GROUNDNUT
ENTOMOLOGY

GROUNDNUT
ENTOMOLOGY

Editors
V. NANDAGOPAL
K. GUNATHILAGARAJ

2008

SATISH SERIAL PUBLISHING HOUSE
403, Express Tower, Commercial Complex, Azadpur, Delhi - 110033 (INDIA)
Phone : 011-27672852 Fax : 91-11-27672046
E-mail : info@satishserial.com, hkjain1975@yahoo.com
Website : www.satishserial.com

Published by :

SATISH SERIAL PUBLISHING HOUSE
403, Express Tower, Commercial Complex, Azadpur, Delhi - 110033 (INDIA)
Phone : 011-27672852 Fax : 91-11-27672046
E-mail : info@satishserial.com, hkjain1975@yahoo.com

ISBN : 81-89304-42-9

Printed at
SALASAR IMAGING SYSTEMS, DELHI-35

Dedication

This book is dedicated to the teachers and considered to be the luminaries of agricultural entomology

Dr.T.N. Ananthakrishnan, a renowned thrips specialist, with lots of contribution both in teaching and research of agricultural entomology. He has authored many research articles and books on Entomology.

Dr.B.V. David, A pioneer of Entomology with enormous contribution both in teaching, authored many books now used for reference books in many of the Agricultural Universities.

Dr.S. Jayaraj, a well known Professor of Agricultural Entomology, formerly VC of TNAU, Coimbatore and now the founder Director of Dr.Jayaraj Resaerch Foundation at Chennai.

Dr.M. Balasubramanian, formerly Professor of Agricultural Entomology, Director of Research of TNAU, and a Agricultural Research Specialist of World Bank.

Dr. Abdul Kareem, formerly Professor of Agricultural Entomology, V.C of TNAU whose contributions in Agrcultural entomology is still a guiding route for the students.

Dr.A. Raghupahty, formerly Professor of Agricultural Entomology, Dean of Agricultural College and Research institute, TNAU Madurai campus, who was a finest teacher and taught the students up to Ph.D.

Dr.R.J. Rabindra, a finest teacher loved by everyone, formerly Professor of Agricultural Entomology, TNAU and now the Project Director of Biological Control (ICAR), Bangalore.

Foreword

Contrary to the spectacular break through witnessed in food grains more particularly wheat and rice since late 1960's, the country's oilseed scenario presented until recently a picture of visual stagnation. Between late 1960's and mid 80's while the food production doubled and the output of wheat and rice went up by 2 and 4 fold respectively, but oilseeds registered a mere 5 lakh tones additional production till 1987-88. Equally unimpressive were the growth rates of productivity of oilseeds when compared to those of food grains in general and wheat and rice in particular. Added to their virtual insulation from the impact of green revolution which transformed rice, the yield levels of oilseeds in general, groundnut in particular suffer from wide fluctuations in space and time. Since 1987-88, the country's vegetable oil scenario has been changed from its decade long inertia due setting up of the TMOP in May, 1986 by the GOI and series of synergistic and innovative farmers-oriented policies it has launched. Thus the country's aggregate output of oilseeds and groundnut which in past never surpassed 13 million tones and 750 kg/ha respectively even under the most favorable seasonal conditions. Simultaneously the quantum of imports of vegetable oils and out flow of scarce foreign exchange reserves on account of it came down from 18 lakh tones (920 crores, Foreign Exchange Equivalent) in 1987-88 to as low as 4.0 lakh tones (250 crores) by 1989-90. These amazing developments in the country's most sensitive and vital sector namely oilseeds have not only capitulated it from a position of one of the major sources of drain on the exchequer to a net foreign exchange earner in recent years but also created much needed sense of optimism and confidence. Realizing the enormous potential of available crop varieties, production and protection technologies for breaking the stagnation in production and productivity of the groundnut and prevailing wide spread skepticism on recommended technology and their reported potentials, vast and extensive research network available in the country for groundnut. The economic yield of groundnut is totally influenced by factors such as soil, climatic conditions, varieties, cultivation practices apart from biotic stress like insects and diseases. Among the insect and non-insect pests recording more than hundred, only about 12 insects are considered to be economically important. The major area under groundnut in the sub-tropics, where the environmental conditions are conducive, consequently a high level of yield loss is recorded. Based on occurrence of one or more number of pests in a given geographic area, the yield loss may vary to an extent of 90%. The authors have taken special interest in compiling the major insect and non-insect pest and their management. I hope that this book with 26 chapters will be of immense use to the researcher, extension personals and for the student of agriculture entomology as well. I compliment the editors Dr.V.Nandagopal, Senior Scientist, NRCG (ICAR), Junagadh and Dr.K.Gunathilagaraj, Professor of Agricultural Entomology, TNAU, Coimbatore for their best efforts in presenting such a book.

Sd/-
Dr. H. David

Date
June, 2007

List of Contributors

A.D. Makwana
Technician (Ent)
National Research Centre for Groundnut
(ICAR) Junagdh-362001

A.Dhandapani
Senior Scientist (Statistics)
NCIPM (ICAR)
Pusa, New Delhi-110012

C.P.S. Yadav
PC (White grubs)
ARS, Durgapura
Jaipur-302018
Rajasthan

C.A. Mahalingam
Asso. Prof. (Entomology)
TNAU, Coimbatore-641003
Tamil Nadu

D.D.R. Reddy
Professor of Entomology (Rtd)
APAU
Rajendra Nagar
Hyderabad
A.P.

D.J. Patel
Principal
Anand Agricultural University
Anand
Gujarat

D.J. Koshiya
Head
Department of Entomology
Junagdh Agricultural University
Junagdh-362001

D.N. Yadav
Professor of Entomology
Anand Agricultural University
Anand, Gujarat

G.D. Satish Kumar
Senior (SS) Extension
NRCG (ICAR)
Junagdh-362001, Gujarat

G. Logiswaran
Professor of Entomology
RRS, TNAU
Coimbatore-641003
Tamil Nadu

G.V. Ranga Rao
Senior Scientist (IPM)
ICRISAT
Patancheru, A.P.

J.G. Bapodra
Professor of Entomology
Junagadh Agricultural University
Junagadh-362001

John Sudhir
Asso. Prof. (Entomology)
APAU, Ananthapur (RRS), A.P.

K.Gunathilagaraj
Professor of Entomology
TNAU, Coimbatore-641003
Tamil Nadu

K. Sahayaraj
Director
Entomological Research Centre
St. Xavier's College (Autonomous)
Palayamkottai-627002
Tamil Nadu

M. Samdur
Senior Scientist
NRCG (ICAR)
Junagdh-362001
Gujarat

M.P. Ghewande
Principal Scientist (Pl. Pathology)
National Research Centre for Groundnut
(ICAR), Junagdh-326001

M.V. Gedia
Technical Officer (Ent)
National Research Centre for Groundnut
(ICAR) Junagdh-362001

P.C. Rath
Senior Scientist (Ent)
CRRI (ICAR)
Cuttack-6
Orissa

P. Manivel
Senior Scientist
NRCG (ICAR)
Junagdh-362001
Gujarat

Rajiv Rathod
Section of Entomology
National Research Centre for Groundnut
(ICAR) Junagdh-362001

Rameswar Rao
Scientist
ICRISAT
Patancheru, A.P.

R. Vishnupriya
Asso. Professor (Ent.)
RRS, TNAU
Tindivanam, Tamil Nadu

S.K. Patel
Professor of Nematology
Agricultural University
Anand, Gujarat

S.P. Singh
Project Director
Directorate of Biological Control
Hebbal, Bangalore

T. Muralikrishna
Asso. Prof. (Entomology)
APAU, Tirupati Centre
Tirupati

T. Senguttuvan
Professor of Entomology
TNAU, Coimbatore-641003
Tamil Nadu

T.P. Trivedi
Principal Scientist (Entomology)
NCIPM (ICAR)
Pusa, New Delhi-110012

T.V. Prasad
Scientist, (Entomology)
National Research Centre for Groundnut
(ICAR) Junagdh-362001

V. Nandagopal
Senior Scientist (Entomology)
National Research Centre for Groundnut
(ICAR), Junagadh-362001

Contents

Insects and Non-insect Pests: Distribution Identification and Symptoms of Damage

V. NANDAGOPAL AND G. V. RANGA RAO

1.1. Introduction

Groundnut is one of the major oilseed crops in India. It is cultivated under diversified agro climatic condition in about 6.6 million ha producing about 5.9 m tonnes of pods (FAO, 2005). Of which about 85% area of cultivation is under rain fed **fig.1**. The pod yield is influenced by soil fertility, climatic condition, varieties grown, cultural practices, insect pests and diseases and environmental stresses. Among the biotic stress are known to inflict considerable losses in pod yield. More than 100 species of insects and mites are known to attack groundnut (Amin, 1988 and Nandagopal, 1992). The total annual loss from these pests has been estimated to be about Rs.1600 million (Amin, 1983). A comprehensive list of insect and non-insect pests of groundnut was given by Nandagopal and Prasad (2004). Most of these insects and mites are distributed throughout the groundnut growing zones of India with one or two as major yield barriers in each zone. Insects like leaf miner endemic in south India has now spread to central India particularly Maharashtra and Gujarat. The zone wise distribution of the major insect pests is given **fig.2**.

1.2. Groundnut insect pests and mites and their distribution

The list of major and minor insects recorded in India on groundnut are given in Table 1.1. Based on the feeding behavior, the insect pests, infesting groundnut may be broadly classified in to those attacking aerial parts of plants, underground parts (subterranean) and in storage. As below:

1.2.1. Defoliators

The groundnut leaf miner, *Aproaerema modicella* Dev., tobacco caterpillar, *Spodoptera litura* F., gram pod borer, *Helicoverpa armigera* F., and the red hairy

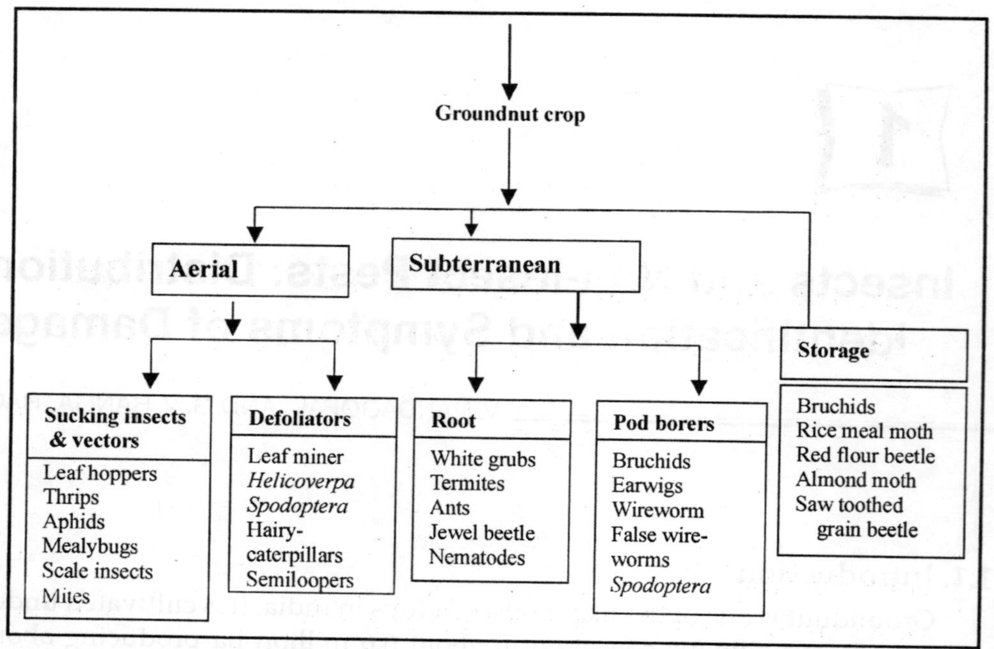

caterpillars, *Amsacta albistriga* (Walker), *A. moorie* (Butler) and Bihar hairy caterpillar, *Spilosoma obliqua* (Walker) are considered to be major insect pests of groundnut. Among the eight Orthopteran insects (Jayanthi *et al.*, 1993; Paras Nath, 1993), so far no major outbreaks have been recorded. These defoliators occur as pests in Andhra Pradesh, Tamil Nadu, Maharashtra and Karnataka.

1.2.2. *Sucking insect pests*

Among the sucking insect pests, the aphid, *Aphis craccivora* Koch, leaf hopper, *Balclutha hortensis* Lindb., *Empoasca kerri* Pruthi, and the thrips, *Caliothrips indicus* Bagnall, *Thrips palmi* Karny are responsible for major yield losses. The aphid is principally a rainy season pest occurring throughout the groundnut growing zones. Leaf hoppers, and thrips occur both in rainy and post rainy seasons. Besides causing direct damage aphids and thrips are vectors of major viral diseases viz., Peanut Stripe Virus (PStV) and Peanut Bud Necrosis Disease (PBND) respectively.

1.2.3. *Subterranean insects*

The root and pod feeding insects belong to the orders Isoptera, Hemiptera, Lepidoptera, Hymenoptera, Coleoptera and Dermaptera. The white grub, *Holotrichiaea consanguinia*, termites, *Odontotermes obesus* (Rambur), wire worms, *Penthicoides seriatoporus* Fair and earwig, *Anisolabis anulipes* Luc. are important. During favorable years, their infestation is wide spread (Senguttuvan and Dhanakodi, 1997) resulting in serious damage.

Figure 1. Rainfed groundnut agriculture in India

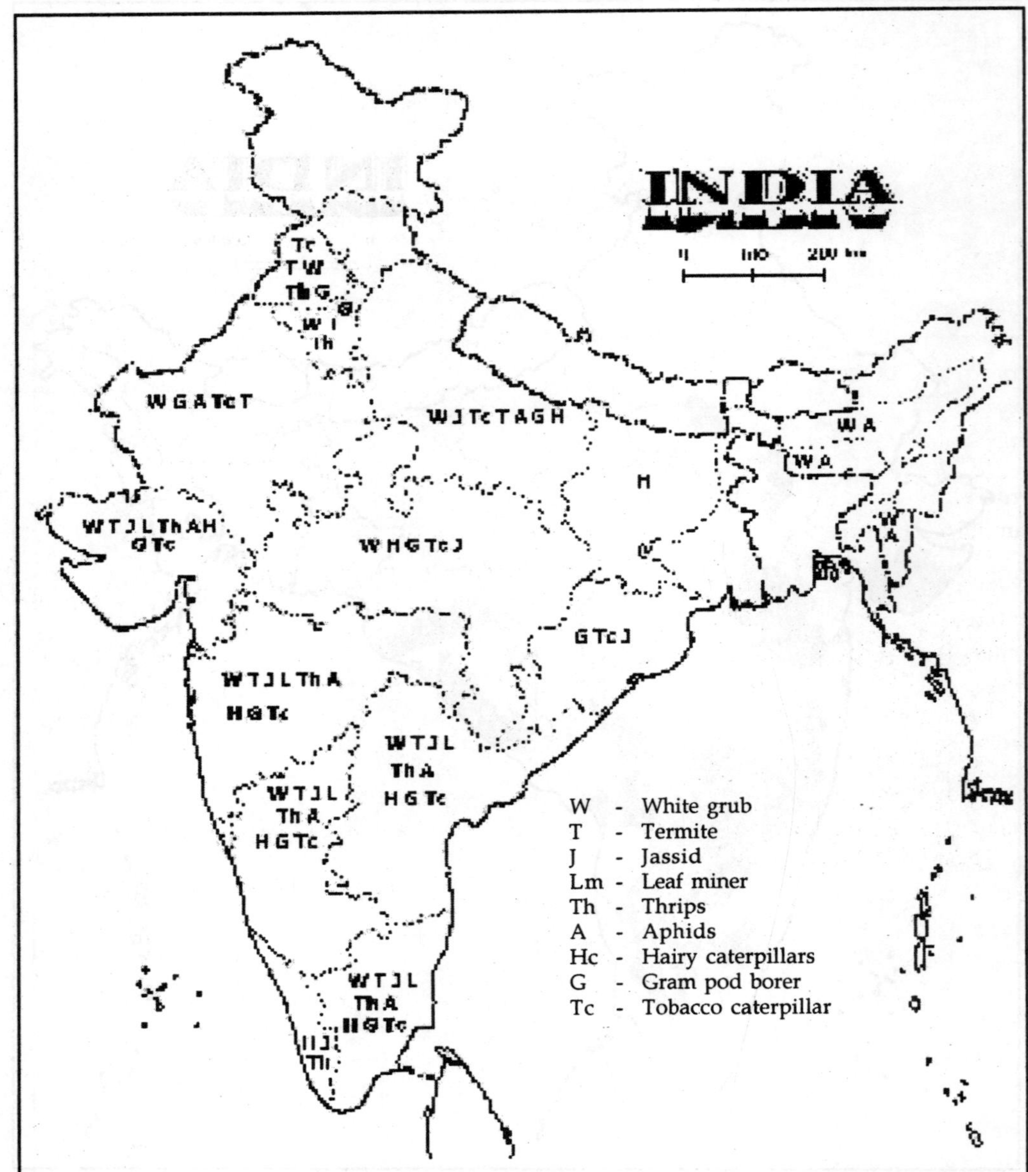

Figure 2. Distribution of insect-pests of groundnut

Table 1.1. List of insect pests of groundnut in India

Order: Orthoptera **Family:Acrididae** *Acrida turrita* Linn. *Acroltytus patruelis* H.S. *Atractomorpha crenulata* Fab. *C. hemipterus* *Chrotogonus hemipterus* Schaum *Chrotogonus trachypterus* (Blanchard) *Colemania sphenarioides* Boliver *Cyrtacanthacris ranacea* (Linn.) *Oedaleus* sp. *Orthacris* sp. *Oxya chinensis* (Thunberg) (=*Velox*) *Oxya velox* (F.) *Trilophida* sp. **Order: Isoptera** **Family: Termitidae** *Microtermes thoracalis* Sjostedt *Odontotermes obesus* (Rambur) *Syntermes* sp. *Trinervitermes biformis* (Wasm.) **Order: Thysanoptera** **Family: Thripidae** *Caliothrips (=Heliothrips) indicus* Bagnall *Frankliniella schulzei* (Tribon) *Frankliniella fusca* (Hinds) *Podothrips bicolor* Seshadri and Ananthakrishnan *Scirtothrips dorsalis* Hood *Taeniothrips distalis* Karny *Taeniothrips longistylus* Karny *Thrips flavus* Schrank *Thrips hawaiensis* Morgan *Thrips Palmi* Karny **Order: Hemiptera** **Family: Miridae** *Calocoris angustatus* Lethierry *Creontiades pallidifer* (Walker)	*Creontiades pallidus* (Rambur) *Megacoelum stramineum* Walker **Family: Lygaeidae** *Aphanus sordidus* (F.) *Aphanus(=Naphius) apicalis* (Dallas) *Elasmolomus sordidus (F.)* **Family:Pyrrhocoridae** *Dysdercus cingulatus* (F.) *Dysdercus koenigii* Fab. **Order: Homoptera** **Family: Cicadellidae (= Jassidae)** *Aconurella indica* (Pruthi) *Amrasca biguttula biguttula* (Ishida.) *Austroasca alfaalfae* (Evans) *Austrogalia omma* (Mkusant and Ray.) *Austrogalia avicula* (Raibaut.) *Balclutha hebe* (Kirkaldy) *Balclutha hortensis* *Cicadulina bipunctata* (Melichar) *Circulifer opacipennis* (Lethierry) *Empoasca dolichi* Paoli *Empoasca fabae* Harris *Empoasca kerri* Pruthi *Empoasca motti* Pruthi *Empoasca parathea* Pruthi *Empoasca* sp. *Exitianus capicola* (Stal.) *Hishimonus phycitis* (Distant) *Macrosteles indrina* (Pruthi) *Nephotettix nigropictus* (Stal.) *Orosius albicinctus* Distant *Orosius argentatus* (Evans) *Zygina equata* Singh **Family: Delphacidae** *Sogatella* sp. **Family: Aleyrodidae** *Bemisia tabaci* (Gennadius) (=*Gossypiperda* Mistra and Lamba)

Contd...

Table 1.1. Contd...

Family: Aphididae

Aphis craccivora Koch
(=*laburni* Kaltenbach)
(=*leguminosae* Theob.)
Aphis gossypi Glover
Myzus persicae Sulzer
Rhopalosiphum maidis (Fitch)
Therioaphis ononidis (Kaltenbach)

Family: Coccidae
Phenacoccus hirsutus (Green)

Family: Pseudococcidae
Dysmicoccus (=Pseudococcus) brevipes (Cockerell)
Ferrisia virgata (Cockerell)
Planococcus(=Pseudococcus) lilacinus Cockerell

Family: Diaspidae
Hemiberelosia latanise (Signoret)

Order: Dermaptera
Family: Labiduridae

Anisolabis annulipes (Lucas)
Euborellia stali Dohrn.
E. annulipes (Luc.)

Order : Coleoptera
Family: Scarabaeidae

Adoretus limbatus
pogonia uniformis
Autoserica atratula
Autoserica insanabilis
Dyscinetus trachipygus
Lachnosterna (=*Holotrichia*) *consanguinea* (Blanchard)
Lachnosterna fissa (Brenske)
Oxycetonia versicolor (F.)
Pogonia cribricollis
Apogonia ferruginea
Apogonia roucca
Autoserica nathani
Autoserica nilgirensis
Holotrichia (=insularis) reynaudi
Holotrichia consanguinea
Holotrichia excisa
Holotrichia fissa
Holotrichia heterotincta
Holotrichia longipennis
Holotrichia nilgiria
Holotrichia problematica
Holotrichia rufoflava
Holotrichia sculpticollis
Holotrichia seticollis
Leucopholis burmeistri
Leucopholis coneophora
Leucopholis lepidophora
Maladera sp.
Schizonycha fuscescens
Schizonycha ruficollis Fab.
Schizonycha sp.

Sub-Family : Rutelidae

Adoretus decanus
Adoretus ladakhensis
Adoretus lasiopygus
Adoretus nitidus
Adoretus punjabensis
Adoretus versutus
Anomala bengalensis
Anomala bicolour
Anomala dimidiata
Anomala dorsalis
Anomala elata
Anomala marginipennis
Anomala polita
Anomala ruficapilla
Anomala varians
Anomala versicolor
Pachynhinadoretus frontatus
Rhinyptia laeviceps Arrow
Rhinyptia meridionalis
Family: Buprestidae
Sphenoptera perroteti (Guer.)

Family: Elateridae

Penthicoides seriatoporus Fair

Contd...

Table 1.1. Contd...

Family: Meloidae	**Family: Arctiidae**
Mylabris balteata (Pall.)	*Cretonotus gangis* Linn.
Mylabris phalerata (Pall.)	*Diachrysia orichalcea* (Fb)
Mylabris pustulata (Thunberg)	*Amsacta albistriga*(Walker)
Mylabris balteata (Pall.)	*Amsacta lineola* (Fabricus)
Family: Chrysomelidae	*Amsacta moorei* (Butler)
Aulacophora sp.	*Diacrisia obliqua* (Walker)
Aulacophora foveicollis (Lucus)	**Family: Noctuidae**
Aulacophora lewesi Baly	*Achaea janata* Linn.
Aulacophora stevensi Baly	*Agrotis ipsilon* (Hufnagel)
Family: Curculionidae	*Chrysodeixis chalcites* (Esp.)
Crytozemia cognata Marshal	*Chrysodeixis erisoma* (Doubleday)
Crytozemia cognata Marshal	*Chrysodeixis orichalcea* (Fab.)
Myllocerus blandus Faust.	*Mymenia recurvalis*
Myllocerus discolor Boheman	*Pericallia ricini* F.
Myllocerus viridanus F.	*Phytometra (=Plusia) orichalcea* F.
Myllocerus discolor Boheman	*Plusia chalcytes* (Esper) (=*Chrysodeixis chalcites* Esper)
Myllocerus dorsatus (F.)	*Plusia signata* (F.)
Myllocerus sp.	*Spodoptera exigua* (Hubner)
Myllocerus undecimpustulatus maculosus Desbr.	*Spodoptera littoralis* (Boisduval)
Myllocerus viridanus F.	*Spodoptera litura* (F.)
Phytoscapus triangularis Olivier	*Trichoplusia ni* (Hubner)
Tenymecus indicus Faust.	**Family : Sphingidae**
Family: Cetonidae	*Agrius convolvuli* (L.)
Oxycetonia versicolor (Fab.)	**Family : Lymantridae**
Order: Lepidoptera	*Agrius convolvuli* Linne
Family: Tortricidae	*Dasychira mendosa* (Hb.)
Cydia ptychora (Meyrick)	*Euproctis ommakas* (Koll.)
Family: Gelechiidae	*Euproctis lunata* Wlk.
Anarsia ephippias (Meyrick)	*Euproctis subnotata* Wlk.
Aproaerema modicella Dev.	*Hedylepta indicata* (F.)
(= *Stomopteryx subsecivella* (Zeller)	**Order: Hymenoptera**
(=*nerteria* Meyrick)	**Family: Formicidae**
Scopula albiflava (Warren)	*Dorylus orientalis* Westwood

1.2.4. Borers

Under the field conditions, the root borer, *Sphenoptera perotetti* Guer, pod borers; *S. litura*, wireworms, *P. seriatoporus*, earwig, *A. anulipes* and ant, *Dorylus orientalis* Westwood are important. A detailed list of borers recorded in India is given in the Chapter 13.

1.2.5. Storage pests

Several insects damage groundnut and its products both in storage and to some extent in the field causing severe damage (Srivastava, 1970). An approximate loss of 6-10% groundnut kernel stored in bags due to insect pests was reported (Amin, 1988). The major pests of stored groundnut in India are *Caryedon serratus* (Ol.), *Tribolium castaneum* (Herbst), *Corcyra cephalonica* (Stainton), *Oryzaephilus mercator* (Fauvel) and *Ephestia cautella* (Walker) (Rai, 1976; Anon., 1983). A detailed list of storage insects recorded in India is given in the Chapter 14.

1.2.6. Non-insect pests

Mites, nematodes, rats, squirrels, jackals and birds are the non-insect pests which occasionally damage the groundnut. A detailed list of non-insect pest recorded in India is given in the Chapter 12.

1.2.7 Nematodes

In India, 38 nematode species have been recorded on groundnut crop and of them, *Meloidogyne* spp. are found to be the major ones causing about 14 to 20% yield lose. A detailed list of nematode species recorded in India is given in the Chapter 15.

1.3. Symptoms of damage

A key for identifying the important pests based on the symptoms at different stages of crop growth is given below:

1.3.1. Seedling stage (up to 30 days)

1. Drying of seedling	i) Including the stem and leaves wither and dry up easy to pull - **Wireworms**
	ii) Entire seedlings dry up and easily pulled out, the roots and stems are hollow inside and generally filled with soil – **Termites**
2. Drying of leaves, papery in appearance	i) A number of feeding spots coalesce into a white patch in the beginning and under severe conditions the total foliage looks deformed – **Thrips**

Contd...

Table Contd...

2. Drying of leaves, papery in appearance	i) A number of feeding spots coalesce into a white patch in the beginning and under severe conditions the total foliage looks deformed – **Thrips** ii) Puckerings, foldings, crinkling, 'V' shaped yellowing on the leaf tips -**Leafhopper** iii) Puckering and folding, crinkling of leaves with lot of nymphs and adults covering the shoot - **Aphids**
3. Partial to complete defoliati on leaving only the stem.	i) No foliage can be seen or with few midribs , look like a cattle grazing field – ***Spodoptera, Helicoverpa*** **and Red hairy caterpillar** ii) Irregular wavy feeding on leaf edges - **Grasshopper** iii) Irregular feedings of mature leaves - **Weevils**
4. Pin head sized brown spots the on leaves	i) Small blister like spots are seen on the upper surface of leaves, as feeding increases, mines increased in size and entire leaflets become brown and rolls, folded, shrivels and dries up - **Leaf miner**
5. Drying of entire plants	i) Entire plant dries, easy to pull with damaged tap roots – **White grubs**

1.3.2. Flowering to maturity

1. Drying of leaves	i) Mature leaves folded and fastened together with 2 or 3 adjacent leaves. The larvae seen feeding inside the folded leaves. - **Leaf miner** ii) Silken webbing and small red coloured mites seen moving in thousands. Ultimately the whole plant withers and dies - **Red spider mite** iii) Silken webbing and small white coloured mites seen moving in thousand. Ultimately the whole plant withers and dies - **White spider mite** iv) Drying of leaves and plants, yellowing and wilting of leaves, roots eaten away –**White grubs** v) Drying of leaves, plant death, root tunneling and filling with soil particles, pod boring and scarification-**Termites** vi) Partially buried, cut pieces of branches at the base of the root zone, heap of soil, presence of burrows- **Rats** vii) Withering leaves and plant death, when roots are split grubs seen inside - **Groundnut root borer**
2. Yellowing of leaves	i) 'V' shaped yellowing on the tip of the leaves - **Leaf hopper**

Contd...

Table Contd...

	ii) Presence of tiny greenish yellow, flat to oval shaped nymphs and puparia with white waxy fringes on the dorsal side of the leaves - **Whitefly**
3. Leaf edges chewed on either side of the midrib	i) Leaf lamina partially or completely eaten up **- Red hairy cater pillar** *-Spodoptera* *-Helicoverpa* ***-Spilosoma*** ii) Leaves rolled up, 2 - 3 leaflets folded white streaks/ patches are seen on the lamina **-Leaf miner** iii) Irregular wavy feeding on leaf edges **- Grass hopper** iv) Irregular feeding of the mature leaves, mostly on the margins- **Weevils**

1.4. Key for identification of major insects and mites

1.4.1. Defoliating pests

- Dark green larvae feeding inside the lamina, folded leaves, measuring about 10 mm - **Leaf miner**
- A small brown pupa (5-6 mm in length) seen inside the folded leaves. - **Leaf miner pupae**
- Smooth bodied larvae with light to dark brown colour body with distinct dark band on the thoracic region about 15 to 25 mm long preferably seen feeding on the foliage after dusky hours. - **Tobacco caterpillar.**
- Larvae with profuse deep red coloured hair seen moving in large numbers measuring 15 to 25 mm long, defoliating the leaves. - **Red hairy caterpillars, Bihar hairy caterpillar.**
- Greenish larvae with sparse setae and longitudinal bands along the body measuring about 15 to 25 mm long feeding on the foliage. - **Gram pod borer.**
- Deep brown to black insects feeding on the foliage from margins. – **Ash weevils.**

1.4.2. Subterranean insects

- Social polymorphic white insects in groups moving in lines during morning and evening hours - **Termites.**
- Roots eaten away, whitish 'C' shaped grubs found near the root zones **-White grubs**

- Dark brownish segmented well skeletonized grubs found inside the pods feeding on the developing kernel - **Wireworms**
- Brownish and blackish insects measuring about 15 to 20 mm long having well developed forceps like structures on the anal end of the body feeding on the kernels inside the pods under field condition - **Earwigs**
- Smooth bodied, larvae with light to dark brown chocolate colour body seen feeding and making irregular holes on pods near soil surface - ***Spodoptera***

1.4.3. Sucking pests

- Deep brownish nymphs and adults having two cornicles from which droplets of honey dew excreted, covering the shots and flowers - **Aphids**
- Yellowish green nymphs feeding from under surface of leaves, on disturbing the plants greenish adults fly away from the plants -**Leafhopper**
- Tiny insects scraping on the upper surface of the leaves causing pin head white patches, on passing a wet hand, the adults and nymphs can be seen trapped - **Thrips**
- Presence of tiny greenish yellow flat oval nymphs and puparia with white waxy fringes on the dorsal side of the leaves - **Whitefly**

1.4.4. Borers

- Whitish about 5 to 15 mm long, robust grubs feeding inside the roots.- **Root borer**

1.4.5. Storage insect pests

- Dark brownish robust adults with large hind legs seen moving on the storage bags in between the pods, whitish grubs feeding inside the kernel under storage condition. This is the largest beetle feeding on the stored pods/kernels. - ***Groundnut bruchid***
- Chesnut brown beetles about 3 to 4 mm long are often first to colonize the stored groundnuts. The larvae are cylindrical in shape with two prominent 'horns' on the last abdominal segments – ***Rust- red flour beetle***
- Tiny beetle with ridged pro-thorax and the larvae cream coloured – ***Merchant grain beetle***
- Oval shaped adult beetles 2 to 3 mm long having dark brown mottling, the dorsal surface is covered with fine hairs. The larvae are straw coloured and the grownup larvae having tuft of hairs on each abdominal and thoracic tergites – ***Khapra beetle***

- Dark brown bugs 10 mm long and 2 mm wide - ***Storage bugs***
- Brown adult moth 12 to 15 mm long. The larvae spin silken webbing together with kernels - ***Rice moth***
- Grayish brown adults with an indistinct pattern on the forewings, the larvae move freely through the produce - ***Almond moth, Tropical warehouse moth***
- Adult moth is having basal third cream coloured and the rest is reddish-brown colour - ***Indian meal moth***

1.4.6. Non-insect pests

- Tiny red coloured mites moving on the foliage, extensive webbings seen on the entire plants - ***Two spotted red spider mites***
- Tiny white coloured two spotted mites moving on the foliage, extensive webbing seen on the entire plants - ***Two spotted white spider mites***
- Birds which damage groundnut either in the field or while drying of the harvested pods and kernels (Table 1.5).
- Rodents which damage the crop under field conditions and in the storage (Table 1.5.3).

References

Amin, P.W. (1983). Major field insect pests of groundnut in India and associated crop losses. Special Issue. *Indian Journal of Entomology*. 2: 337-344.

Amin, P.W. (1988). Insect and mite pests and their control. In *Groundnut* (Ed. P.S. Reddy). pp. 393-452. Indian Council of Agricultural Research, New Delhi.

Anonymous, (1983). Annual Report 1983-84. International Crops Research Institute for the Semi-Arid Tropics, Patancheru, Andhra Pradesh, India.

Anonymous, (1993). Annual Report (1st April, 1992 to 31st March, 1993). All India Co-ordinated Research Project on Biological Control of Crop Pests and Weeds. Biological Control Centre (ICAR-NCIPM)., Bangalore, India. Tech. Doc. No. 40, 1-313.

FAO, (2005). Statistics of area and production of groundnut in India.

Jayanthi, M., Singh,K., M. Singh,R. N. (1993). Population build up of insect pests on MH 4 variety of groundnut influenced by abiotic factors. *Indian Journal of Entomology*. 55(2):109-123.

Nandagopal, V. (1992). Studies on Integrated pest management in Groundnut in Saurashtra, Ph. D. thesis, Saurashtra University, Rajkot, pp 246.

Nandagopal,V and T.V.Prasad.(2004). World list of insect and non-insect pests of groundnut and their natural enemies. National Research Centre for Groundnut (ICAR), Junagadh, Gujarat. 140 pp.

Paras Nath. (1993). Studies on the groundnut insect pest and their management in east utter pradesh. Final technical report of ICAR AD-HOC Project, Dept. of Entomology and Agril. Zoology, Institute of Agril. Sciences, Banaras Hindu University, Varanasi. pp114.

Patel, M.H.(1988). Studies on predation of *Heliothis armigera* Hubner and *Spodoptera litura* Fab. By insectivorus birds with special reference to *Acridotheres tritis* L. *M.Sc. thesis submitted to the Gujarat Agri. Univ. Anand,* 77 pp.

Rai, K. (1976). Pests of oilseed crops in India and their control. Indian Council of Agricultural Research (ICAR)., New Delhi. 121 Pages.

Senguttuvan, T. and Dhanakodi, C.V. (1997) . Control of pod borers of groundnut by soil application of insecticides in alfisols. *Journal of Oilseed Research*. 14(2):274-277.

Srivastava, A.S. (1970). Important insect pests of stored oil seeds in India. *International Pest Control*. 12:18 – 20.

□□□

Ecology of Groundnut Pests

V. NANDA GOPAL

2.1. Introduction

In India groundnut cultivation is mostly monoculture in major groundnut growing states likes Andhra Pradesh, Gujarat, Tamil Nadu, Karnataka, Maharashtra and Orissa. However, there is a great diversity in the cultural practices followed in different agro climatic regions. Extensive cultivation in groundnut with introduction of new varieties and adoption of modern agrotechniques have brought a great deal of changes in the abundance of cosmopolitan and regional pests and their associated natural enemies complex. Three decades earlier, leaf miner and sucking insects (Aphids) and hairy caterpillars were considered to be the major insect pests of groundnut and pests like tobacco caterpillar and gram pod borer were consider to be minor and sporadic. But, as of now, the tobacco caterpillar and the gram pod borer are the major threat to the cultivation of groundnut in states like Karnataka, Tamil Nadu and Gujarat. There also exist a spatial variations in the distribution and intensity of insects. This is evidenced from the fact that groundnut leaf miner (*Aproaerema modicella* Dev.) is confined only to southern India. Similarly, Red hairy caterpillar (*Amsacta albistriga* Wlk.) occurs only in states like Tamil Nadu & Andhra Pradesh. Amin (1983) gave a clear picture of the pests which occurred in 1968 (Rai, 1976) and the later period witnessed an increase in importance of pests like white grubs, thrips, tobacco caterpillar and leaf hopper.

2.2. Groundnut Agro ecosystem

The agro ecosystem in India where groundnut is grown is diverse. The groundnut areas are divided into northern, western, central, south eastern, peninsular and southern zones. The majority of groundnut cultivation is in the southern and peninsular zone. Cultivation of groundnut in large and continuous areas results in the availability of continuous and abundant food supply for the various species of insect pests attracted to the different phenophases of crop growth. In addition to that, introduction of rabi-summer groundnut cultivation in Gujarat in 1977 –78 and in Maharashtra since 1978-79 covering 0.15 million hectares in Gujarat and Maharashtra respectively aggravated the situation, augmenting the multiplication of insect pests of groundnut. Availability of alternate host plants throughout the year such as *Psoralea corelifolia* (L.) for leaf miner, continuous stand of *kharif* crops, such as castor for *S. litura* and pigeon pea for *H. armigera* favoured the sustainability of the pest population in the field.

2.3. Distribution Pattern of Important Pests

Insect pests like leaf miner (*A. modicella*) is cosmopolitan in occurrence and distribution, infesting the groundnut crop from vegetative stage to even just before harvest both in peninsular and southern India. The next category of insects includes those that occur more predominantly in one zone but are noticed only sporadically in north zone. For instance, *S. litura* is a major pest in southern India but noticed only sporadically in other zones Tables 2.1. and 2.2 show the importance of insect pests and their associated yield loss in different states of India.

Table 2.1. *Spodoptera litura* male moths catches in pheromone traps over a period (1985 to 1993) in various locations

Stations	85-86	86-87	87-88	88-89	89-90	90-91	91-92	92-93
ICRISAT (Hyderabad) South Zone	170	158	239	283	233	272	172	391
Gwalior M.P.,Central	148	169	176	206	154	253	-	-
Hissar, Northen Temparate	104	210	68	6	22	32	-	-
Ludhiana, Western Zone	84	142	190	152	84	141	132	136
Padappai, Chennai South zone	64	36	22	31	28	16	46	35
Jalgaon, Maharashtra	153	51	46	53	22	31	39	50

(after Ranga Rao and Wightman, 1999)

A third category of insect pest occurring both in tropical and temperate zones includes white grub (*Holotrichia consanguinea*). It occurs both in Andhra Pradesh which is in south zone (Rao Deo,1974) and in Uttar Pradesh, Punjab and Rajasthan which are in northern zone. Loose sandy and well drained

Table 2.2. Yield loss associated with insect pests of groundnut in India

State	Area ('000 ha)	Major pests	Secondary pests	Yield loss associated with pests (%)	Loss in million Rs.
Gujarat	2150	White grub	Leaf hopper	40-7	95
Maharastra	737	Leafminer	Thrips, aphid, white grub	5-10	Not known
Madhya Pradesh	325	IN	Termites	---	Not known
Andhra Pradesh	1695	Leafminer Red hairy caterpillar	Leaf hopper	15	505
Tamil Nadu	1151	Leafminer Hairy caterpillar	Leaf hopper	42-70	737
Orissa	310	Thrips	Leafminer	35	Not known
Karnataka	863	Leafminer Thrips Red hairy caterpillar	Semilooper, gram pod borer	22	125
Punjab	45	White grub Termite	None	49	82
Uttar Pradesh	186	Termite , White grub, *Heliothis armigera*	None	20	Not known
Rajasthan	252	Termite	None	50	40
Haryana	9	Termite, White grub, Tobacco caterpillar	None	Not known	Not known

soil was found to be suitable for survival and multiplication of white grubs. The population build up of leafminer depends mostly on the increase in temperature. The egg production has been shown to be temperature dependent at 15°C and at 55°C, low egg production was recorded. The ambient temp. under the field condition, favours the larval development lasting for 9 to 28 days (Cherian and Basheer, 1942; Kapadia, 1982; Shanower *et al.*, 1992).

2.4. Role of abiotic Factors

2.4.1.1. Temperature and Humidity

The increase in the species complex depends on two broad categories of physical environment namely geographic barriers and climatic restriction. Because of the unlimited human activities there existed a change in the bio-climatic environment controlling the abundance of insect pests.

Temperature has been the most dominant factor influencing the growth, development, feeding, survival, fecundity, dispersal, distribution and abundance of insects. Insects generally grow faster in warmer conditions in a given zone of effective temperature. The development of insect species shows a linear relationship with temperature. The required optimum humidity for each stage of insect species varies. Hence these two physical factors are highly interactive in affecting the activity, development and survival of insect species in nature.

2.4.1.2. Defoliators

The groundnut leafminer is active both in kharif and rabi-summer condition (Nandagopal, 1992; Ranga Rao *et al.*, 1993). Sandhu (1978) reported a total duration of GLM as 37-45 days on Lucerne in the laboratory at 14-32°C and 67-87% Rh. For egg development 60 degree days above 12.4°C are required (Shanower, 1989). Under field conditions eggs generally hatch in 3 to 4 days but may require 6-8 days at lower temperature (Kapadia *et al.*, 1982). In the case of *H. armigera* a negative correlation with minimum temperature and sunshine hour was observed (Bapodra, 1988). *Spodoptera litura,* required a developmental threshold temperature of 37°C, if it goes above it was found lethal. The lower mean developmental threshold for egg development was 8.2°C, for larvae 10°C with a range of 6.6°C to 13.14°C and for pupae 10.2°C with a range of 9.6 °C to 10.8°C. The adult during its pre-oviposition period had a slightly higher threshold of 10.8°C (ranging 7.5°C -14.1°C). There was no oviposition at 35°C -37°C (Ranga Rao *et al.*,1989).

2.4.1. Sucking insect pests

The weather conditions found during November 1st week to February last week favoured a fast multiplication of the aphid species, especially *A. craccivora* and *R. maidis.* The pest-weather interaction study revealed that population build up was directly related to minimum temperature. If the temperature falls below 20°C, the aphid population builds up and reached the peak population when the weekly average minimum temperature is around 15°C. There was no clear relationship between Rh and aphid population (Anon., 1997).

Both minimum and maximum temperature, average vapour pressure as well as soil temperature at 10cm and 20cm depth were significantly correlated with the population of *B. hortensis.* There was no relationship with humidity. The length of life cycle of groundnut leaf hopper *E. kerri* decreased with decrease in temperature (Singh *et al.*, 1989). *E. kerri* showed a positive correlation with maximum daily temperature, sunshine and negative correlation with minimum daily temperature, morning Rh and wind speed (Jayanti *et al,*. 1993). High humidity more than 70.5 % coupled with dry spell for about a week favoured heavy build up of thrips *C. indicus* population. The population of *C. indicus* had positive correlation with Rh and a negative correlation with evening Rh and wind speed. (Jayanthi *et al.*, 1993). The physical factors such as temperature, Rh, sunshine hour did not directly influence the population build up of thrips. The population of white fly *B. tabaci* were positively correlated with temperature, evening Rh and wind speed, and negatively correlated with morning Rh and sunshine (Jayanthi *et al.*, 1993).

2.4.2. *Rainfall*

2.4.2.1. Defoliators

The abundance of insect pests in an area is affected by the total rainfall and distribution. Khan and Reodeo (1987) claimed that high rain fall was a key factor in regulating the leaf miner population. (Amin, 1987) suggested that heavy rain fall reduced leaf miner population, although Wheatley *et al.*, (1989) found that water from an over head irrigation system did not lower leafminer density. Lewin *et al.* (1979) found a significant negative correlation between leafminer population and rainfall. In an another interesting observation, with a rain simulator physical impact of rain fall on leafminer egg and larvae did not significantly increase the mortality. There was no linkage between the rain fall and the appearance of leafminer male moths in pheromone traps (Anon., 1993 and 1994).

H. armigera populations are reduced each year by the hot dry summer in the south and cold winter in the north. With increase in irrigation, in South India there is a possibility of increase in the pest population due to continuous availability of food. There appears to be no correlation between rainfall and *H. armigera*, incidence. Though there was continuous rainfall from 1st week of June in the year 1992, the moths emerged from august 1st week and continued up to 3rd week of September at Latur (Anon., 1993). Similarly in the year 1993 in the same center, the population of *H. armigera* continued to appear ranging 1 to 31 males/trap/week even though the rainfall occurred during June 2nd week to October last week.

Rainfall ranging 5 to 26mm favoured *S.litura* moth emergence, rather than rainy days (Anon.,1998) at Dharwad. In the same centre, in the case of *S. litura,* though rainfall was confined between second week of June to 2nd week of September, the male moths of *S.litura* appeared throughout the season (Anon., 1993). In a similar situation *S. litura* also appeared from June to October 1st week irrespective of intensity of rain during that period.

2.4.2.2. Sucking insect pests

The aphid, *A. craccivora* is cosmopolitan in distribution and infests many host plants belonging to leguminous crop. This is by and large a rainy season insect and attains pest status, at times during the post rainy season. In a water stress experiment, it was observed that 40 days after sowing, the population of *A. craccivora* was much higher in plants that had sufficient moisture (Rang Rao *et al.*, 1991). Rainfall of about 40mm in August and rapid build up of predators wipe out the population of aphids. Serious aphid damage can also occur during drought period (Amin, 1988). There was no relationship between rainfall and *A. craccivora* population trapped in sticky traps over six years. Similar non-relation was exhibited in the case of leaf hopper (*B. hortensis*) and thrips (*C. indicus*). There was a positive correlation with rain fall in *E. kerri*, as well as *C. indicus* and *B. tabaci* (Jayanti *et al.*, 1993).

2.4.2.3. Subterranean insects

Adults of *H. serrata* and *H. consanguinea* emerged from the soil after 25 mm of pre-monsoon or monsoon rain during mid-May or later. Any rain-received prior to mid-May does not initiate the emergence of beetles as they do not become sexually matured. If there is pre-monsoon prior to monsoon rain, two emergence peaks of beetle are observed (Desai and Patel, 1965; Rai *et al.*, 1969). Mass emergence of beetle takes place during the 1st summer shower (Gupta and Avashti 1976; Karla and Kulshreshta 1961). In southern India the beetle emergence commences during March – April, even without any rains but there is no much enhancement with heavy rainfall (David and Ananthanarayana, 1974). Termites generally produces sexual forms, which is triggered by the receipt of rainfall.

2.4.3 *Other Climatic Factors*

The groundnut crop occupies to a greater extent of area in southern states of India and in Maharashtra, Gujarat and Orissa. There was no drastic climatic changes like too much winter as in the case of northern India, frost, clouds have no effect on the insect pest of groundnut grown in south India. It is clearly observed in the case of Aphid (*A craccivora*) Leaf hopper (*B. hotensis*), Thrips (*C. indicus*) where, the sunshine hours, have no correlation with the population of above insects. However, the wind velocity is related to the migration of the passive flyers such as thrips and aphids which are usually carried by wind. Light loamy soils are highly suitable for termite activity (Leela David *et al.*, 1964), whereas, sandy loam soils are suitable for white grubs (Gupta and Avashti 1957; Sunil Kumar, 1971). *H. consanguinea* has been reported form sandy track of south Bihar, while, the Dynastid (*Heteronydis* sp.) is reported from light loamy soil of North Bihar (Prasad, 1961). However, Veeresh (1977) contradicted as white grubs are found in different types of soil, except in hard soil where the soil moisture play a dominant role in the mobility of the grubs.

The intensity of moon light associated with the daily change in the degree of moon phase in the lunar cycle, because of its interference in photo tactic response, should affect the trap catch considerably in the case of both *S.litura* and *H. armigera*. Many workers in the past have reported a significant reduction in trap catches of noctuid moth on full moon and around compared to the no moon period (Nemack, 1971). Most of them compared the response of full moon and no moon. Bowden and Church (1973) and Persoon (1976) have made very extensive studies examining responses of large number of insect species in relation to the regular changes in light illumination. On different days of lunar cycle Vaishampayan and Shrivastara (1978) studied the response of *S. litura* covering 12 lunar cycle of a year and reported, significant inverse correlation between the trap catches and degree of moon phase.

Vaishampayan and Rajesh Verma (1982) analysed four years data of light trap (using mercury vapour lamp) catches of *H. armigera* at Jabalpur on the effect of moon light on the trap catches. In all the lunar cycle, in all the years (1975-79) the trap catch was constantly very low on moon light around full moon nights. Result revealed a significant linear and negative correlation between the degree of moon phase (moon illumination) and light trap catches with a value exceeding –0.85 in all the years. The theory of suppression of response of males on full moon and around due to brightness of moon light was appeared beyond doubt. However all these studies were carried out using light trap not the pheromone trap because in the case of leafminer (*A. modicella*) there was a highly significant difference in the moth catch in a light trap and in a pheromone trap (Ranga Rao and Whightman, 1994). In a pheromone trap, the moths trapped were to a maximum of 2500 males/ week compared to only 50 males in a light trap. This trend was repeated in several broods of leafminer moth over years.

2.5. Impact of cultural practice

2.5.1. Date of sowing

The effect of sowing date on leafminer infestation was contradictory, Lewin *et al.* (1979) found that early sowing of groundnut led to higher infestation of GLM, and Logiswaran *et al.* (1982) clarified that late planting was more heavily attacked. The observation made at ICRISAT on the early sown rainy season crop where heavy leaf miner infestation were observed at 15 days after emergence of crop indicating the possibility of manipulating sowing date as a control method has a remote possibility.

The effect of sowing date on the incidence of *S. litura* on different groundnut cultivars has been reported at Dharwad (Patil *et al.*, 1996). Pod yield and incidence of *S. litura* tend to decrease with delay in sowing date from 15th June to 30th July, irrespective of cultivars. Mean pod yield of 2684 kg/ha. was obtained from 15th June sown crop which was calculated to be 15%, 48% and 63% higher than that of the yield obtained from the crop sown on 30th June, 15th July and 30th July respectively. However, a lower % of leaf damage due to *S. litura* was noticed in 30th July sown crop (19%) as compared to 52% foliage damage in the 1st June sown crop and 41.7 % and 31% in crops sown at 30th June and 15th July respectively.

2.5.2. Effect of Intercropping

In India, groundnut is grown in between the rows of cassava (*Manihot esculenta),* a crop of 6 to 8 month period and sunhemp (*Crotalaria juncea*) which has been reported to be producing phenolic compound (Kandaswamy and Prasad, 1979) repelling soil insects (CIAT, 1987).

In Meghalaya groundnut intercropped with maize had a significantly lower leaf miner damage compared to sole groundnut (Gangwar *et al.*, 1994). Singh *et al.*(1991) investigated the influence of inter cropping of red gram, green gram, sorghum, soybean and sesamum with groundnut on the incidence of different insects in Delhi. When red gram was intercropped in groundnut, the leaf hopper population was found to be more on groundnut than on sole crop. But there was considerable reduction in the population of thrips, *C. indicus* when compared to sole groundnut.

The population of *S.litura* was slightly increased with the intercropping of red gram as compared to sole groundnut crop. The groundnut-pearlmillet intercropping system had significantly higher population of predatory coccinellad beetles than the sole groundnut crop (Kennady *et al.*, 1994). The population of leafhopper and thrips on groundnut were less when groundnut was grown with pearl millet and sorghum.

2.6 Natural enemies

Natural enemies play a vital role in minimizing the population of leafminer (*A. modicella*). About 47 hymenopteron parasites have been recorded on groundnut leaf miner (Shanower, 1993). Predominantly *Goniozus* sp. is the parasitoid recorded in the post rainy season, while *Stemomisius japonicus* and *Apanteles* sp. during the rainy season. A detailed list of natural enemies of the leafminer is being given in the chapter 3. Data collected by Ranga Rao (Personal communication) over 13 years on the parasites in leafminer during rainy and post rainy season at Hyderabad is given in table 2.3. The tropic relationship of the groundnut leaf miner and their natural enemies depicted by Shanower *et al.*, (1992) is given in Fig 1.

Table 2.3. Leafminer larval parasitism^{-1} in different seasons at ICRISAt centre

Year	(%) Parasitism	
	Rainy season	Postrainy season
1984	-	42
1985	90	58
1986	21	62
1987	-	45
1988	44	48
1989	10	32
1990	36	6
1991	16	27
1992	33	40
1993	21	-
Mean	34	40

1.Sample size 100 larvae season^{-1}

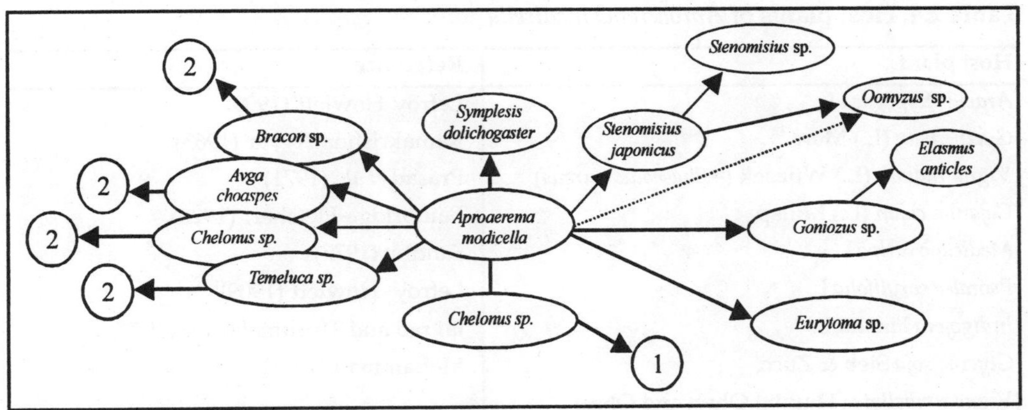

1. *Pteromalus* sp., *Eupelmus* sp., *Tetrastichus* sp.
2. *Oomyzus* sp., *Aphanagmus fijiensis*

Fig.1. Trophic relation between *Aproaerema modicella* and its parasites at ICRISAT, India

Goniozus sp. is the dominant parasitoid in the postrainy season and *Stenomesius japonicus* and *Apantles* sp. in the rainy season.In sprayed areas where population was low during April the parasitism was maximum (100%).Similarly in the intercrop system, there was an increased parasitism of leafminer larvae with cowpea as intercrop (28%), pearl millet (26%), sorghum (25%) and black gram (26%) compared to sale groundnut crop (22%) (Logiswaran and Mohanasundaram, 1985). There were about 44 species of parasitoids, 24 predators, 4 nematodes and 13 entomopathogens recorded on *S. litura* in India (Ranga Rao *et al.*, 1993). The egg parasitoids belong to Trichogrammatidae, Scelionidae and Braconidae. Among the parasitoids, 47% of the parasitoids are braconids, 19% ichneumonids, 16% techanids, 10% eulophids, 3% chalcids and 2% by selionids, encrytids and muscids. Altogether it was 84% by hymenopteron parasitoids and 16% by dipteran parasitoids, the parasitoids recorded on *H. armigera* were mostly belong to Diptera and Hymenoptera in the families of Ichneumonidae, Braconide and Trichogammatidae. The egg parasitoid, *Trichogramma chilonis* has been reported to parasitise upto 79% (Greathead and Girling 1982).

2.7 Alternate hosts

The leafminer being oligophagous, feeds mainly on leguminous host plants with an exception of *Boreria hispida,* (Rubiacae) (Shrinivasan and Shivarao, 1984). The list of alternate host plant is given in Table 2.4. It feeds to a smaller extent on these hosts compared to groundnut and soybean, which are main hosts of leafminer. Exceptionally more than 3000 leafminer larvae were recorded on a single shrub of *Psoralea corylifolia* L. (Mohansundaram and Chandramohan, 1986) indicating the potential of this plant serve as an alternate host for leafminer.

Table 2.4. Host plants of *Aproaerema modicella*

Host plant	Reference
Arachis hypogaea L.	Lefroy Howlett (1909)
Gycine max (L.) Merr.	Ramakrishna Ayyar (1963)
Vigna radiata (L.) Wilzeek (=*Phaseolus aureus*)	Prasad *et al.*, (1971)
Cajanus cajan (L.) Millsp.	Bainbridge-Fletcher, (1920).
Medicago sativa L.	Sandhu (1977)
Psoralea corylifolia L.	Lefroy Howlett (1909)
Indigofera hirsuta L.	Jai rao and Thirumalachar (1977)
Glycine soja Sieb & Zucc.	Mohammad (1981)
Vigna umbellate (Thunb.) Ohwi and Ohasi (=*Phaseolus calacaratus*)	Jai Rao and Thirumalachar (1977)
Trifolium alexandrium L.	Thintadarya *et al.*, (1979)
Teramnus labiolis (L.) Spring	Das and Misra (1984)
Lablab purpureus L.	Das and Misra (1984)
Rhynchosia minima Dc.	Srinivasan and Siva Rao (1984)
Boreria hipda K.Sch.	Srinivasan and Siva rao (1984)
Boerhavia diffusa L.	Kennady *et al.*, (1992)
Amaranthus viridis L.	Kennady *et al.*, (1992)

Mule and Chari (1978) reported 96 crops and 61 weeds damaged by *H. armigera*. Among them, in recent years groundnut crop has become one of the major host plants for this pest. Singh and Jalali (1997) compiled the alternate hosts for *S. litura*. This polyphagous species feeds on more than 120 host plants in India. In recent years, under storage conditions, the groundnut bruchid *Caryedon serratus* has been cause of concern (Motka, 1997). There were about 49 plant species recorded as different host for this insect. About 33 different plants under the field conditions have been recorded as hosts for the bruchid.

2.8. Varieties

The introduction of new varieties, which have good response to synthetic fertilizers in a few cases, has increased the pest status of few species. For instance, when the variety Girnar 1 was introduced, it showed high resistance to sucking insect such as leaf hopper and thrips with an additional character of resistance to leaf spot disease but high susceptibility to leaf miner causing 100% foliar damage.

POL-2, S-206, Karod 4-1, TMv-11 and GG-2 in an alternative choice test, Kadiri-3, Girnar-1, BG-2 were found to be highly resistance to *S. litura*. Based on oviposition, pod damage, loss of pod weight and developmental period, the cv. ICGS-11, SB-11 were classified as least susceptible to *C.serratus*

(Gorpade et. al., 1998). Similarly in the oviposition preference RSB-1, TMV-3 and RS-1 were found to be least preferred for oviposition compared to TMV-2 with 205 eggs (Motka, 1997). There are some high yielding cultivars released against multiple pest situation from ICRISAT given in table 2.5.

Table 2.5. Some multiple pest resistant genotypes

Genotype	Resistant to
NC AC 343*	Termite, Pod borer, Thrips, Leaf hopper, Aphids (Tolerance), Leafminer, Bud necrosis diseases
Myria PC	Thrips, Leaf hopper, Leafminer, Bud necrosis diseases
Ah 7215	Thrips, Leaf hopper, Leafminer, Bud necrosis diseases
APAU C4	Thrips, Leaf hopper, Leafminer, Bud necrosis diseases
EC 36892*	Thrips, Leaf hopper, Aphids, Bud necrosis diseases
M 13*	Termite, Thrips, Leaf hopper, Leafminer, Bud necrosis diseases
ICGPRS 312*	Thrips, Leaf hopper, *Spodoptera litura, Spodoptera exiua*
ICGV 86687*	Leaf hopper, Leafminer, Bud necrosis diseases, Rust
ICGV 87162*	Leaf spot

* High yielding cultivars

The high yielding cultivars released such as NCAc 343, M 13, ICGPRS 312, ICGV 86687 were multiple pest resistant cultivars. BG-2 was found to be highly resistant to leaf miner even though, 100% incidence was recorded on this variety, there was no further development of larvae. Several genotypes have been identified in field condition as well as under laboratory choice test which were resistant to *S. litura* (Nandagopal *et al.*,1997).

2.9. Life Tables

Among the ecological techniques, the life table studies are one of the very useful tools to bring out the population fluctuation in a given ecosystem. Such life table can reveal the real population changes throughout the lifecycle of a species in its natural environment. When this observations are related to several causes of key factors for the mortality, the life tables from a budget of the successive processes which operate in a given population in that environment. A stable age distribution of *A. modicella* has been reported by Madhumati and Thippuswamy (1997). It shows clear broods in both summer and *kharif* crops. The intrinsic rate of increase (rm) and the life table indicated that in the 1st generation of kharif for larval phase I, larval phase II, pupae and adults were 44.5, 29.7, 88.3, 3.1 and 4.5% respectively for the 1st generation. While, in 2nd generation 39.2, 28.1, 22.3, 3.6, 3.7% respectively. Life tables for *H. armigera* on groundnut has been reported by (Koshiya and Patel, 1987). The mortality in younger larvae (1st and 2nd instar) was high (25.6 and 26%) due to unknown causes. The population of older larvae (III, IV, V and VI) declined by 22.8 and 23.1% due to NPV infection (Table 2.6 and 2.7). The negative values of population trends index (1.1 and 1) indicated that the mortality factors operating during summer were not effective in suppressing the *H. armigera* population on groundnut crop. Pupal mortality due to failure to complete development or inability of the imago to eclose was 7.6 and 6.7% during the two summer seasons. A few pupae also died due to larval/pupal parasitoid, *Eucarcelia illota* (Curr.).

Table 2.6. Life table ot *H. armigera* during the summer (February to may, 1982) on groundnut Population/ha).

x	lx	dxf	dx	100qx	sx
Expected Eggs	36513	Mortality/sterility	4763	15.00	0.87
Viable eggs	31750				
Younger larvae		Parasitoid	0000	0.00	
(N1)	31750	Disease	0000	0.00	0.74
(1st & 2nd instars)		Unknown	8125	25.59	
		Parasitoid	875	3.70	
Older larvae	23625	Disease	5375	22.75	0.69
(3rd to 6th instar)		Unknown	1000	4.23	
			7250	30.68	
		Parasitoid	625	3.82	
Pupae	16375	Disease	500	3.05	0.87
		Unknown	1250	7.63	
			2375	14.50	
Moths	14000	(Sex 50% Female)			1.00
Female X_2 (N_3)	14250				1.00
Normal Female X_2	14250				1.00
General total	22513			60.97	

Expected eggs	4624125
No. of dead sterile eggs	693619
Viable eggs	3939506
Expected no. of younger larvae	3930506
Actual no. of younger larvae (N_2)	33163
Trend index = N_2/N_1	1.4
Generation survival (SG) = N_3/N_1	0.45

Table 2.7.Life table ot *H. armigera* during the summer (February to may, 1982) on groundnut (Population/ha).

x	lx	dxf	dx	100qx	sx
Expected Eggs	30659	Mortality/sterility	4292	14.00	0.86
Viable eggs	26367				
		Parasitoid	750	2.84	
Younger larvae (N1)	26367	Disease	000	0.00	0.74
(1st & 2nd instars)		Unknown	6125	23.23	
			6875	26.07	
		Parasitoid	250	1.28	
Older larvae	19500	Disease	4500	23.08	0.65
(3rd to 6th instar)		Unknown	2000	10.28	
			6750	34.64	
		Parasitoid	250	1.96	
Pupae	12750	Disease	750	5.88	0.85
		Unknown	875	6.68	
			1875	14.52	
Moths	10875	(Sex 50% Female)			1.00
Female X_2 (N_3)	10875				1.00
Normal Female X_2	10875				1.00
General total	19792			64.53	
Expected eggs			3726155.63		
No. of dead sterile eggs			521661.79		
Viable eggs			3204493.84		
Expected no. of younger larvae			3204493.84		
Actual no. of younger larvae (N_2)			27136		
Trend index = N_2/N_1			1.03		
Generation survival (SG) = N_3/N_1			0.41		

2.10. Forecasting

Forecasting in groundnut is still in the infancy state. Only a few reports are available, that too on the abiotic factors on the incidence of insect pest in groundnut. At Maharashtra, high, maximum and minimum temperature and low humidity in the pre-monsoon period followed by continuous rain for about 3 weeks with maximum and minimum temperature in the range of 27°C, 32.7°C and 21.3 to 23.7°C respectively, during mid June and mid July, followed by monsoon break for a week or two favoured outbreak of pests (Dubey *et al.,* 1995). Madhumati and Thippeswamy (1997) calculated net reproductive rate and different rates of increase in abiotic factors favored rate of multiplication, highest at 31.9°C and 64.16 mm rainfall. Shanower *et al.*(1993) calculated different degree days requirement for different phonological phages of leaf miner. Rao *et al.* (1989) in a laboratory studies calculated the developmental threshold and thermal requirement for different stages of *S. litura*.i.e. 37°C for all stages. Anon (1997) worked out a stepwise correlation between aphid and abiotic factors likes maximum, minimum and average temperature had significantly negative relation. The relation worked out between the temperature and aphid was y = 1839.57 + (-60.65) x (Average).

References

Amin, P.W. (1983). Major field insect pests of groundnut in India and associated crop losses. Special Issue, *Indian Journal of Entomol.,* 2:337-344.

Amin, P.W. (1987). Insect pests of groundnut in India and their management. In: *Plant Protection in Field Crops,* pp. 219-233. Veerabhadra Rao, M. and Sithanantham, S. (Eds). Plant Protection Association of India, Rajendranagar, Hyderabad, India.

Amin, P.W. (1988). Insect and mite pests and their control. In *Groundnut* (Ed. P.S. Reddy). pp. 393-452. Indian Council of Agricultural Research, New Delhi.

Anonymous, (1998). Proceedings of the All India Co-ordinated Research Project on Groundnut.

Anonymous, (1997). Annual Report, National Research Center for Groundnut (Indian Council of Agricultural Research), Junagadh, Gujarat. pp,69.

Anonymous, (1993). Annual Report (1st April, 1992 to 31st March, 1993). All India Co-ordinated Research Project on Biological Control of Crop Pests and Weeds. Biological Control Centre (ICAR-NCIPM)., Bangalore, India. Tech. Doc. No. 40, 1-313.

Anonymous, (1994). Elite groundnut germplasm lines ICGV 86252, ICGV 86393, ICGV 86455, ICGV 86462. *Plant-Material-Description.* 1994, No. 47, 4 pp. International Crops Research Institute for the Semi-Arid Tropics. ICRISAT, Patancheru, Andhra Pradesh 502 324, India.

Bainbridge-Fletcher, T. (1920).. Life histories of Indian insects: Microlepidoptera. Memoirs, Department of Agriculture, India, Pusa. *Entomology Series.* 6:77-79.

Bapodra, J.G. (1988). Studies on biometrical analysis, biology, population dynamics and chemical control of *Heliothis armigera* (Hubner) on different hosts. Ph.D thesis submitted to G.A.U., Sardar Krishinagar, Guajarat.

Bowden, J and Church, B. M. (1973). The influence of moon light on the catches of insect in light trap in Africa. Part II. The effect of moon phase on light trap catches. *Bulletin of Entomolgical Research.* 63:129-142.

Cherian, M.C. and Basheer, M. (1942). Studies on *Stomopteryx nerteria* Meyr., a pest of groundnut in the Madras Presidency. *Madras Agricultural Journal.* 30:379-381.

CIAT, (1987). Research highlight for 1986, Cali, Columbia.

Das, B.B. and Misra, D.P. (1984). New records of alternate hosts of groundnut leafminer, *Aproaerema modicella* Deventer (Gelechiidae: Lepidoptera). *Indian Journal of Agricultural Science.* 54:776.

David, H. and Anthanarayana, K. (1974). Some aspects of biology of whitegrubs in sugarcane in Tamil nadu. *Whitegrubs Newsletter.* 1:14-15.

Desai, and Patel, (1965). Some observation on the biology and control of white grubs in soil (*Hololotricha* near *consanguinea* Blanch) affecting groundnut and cereals in Gujarat. *Indian Journal of Entomology.* 27:89-94.

Dubey *et al.,* (1995). Life histories of Indian insects: Microlepidoptera. Memoirs, Department of Agriculture, India, Pusa, *Entomology Series* 6, 77-79.

Gangwar, S.K., Singh, Y.P. and Patel, C.S. (1994). Influence of intercropping on infestation by insect pests of crops at medium-high altitude of Meghalaya. *Indian Journal of Agricultural Sciences.* 64(2):137-140.

Ghorpade, S.A., Ghule, S.L. and Thakur, S.D. (1998). Relative susceptibility of some groundnut cultivars to pod borer *Caryedon serratus* Olivier in storage. Seed Research. 26 (2):174-177.

Greathead, D.J. and Girling, D.J. (1982). Possilities for Natural Enemies in *Heliothis* Management and the Contribution of the Commonwealth Institute of Biological Control.pp. 147-158 *In* Proceedings of the International Workshop on *Heliothis* Management, 15-20 November1981, ICRISAT, Patancheru, AP., India.

Gupta, B. D. and Avashti, P. N. (1957). Observation on a new beetle pest of sugarcane crop in Bihar. *Indian Sugar.* 7(9):587-594.

Gupta, H.C.L., Misra,R.P. and Mittal, V.P. (1976). Evaluation of insecticides against *Caryedon serratus* (Ol.) (Bruchidae: Coleoptera). *Buletin of Grain technology.* 14 (1) : 68-69.

Jai Rao, K. and Thirumalachar, D.K. (1977). New record of alternate host plants of groundnut leafminer *Stomopteryx subsecivella* Zeler (Syn.: S.neteria Meyrick). (Lepidoptera: Gelechiidae). *Current Scienceence.* 46:91-92.

Jayanthi, M., Singh, K.M. and Singh,R.N. (1993). Pest complex of a high yielding groundnut variety Mh4 under delhi conditions. *Indian Journal of Entomology.* 55(1):30-33.

Jayanthi, M., Singh,K., M. Singh,R. N. (1993). Population build up of insect pests on MH 4 variety of groundnut influenced by abiotic factors. *Indian Journal of Entomology.* 55(2):109-123.

Kandaswami & Prasad (1979). *Soil Biology and Biochemistry.* 11:7375.

Kapadia, M.N., Bharodia, R.K. and Vora, V.J. (1982). Biology and estimation of incidence of groundnut leafminer, *Stomopteryx subsecivella* Zell (Gelechiidae: Lepidopter). *GAU. Research. Journal.* 8:37-39.

Karla, A.N. and Kulshreshtha, J.P. (1961). Studies on the biology and the control of *Lachnosterna consanguinea* (Blanch.), a pest of sugarcane in Bihar(India). *Bulletin of Entomological Research.* 51:577-587.

Kennady, F.J.S., Balagurunathan. R., Christopher,A and Rajamanickam , A. (1994). Insect pest management in peanut: a cropping system based approach. *Tropical Agriculture* 71 (2), 116-118.

Kennedy, F.J.S., K. Rajamanickam, R.Balagurunathan, and Lourduraj, A.C. (1992). Yield loss due to insect pests and their control in groundnut. *Groundnut News.* 4(2):7.

Khan, M.I. and Rao Deo. A.K. (1987). Importance of larval parasites in the control of *Stomopteryx subecivella* Zeller. *Journal of Maharashtra Agricultural University.*, 3, 261-263.

Koshiya, D.J. and Patel, H.K. (1987). Life table and population dynemics of *Heliothis armigera* Hubn. on groundnut. *G.A.U. Research Journal.* 12:11-14.

Leela David, A., V.M.kalyanaraman and P.S.Narayanaswamy. (1964). Distribution,status and seasonal variations in population density of the pests of sugarcane in Madras state. *Indian Journal of Sugarcane Research.* 8:367-377.

Lefroy Hawlett. (1909). Manual of Entomology with special reference to economic Entomology. *Agricole Reprints Corporation*, New Delhi. Pp.539.

Lewin, H.D., Saroja, R., Leela David, A., and Padmanabhan, M.D. (1979). Influence of sowing time and weather on the incidence of groundnut leafminer. *Indian Journal of Agricultural Science.* 49:886-891.

Logiswaran, G, Madhava Rao, G.vasudevan and V.Kannan. (1982). Influence of time of sowing and weather factors on the infestation of leaf miner *Aproaerema modicella* Deventer and yield in rainfed groundnut. *Madras Agricultural Journal.* 69:359-363.

Logiswaran, G. and Mohanasundram, M. (1985). Effect of intercropping, spacing and mulching in the control of groundnut leafminer, *Aproaerema modicella* Deventer (Gelechiidae:Lepidoptera). *Madras Agricultural Journal.* 72:695-700.

Madhumathi, T. and Thippeswamy, M. (1997). Stable age distribution of *Aproaerema modicella* Deventer (Lepidoptera: Gelechiidae) in groundnut ecosystem. *Journal of Research ANGRAU.* 25 (4):29-35.

Mohansundaram and Chandramohan, N. (1986). *Psoralea corylifolia*, an alternate host for carry over of groundnut leafminer (*Aproaerema modicella* Deventer). *Madras Agricultural Journal.* 73:591.

Maxwell-Lefroy, H. and Howlett, F.M. (1909). Indian insect life: a manual of the insects of the Plain (Tropical India), reprinted 1971 by Today and Tomorrows' Publishers, New Delhi.

Mohammad, A. (1981). The groundnut leaf miner (*Stomopteryx subsecivella* Zeller) (Lepidoptera: Gelechiidae). A Review of world literature. Occasional paper 3, Groundnut Improvement Program, Interational Crops Research Institute for Semi-Arid Tropics (ICRISAST), Patancheru, A.P., India. pp.21.

Motka, M.N. (1997). Biology, Population Dynamics, Varietal screening and Management of Groundnut beetle *Caryedon serratus* (Olivier). Ph.D.thesis submitted to the Gujarat Agricultural University, Junagadh pp260.

Motka, M.N., Bhalani, P.A. and Bharodia, R.K. (1985). Growth and development of *Aproaerema modicella* (Deventer). (Lepidoptera: Gelechiidae). on certain groundnut varieties. *Agricultural Science Digest* (India). 5:162-164.

Mule, N.M. and Chari, M.S. (1978). Chemo sterilization of male moths of tobacco capsule borer (*Heliothis armigera* Hubn.). *Tobacco Research.*, 4, 16-23.

Mule, N.M. and Chari, M.S. (1976). Chemosterilization of male moths of tobacco capsule borer (*Heliothis armigera* Hubn.). *Tobacco Researchearch.* 4:16-23.

Nandagopal,V. S.K.Yadav, J.B.Misra, A.Bandyopadhyay, V.Singh, M.S.Basu and Premnarayan.(1997). Status of *Spodoptera litura* resistance in Groundnut in India. Proceedings of the Natioanl Scientist Forum on *Spodoptera litura* (F.). 2-4, August,1996. ICRISAT Asia Centre, India Patancheru, A.P. and 94,Victoria Street, London,SW1 SJL,U.K., Overseas Development Administration pp158.

Nandagopal, V. (1992). Studies on Integrated pest management in Groundnut in Saurashtra, Ph. D. thesis, Saurashtra University, Rajkot, pp 246.

Nemac, S. J. (1971). Effect of lunar phase on light trap collection and population of bollworm moth. *Journal of Economic Entomology*. 64:860-863.

Persoon, B. (1976). Influence of weather and nocturnal illumination on the activity and abundance of populations of noctuids (Lepidoptera) in south coastal Queensland. *Bulletin of Entomological Research*. 66(1):33-63.

Prasad, S. K. (1961). Sugar cane and its problem. White grub injuring sugarcane and there control. *Indian Sugar* 11:1-4.

Prasad, S.K., Chawla, M.L., Kumar, S. and Saxena, H.P. (1971). Root-knot nematode, *Meloidogyne javanica* (Treub, 1885) Chitwood, 1949 and stem borer, *Stomopteryx nertaria* Meyrick affecting green gram, *Phaseolus aureus*. *Indian Journal of Entomology*. 33:55-60.

Rai, B. K., Joshi., H. C., Rathore, Y. K., Dutta, S. M. and Shinde, V. K. R. (1969). Studies on bionomics and control of *Holotrichia consanguinea* Blanch., in Lalsot, Jaipur Distict, Rajasthan. *Indian Journal of entomology*. 31:132-142

Rai, K. (1976). Pests of oilseed crops in India and their control. Indian Council of Agricultural Research (ICAR)., New Delhi. 121 Pages.

Ramakrishna Ayyar, T.V. (1940). A handbook of economic entomology for South India. McMillan Publ. (India), New Delhi.

Ranga Rao, D.V., Singh, K.N., Wightman, J.A., and Ranga Rao, G.V. (1991). Economic Status of neem cake for termite control in groundant. International Arachis Newsletter, 9-12-13.

Ranga Rao, G.V., and Whightman J.A. (1994). First Annual Rabi/Summer Groundant Research Worker Group Meeting held at Indian Institute of Technology, Karagpur, West Bengal from 12-15 September, 1994.

Ranga Rao, G.V. and Whiteman, J.A. (1997). Techniques for screening groundnut genotypes, pp.67-75. In: *Spodoptera litura* in India: *Proceedings of the National Scientists Forum on Spodoptera litura* (F.). (April 2-4, 1996). ICRISAT Asia Centre (J.A.Wightman and G.V. Ranga Rao), ICRISAT, Patancheru, Andhra Pradesh, India, 146+6 pp.

Ranga Rao, G.V., Reddy,D.D.R., and Shanover, T.G. (1993). Bio-Ecology and management of groundnnut leaf miner, *Aproaerema modicella*, Paper presented in Group discusssion on Integrated Pest Management Strategies in Oilseeds in India, Held on 23-24 December,1993, Punjab Agricultural University, Ludhiana.

Rao Deo, A.K. (1974). A note on white grub control in Maharashtra State. Whitegrub Newsletter. 1:19-20.

Ranga Rao, G.V.R., Wightman, J.A. and Rao, D.V.R.(1989). Threshold temperatures and thermal requirements for the development of *Spodoptera litura* (lepidoptera: Noctuidae). *Environmental Entomology*. 18:548-551.

Sandhu, G.S. (1977). Some insects recorded as pests of fodder crops at Ludhiana (Punjab). *Punjab Agriculture University Research Journal*. 14:449-459.

Sandhu, G.S. (1978). Note on the incidence of *Stomopteryx subsecivella* (Zell.). (Lepidoptera: Gelechiidae). on lucerne at Ludhiana. *Indian Journal of Agricultural Sciences*. 48:53-54.

Shanower, T. G., Das, V. S. R. and Pal, S. K. (1993). A typical oviposition behaviour in *Helicoverpa armigera* (Hubner). *International Chickpea Newsletter*. 28:16-17

Shanower, T.G. (1989). The biology, population dynamics, natural enemies and impact of the groundnut leafminer, *Aproaerema modicella* (Deventer). (Lepidoptera Gelechiidae) on groundnut in India. Ph.d. dissertation, Univ. of California, Berkeley, USA.

Shanower, T.G., Wightman, J.A., Gutierrez, A.P. and Ranga Rao, G.V. (1992). Larval parasitoids and pathogens of the groundnut leaf miner, *Aproaerema modicella* (Lep.: Gelechiidae)., in India. *Entomophaga*. 37(3):419-427.

Singh, H., Rohilla, H.R., Kalra, V.K. and Singh, H. (1989). Control of termites (*Odontotermes obesus* Ramb.) in groundnut through seed treatment with aldrin. *Indian Journal of Plant Protection*. 17:263-264

Singh, S.P. and Jalali, S.K. (1997). Management of *Spodoptera litura* (Fabricius) (Lepidoptera: Noctuidae). pp.27-65. In: *Spodoptera litura* in India: *Proceedings of the National Scientists Forum on Spodoptera litura* (F.). (April 2-4, 1996). ICRISAT Asia Centre (J.A.Wightman and G.V. Ranga Rao), ICRISAT, Patancheru, Andhra Pradesh, India, 146+6 pp.

Singh, T.V.K., Singh, K.M. and Singh, R.N. (1991). Influence of intercropping. III. Natural enemy complex in groundnut. *Indian Journal of Entomologyomol*. 53(3):333-368.

Srinivasan, S. and Siva Rao, D.V. (1984). New off-season summer hosts of the groundnut leafminer, *Aproaerema modicella* Deventer, (=*Stomopteryx subsecivella* Zell.). (Lepidoptera: Gelechiidae). in Andhra Pradesh. *Entomon*. 9:11-12.

Sunil Kumar (1971). Integrated control of white grub, *Lachnosterna consanginea* Bl. Whitegrubs Newsletter. 1:28-33.

Thontadarya, T.S., Jai Rao, K. and Kumar, N.G. (1979). Occurrence of the groundnut leafminer *Biloba* (*Stomopteryx*) *subsecivella* (Zeller) on berseem (*Trifolium alexandrinum* L.). in Karnataka. *Current Research*. 8:65.

Twine, P. H. (1973). Egg parasites of *Heliothis punctigera* and *H. armigera* in south-eastern Queensland. *Queensland Journal of Agricultural and Animal Sciences*. 30(4):331-336.

Vaishampayan, S. M. and Rajesh Verma.(1982). Influence of moon light and lunar periodicity on the light trap catches of gram podborer, *Heliothis armigera* (Hubner) moths. *Indian Journal of Entomology*. 44(3):206-212.

Vaishampayan, S. M. and Shrivastava, S. K. (1978). Effect of moon phase and lunar cycle on the activity of light trap catch of tobacco caterpillar *Spodoptra litura* (F). *Journal of Bombay Natural History Society*. 75(1):83-87.

Vanhall, C. J. J. (1922). (Disease and pests of cultivated plants in the Dutch East Indies in 1921). Zeikten en Plagen der cultuurge–wassen in Nederlandsche– Indie in 1921. Mededelingen van het Institut voor Plantenziekten. Buitenzorg No.53. pp 46.

Veeresh, G. K. (1977). Root grub controlling campaign in Karnataka. *Whitegrub Newsletter*. 1:28-33.

Veeresh, G.K. (1977). Bionomics and control of white grubs in India. *Sugarcane Newsletter*. 9:44 -56.

Wheatley, A.R.D., Wightman, J.A., Williams, J.H. and Wheatley, S.J. (1989). The influence of drought stress on the distribution of insects on four genotypes grown near Hyderabad, India. *Bulletin of Entomological Research*. 79:567-577.

□□□

3

Aproaerema modicella (Deventer)

G. V. RANGA RAO, V. NANDAGOPAL AND D. D. R. REDDY

3.1. Introduction

The groundnut leaf miner (GLM), *Aproaerema modicella* (Deventer) (Lepidoptera: Gelechiidae), is a major insect pest of groundnut and soybean in many south Asian and south-east Asian countries (Wightman *et al.*, 1990). It is the most important insect pest causing considerable damage to groundnut in India (Amin, 1983). Originally described as *Anacampsis nerteria* Meyr. from specimens collected in India (Meyrick, 1906), four other binomials have referred to the same pest : *Biloba subsecivilla* Zeller, *Stomopteryx nerteria* Meyrick, *Stomopteryx subsecivella* Zeller, and *Aproaerema nerteria* Meyrick. According to a recent opinion, a species from South Africa is now called *Stomopteryx subsicivella* (Zeller); the second is the Indian – Indonesian groundnut leaf miner, *Aproaerema modicella* (Deventer) (Mohammad, 1981). Mohammad (1981) originally described *A.modicella* from a moth collected in Java, Indonesia.

Various common names have also been applied to this insect : Arachis leaf miner (Van der Laan and Ankersmit, 1951), madu puchi (Reddy, 1968), surul - poochi (Bainbridge Fletcher, 1914; Ramakrishna Ayyar, 1963), the groundnut leaf webber (Sadakatulla *et al.*, 1976) and the soybean leaf miner (Gujarati *et al.*, 1973).

3.2 Distribution

The geographical distribution of GLM is restricted to south and south-east Asia, from Pakistan to China and as far south as the Philippines and Sri Lanka. It has been reported from Burma (Frohlic and Rodewald, 1963), China (Yang and Liu, 1966) India (Meyrick, 1906), Indonesia (Vanhall, 1922); Pakistan (Atwal, 1976); The Philippines (Litsinger *et al.*, 1978), Sri Lanka (Lefroy and Howlett, 1909; Nayyar *et al.*, 1976) and Thailand (Keeratikasikorn and Hiryansaree, 1975). In India, GLM is found in Andhra Pradesh

(Channabasavanna, 1957), Gujarat (Nandagopal, 1992), Karnataka (Channabasavanna, 1951) Madhya Pradesh (Gujarati, *et al.*, 1973), Maharashtra and Tamil Nadu (Anon. 1941, 1950, 1963; Barber, 1906; Lefroy, 1907; Cherian and Basheer, 1942; Ramakrishna Ayyar, 1963), Punjab (Sandh, 1978) and Rajasthan (Srivastava *et al.*, 1962).

3.3. Host range

GLM is an oligophagous pest and feeds only on leguminous host plants (Table 3.1) with the exception of *Boreria hispida* of Rubiaceae (Shanower *et al.*, 1993b). Several hosts are crop plants, including groundnut, soybean, pigeon pea and alfalfa.

Table 3.1. Host plants of *Aproaerema modicella*

Host plant	Reference(s)
Arachis hypogaea L.	Lefroy and Howlett, (1909)
Gycine max (L.) Merr.	Ramakrishna Ayyar, (1963)
Vigna radiata (L.) Wilzeek (=*Phaseolus aureus*)	Prasad *et al.*, 1971
Cajanus cajan (L.) Millsp.	Bainbridge Fletcher, (1920)
Medicago sativa L.	Sandhu, (1977)
Psoralea corylifolia L.	Lefroy and Howlett, (1909)
Indigofera hirsute L.	Jai rao and Thirumalachar, (1977)
Glycine soja Sieb & Zucc.	Mohammad, (1981)
Vigna umbellate (Thunb.) Ohwi and Ohasi (=*Phaseolus calacaratus*)	Jai Rao and Thirumalachar, (1977)
Trifolium alexandrium L.	Thintadarya *et al.*, 1979
Teramnus labiolis (L.) Spring	Das and Misra, (1984)
Lablab purpureus L.	Das and Misra, (1984)
Rhynchosia minima Dc.	Srinivasan and Siva Rao, (1984)
Boreria hipda K.Sch.	Srinivasan and Siva rao, (1984)
Boerhavia diffusa	Kennedy *et al.*, (1992)
Amaranthus viridis	Kennedy *et al.*,1992

3.3 Nature of injury

The young larvae mines into the leaves as soon as they hatch. The presence of small brown blotches on the leaf can be seen. **(Plate 1),** The mines are about 1mm long when noticeable. The mines enlarge as the larvae grow. When they become too large and the larvae completes the different instars, emerge out and web the adjacent leaflet's together, and continuously feed on leaf tissue from inside the webbed leaves. A severely attached field presents a burnt up appearance **(Plate 2).**

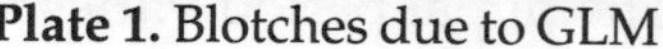

Plate 1. Blotches due to GLM

Plate 2. Burnt up appearance

3.4. Life cycle

The life cycle of GLM has been studied by several workers (Lefroy and Howelt, 1909; Bainbridge Fletcher, 1920; Cherian and Basheer, 1942; Narayanan, 1961, Ramakrishna Ayyar, 1963; Reddy, 1968; Bindra, 1970; Gujrati *et al.*, 1973; Kothai, 1974; Atwal, 1976; Rai, 1976; Sanchu, 1978; Kapadia *et al.*, 1982, and Phisitkul, 1985). Small (<1.0 mm) oval eggs are laid on the undersides of the groundnut leaflets, petioles and stems. The pre-oviposition period is 1-3 days and fecundity averages between 86.6 and 185.8 eggs per female, the maximum being 473 eggs (Cherian and Basheer, 1942). Egg production is significantly lower below 15° C and above 35° C (Shanower *et al.*, 1993a). Eggs hatch in 3-4 days but may require 6-7 days at lower temperatures (Kapadia *et al.*, 1982) **(Plate 3).**

First instar creamy yellow larvae typically chew through the epidermis to reach the leaf mesophyll upon hatching. Early instars mine the leaflets usually near the midrib and proceed towards the margins by feeding on the

Plate 3. Adult & Eggs of leaf mines

green tissues in between the epidermal layers. After about a week, the larvae leave the mine, fold a leaflet or web together two or more leaflets and continue to feed. In peninsular India, the larval has five instars (Shanower *et al.*, 1993a) but 3-6 larval instars have been reported in the literature (Gujarati *et al.*, 1973; Kapadia *et al.*, 1982; Islam *et al.*, 1983; Amin, 1987). Larval period lasts between 9 and 28 days (Cherian and Basheer, 1942; Sandhu, 1978; Kapadia *et al.*, 1982) and its development requires approximately 325 degree days above a threshold temperature of 11.3°C (Shanower *et al.*, 1993a). Males and females can be distinguished in the larval stage by the distinctive pink gonads of the male which are visible through the cuticle. Final instar larvae are approximately 6.0 mm long and very active **(Plate 4)**.

Pupation occurs within the webbed leaflets in white silken cocoons and the pupae are yellowish to reddish brown. The pupal period varies from 3 to 10 days (Cherian and Basheer 1942; Sandhu, 1978) and requires 72 degree days above a threshold of 14.7° C (Shanower *et al.*, 1993a). Males emerge two days earlier than females and the sex ratio is approximately 1:1 (Lalita Kumari, 1989). The adults live for five to 20 days. The moths are small 8-10 mm in length and grayish, the forewings are darker than the hindwings and have a white spot on the upper margin towards the apical end. The life cycle is completed in 13-28 days in southern India (Cherian and Basheer, 1942), 37-45 days in northern India (Sandhu, 1978) and roughly requires 450 degree days (Shanower *et al.*, 1993a).

Plate 4. Larva of GLM

3.5. Population dynamics

GLM has been an important pest of groundnut in India (Amin and Mohammad, 1980). Continuous cultivation or multiple cropping of groundnut (Krishna Murthy Rao *et al.*, 1962) with the introduction of *Rabi*-summer groundnut crop (irrigated groundnut) and a groundnut/soybean rotation (Van der Laan and Ankersmit, 1951; Feakin, 1973; Wightman and Amin,

1988), allow GLM populations to build up. GLM populations can persist on several wild hosts, even in the absence of groundnut or soybean and more than 3000 GLM larvae have been found on a single *Psoralea corylifolia* L. shrub, indicating the potential of this plant as an alternate host (Manoharan and Chandramohan, 1986). In China the pest over-winters in the pupal stage within the foliage of buried groundnut plants in the soil (Yang and Liu, 1966). However, Kothai (1974) did not observe any adult emergence from the dried groundnut and soybean leaves along the pupae buried and mixed up with soil. Alternatively, GLM may survive the extreme hot, dry Indian summer in pupal aestivation (Jagtap *et al.*, 1985). Alternate host plants may, therefore be an important source of survival. Given the small size of the moths it is very likely that moths are carried by wind across locations and studies on movement of the pest across seasons and locations are needed.

GLM populations fluctuate widely between seasons. Population peaks have been reported in July and August in Thailand (Campbell, 1983) at the end of the rainy season (September and October) specially in drought or low rain fall years (Amin, 1983) and also in November and December (Mohammad, 1981) as well as at the end of the post-rainy season i.e. March and April In Bangladesh and India (Amin and Mohammad, 1980; Islam *et al.*, 1983). Light trap data at International Crop Research Institute for Semi-Arid Tropics (ICRISAT) center during 1989-90 and 1990-91 also indicated wide population fluctuations during different seasons even though population peaks varied between the years.

With the demonstration of presence of sex pheromone in the GLM (Nandagopal and Reddy, 1990; Lalita Kumari and Reddy, 1991) and availability of a synthetic sex pheromone (Hall *et al.*, 1993), it would be a distinct possibility to monitor GLM adult populations using pheromone traps. Pheromone trap data during 1992-93 at ICRISAT center indicated peaks of GLM populations with two highest peaks in the 30-33 standard weeks (July-August) during rainy season and 6-11 standard weeks (February-March) during post-rainy season. Several years of pheromone trap data is required to indicate population fluctuations of GLM over seasons. From our observations under field conditions, distinct larval/pupal stages occur in a given locality and overlapping generations are uncommon. At ICRISAT center larval populations have been recorded regularly from 1980 to 1993 in unsprayed groundnut (cv. Kadiri 3) trials. Population densities ranged from 1 to >320 larvae per plant. Extremely high densities (>50 larvae per plant) were recorded in rainy seasons (1984 and 1987) and post-rainy seasons (1981 and 1982). In other seasons, densities ranged between 10 and 20 larvae per plant.

Three to four GLM generations per season are common on groundnut in India, although five generations have been reported during the rainy season in south India (Logiswaran and Mohana Sundaram, 1986). Two generations per crop are typical in Thailand (Campbell, 1983), whereas in China seven generations have been reported on a single soybean crop (Yang and Liu, 1966). Recently pheromone trap data at ICRISAT center indicated 4-5 generations during the rainy and 6 generations during the post-rainy seasons. Abiotic factors, principally rainfall, humidity and temperature, are frequently suggested as causes of population fluctuations. Lewin *et al.*(1979) found a significant negative correlation between GLM incidence and rainfall: higher GLM incidence was correlated with lower rainfall. Amin (1987) and Khan and Raodeo (1978) suggested that high rainfall reduces leaf miner populations even though data from the latter do not support this conclusion. They observed the high populations in August-September during a high rainfall period, and populations declined in March when no rain was recorded (Khan and Raodeo, 1978). Wheatley *et al.*(1989) also found that water from an overhead irrigation system did not lower GLM density. Reduction in the GLM populations may depend upon the stage present during heavy rainfall period. It is likely that adults are easily washed away by rain resulting in reduced incidence of GLM. Experiments using rain simulator indicated that the physical impact of rainfall on GLM eggs and larvae does not significantly increase mortality in these two stages (Shanower, 1989). Temperature was also positively correlated with GLM incidence and accounted for more of the variation than rainfall (Lewin *et al.*, 1979).

3.6. Yield loss and threshold populations

The GLM reduces yields by feeding on leaves, thereby reducing the photosynthetically active leaf area. Shanower (1989) measured the consumption of individual larvae and calculated that on an average 179.3 mm^2 of leaf area is eaten. The estimation of Islam *et al.*(1983) of 348 mm^2 of leaf tissue (equivalent to 6-10 groundnut leaflets, depending on the genotype) appears to be too large for such a small caterpillar.

Reduced groundnut yields were associated with severe infestations in south India (Kulshreshta, 1964). Yield losses due to GLM were reported to be more than 50 per cent in Tamil Nadu by GLM alone (Logiswaran and Mohanasundaram, 1985) and 16 per cent pod dry weight (equivalent to 303 kg ha-1) by GLM and *Aphis craccivora* Koch. (Jagtap *et al.*, 1984). Yield reductions in the range of 20-30 per cent in groundnuts have been reported from China (Yang and Liu, 1966). Most of the yield reduction studies were based on insecticidal protection of groundnuts (compared to unprotected groundnuts) resulting in increased pod yields. Yield increases up to 24 per cent (Palaniswamy and Ramachandran, 1978); 36 per cent (Lewin *et al.*, 1973); 65 per cent (Sivasubramanian and Palaniswamy, 1983; Rajput *et al.*, 1985); 71 per cent (Lal *et al.*, 1974); 76 per cent (Vittal *et al.*, 1964); 85 per cent (Sangappa and Ali, 1977); 88 per cent (Vittal and Saroja, 1965); 89% (Abdul-Kareem and Subramanian, 1976); 92 per cent (Krishnanand and Kaiwar,

1965); represent yield increases compared to insecticidal protection and include the total effects of the insecticides on the groundnut arthropod pest fauna, rather than their effects on the leaf miner populations alone. Care should be taken in drawing conclusions from such data. Data from screen cages and artificially infested groundnut with different levels of GLM (Tajkumar and Devraj Urs, 1983) which resulted in 1.2% yield loss with each 1% infestation of GLM (based on regression of yield loss versus infestation) can be misleading because the cages reduce sunlight, thus possibly confounding the results.

Ghewande *et al.* (1987) recommended an action threshold of 61-70 larvae per 100 leaflets in India. The impact of GLM infestation on groundnut growth and yield is determined by the time and stages of the crop. An infestation of five larvae per plant after 10 days of emergence (DAE) has a much greater impact than 20 larvae per plant at 75 DAE (Shanower *et al.*, 1993b). Ghule *et al.*(1987b) suggested that groundnut needs protection from GLM between 45 and 75 days after emergence and this is true only if GLM populations are low, early in the season. However they cautioned that one need to look for parasites and if 50 per cent of the larvae/pupae are parasitized and the population is close to the threshold, no protection is necessary.

3.7. Management

3.7.1. Physical and cultural control

GLM moths are attracted to light and petromax lights placed at ground level attracted more moths than those placed 5-10 feet above the ground (Cherian and Basheer, 1942). Collection and destruction of moths by setting light traps early in the season, has been reported to reduce the pest incidence (Narayanan, 1961; Ramakrishna Ayyar, 1963; Reddy, 1968). With the demonstration of a sex pheromone in the GLM (Nandagopal and Reddy, 1990; Lalita Kumari and Reddy, 1991), the emergence of males two days earlier than females (Lalita Kumari, 1989) with distinct broods of GLM in a given location and the availability of synthetic pheromone (Hall *et al.*, 1993) there is a possibility of mass trapping and mating disruption in GLM which needs investigation. Though several cultural methods have been recommended for control of GLM, rotation of groundnuts with non-leguminous crops has been suggested to considerably reduce the GLM populations and rotation of groundnuts with soybeans should be avoided to prevent outbreaks of the pest (Van der Laan and Ankersmit, 1951). However, intercropping soyabean in groundnut reduces the GLM incidence considerably (Nandagopal,1992). Lower GLM larval densities have been found when groundnut was intercropped with sorghum, millet or cowpea than in monoculture groundnut at 30x10 cm spacing (Logiswaran and Mohanasundaram, 1985). However, the lowest larval densities were recorded in monoculture groundnut at close spacing (15x10 cm). Mulching with rice straw had no effect on GLM levels but did have a positive effect on parasitism levels: monocropped groundnut at 30x10 cm spacing had the lowest level of parasitism whereas a similar monocrop at 15x10 cm spacing had the highest; intercrop treatments all had intermediate levels of parasitization (Logiswaran and Mohanasundaram, 1985). Studies on the effect of sowing date on GLM infestations were contradictory. For example Lewin *et al.* (1979) showed that early sowing led to higher infestations of GLM but Logiswaran *et al.*(1982) concluded that late plantings were more heavily attacked. Our observations on the early sown rainy season crop revealed heavy infestations at 15 DAE itself, indicating that the possibility of manipulating planting date as a control method has a remote possibility.

3.7.2. Host plant resistance

Considerable progress has been made in developing GLM – resistant or tolerant cultivars. Most of the earlier studies (Lewin *et al.*, 1971; Sathiamoorthy *et al.*, 1978; Jai Rao; Sindagi, 1973 and 1974) were field screening of groundnut germplasm. Bunch varieties are generally considered to be less susceptible to GLM except in the studies of Motka *et al.*, 1985. Lewin *et al.* (1971) also found that spreading varieties were less susceptible to GLM than semi- spreading or bunch types. Vikram Singh (1979) compiled the list of the trails under taken by the All India Coordinated Research Project on Oilseeds several resistant and tolerant lines to GLM and identified ICG 6022, ICG 7887 and 7890 as the prominent. Recently GLM resistance has been demonstrated at ICRISAT center in a wide range of genotypes, including spreading, Spanish bunch and Valencia growth habits (ICRISAT, 1986). The incidence of GLM on selected genotypes under high pest pressure is given in Table 3.2. One variety, ICGV 86031, has shown good tolerance to GLM as well as to other defoliators (ICRISAT, 1991). ICGV 86031 is an improved genotype that may be grown in GLM endemic areas and may be used as a suitable parent for germplasm enhancement by national agricultural research centers. A cultivar

BG 2 was also found promising, even though it recorded cent percent incidence there was no further development of the larvae indicating the antibiosis nature of resistance to GLM (Nandagopal, unpublished data).

Table 3.2. Groundnut leaf miner incidence^{-1} on some selected genotypes under high pest pressure at ICRISAT, India (post rainy season 1990/91)

Genotype	leaf area (cm^2 plant^{-1})	Total leaf miner plant^{-1} (larvae +pupae)	leaf miner population (100 cm^{2})
ICG 1697	380	46	12
ICG 86162	236	66	27

Contd..

Table 3.2. Contd..

ICG 87160	404	103	24
ICG 86031	441	68	16
ICG 87495	407	87	20
ICG 2271	335	82	26
ICG 5040	328	77	24
TMV 2	228	50	24
SE	±57.2	±17.4	±4
CV (%)	29	42	32

1.Mean of 5 plants per sample, and 3 replications under high –pressure situation

Attempts to find GLM resistant soybean varieties have not been very successful (Mundhe, 1980; Shetgar and Thrombre, 1984). Shrivastava *et al.* (1988) found three soybean varieties with lower GLM larval populations evaluated during two rainy seasons even though all were attacked by GLM.

3.7.3. Natural enemies

GLM is mainly suppressed by parasitoids, predators and diseases. Several parasitoids have been identified on GLM mainly in south India (Cherian and Kalayasam, 1941; Cherian and Basheer, 1942; Krishnamurthy and Usman, 1954; Subba Rao *et al.*, 1965; Subba Rao and Sharma, 1966; Atma Ram and Subba Rao, 1967; Kothai, 1974; Khan and Raodeo, 1978; Bhatnagar and Davies, 1979; Nandagopal,1992), but their impact has not been fully quantified. The most important and abundant GLM natural enemies, mainly of parasitic Hymenoptera attacking the larval stage are listed (Shanower *et al.*, 1992). The parasitoid community associated with GLM is large and complex, involving at least two trophic levels, which includes GLM and primary and secondary parasitoids. Shanower *et al.* (1992) found nine primary parasitoids and seven secondary parasitoids, which killed up to 50 per cent of the GLM larvae in a generation The parasitoid community is very dynamic with one or two primary parasitoids usually predominating although not always the same ones (Shanower *et al.*, 1992). The relative abundance of the predominant primary parasitoids changed during the five GLM generations studied (Shanower *et al.*, 1992). In the first sampling period *Sympiesis dolichogaster* was dominant, emerging from more than 25 per cent of the parasitized GLM larvae. Three braconids mistakenly considered a single species, combined to kill 28 per cent of the GLM larvae collected in the first sampling period. In subsequent sampling periods, the proportion attacked by *S. dolichogaster* declined while the fraction killed by *Stenomesius japonicus* increased (Table 3.3).

The most important and abundant GLM natural enemies, attacking the larval stage are listed table 3.4, predators (table 3.5) and entomopathogens (table 3.6).

Table 3.3. Groundnut leaf miner larval parasitism[1] in different seasons at ICRISAT centre

Year	Parasitism (%)	
	Rainy season	Post-rainy season
1984	-	42
1985	90	58
1986	21	62*
1987	-	45
1988	44	48
1989	10	32
1990	36	6
1991	16	27
1992	33	40
1993	21	-
1994	28	45
1995	-	20
1996	28	48
Mean	33	39

1. Sample size 100 larvae season^{-1}
2. *Goniozus* sp. is the dominant parasitoid in the post-rainy season, *Stenomesius japonicus* and *Apantles* sp. in the rainy season.
3. In sprayed areas where population was low during April the parasitism was maximum (100%).

Parasitism by the three braconids (*Apantelus* sp., *Avga choapes*, *Bracon* sp.) also decreased. Two species, *Chelonus* sp. and *Goniozus* sp. never emerged from more than 15 per cent of the parasitized larvae in any sampling period. Secondary parasitoids emerged from between 19 per cent and 40 per cent of the hosts collected within any sampling period (Shanower *et al.*, 1992). In other reports, parasitization rates varied from 25 to 90 per cent of the available GLM larvae (Khan and Raodeo, 1978; Srinivasan and Siva Rao, 1986; Yadav *et al.*, 1987). Parasitization rates are highly variable and insufficient data are available to understand natural enemy population dynamics, nor is it very clear why the effectiveness of the natural enemies is so varied from year to year and season to season. GLM larval parasitism data from 1984 to 1992 during both rainy and post-rainy seasons at ICRISAT center indicted a wide variation (6 to 90%) with an average of 34.6 per cent during rainy and 40 per cent during post-rainy seasons.

Although the largest number of parasitoids have been reported from India, little is known of their distribution and seasonal abundance. Khan and Raodeo (1978) reported that six larval parasitoids were most abundant in August and September in Maharashtra. Nine parasitoids were active in both rainy and post-rainy seasons in southern Andhra Pradesh. Parasitism was highest in the September-November and February-March periods (Srinivasan and Siva Rao, 1986). Of the eight parasitoid species recorded in Gujarat, six were present in the rainy season (July-October) and four were active at the end of the postrainy season (March-May). Two species, *Goniozus* sp. and *S. japonicus* Ashm. were active in both seasons (Yadav *et al.*, 1987). However observations at ICRISAT center over a number of years indicate that *Goniozus* sp. is more prevelant in the postrainy and *S. japonicus* and *Apantles* sp. in the rainy season (Table 4). Six species of natural enemies were active in Gujarat (Nandagopal, 1992).

The number of parasitoid species is lower in soybean than in groundnut, as crop is not fully studied yet. In soybean seven parasitoid species *viz. Bracon gelichiae* Ashm. and *Elasmus brevicornis* Gahan, *Chelonus* (*Microchelonus*) sp., *Apanteles* sp., *A. litae* Nixon, *Stenmesioideus ashmeadi* Subba Rao and Sharma and *Goniozus* sp. (Gujarati *et al.*, 1973; Shetgar and Thombre, 1984) have been recorded. The parasitism was 84 per cent in the rainy season and 44 per cent in the post-rainy season.

Several predators have been identified that attack GLM larvae but their impact has not been quantified. *Odynerus punctum* Fabr. (Hymenoptera: Eumenidae) (Lefroy and Howlett, 1909). *Chlaenius* sp. (Coleoptera : Carabidae) have been observed attacking GLM larvae (Shanower and Ranga Rao, 1990). Predation by spiders and robber flies (Deptera: Asilidae) has also been reported (Srinivasan and Siva Rao, 1986).

Pathogens have also been found infecting GLM larvae in India. Unidentified mermithid nematodes (Kothai, 1974; Srinivasan and Siva Rao, 1986), nuclear polyhedrosis virus (Godse and Patil, 1981) and a fungus *Aspergillus flavus* have been recovered from GLM larvae (Oblisami *et al.*, 1969; Shanower *et al.*, 1992) highlighted the importance of pathogens on the population dynamics of *A. modicella*; up to 30 per cent of the larvae were infected by viral and fungal pathogens.

3.7.4. Chemical control

Many synthetic insecticides have been tested as foliar sprays, dusts, seed treatments and also as granules against GLM. DDT was the first synthetic insecticide recommended for GLM control (Ramakrishna Ayyar, 1963). Carbaryl, dieldrin, endrin and parathion (Krishna Murthy Rao *et al.*, 1962; Vittal *et al.*, 1964; Krishnananda and Kaiwar, 1965; Vittal and Saroja, 1965). (Lewin *et al.*, 1973; Rajput *et al.*, 1985; Ghule, *et al.*, 1987a). During the 1970s, organochlorines were continued to be used even as the newer organophosphates and carbamates were recommended (Lewin *et al.*, 1973; Kapoor *et al.*, 1975; Devraj Urs and Krishna Kothai, 1976). Most organophosphates and carbamates with systemic and strong contact action were quite effective against GLM and the dosages that have been reported to be effective are: carbaryl dust 10% (Lewin *et al.*, 1973); carbaryl W.P. 0.2% spray (Krishnananda and Kaiwar, 1965); carbofuran EC 0.025 and 0.05% (Singh, 1978) and carbofuran WP 0.05% (Sangappa and Ali, 1977); chloropyriphos EC 0.05% (Sadakathulla *et al.*, 1976); dimethoate EC 0.03% (Devraj Urs and Kothai, 1976) and 0.05% (Sangappa and Ali, 1977); monocrotophos EC @ 1.6 1/ha (Abdul-Kareem *et al.*, 1974; Abdul-Kareem and Subramaniam, 1976) and 0.04% (Sangappa and Ali, 1977); phosalone EC 0.025 and 0.05%(Singh, 1978); quinalphos EC 0.025% (Abdul-Kareem and Subramanianm, 1976; Devaraj Urs and Kothai, 1976) and 0.05% (Singh, 1978). Geh and Lim (1975) in Malaysia protected rainfed groundnuts with four sprays of monocrotophos 0.075% at fortnightly intervals.

Pyrethroid insecticides were evaluated in the mid-1980s (Sivasubramanian and Palaniswamy, 1983; Rajput *et al.*, 1985) and found to be highly effective, but are more expensive than many other insecticides. Srinivasan *et al.*(1986) compared the cost and efficacy of seven insecticidal dusts for control of GLM: a single application of lindane cost one-third of the price an application of carbaryl or malathion and was as effective.

Furrow applications of carbofuran, disulfoton or phorate granules at sowing are effective in protecting soybeans against the leaf miner damage in Thailand (Arunin, 1978). Seed treatments of groundnuts with 50% carbofuran @ 3, 5 and 7 g a.i./100g seed was effective in suppressing the leaf miner populations for 4 to 6 weeks and resulted in 46, 68, 71 per cent increased pod yields over untreated groundnuts in south India (Lal *et al.*, 1974). Two anti-feedents i.e. TPTA and TPTH (WP) 0.05% also gave significantly increased pod yields with effective control of the GLM (Abdul-Kareem, 1976; Abdul-Kareem and Subramaniam, 1976).

The only botanical insecticide tested for GLM activity was an extract from neem seed (Sadakathulla *et al.*, 1976) and four sprays of 1% extract failed to reduce the incidence of GLM relative to the unsprayed check. Neem-based insecticides namely, Jawan and Margoside OK and CK were compared with crude neem oil or its blending with hostothion. The immediate toxicity of crude neem oil at lowest concentration of 2% either alone or with hostathion deterred 100 per cent oviposition of leaf miner and the residual toxicity persisted up to 15 days, while other treatments had no residual effect. (Nandagopal,1992). Crude neem oil suppressed the oviposition, but, had little influence on larval and pupal mortality of leaf miner and did not deform pupae or adults in the filed conditions. Cent per cent suppression in the hatching of eggs of leaf miner was probably due to the choking effect of oil. Hence, to control this insect, neem oil may be used to prevent its oviposition and it had no advance effects on parasite emergence.

The crude neem oil toxicity persisted for 12 days against leaf miner. The longer persistence against leaf miner might be attributed to the shiny coating on the leaflets which might have prevented the orientation of the ovipositing females. Since the residual effect was more than 12 days three sprays coinciding with emergence of each brood could be sufficient to contain this pest below the economic injury levels.

The deleterious effect of insecticides on natural enemies is widely known and is the cause of secondary pest outbreaks and resurgence in primary pest species (Reynolds, 1971; Luck *et al.*, 1977; van den Bosch, 1978). Shanower *et al.* (1992) demonstrated that the effectiveness of parasitoids was constrained by the use of insecticides, resulting in lower parasitization rates in insecticide treated fields. GLM populations were slightly lower on sprayed plots because the insecticide killed both GLM larvae and the parasitoids. Parasitism rates were reduced less in the systemic insecticide treatment than in the foliar spray treatment. Ranga Rao and Shanower (1988) noted the increased incidence of *Spodoptera litura* (Fab.) in groundnut fields treated for GLM. These results should serve as a warning against the excessive use of chemical insecticides for groundnut pest management. Several studies have evaluated systemic insecticides including phorate, carbofuran and isophenphos, and found them to be highly effective in reducing GLM populations (Lal *et al.*, 1974; Khan and Raodeo, 1978; Logiswaran *et al.*, 1982; RadhaKrishnan *et al.*, 1983). Systemic insecticides provide a means of reducing the negative impact of chemical pesticides on natural enemies. Shanower *et al.* (1992) also showed no impact of systemic insecticide on the pathogens and up to 30 per cent of the larvae in a sample found infected by viral or fungal pathogens in both sprayed and unsprayed plots.

Dudley *et al.* (1989) used a simulation model to investigate the interactions between host-plant resistance, natural control and chemical control for GLM in India. The number of sprays required to protect the crop was high when host plant resistance and natural enemies contributed little to the overall control. However, the efficacy of chemical control, which is unknown, has a fundamental influence on this relationship. A recent survey of groundnut production areas in Andhra Pradesh indicated that pesticide applications for the control of GLM appeared to be upsetting the natural control of *Helicoverpa armigera* Hub. and *Spodoptera litura* L (Ranga Rao and Shanower, 1988). Both of these defoliators were developing into pests of equal importance to GLM in areas of heavy insecticide applications.

3.8. Conclusions

The groundnut leaf miner is a serious pest of groundnut and soybean throughout south and south-east Asia, and so far management is mainly through chemical control. A more integrated approach to GLM control is needed because of the limited resources available to subsistence farmers and the negative aspects of excessive reliance on chemicals. The approach would include monitoring and mass trapping with pheromone traps, use of resistant/ tolerant genotypes, agronomic manipulations to minimize GLM infestation and conservation and enhancement of natural enemies. Data on the population fluctuation of GLM and also its natural enemies, specially quantifiable data on effects of rainfall and other climatic factors on GLM and natural enemies are very much essential before any attempt is made to develop IPM based on natural control. In addition, the role of pathogens on GLM population fluctuations is very little understood and detailed investigations are needed. Chemical control of GLM will continue to be an integral part of its management for some more years to come. Economic threshold for different seasons, locations and crop growth stages are essential to improve the economics and efficiency of chemical control. Except for identification of limited sources of resistance, progress in the development of GLM resistant/ tolerant varieties which are commercially viable is an immediate necessity, especially in developing countries where GLM is a serious pest. In the event of availability of resistant/tolerant variety and with a little management of natural enemies it may be possible to avoid the use of insecticides for GLM control.

References

Abdul Kareem, A. (1976). A study on the anti-feeding effects of two organotin compounds TPTA and TPTH on *Spodoptera litura* F., *Pericallia ricini* F., and *Stomopteryx subsecivella* Zell. *Pesticides* 10: 57.

Abdul Kareem, A., and Subramaniam, T.R. (1976). Antifeeding effect of two organotin compounds on *Stomopteryx subsecivella* Zell. (Lepidoptera). *Madras Agri. J.*, 63: 354-357.

Abdul Kareem, A., S. Sadakathulla and Subramaniam. T. R. (1974). Control surulpoochi on groundnut. *Farm and Factory* 8(4) : 35-36.

Amin, P.W. and Mohammad, A. (1980). Groundnut research at the International Crops Research Institute for the Semi-arid Tropics. Proceedings of the International Workshop on Groundnuts held at the ICRISAT center, Andhra Pradesh, during 13-17 October, 1980.

Amin, P.W . (1987). Insect pests of groundnut in India and their management. In: *Plant Protection in Field Crops,*. Veerabhadra Rao, M. and Sithanantham, S. (Eds). Plant Protection Association of India, Rajendranagar, Hyderabad, India. 219-233 pp.

Amin, P.W. (1983). Major field and storage insect pests of groundnut in India and their control. Paper presented at the Summer Institute on *'Recent Advances in Groundnut Productivity* held at Junagadh, 9-30, April 1983.

Anonymous, (1941). Agriculture and Animal Husbandry in India. 1938-39, Delhi. 422 pp.

Anonymous, (1963). Groundnut leaf miner. *Plant Protection Bulletin* (pest and disease number). 11:121.

Anonymous, (1993). Annual Report (1st April, 1992 to 31st March, 1993). All India Co-ordinated Research Project on Biological Control of Crop Pests and Weeds. Biological Control Centre (ICAR-NCIPM)., Bangalore, India. Tech. Doc. No. 40: 1-313.

Arunin, A. (1978). Pests or soybean and their control in Thailand. Pages 43-46 in *Pests of Grain Legumes: Ecology and Control.* Eds. S.R.Singh, H.F.van Emden, and T. Ajibola Taylor. London, U.K.: Academic Press.

Atma Ram and Subba Rao, B. R. (1967). Descriptions of *Goniozus stomopterycis* nr. sp. (Bethylidae: Hymenoptera). a primary larval parasite of *Stomopteryx nerteria* (Meyrick). in south India and a revised key to oriental species. *Bulletin of Entomology.* 8(2): 73-78.

Atwal, A. S. (1976). *Agricultural pests of* India *and south east Asia.* Kalyani Publishers (India)., Delhi 502 p.

Bainbridge Fletcher, T. (1914). Some south Indian insects and other animals of importance considered especially from an economic point of view. Government Press, Madras – 565 p.

Bainbridge Fletcher, T. (1917). Groundnut pests. *Proceedings of the Second Entomological Meeting,* Pusa 43 p.

Bainbridge Fletcher, T. (1920). Life histories of Indian insects: Microlepidoptera. Memoirs, Department of Agriculture, India, Pusa, *Entomology Series* 6: 77-79.

Barber, C. S. (1906). The pests of groundnut in Madras. *Agricultural Journal of India.*1: 170.

Bhatnagar, V. S. and Davies, J. C. (1979). Arthropod endoparasitoids of insect pests (excluding *Heliothis* spp.). recorded at ICRISAT centre, Andhra Pradesh, India, 1974-79. *Cropping Entomology Progress Report* –3. 7-8 pp, International Crops Research Institute for the Semi-Arid Tropics (ICRISAT)., Patancheru – India.

Bindra, O. S. (1970). Entomological research provides hope for green revolution in oilseeds. *Oilseeds Journal* 2(3) : 24-43.

Campbell, W. V. (1983). Management of arthropods on peanut in Southeast Asia Annual Report. Peanut Collaborative R*esearch Support Program* (CRSP)., University of Georgia, Georgia Experiment Station, USA.

Channabasavanna, G. P. (1951). 'The surulpuchi' (leaf miner). of groundnut, *Mysore Agricultural Journal.* 26: 67-68.

Channabasavanna, G. P. (1957). The groundnut leaf miner and its control. *Mysore Agricultural College Year Book* (1956-57). 123-124.

Cherian, M. C. and Basheer, M. (1942). Studies on *Stomopteryx nerteria* Meyr., a pest of groundnut in the Madras Presidency. *Madras Agricultural Journal,* 30: 379-381.

Cherian, M. C. and Kalayasam, M. S. (1941). Preliminary notes on the parasites of the spotted and the pink bollworms of cotton in Coimbatore. Proceedings of the Indian Academy of Sciences 14(6): 517-28.

Das, B. B. and Misra, D. P. (1984). New records of alternate hosts of groundnut leaf miner, *Aproaerema modicella* Deventer (Gelechiidae: Lepidoptera). *Indian Journal of Agricultural Science.* 54 (9) : 776.

Devraj Urs, K.C. and Krishna Kothai. (1976). Evaluation of some selected insecticides for the control of the groundnut leafminer, *Stomopteryx subsecivella* Zeller. *Madras Agriultural Journal.* 63: 371-372.

Dudley, N. J., Muller, R. A. and Wightman, J. A. (1989). Application of dynamic programming for guiding IPM on groundnut leaf miner in India. *Crop Protection.* 8: 349 - 357.

Feakin, S.D. (1973). Pest control in groundnuts. *PANS mannual* No.2, 3rd edn., Centre for Overseas Pest Research, London, pp. 155-175.

Frohlich, G. and Rodewald, W. (1963). Planzenschutz in den Tropen. Special Volume 3. Karl Marx University, Leipzig. 136-141 p.

Gangrade, G. A. (1976). Terminal technical report of the project on "Assessment of effects on yield and quality of soybeans caused by major arthropod pests". Department of Entomology, Jawaharlal Nehru Krishi Vishwa Vidyalaya, Jabalpur, M.P., India, 144 pp.

Geh, S. L. and Lim, G. S. (1975). Current situation of pests, disease and weed problems rainfed legumes in Peninsular Malaysia. Pages 55-68 in Review on Pest, Disease and Weed problem in Rainfed crops in Asia and the Far East (Eds.)D. B. Reddy. Bangkok.

Ghewande, M. P., Nandagopal, V. and Reddy, P. S. (1987). Plant protection in Groundnut. N.R.C.G. (I.C.A.R), Junagadh (Gujarat State). *Technical Bulletin* No.1: 35.

Ghule, B. D., Dhumal, V. S. and Deokar, A. B. (1987a). Chemical control of groundnut leaf miner. *Journal of Maharashtra Agricultural Universities.* 12: 257-259.

Ghule, B.D., Jagtap, A.B., Dhumal, V.S. and Deokar, A.B. (1987b). Determination of critical time for protection against leafminer. *Journal of Maharashtra Agricultural Universities.* 12: 292-293.

Godse, D.B. and Patil, R.B. (1981). New nuclear polyhedrosis of Lepidopterous insects in Karnataka, India. *Proceedings of Indian Academy of Science* (Animal Science) 90: 93-97.

Gujrati, J. P., Kapoor, K. N. and Gangrade, G. A. (1973). Biology of soybean leafminer, *Stomopteryx subsecivella* (Lepidoptera: Gelechiidae). *The Entomologist.* 106: 187-191.

Hall, D. R., Peter S. Beevor, Derek G. Campion, David J. Chamberlain, Alan Cork, Rosemary White, Aurelio Almetre, Thomus J. Henneberry, Nandagopal, V., Wightman, J. and Ranga Rao, G. V. (1993). Identification and synthesis of new pheromones. *IOBC Wprs Bulletin,* 16, 1-10.

ICRISAT. (1986). Annual Report, International Crops Research Institute for the Semi-Arid Tropics, Patancheru, A.P., 502 324, India.

Islam, W., Ahmed, K. N., Nargis, A., and Islam, U. (1983). Occurrence, abundance and extent of damage caused by insect pests of groundnuts (*Arachis hypogaea*). *Malaysian Agricultural Journal.* 54: 18-24.

Jagtap, A. B., Bothe, N. N. and Deokar, A. B. (1985). Carryover of groundnut leafminer, *Aproaerema modicella* Dev. across the season. *Indian Journal of Entomology.* 47: 463.

Jagtap, A. B., Ghule, B. D., and Deokar, A. B. (1984). Assessment of losses in yield of 'Phule Pragati' groundnut caused by insect pests. *Indian Journal of Agricultural Science.* 54: 697-698.

Jai Rao, K. (1992). Final Progress Report. (1989-92). ICAR project -Bio-ecology and natural enemies of groundnut leaf-miner and their utilisation in the integrated pest management. University of Agricultural Sciences, Dharwad, Karnataka, India, 87 pp.

Jai Rao, K. and Sindagi, S. S. (1973). Preliminary observations on the incidence of leaf miner, *Stomopteryx subsecivella* Zell. on irradiated groundnut material. *Current Research.* 12: 106-108.

Jai Rao, K. and Sindagi, S.S. (1974). Screening of varieties for resistance to groundnut leaf miner, *Stomopteryx subsecivella* Zell. (Lepidoptera: Gelechiidae). *The Mysore Journal of Agricultural Sciences.* 8(1):133-137.

Jai Rao, K. and Thirumalachar, D. K. (1977). New record of alternate host plants of groundnut leafminer *Stomopteryx subsecivella* Zeler (Syn.: *S. neteria* Meyrick). (Lepidoptera: Gelechiidae). *Current Science.* 46: 91-92.

Kapadia, M. N., Bharodia, R. K. and Vora, V. J. (1982). Biology and estimation of incidence of groundnut leafminer, *Stomopteryx subsecivella* Zell). (Gelechiidae: Lepidoptera). *GAU. Research Journal* 8: 37-39.

Kapoor, K. N., Gujrati, J. P. and Gangrade, G. A. (1975). Chemical control of soybean leafminer, *Stomopteryx subsecivella* Zeller (Lepidoptera: Gelechiidae). *Indian Journal of Entomology,* 37: 286-291.

Keeratisasikorn, M. and Hiranyasree, A. (1975). Soybean and peanut insects. KKU-IDRC Semi Arid Crops Project – Annual Report (1975). – Khon Kaen University, Thailand.

Kennedy, F. J. S., Lourduraj, A. C. and Rajmanickam, K. (1992). Weeds as alternate hosts for groundnut leaf miner. Groundnut News 4 (2): 7.

Khan, M. I. and Raodeo, A. K. (1978). Importance of larval parasites in the control of *Stomopteryx subsecivella* Zeller *Journal of Maharashtra Agricultural University* 3: 261-263.

Kothai, K. (1974). Bionomics and chemical control of the groundnut leaf miner *Stomopteryx subsecivella* (Zeller) (Lepidoptea: Gelechiidae), Msc (Ag) thesis, University of Agricultural Sciences, Bangalore, India, pp.78.

Krishnamurthy Rao, S., Rangacharulu, P. and Yesudass, T. (1962). Aerial spraying against surulpoochi (*Stomopteryx nerteria*) on summer irrigated groundnut. *The Andhra agricultural Journal.* 9: 202-206.

Krishnamurti, B. and Usman, S. (1954). Some parasites of economic importance noted in Mysore State. *Indian Journal of Entomology.* 16: 327-344.

Krishnananda, N. and Kaiwar, S. R. (1965). A preliminary study on the control of leafminer of groundnut. *Indian Oilseeds Journal.* 9: 180-185.

Kulshreshta, J. P. (1964). Larval feeders of castor and groundnut and methods of their control. Indian Central Oilseeds Committee Third Conference of Oilseed Research Workers in India. 72-73 p.

Kuntha, C. and Jamjanya, T. (1976). Soybean and peanut insects. KKU-IDRC Semi-Arid Crops Project Annual Report (1976). – Khon Kaen University, Thailand. Not in the text

Lal, S. S., Varma, B. K., Sanyal, P. K. and Sagar, R. D. (1974). A new approach to the control of groundnut leaf miner, *Stomopteryx nerteria* M. by seed treatment with carbofuran. *Indian Journal of Plant Protection*. 2: 111-116.

Lalita Kumari, V.L. (1989). Sex pheromone systems of selected Lepidopterous pests of groundnut, Ph.D. thesis submitted to Andhra Pradesh Agricultural University, Rajendranagar, Hyderabad, India.

Lalita Kumari, V. L. and Reddy, D. D. R. (1991). Sex pheromone in groundnut leaf miner *Aproerema modicella*. *International Arachis Newsletter*. 9: 1-16.

Leefmans, S. (1927). Zeikten en plagen der culturregewassen in Nederlandsch – Indie in 1926. Mededelingen van het Institut voor Plantenzeikten. Buitenzorg.73: 60.

Lefroy, H. (1907). The groundnut leaf miner. Agricultural *Journal of India*., 2, 292.

Lefroy, H. (1909). Manual of Entomology with special reference to economic Entomology. *Agricole Reprints Corporation*, New Delhi. 539 pp.

Lewin, H. D., Saroja, R., Leela David, A. and Padmanabhan, M. D. (1973). Chemical control of groundnut leafminer (*Stomopteryx subsecivella* Zell). *Agriculture and Agro-Industries Journal*. 6: 12-16.

Lewin, H. D., David, A. L., Saroja, R., Natarajan, S. and Janarthanan, R. (1971). A comparative study of pests and diseases of different strains of groundnut. *Madras Agricultual Journal*. 58: 15-20.

Lewin, H. D., Saroja, R., Leela David, A., and Padmanabhan, M.D. (1979). Influence of sowing time and weather on the incidence of groundnut leafminer. *Indian Journal of Agricultural Science*. 49: 886-891.

Listinger, J.A., Quirino, C.B., Lumann, M.D., and Bandong, J.P. (1978). The grain legume pest complex of rice-based cropping systems at three locations in the Philippines. In Pests of Grain Legumes. Ecology and Control, pp. 309-320. (Eds. Singh,S.K., Van Emdon,H.F., and Taylor, T.), Academic Press, London, U.K.

Logiswaran, G., Madhava Rao, G., Vasudevan and Kannan, V. (1982). Influence of time of sowing and weather factors on the infestation of leaf miner *Aproaerema modicella* Deventer and yield in rainfed groundnut. *Madras Agricultural Journal*. 69:359-363.

Logiswaran, G. and Mohanasundram, M. (1985). Effect of intercropping, spacing and mulching in the control of groundnut leafminer, *Aproaerema modicella* Deventer (Gelechiidae:Lepidoptera). *Madras Agricultural Journal*. 72: 695-700.

Logiswaran, G. and Mohanasundaran, M. (1986). Influence of weather factors on the catches of moths of groundnut leafminer (*Aproaerema modicella* Deventer). in the light trap. *Entomon*. 12: 147-150.

Luck, R. F., Van deo Bosch, R. and Garcia, R. (1977). Chemical insect control. A troubled pest management strategy. *Bioscience.* 27: 606-611.

Manoharan, T. and Chandramohan, N. (1986). *Psoralea corylifolia*, an alternate host for carry over of groundnut leafminer (*Aproaerema modicella* Deventer). *Madras Agricultural Journal.* 73: 591.

Meyrick, E. (1906). Description of Indian microlepidoptera. Journal *of Bombay Natural History Society.* 17: 139-140.

Mohammad, A. (1981). The groundnut leaf miner (*Stomopteryx subsecivella* Zeller) (Lepidoptera: Gelechiidae). A review of world literature. Occasional paper 3, Groundnut Improvement Program, International Crops Research Institute for Semi- Arid Tropics (ICRISAST), Patancheru, A.P., India. pp.21.

Motka, M. N., Bhalani, P. A. and Bharodia, R. K. (1985). Growth and development of *Aproaerema modicella* (Deventer). (Lepidoptera: Gelechiidae). on certain groundnut varieties. *Agricultural Science Digest.* 5: 162-164.

Mundhe, D.R. (1980). Studies on the susceptibility of twenty varieties of soybean to the leafminer, *Stomopteryx subsecivella* Zeller. *Research Buletin Marathwada Agricultural University.* 4: 10-11.

Muthiah, C. (1991). Some new parasites of groundnut leaf miner (*Aproaerema modicella* Dev.). *Indian Journal of Plant Protection.* 19(1): 105.

Muthiah, C. and Abdul Kareem, A. (2000). Survey of groundnut leaf miner and its natural enemies in Tamil Nadu, India. *International Arachis Newsletter.* 20: 62 – 63.

Nair, M. R. G. K. (1975). Insects and mites of crops in India. ICAR, New Delhi, pp 404.

Nandagopal, V. and Reddy, P. S. (1990). Report of sex pheromone in groundnut leaf miner *Aproaerema modicella* Dev. *International Arachis Newsletter.* 7: 25-26.

Nandagopal. V. (1992). First record of insect pest and predators of thrips and jassids in groundnut. *International Arachis Newsletter.* 11: 26.

Narayanan, E. S. (1961). Insect pests of groundnuts and methods of their control. In Groundnut - A Monograph - C. R. Sheshadri (Ed.). Indian Central Oilseeds Committee, Hyderabad, pp 222-228.

Nayyar, K. K., Ananthakrishnan, T. N. and David, B. V. (1976). General and applied Entomology, Tata McGraw Hill Publishing Co. pp. 586 .

Oblisami, G., Ramamurthy, K. and Rangaswami, G. (1969). Studies on the pathology of some crop pests of South India. *Mysore Journal of Agriculture Science.* 3: 86-98.

Palanisamy, M. S. and Ramchandran, T. K. (1978). Chemical control of groundnut leafminer *Stomopteryx subsecivella* Zell. (Gelechiidae: lepidoptera). *Pesticides.* 12: 24-26.

Phisitkul, S. (1985). The biology and natural enemies of the leafminer, *Aproaerema modicella* Deventer (Lepidoptera: Gelechiidae) on groundnut (*Arachis hypogaea* L.) Ph.d. dissertation, Univ. of the Philippines, Los Banos.

Prasad, S. K., Chawla, M. L., Kumar, S. and Saxena, H. P. (1971). Root knot nematode, *Meliodogyne javanica* (Treub, 1885) Chitwood, 1949 and stem borer, *Stomopteryx nertaria* Meyrick affecting green gram, *Phaseolus aureus. Indian Journal of Entomology.* 33: 55-60.

Radhakrishnan, S., Mohanan, M. and Vittal, S. M. (1983). A preliminary note on the efficacy of isofenphos 40 SD as seed treatment against groundnut surulpoochi. *Pesticides*. 17: 30-31.

Rai, K. (1976).Pests of oilseed crops in India and their control. Indian Council of Agricultural Research (ICAR)., New Delhi. 121 pp.

Rajagopal, D, S. Mallikarjunappa, and Gowda, J. (1988). Occurrence of nautral enemies of the groundnut leaf miner, *Aproaerema modicella* (Lepidoptera: Gelechiidae). *Journal of Biological Control*. 2: 129-130.

Rajgopal. K, V. Nandagopal and N. R. Bhagat. (1988). Preliminary screening of Virginia groundnut genotypes against tobacco caterpillar *Spodoptera litura* (F). *Journal of Oilseeds Research*. 5:162-165.

Rajput, S. G., Dalaya, V.P., and Awate, B. G. (1985). Field evaluation of synthetic pyrethroids and other insecticides against groundnut leaf miner (*Aproaerema modicella*, D.). *Pesticides*. 19: 34-35.

Ramakrishna Ayyar, T.V. (1963). A handbook of economic entomology for South India. McMillan Publ. (India)., New Delhi.

Rang Rao, G. V. and Shanower, T. G. (1988). A survey of groundnut pests in Andhra Pradesh, India, Post Rainy season 1987/88. *International Arachis Newsletter*. 4: 8-10.

Ranga Rao, G. V. (1996). Groundnut entomological work during winter/ spring 1995-96 in Vietnam, 108pp. ICRISAT, Patancheru, Andhra Pradesh, India.

Reddy, D. B. (1968). Plant Protection in India, 454 pp. Applied Publishers, New Delhi.

Reddy, P.S. (1988). Genetics, breeding and varieties. In Groundnut, (ed. P.S.Reddy) pp. 508-525.

Reynolds, H. T. (1971). A world review of problems of insect populations upsets and resurgences caused by pesticide chemicals. In Agricultural Chemicals -Harmonyor Discord for food - People environment. Ed. J.E. Smith - University of California, Division of Agricultural Sciences, pp 108-112.

Sadakathulla, S., Abdul Kareem, A., Srinivasan P. M. and S. Jayaraj. (1976). Chemical conrol of groundnut leaf webber, *Stomopteryx subsecivella* Zell. (Gelechiidae: Lepidoptera). in Tamil Nadu. *Pesticides* 10: 30-31.

Sandhu, G. S. (1977). Some insects recorded as pests of fodder crops at Ludhiana (Punjab). *Punjab Agriculture University Research Journal*. 14: 449-459.

Sandhu, G. S. (1978). Note on the incidence of *Stomopteryx subsecivella* (Zell.). (Lepidoptera: Gelechiidae). on lucerne at Ludhiana. *Indian Journal of Agricultural Sciences*. 48: 53-54.

Sangappa, H. K. and Ali, T. M. M. (1977). Evaluation of some newer insecticides in the control of the groundnut leaf miner, *Stomopteryx subsecivella* Zell. (Gelechiidae: Lepidoptera). *The Mysore Journal of Agricultural Sciences*. 11(4): 559- 561.

Sathiamoorthy, M. R., Palaniswami, M. S., Ramachandran, T. K., and Natarajan, S. T. (1978). Screening of groundnut bunch varieties against the leaf miner *Stomopteryx subsecivella* Zell. (Gelechiidae: Lepidoptera). *Agriculture and Agro Industries Journal*.11(9): 35-36.

Shanower, T. G. and Ranga Rao, G. V. (1990). *Chlaenius* sp. (Col.: Carabidae): a new predator of groundnut leaf miner larvae. *International Arachis Newsletter*., 8: 19- 20.

Shanower, T. G., Gutierrez, A.P. and Wightman, J.A. (1993a). Effect of temperature on development rates, fecundity and longevity of the groundnut leaf miner, *Aproaerema modicella* (Lepidoptera: Gelechiidae)., in India. *Bulletin of Entomological Research.*, 83, 413-419.

Shanower, T. G., Wightman, J. A., Gutierrez, A. P.,and Ranga Rao, G. V. (1992). Larval parasitoids and pathogens of the groundnut leaf miner, *Aproaerema modicella* (Lep.: Gelechiidae)., in India. *Entomophaga.* 37(3): 419-427.

Shanower, T. G. (1989). The biology, population dynamics, natural enemies, and impact of the groundnut leafminer, *Aproaerema modicella* (Deventer). (Lepidoptera Gelechiidae)., on groundnut in India. Ph.d. dissertation, Univ. of California, Berkeley, USA.

Shanower, T. G., Wightman, J. A. and Gutierrez, A. P. (1993b). Biology and control of the groundnut leaf miner, *Aproaerema modicella* (Deventer). (Lepidoptera: Gelechiidae). *Crop Protection.* 12: 3-10.

Shekharappa, Somasekhar and Patil B. V. (1990). Parasitoid complex of the groundnut leaf miner, *Aproaerema modicella* Deventer (Lepidoptera: Gelechiidae). *Journal of Biological Control.* 4 (1): 55-56.

Shetgar, S. S. and Thombre U. T. (1984). Occurrence of natural enemies on soybean leafminer and relative susceptibility of some soybean varieties to its attack. *Journal of Maharashtra Agricultural University.* 9: 218-219.

Shetgar, S. S., Bilapate, G. G., Patil, V. V. and Londhe, G. M. (1993). A note on the natural enemies of Bihar hairy caterpillar, *Spilosoma obliqua* (Walker). *Indian Journal of Entomology.* 52 (1): 158.

Shrivastava, K. K., Srivastava, B. K., and Deole, J. Y. (1988). Studies on chemical control and varietal resistance of soybean against leafminer, *Stomopteryx subsecivella* Zeller. *Indian Journal of Plant Protection* 176: 147-151.

Singh, K.(1978). Evaluation of various insecticides against groundnut leaf miner , stomopteryx nerteria (lepidoptera: Gelechiidae). *Indian Journal of Plant Protection* 6: 27-29.

Sivasubramanian, P. and Palaniswamy, G. A. (1983). Studies on the chemical control of groundnut leaf miner, *Aproaerema modicella* Dev. *Madras Agricultural Journal.* 70: 485-486.

Srinivasan, S. D. V., Siva Rao and Sundrarajan, M. (1986). Influence of crop management practices on incidences of leaf miner (*Aproaerema modicella* dev.) in groundnut. *Journal of Agriculture Research* 33: 381-382.

Srinivasan, S. and Rao, D. V. S. (1987). New report of parasites of groundnut leaf webber, *Aproaerema modicella* Deventer (Lepidoptera: Gelechiidae). *Entomon.* 12 (2): 117-119.

Srinivasan, S. and Siva Rao, D.V. (1984). New off-season summer hosts of the groundnut leafminer, *Aproaerema modicella* Deventer, (=*Stomopteryx subsecivella* Zell.). (Lepidoptera: Gelechiidae). in Andhra Pradesh. *Entomon.* 9: 11-12.

Srinivasan, S. and Siva Rao, D. V. (1986). New report of parasites of groundnut leaf webber, *Aproaerema modicella* Deventer (Lepidoptera:Gelechiidae). *Entomon* 12: 117-119.

Srinivasan, S. D. V., Siva Rao. and Sundrarajan, M. (1986). Influence of crop management practices on incidences of leaf miner (*Aproaerema modicella* Dev.) in groundnut. *Journal of Agriculture Research* 33: 381-382.

Srivastava, A. S., Vyas, H. K., Rathore, Y. K. and Khan, R. M. (1962). Survey of injurious insects of oilseed crops in Rajasthan and suggestions for control of the major pests. *Uttar Bharati Journal of Research* 8: 111-117.

Subba Rao, B. R. (1955). A new species of *Chilonis* on *Heliothis armigera* (Fab.). *Indian Journal of Entomology* 17: 63-64.

Subba Rao, B. R. and Sharma, A. K. (1966). Two new species of parasites reared from *Stomopteryx nerteria* (Meyrick) (Lepidoptera: Gelechiidae). *Indian Journal of Entomology*. 28: 299-304.

Subba Rao, B.R. , Kundu, G.G., Sharma, Y.K., Anand, R.K. and Rai S. (1965). New records of parasites of groundnut leafminer, *Stomopteryx nerteria* (Meyrick). (Gelechidae: Lepidoptera). *Indian Journal of Entomologyomol*. 27: 355-357.

Subba Rao, B. R. and Sharma, A. K. (1966). Two new species of parasites reared from *Stomopteryx nerteria* (Meyrick) (Lepidoptera: Gelechiidae). *Indian Journal of Entomology*. 28: 299-304.

Sundarababu, P.C. and Rabindra , R. J. (1993). Scope of utilizing microbial pesticides in the management of key pests of oilseed crops, In: Group Discussion on IPM strategies in oilseeds in India. Summary of papers (Bakhetia, D.R.C. and Balraj Singh, Eds.). Department of Entomology, Punjab Agricultural University, Ludhiana, Punjab, India (December 23-24, 1993). pp 4-5.

Tej Kumar, S. and Devraj Urs, K. C. (1983). Estimation of crop loss in groundnut due to the leafminer *Aproaerema modicella* (Lepidoptera: Gelechiidae). Special Issue, *Indian Journal Of Entomoogy*. 2: 345-351.

Thontadarya, T. S., Jai Rao, K., and Kumar, N. G. (1979). Occurrence of the groundnut leafminer *Biloba* (*Stomopteryx*) *subsecivella* (Zeller). on berseem (*Trifolium alexandrinum* L.). in Karnataka. *Current Research*. 8: 65.

Van den bosch, R. (1978). The pesticide conspiracy. Doubleday Publ. New York, USA. 135 pp.

Van der Laan, P. A. and Ankersmit, G. W. (1951). Chemical control of arachis leaf miner, *Stomopteyx subsecivella* Zell. Contribution from the General Agricultural Research Station, Bogor (Indonesia). No.119. pp 13.

Vanhall, C. J. J. (1918). Diseases and pests of cultivated plants in the Dutch East Indies. Meded, Laboratarium voor plantenziekten, Buitnzorg, No.33, 42 pp. Not in the text

Vanhall, C. J. J. (1922). Disease and pests of cultivated plants in the Dutch East Indies in 1921). Zeikten en Plagen der cultuurge–wassen in Nederlandsche– Indie in 1921. Mededelingen van het Institut voor Plantenziekten. Buitenzorg No.53. 46 pp.

Vikram Singh. (1979). Current status of oilseeds research in India – ICAR, Directorate of Oilseeds Research – Hyderabad, 36 pp.

Vittal, S. M. and Saroja, R. (1965). Insecticidal control of surulpoochi (*Stomopteryx subsecivella* Zell.) on grounnsut 19-24. Madras Agricultural Journal. 52 : 275-279.

Vittal, S. M., Azeez Basha, A. and Saroja, R. (1964). A note on the control of 'surulpoochi' (*Stomopteryx subsecivella* Zell.) on groundnut. *Madras Agricultural Journal.* 51: 475-476.

Wheatley, A. R. D., Wightman, J. A., Williams, J. H., and Wheatley, S. J. (1989). The influence of drought stress on the distribution of insects on four genotypes grown near Hyderabad, India. *Bulletin of Entomological Research* 79: 567-577.

Wightman, J. A and Amin, P. W. (1988). Groundnut pests and their control in the Semi-Arid Tropics. *Tropical Pest Management*. 34: 218-226.

Wightman, J. A., Dick, K. M., Ranga Rao, G. V., Shanower, T. G. and Gold, C. S. (1990). Pests of groundnut in the semi-arid tropics. In: *Insect Pests of Food Legumes* (Ed. by Singh S. R.) John Wiley & Sons, New York. pp.243-322.

Yadav, D. N., Patel, R. R. and Patel, R. C. (1987). Natural enemies of groundnut leaf-miner, *Aproaerema modicella* Deventer (Lepidoptera: Gelechiidae) and their impact on its infestation at Anand (Gujarat). *Gujarat Agricultural University Research Journal.* 13(1): 13-16.

Yang, C. and Liu, H. (1966). Biological observations on *Stomopteryx subsecivella* Zell. in Den Bei District, Kwantung. (In Chinese). *Acta Entomol. Sin.*, 15: 39-46.

Table 3-4 contd.

Table 3.4. List of parasitoids of *Aproaerema modicella* (Deventer)

Species	Order/Family	Country	Host stage attacked	Reference
Goniozus sp.	Hymenoptera Bethylidae	India	Larva	Subba Rao *et al.*, 1965; Yadav *et al.*, 1987; Subba Rao *et al.*, 1965; Khan and Raodeo 1978; Cherian and Basheer, 1942
G. stomopterycis	Hymenoptera Bethylidae	India	Larva	Atma Ram and Subba Rao, 1967 Shetgar *et al.*, 1993
Perisierola sp.	Hymenoptera Bethylidae	India	Larva	Krishnamurthi and Usman, 1954 Subba Rao *et al.*, 1965 Subba Rao and Sharma, 1966; Mohammad 1981; Phisitkul, 1985; Srinivasn and Siva Rao, 1986; Shanower, 1989
Apanteles sp.	Hymenoptera Braconidae	India	Larva	Srinivasan and Rao, 1987; Khan and Raodeo, 1978; Cherian and Basheer, 1942
A. agilis	Hymenoptera Braconidae	India	Larva	Jai Rao, 1992
A. javensis	Hymenoptera Braconidae	India	Larva	Amin, 1987 Subba Rao *et al.*, 1965
Apanteles sp. (laevigatus group)	Hymenoptera Braconidae	India	Larva	Yadav *et al.*,1987
A. singaporensis	Hymenoptera Braconidae	India	Larva	Subba Rao *et al.*, 1965; Subba Rao and Sharma, 1966
A. litae	Hymenoptera Braconidae	India	Larva	Krishnamurthi and Usman, 1954; Subba Rao *et al.*, 1965; Subba Rao and Sharma, 1966; Mohammad, 1981; Phisitkul, 1985; Srinivasn and Siva Rao, 1986; Shanower, 1989
Avga choaspes	Hymenoptera Braconidae	India	Larva	Anonymous, 1993; Shekharappa *et al.*, 1990
A. nixoni	Hymenoptera Braconidae	India	Larva	Subba Rao and Sharma, 1966
Bracon sp.	Hymenoptera Braconidae	India	Larva	Subba Rao *et al.*, 1965; Cherian and Basheer, 1942

Contd...

Table 3.4. Contd...

B. gelechiae	Hymenoptera Braconidae	India	Larva	Yadav *et al.*, 1987; Shekharappa *et al.*, 1990; Khan and Raodeo, 1978
B. brevicornis	Hymenoptera Braconidae	India	Larva	Subba Rao *et al.*,1965
B.(Microbracon) hebetor	Hymenoptera Braconidae	India	Larva	Amin, 1987; Cherian and Kylasam, 1941
Chelonus sp.	Hymenoptera Braconidae	India	Egg	Subba Rao *et al.*,1965; Yadav *et al.*, 1987
Chelonus sp. nr. *malayana*	Hymenoptera Braconidae	India	Egg	Cherian and Basheer, 1942
C. blackburni	Hymenoptera Braconidae	India	Egg	Srinivasan and Rao, 1987
C. curvimaculatus	Hymenoptera Braconidae	India	Larva	Khan and Raodeo, 1978; Amin, 1987
Dolichogenidea sp.	Hymenoptera Braconidae	India	Larva	Shetgar *et al.*,1993
Phanerotoma sp.	Hymenoptera Braconidae	India	Egg/ Larva	Krishnamurthi and Usman, 1954; Subba Rao *et al.*,1965; Subba Rao and Sharma, 1966; Mohammad, 1981; Phisitkul, 1985; Srinivasn and Siva Rao, 1986; Shanower, 1989
Ceraphron sp.	Hymenoptera Ceraphronidae	India	Egg/ Larva	Krishnamurthi and Usman, 1954; Subba Rao *et al.*,1965; Subba Rao and Sharma, 1966; Mohammad, 1981; Phisitkul, 1985; Srinivasn and Siva Rao, 1986; Shanower, 1989
Aphanagmus fijiensis	Hymenoptera Ceraphronidae	India	Egg/ Larva	Krishnamurthi and Usman, 1954; Subba Rao *et al.*,1965; Subba Rao and Sharma, 1966; Mohammad, 1981; Phisitkul, 1985; Srinivasn and Siva Rao, 1986; Shanower, 1989
Brachymeria sp.	Hymenoptera Chalcididae	India	Larva/ pupa	Amin, 1987; Yadav *et al*,.1987; Shekharappa *et al.*, 1990; Muthiah, 1991; Cherian and Basheer, 1942; Khan and Raodeo, 1978

Contd...

Table 3.4. Contd...

Copidosoma sp.	Hymenoptera Encyrtidae	India	Larva	Subba Rao *et al.*, 1965; Krishnamurthi and Usman, 1954; Subba Rao and Sharma, 1966; Mohammad, 1981; Phisitkul, 1985; Srinivasn and Siva Rao, 1986; Shanower, 1989
Cerchysiella sp.nr. *laevigatus*	Hymenoptera Encyrtidae	India	Larva	Jai Rao, 1992
Homalotylus eytelweini	Hymenoptera Encyrtidae	India	Larva	Jai Rao, 1992
Asympiesiella sp.	Hymenoptera Eulophidae	India	Larva	Amin, 1987
Euryscotolynx coimbatorensis	Hymenoptera Eulophidae	India	Larva	Krishnamurthi and Usman, 1954; Subba Rao *et al.*, 1965; Subba Rao and Sharma, 1966; Mohammad, 1981; Phisitkul, 1985; Srinivasn and Siva Rao, 1986; Shanower, 1989
Oomyzus sp.	Hymenoptera Eulophidae	India	Larva	Krishnamurthi and Usman, 1954; Subba Rao *et al.*, 1965; Subba Rao and Sharma, 1966; Mohammad, 1981; Phisitkul, 1985; Srinivasn and Siva Rao, 1986; Shanower, 1989
Pediobius sp.	Hymenoptera Eulophidae	India	Larva	Krishnamurthi and Usman, 1954; Subba Rao *et al.*, 1965; Subba Rao and Sharma, 1966; Mohammad, 1981; Phisitkul, 1985; Srinivasn and Siva Rao, 1986; Shanower, 1989
Stenomesioideus ashmeadi	Hymenoptera Eulophidae	India	Larva	Krishnamurthi and Usman, 1954; Subba Rao *et al.*, 1965; Subba Rao and Sharma, 1966; Mohammad, 1981; Phisitkul, 1985; Srinivasn and Siva Rao, 1986; Shanower, 1989
Stenomesius japonicus	Hymenoptera Eulophidae	India	Larva	Krishnamurthi and Usman, 1954; Subba Rao *et al.*, 1965;

Contd...

Table 3.4. Contd...

				Subba Rao and Sharma, 1966; Mohammad, 1981; Phisitkul, 1985; Srinivasn and Siva Rao, 1986; Shanower, 1989
***Stenomesius* sp.**	**Hymenoptera Eulophidae**	**India**	**Larva**	**Krishnamurthi and Usman, 1954; Subba Rao *et al.*, 1965; Subba Rao and Sharma, 1966; Mohammad, 1981; Phisitkul, 1985; Srinivasn and Siva Rao, 1986; Shanower, 1989**
***Sympiesis (Asympiesiella)* sp.**	**Hymenoptera Eulophidae**	**India**	**Larva**	**Krishnamurthi and Usman, 1954; Subba Rao *et al.*, 1965; Subba Rao and Sharma, 1966; Mohammad, 1981; Phisitkul, 1985; Srinivasn and Siva Rao, 1986; Shanower, 1989**
Sympiesis dolichogaster	**Hymenoptera Eulophidae**	**India**	**Larva**	**Krishnamurthi and Usman, 1954; Subba Rao *et al.*, 1965; Subba Rao and Sharma, 1966; Mohammad, 1981; Phisitkul, 1985; Srinivasn and Siva Rao, 1986; Shanower, 1989**
S. indica	**Hymenoptera Eulophidae**	**India**	**Larva**	**Krishnamurthi and Usman, 1954; Subba Rao *et al.*, 1965; Subba Rao and Sharma, 1966; Mohammad, 1981; Phisitkul, 1985; Srinivasn and Siva Rao, 1986; Shanower, 1989**
***Tetrasticus* sp.**	**Hymenoptera Eulophidae**	**India**	**Larva**	**Krishnamurthi and Usman, 1954; Subba Rao *et al.*, 1965; Subba Rao and Sharma, 1966; Mohammad, 1981; Phisitkul, 1985; Srinivasn and Siva Rao, 1986; Shanower, 1989**
***Elachertus* sp.**	**Hymenoptera Eulophidae**	**India**	**Larva**	**Shekharappa *et al.*, 1990**
***Elasmus* sp.**	**Hymenoptera Elasmidae**	**India**	**Larva**	**Yadav *et al.*, 1987**
Elasmus* sp. nr. *luteus	**Hymenoptera Elasmidae**	**India**	**Larva**	**Krishnamurthi and Usman, 1954;**

Contd...

Table 3.4. Contd...

				Subba Rao *et al.*, 1965; Subba Rao and Sharma, 1966; Mohammad, 1981; Phisitkul, 1985; Srinivasn and Siva Rao, 1986; Shanower, 1989
Elasmus anticles	Hymenoptera Elasmidae	India	Larva	Krishnamurthi and Usman, 1954; Subba Rao *et al.*, 1965; Subba Rao and Sharma, 1966; Mohammad, 1981; Phisitkul, 1985; Srinivasn and Siva Rao, 1986; Shanower, 1989
Elasmus brevicornis	Hymenoptera Elasmidae	India	Larva	Amin, 1987
Eupelmus sp. (*urozonus group)*	Hymenoptera Eupelmidae	India	Larva/ pupa	Subba Rao *et al.*, 1965
Eupelmus sp.nr. *anpingensis*	Hymenoptera Eupelmidae	India	Larva/ pupa	Cherian and Basheer 1942; Amin, 1987
Eurytoma sp.	Hymenoptera Eurytomidae	India	Larva	Srinivasan and Rao, 1987; Shekarappa *et al.*, 1990; Shetgar *et al.*, 1993
Plutarchia giraulti	Hymenoptera Eurytomidae	India	Larva	Krishnamurthi and Usman, 1954, Subba Rao *et al.*, 1965; Mohammad, 1981; Phisitkul, 1985; Srinivasn and Siva Rao, 1986; Shanower, 1989
Temelucha sp.	Hymenoptera Ichneumonidae	India	Larva	Shekharappa *et al.*, 1990
Anagrus sp.	Hymenoptera Mymaridae	India	Larva	Shetgar *et al.*,1993
Pteromalus sp.	Hymenoptera Pteromalidae	India	Larva	Srinivasan and Rao, 1987; Shekharappa *et al.*, 1990
Dibrachys sp.	Hymenoptera Pteromalidae	India	Larva	Krishnamurthi and Usman, 1954 Subba Rao *et al.*, 1965; Subba Rao and Sharma 1966 Mohammad, 1981; Phisitkul, 1985; Srinivasn and Siva Rao, 1986; Shanower, 1989
Habrocytus sp.	Hymenoptera Pteromalidae	India	Larva	Krishnamurthi and Usman, 1954; Subba Rao *et al.*, 1965; Subba Rao and Sharma, 1966; Mohammad, 1981; Phisitkul, 1985;

Contd...

Table 3.4. Contd...

				Srinivasn and Siva Rao, 1986; Shanower, 1989
Idris sp.	Hymenoptera Scelionidae	India	Larva	Yadav *et al.*, 1987

Table 3.5. List of predators of *Aproaerema modicella*

Predator	Order/Family	Country	Host stage attacked	Reference
Apertochrysa sp.	Neuroptera Chrysopidae	India	Egg/larva	Jai Rao, 1992
Chlaenius sp.	Coleoptera Carabidae	India	Larva	Shanower and Ranga Rao, 1990; Ranga Rao, 1996
Spiders	Araneae	India	Larva	Srinivasan and Rao, 1987
Robber flies	Diptera Asilidae	India	Larva	Srinivasan and Rao, 1987

Table 3.6. List of entomopathogens of *Aproaerema modicella*

Species	Order/Family	Country	Host stage attacked	Reference
Bacillus thuringiensis	Eubactariales Bacillaceae	India	Larva	Rajgopal *et al.*, 1988
Aspergillus flavus	Moniliales Moniliaceae	India	Larva	Srinivasan and Rao, 1987
Beauveria bassiana	Moniliales Moniliaceae	India	Larva	Rajgopal *et al.*, 1988
Baculo virus	Virales Baculoviridae	India	Larva	Sundrababu and Rabindra, 1993; Rajagopal *et al.*, 1988
Entomophilic nematodes	Nematoda Steinernematidae	India	Larva	Srinivasan and Rao, 1987

□□□

Spodopetera litura (F.)

G. V. RANGA RAO, R.J. RABINDRA,
V. NANDA GOPAL AND RAMESWAR RAO

4.1. Introduction

Spodoptera litura (F.) commonly known as the tobacco caterpillar is one of the serious insect pests of groundnut. In India it is considered as a pest of national importance since it is polyphagous attacking several agricultural and horticultural crops. It has been reported from 51 countries causing damage to more than 120 species of plants belonging to 44 families. In India it feeds on 74 species of cultivated crops and some wild plants. Besides groundnut, it attacks tobacco, cotton, pulses and several vegetable crops (Singh and Jalali, 1997).

4.2. Economic importance

The larvae feed voraciously on the leaves causing defoliation. It has been reported that an infestation level of one larva/plant during the seedling or flowering stage can result in 20% yield loss in groundnut. Severe outbreak of the pest can result in 30-40% loss in pod formation. At Dharwad, the pod loss was to an extent of 66.6% due to *S. litura* (Kulkarni, 1989). Studies at ICRISAT Centre and at-on farm stations indicated that the groundnut crop is vulnerable to defoliation during the early phase of the crop. In the first 50 days of seedling emergence, if the defoliation is <25% and after 50 days of crop age if the defoliation is around 50%, there would not be any effect on yield. However, there were a number of instances where total crop failures were reported despite intensive pest management practices (Wightman and Ranga Rao, 1993).

4.3. Nature of Injury

The larvae are gregarious in feeding. The first instar feeds by scrapping and the grown up larvae feed the lamina the under surface of the leaves leaving the vein and upper epidermis intact giving a fabric surface **(Plate 1).**

When the damage is severe, they defoliate totally leaving only the stems **(Plate 2)**. The 2nd and 3rd instar larvae feed by making small holes here and there on the leaf. The subsequent instars feed voraciously on the entire lamina, petioles and some times on the tender twigs on the terminal shoot of the plant. In certain cases they feed on the flower and bore the tender groundnut pods, which are just below the soil surface.

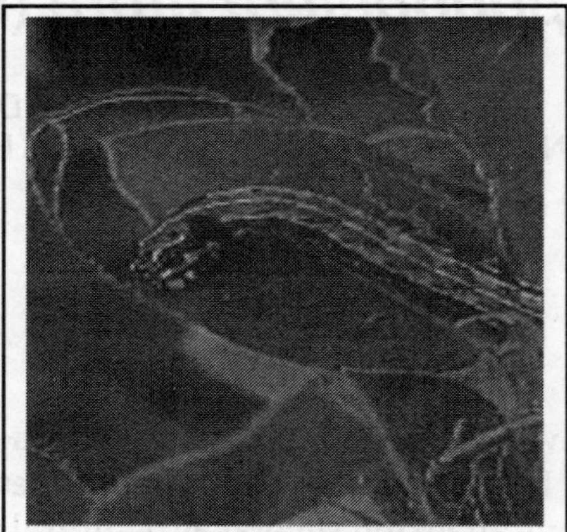

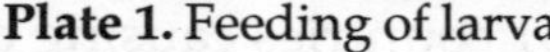

Plate 1. Feeding of larva

Plate 2. Total defoliation *S.litura*

4.4. Biology

A female moth on an average mates 3-4 times and males 10 times in a life span. Males mate only once in night. A moth lays on an average about 300-700 eggs in clusters covered by scales **(Plate 3)**. The fecundity of the female moth varies from 2000 to 2600 eggs and oviposition period spread over 6 to 8 days. Eggs hatch in about 3-4 days at 26°C and in 3 days at 30°C (Bilapate *et al.*, 1980) and feed gregariously by scraping the leaf surface. As they grow older, the larvae tend to disperse. The larvae feed for about 2 to 3 weeks and when fully grown they are stout, cylindrical and pale greenish brown with dark marking. The colour of the larvae also varies according to the host plant and season. There may be rows of dark spots or transverse and longitudinal gray and yellow band on the body of the larvae. As they grow in size, the colour changes from light green to dark gray or black. Early instars are tiny blackish green with a distinct black band on the first abdominal segment. Grown up caterpillars are 40-50mm long, stout, smooth but having scattered short setae. They have dull blackish and greenish body with yellow dorsal and lateral stripes, the lateral yellow stripes are bordered dorsally with a series of semi lunar black marks and a 'V' shaped white mark on the front portion of the black head. After passing through 6 larval instars in a span of about 25 days they pupate inside the soil in an earthen cell. The pupa is dark brownish black in colour. Pupation is maximum under fallen leaves in wet sandy loam soil and clay soil is unfavorable for pupation. Most of the larvae pupate at 4cm depth in any soil, very rarely the pupation can be noticed up to 12 cm. After completing the pupal period, the adult moth emerge during early hours of the night. The egg, larval and pupal period is 3, 12-26 and 6 to 16 days respectively with a total lifespan of 49 to 63 days for Females and 50 to 66 days for males at 20°C. A detailed biology of this insect is reported (Anon., 1993) **(Table 4)** The life cycle of the insect is shown in. The female moth produces sex pheromone that attracts male prior to courtship (Tamaki *et al.*, 1973; Yashima *et al.*,1974). Males have short and pointed abdomen, measuring about 6mm and females 8mm long abdomen with tuft of hair measuring 3mm long. The fore wings of both male and female have similar pattern of coloration i.e. black patches with yellow marking in both the sexes, the apical margin up to anal angle, there is a band of about 8 intermittent with yellow lines which measure 1mm. The remigium area is marked with black scale. There is yellow band starting in between costal margin and pterostigma ending just 2mm above anal region. The hind wing is hyaline with a small black lining starting from humeral angle passing through costal margin, pterostigma, apical angle, apical margin, anal angle and ends with light marking just below anal margin.

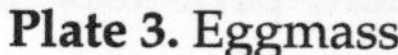

Plate 3. Eggmass

Plate 4. Life stages

Table 4.1. Biometrical observation on *Spodoptera litura*

Stages	Duration (Days) Mean ± SEm	Range (days)
eggs	4.28 ± 0.24	3-8
larvae		
1st Instar	2.96± 0.2.	2-5
2nd Instar	3.00± 0.15	2-4
3rd Instar	3.28± 0.20	2-5
4th Instar	3.96± 0.21	2-5
5th Instar	4.36± 0.21	3-6
6th Instar	3.84± 0.24	2-6

Contd...

Table 4.1. Contd...

Total larval period	21.48± 0.83	14-28
Pre-pupa	1.84± 0.13	1-3
Pupa	8.44± 0.25	7-11
Pre-oviposition	1.92± 0.14	1-3
Oviposition	5.72± 0.42	3-11
Post-oviposition	3.72± 0.24	2-6
adult		
Male	5.96± 0.42	3-11
Female	11.36± 0.40	8-17
Total life	42.12± 1.36	32-53
Fecundity	431.64± 22.54	250-600

4.5. Seasonal history

The brood of *S. litura* overlap and all the stages of the insect are found at a time. However, there are effective brood when moth emergence take place in large numbers at the favourable period. The outbreak initiated by heavy rainfall after the dry spell lasting for 4-7 days under north Indian condition. The light trap catches in south India reveal peak population during November-December and again in April and October (Ranga Rao and Wightman, 1994).

The moth catches are usually high (186moth/night) around new moon night and low (96moth/trap/night) in full moon night. Over 90% moths are caught before midnight and if there are any drizzles in the early hours of night, the trap catches are low. The pest population showed a negative correlation with mean maximum temperature and positive correlation with relative humidity. In general, the population level is often influenced by the availability of the favourable crop plants in large area coupled with suitable weather conditions. In north India the population of the male moth was low from November last week to February third week, probably due to heavy winter during this period and in South India, the population was low, from last week of April to second week of June. Otherwise the population goes up to maximum of 2300moth/trap/week during July and October (Ranga Rao and Wightman, 1994). Sridhar and Prasad (1996) studied the life table of *Spodoptera litura* and reported that generation mortality due to known and unknown causes was 1.0314 in the 1st generation, 0.9355 in the 2nd generation and 1.0094 in the overlapping generation. Of 4 known mortality factors, *Nomuraea rileyi*, an entomogenous fungus, caused a maximum mortality of 36.9%. A protozoan (*Nosema* sp.) caused 13.8% and larval parasites, *Peribaea orbata* and *Apanteles ruficrus* [*Cotesia ruficrus*], caused 13.7 and 8.2% mortality respectively. Bhanukiran *et al.*(1998) studied the life tables of *Spodoptera litura* on groundnut. Key factor analysis showed that the entomogenous fungus *Nomuraea rileyi* was the main larval mortality factor. Also, birds (*Acridotheres tristis* and *Dicrurus adsimilis*) recorded were important mortality factors followed by *Vespa cincta* [*Vespa tropica*] and *Eumenes* sp., and spiders. The coccinellids *Verania vincta* [*Micraspis vincta*], *Coccinella transversalis* and *Menochilus sexmaculata* [*Cheilomenes sexmaculata*] played a major role in reducing the pest population. Negative trend indices of the first (I=0.4855) and second (I=0) generation indicated the effect of natural factors in reducing the population of *S. litura*. Generation survival of the first generation (0.5745) was more than that of the second generation (0.3615).

4.6. Alternate hosts

S. litura is a polyphagous insect infesting varied host plants, which are distributed throughout India. **(Table 4.2).** The continuance of the castor crop up to February- March in the field grown either as sole crop or as intercrop in groundnut facilitates the population build up and sustainability.

Table 4.2. Host plants of *Spodoptera litura*

Host plant	Reference
Agathi *(Agati grandiflora)*	David & Kumarswami, 1975
Alfalfa	Okada, 1993
Amaranth *(Amaranthus hybridus)*	Ayyar, 1963
Banana (*Musa paradisiaca*)	Rao, 1956
Piper nigrum	Singh, 1980
Blackcumin	Nagalingam *et al.*, (1979)
Blackgram (*Phaseolus mungo*)	Ayyar, 1963
Brinjal *(Solenum melongena)*	Pruthi, 1969
Cabbage*(Brassica oleracea var. capitata)*	Ayyar, 1963
Castor (*Ricinu communis)*	Rao, 1956
Cauliflower*(Brassica oleracea var.capitata)*	Ayyar, 1963
Chilli (*Capsicum annuum*)	David & Kumarswami, 1975
Chow chow	Ayyar, 1963
Chromolaena	Venkatesha, 1993
Clove	Thomas *et al.*, 1969
Coffee *(Coffea arabica)*	Venkatesha, 1993
Cotton *(Gossypium* spp.*)*	Ayyar, 1963
Cowpea (*Vigna unguiculata*)	Yadav & Yadav, 1983
Cynara	Krishnaiah *et al.*, 1975
Fennel	Nagalingam *et al.*, 1979
Fenugreek *(Trigonella foenumgraceum)*	Nagalingam *et al.*, 1979
Gladiolus	Krishnaiah *et al.*, 1975

Contd...

Table 4.2. Contd...

Gram (*Cicer arientimum*)	Pruthi, 1969
Green gram	Nagalingam *et al.*, 1979
Groundnut	Rao, (1956)
Jatropha curcas	Meshram & Joshi, 1994
Jute (*Corchorus capsularis*)	Patel, 1943
Linseed (*Linum utitatissimum*)	Pruthi, 1969
Long bean	Singh & Hoi, 1972
Lucerne	Pruthi, 1969
Maize (*Zea mays*)	Rao, 1956
Marsilea	Sain *et al.*, 1983
Okra	Thomas *et al.*, 1969
Onion *(Allium cepa)*	David & Kumarswami, 1975
Opium	Singh & Dabral, 1977
Poppy	Pruthi, 1969
Potato (*Solanum tuberosum*)	Pruthi, 1969
Oryza sativa	Im *et al.*, 1989
Rosa sp.	Pruthi, 1969
Sorghum (*Sorghum vulgare*)	Singh & Byas 1975
Soybean (*Glycine max*)	Srivastava *et al.*, 1971
Sugar beet (*Helianthus annuus*)	Pruthi, 1969
Sunflower	Rao, 1956
Sun hemp (*Crotalaria juncea*)	Ramamurthy *et al.*, 1967
Sweet potato (*Ipomoea batatas*)	Pruthi, 1969
Taro	Macfarlane *et al.*, 1987
Tobacco (*Nicotina tabacum*)	Rao 1956
Tomato (*Lycopersicon lycopersicum*)	Ayyar 1963
White Clover	Okada, 1993
Radish (*Raphanus sativus*)	Vaish and Agrawal, 1978
Dhaincha (*Sesbanium aculeata*)	Dhandapani *et al.*, 1982
Apple (Malus pumila)	Thakur *et al.*, 1976
Rami	Mustafee, 1977
Tea (*Camellia thea*)	Muraleedharan and Kandasamy, 1981
Ommo	Nagalingam *et al.*, 1979
Citrus	Nagalingam *et al.*, 1979
Coriander	Butani, 1984
Artichoka	Krishnaiah *et al.*, 1975
Mulbery	Anon., 1984
Maize (*Zea mays*)	Patel and Vyas, 1984

Contd...

Table 4.2. Contd...

Bajra *(Pennisetum typhoides)*	Patel and Vyas, 1984
Verseem	Pruthi, 1936
Winged bean	Ahmed and Gangwar, 1984
Ridge gourd	Chhibber *et al.*, 1978
Coccinia sp	Panday and Srivastava, 1967
Zinnias	Chockalinga and Somasundaram, 1983
Strawberry (*Fragaria* sp.)	Oikawa *et al.*, 1999
Perilla nankinensis	Oikawa *et al.*, 1999
Sweet pepper	Oikawa *et al.*, 1999
Grape (*Vitis vinifera*)	Jagginavar *et al.*, 1999
(Grapevine)	Balikai *et al.*, 1999
Teak	Roychoudhury *et al.*, 1998
Lilly (*Lillium* sp.)	Liu-Tashiu *et al.*, 1998
Paulownia	Manoj-Kumar and Mukhtar-Ahmad, 1998
Piper nigrum	Vos and Frinking, 1998
Hevea brasiliensis	Duong-Nguyen Hai *et al.*, 1998
Sesbaia spp	Roychoudhury *et al.*, 1997
Acacia mangium	Intachat and Kirton, 1997
Brahmi	Jyoti-Tripathi *et al.*, 1997

4.7. Causes of outbreak

The primary cause of the outbreak, though appearing once in 3-4 years in large numbers is due to resistance developed against the commonly used pesticides such as cypermethrin ranging from 0.22 to 197 fold, fenevelerate 8 to 12 folds, endosulfan 1 to 13 folds, methomyl 0.7 to 19 folds; quinolphos 1 to 29 folds and monocrotophos 2 to 362 folds among the 22 strains of *S. litura* collected from different locations from Andra Pradesh (Arms *et al.*, 1994)

- Favorable weather conditions : Heavy rainfall following a prolonged dry spell can contribute to outbreak (Chelliah, 1985)
- Indiscriminate use of chemical insecticides resulting in
  a. Destruction of natural enemies
  b. Development of insecticide resistance

  The use of substandard insecticides and improper appliance leading to in control failures.
- Excessive application of nitrogenous fertilizers
- Frequent irrigation resulting in high humidity
- Availability of alternate weed hosts and crop plants in the ecosystem.

Twenty-two strains of *Spodoptera litura* collected from groundnut crops of eight locations in Andhra Pradesh, India, between 1991 and 1996 were assayed in the F1 generation for resistance to commonly used insecticides. Resistance levels ranged as follows: cypermethrin, 0.2- to 197-fold; fenvalerate, 8- to 121-fold; endosulfan, 1- to 13-fold; quinalphos, 1- to 29-fold; monocrotophos, 2- to 362-fold and methomyl, 0.7- to 19-fold. In nearly all strains pre-treatment with the metabolic inhibitor, piperonyl butoxide, resulted in complete suppression of cypermethrin resistance (2- to 121-fold synergism), indicating that enhanced detoxification by microsomal P450-dependent monooxygenases was probably the major mechanism of pyrethroid resistance. Pre-treatment with the synergist DEF, an inhibitor of esterases and the glutathione S-transferase system, resulted in a 2- to 3-fold synergism with monocrotophos indicating that esterases and possibly glutathione S-transferases were at least to some extent contributing to organophosphate resistance Armes et al.(1997). Kumar and Regupathy (2000) estimated the LD50s of endosulfan, chlorpyrifos, profenofos, fenvalerate and deltamethrin were 0.7628, 0.0355, 0.3710, 0.0672 and 0.0647 ppm, respectively, for 3rd-instar larvae of *S. litura* from a field population on rainfed groundnut in the Madurai district of Tamil Nadu.

4.8. Natural enemies

Rao *et al.* (1993) listed 71 species of parasitoids in 7 families of Hymenoptera and 2 families of Diptera. Thirty six predators belonging to 14 families and 12 species of spiders from six families have been reported to feed on this species (Table 4.3). The persistence of *N. rileyi* conidia, similarly applied to the soil in a groundnut field in India, was high, even at 2 weeks after application (Devi, 1995). The predator, *Rhynocoris kumarii* preferred *S. litura* (48%), followed by *H. armigera* (34%) and *B. subsecivella* (18%) (Sahayaraj and Sivakumar, 1995). Populations of *S. litura* were significantly reduced following release of predator, *Rhynocoris marginatus* at 5000 predators/ha at 15-day intervals Pod yield was also highest (1867.77 kg/ha) in plots where the predator was released (Sahayaraj, 1999).

Four protozoan, four fungal, seven bacterial, four each of viral and nematode species were also reported as pathogens infecting the larvae. Published information suggest that periodic releases of large number of egg parasites could help in suppressing this pest. Mass releases of an indigenous egg-larval parasite *Chelonus heliopae* against *S. litura* in cauliflower proved ineffective in controlling the pest. However, five weekly releases of 50000 parasites per 0.2 ha and two releases of 15 000 parasitoids per 0.2 ha in cauliflower resulted in 60% parasitism (Patel *et al.*, 1979). *Telenomus remus* was introduced to Western Samoa and was recorded by Braune (1982) as a common egg-larval parasitoid of *S. litura,* with parasitism averaging 54%. Complete parasitisation was observed only in small egg masses and the percentage of parasitisation decreased with an increase in size of egg mass. *T. remus* could oviposit only on the surface of the host egg mass. Thus the effectiveness of *T. remus* was limited to the large compact egg masses of *S. litura*. *Rhynocoris marginatus* fourth and fifth instar nymphs preferred *H. armigera*, followed by *S. litura*. Adult predators preferred *S. litura* the most, followed by *H. armigera*. Predators at all life stages showed the least preference for A. modicella. (Sahayaraj and Paulraj, 2001). Two applications of *Spodoptera litura* NPV at 250 larval equivalents (LE)/ha with the adjuvant crude sugar at 2.5 kg/ha with an interval of 10 days significantly reduced the larval population of *Spodoptera litura* and increasing the pod yield in comparison with untreated controls (Dhandapani and Babu, 1995).

The nuclear polyhedrosis virus (NPV) of *Spodoptera litura* was not cross infective to other groundnut pests such as *Heliothis armigera* [*Helicoverpa armigera*], *Aproaerema modicella, Empoasca kerri* and *Aphis craccivora* and was safe to their predators (*Coccinella septempunctata, Menochilus sexmaculatus* [*Cheilomenes sexmaculata*] and *Chrysoperla carnea*) and parasitoids (*Bracon gelechiae* and *Tetrastichus* sp). Laboratory experiments were conducted to determine the effect of biological pesticides, i.e. Vijayneem and Nimbicidine (neem [*Azadirachta indica*] formulations), and nucleopolyhedroviruses (*Spodoptera*; NPV(S)), on the behaviour and biological control potential of *R. marginatus*. First, second and third instars of *R. marginatus* were highly deterred by Nimbicidine, and the results were statistically significant compared to the other biological pesticides tested. Late instars (fourth and fifth) were deterred by Vijayneem and its effect was significant (except fourth instars) compared to Nimbicidine and NPV(S) treatments. The first, second, third, fourth and fifth nymphal instars and adults of *R. marginatus* consumed 1.3, 2.6, 3.9, 5.2, 6.7 and 7.5 *S. litura* third instars, respectively. In the untreated control, the mean consumption was 28.8 S. litura larvae per predator. *R. marginatus* consumed 26.3 *S. litura* larvae when treated with NPV and the value was statistically not significant when compared to the untreated control. The predation rate was reduced in larvae fed on leaves treated with Vijayneem (21.9 prey per predator) and Nimbicidine (23.7 prey per predator) treatments Sahayaraj (2001).

Ramegowda and Basavanagoud (2001) tested the efficacy of microbial agents (*S. litura* nuclear polyhedrosis virus, SlNPV at 1 ml/litre; *Bacillus thuringiensis* subsp. kurstaki, Btk at 1 ml/litre; and *Nomuraea rileyi* at 1 g/ litre) against insecticide susceptible and resistant *S. litura* infesting groundnut. The new compounds, lambda cyhalothrin, carbosulfan, profenofos and polytrin C [cypermethrin+profenofos], and test insecticides MPO 62, RH 2485, spinosad and fipronil were also evaluated against the pest. The control insecticides used were monocrotophos at 0.036%, quinalphos at 0.05%, endosulfan at 0.07% and cypermethrin at 0.005%. Mortality by the bioagents was evident on the 3rd day after treatment. SlNPV and *N. rileyi* did not cause significant effects on the pest, while *Btk* caused 4 and 26% mortality in the resistant and susceptible insects, respectively. The new compounds were highly effective against the insecticide susceptible strain. However, against the resistant strain, only carbosulfan, profenofos and plolytrin-C were highly effective. Naik and Ramaswamy (2005) reported that application of *S. litura* nucleopolyhedrosis virus (SLNPV) SLNPV + vegetable oil + jaggery + boric acid + tinopal was the most effective against *S. litura* on groundnut cv. TMV-2 registering the least mean percent of infestation (4.81%) and the highest pod yield (4071 kg/ha).

The incidence of *Nomuraea rileyi* on *Spodoptera litura* in groundnut was observed between thirty-second to fortieth standard week (SW) with peak activity in the thirty-seventh standard week (10-16 September) during 1996 and 1997 and in the thirty-fourth (20-26 August) during 1998. *Nomuraea rileyi* was recorded infecting 11% of the larval population of *Spodoptera litura* on groundnut at Patancheru, Andhra Pradesh, in September 1999. (Venkatesan *et al.,* 2000). The most favourable period of the season for the fungus was from the thirty-second to the fortieth week namely, first week of August to first week of October. Maximum incidence was noticed in groundnut (27.12%) and correlation studies between weather parameters and disease incidence of *N. rileyi* on S. litura indicated positive significant influence of afternoon relative humidity in groundnut (0.87) (Kulkarni and Lingappa, 2002). Patil *et al.*(2003) observed the natural incidence of the entomopathogen, *N. rileyi,* on *S. litura* in a groundnut The first appearance of *N. rileyi* was recorded in the 30th standard week, which continued up to the 39th standard week and peaked between the 33rd and 37th standard weeks. During this period, the temperature ranged between 19.64 and 29.97°C and the relative humidity ranged from 72.49 to 94.19%, with mean weekly rainfall fluctuating from 2.0 to 62.30 mm. Groundnut recorded 100% mycosed larvae of *S. litura* due to infection by *N. rileyi*. Mycosed larvae were low in sweet pepper. The incidence of *N. rileyi* showed positive association with morning relative humidity. *H. armigera* recorded higher infection compared to *S. litura* in groundnuts. However, higher infection was observed in *S. litura* compared to *H. armigera* in black gram (Manjula *et al.,* 2003).

The list of parasitoids (Table 4.3), predators (Table 4.4), Entomopathogens (Table 4.5) and bird predators (Table 4.6) recorded on *Spodoptera litura* in India are given.

4.9. Host plant resistance

Patil *et al.* (1991) reported two lines ICGV87264 and ICGV 86590 recording the least foliar damage as compared to 37 per cent damage in ICGV 86600. Tiwari *et al.* (1988) classified released varieties for there susceptibility based on the larval growth, pupal period and weight and the longevity of the adult. The variety dwarf mutant registered a lowest larval period (15 days), while, M-13 and C501 had the longest larval period (19 days). Rajgopal *et al.* (1988) reported significant varietal difference both in larval orientation as well as leaf damage in the host preference studies. The varieties having less damage were V40 and AH629 in Virginia bunch and NCAC17840, NFG79, EC21989 in Virginia runner. Stevenson *et al.*(1993) identified and characterized some of the wild species of the *Arachis* resistance to *S. litura* in terms of larval mortality and weight gain. *A. major* and *A. kemphmercadoi*, were rejected by the larvae for feeding indicating their resistance to *S. litura.* Sreenivasa *et al.*(1997) studied the larval mortality of *Spodoptera litura* which was highest on the groundnut variety ICGV 86031 (82.8%) followed by ICGV 86350 (53.32%) and Dh-3-30 (43.02%). Development was shortest on Dh-3-30 (32.25 days, adults surviving for a maximum of 10.5 days), and longest (37.5 days) on ICGV 86031 with adults surviving a maximum of 9.50 days. ICGV 86350 was intermediate.

Out of 32 genotypes the lowest *S. litura* leaf damage (5%) was recorded in ICGV 86156, ICGV 86400, ICGV 86528, ICGV 87128, ICGV 87141, ICGV 87290, ICGV 87411 and ICGV 91214 (Dharne and Patel, 2000). Accessions belonging to *Arachis cardenasii* (ICG 8216), *Arachis ipaensis* (ICG 8206), *A. paraguariensis* (ICG 8130) and *Arachis appressipila* (ICG 8946) showed resistance to leaf feeding and antibiosis to *Spodoptera litura* under no-choice conditions. Six lines, derived from wild relatives, showed resistance to H. *armigera* and *S. litura*, and/or leaf miner. Plant morphological characteristics such as main stem thickness, hypanthium length, leaflet shape and length, leaf hairiness, standard petal length and petal markings, basal leaflet width, main stem thickness and hairiness, stipule adnation length and width, and peg length showed significant correlation and/or regression coefficients with damage by *H. armigera, S. litura*, and leafhoppers, and these traits can possibly be used as markers to select for resistance to these insect pests (Sharma *et al.*, 2003).

4.10. Management of *S. litura*

The IPM practices suggested for *S litura* (Rama Prasad, 1997) are as follows.

4.10.1 Cultural methods

- Summer ploughing to expose the pupae for predation by birds and desiccation by sunlight
- Optimized irrigation and fertilization to reduce luxuriant and succulent vegetative growth which can favour multiplication of *S. litura*
- Destruction of weed hosts to prevent the carry over to the next generation

- Raising of castor s a trap crop on the bunds or on the borders to attract adults for oviposition
- Inclusion of less preferred crop plants like sorghum and finger miller in cropping system

Nath and Singh (1998) studied the effect of groundnut intercropping with millet [*Pennisetum glaucum*] and pigeon pea on the relative infestation of insect pests and concluded that the populations of *Spodoptera litura* and *Aulacophora* spp. were the highest in the groundnut + maize intercrop, The populations of *Aulacophora* and *A. crenulata* were lowest in the groundnut + *P. glaucum* and the groundnut + pigeon pea intercrops. The incidence of S. litura was low in the groundnut + *P. glaucum* intercrop and in the groundnut + pigeon pea intercrop. The effect of intercropping groundnut (cv. Chitra) with *Cajanus cajan* cultivars (Bahar and UPAS-120), *Phaseolus aureus* [*Vigna radiata*] (cv. Pant moong-1), *P. mungo* [*Vigna mungo*] (cv. Type-9), *Glycine max* (cv. Ankur), *Crotalaria juncea* (cv. Local), *Sesamum indicum* (cv. Gujarat-1), *Guizotia abyssinica* (cv. Local), *Abelmoschus esculentus* (cv. Pusa Sawani), *Sorghum vulgare* [*Sorghum bicolor*] (cv. HOS-1), *Pennisetum typhoides* [*Pennisetum glaucum*] (cv. Local) and *Zea mays* (cv. Jaunpur Local) on the population of and groundnut leaf damage caused by Spodoptera litura was studied by Singh and Paras-Nath, (2003). Among the groundnut-based intercrops, the intercrop groundnut + niger had the lowest incidence of the pest, while the lowest leaf damage (%) was observed with groundnut intercropped with pigeon pea. The intercrop groundnut + black gram showed the highest incidence and percent leaf damage by this pest.

4.10.2 Mechanical methods

- Hand collection of egg masses and gregarious early instar larvae from the trap crop of castor as well as the groundnut crop on a community basis.
- Digging trenches around the fields to trap migrating *S. litura* larvae.

4.10.3. Behavioral methods

Setting up of black light traps as well as pheromone traps for monitoring the adult activities. The sex pheromone of *S. litura* contains, 2(E)9,11-tetra decadienyl acetate 2, (E), 9,12-tetra-decadienyl acetate in 10:1 ratio and was reported to be effective in trapping male moths. Two traps can monitor 100 ha and the ETL is 100 moths/trap/day over a period of one week. Srinivas and Rao (1999) conducted experiments on the mating disruption of *Spodoptera litura* (F.) infesting groundnuts by using synthetic sex pheromone of S. litura (comprising (9Z,11E)-9,11-tetradecadienyl acetate and (9Z,12E)-9,12-tetradecadienyl acetate in a 10:1 ratio at _10 mg dispenser-1 (50 gm a.i. ha-

1)). Observations showed that moth catch in pheromone traps, egg masses, and larvae were significantly lower when compared to the control treatment throughout crop growth. High concentrations of synthetic sex pheromone released from many sources led to effective mating disruption in *S. litura* adults. The linear regression analysis indicated that the total rainfall had no effect on the trap catches throughout the study period. The maximum temperature had a positive association with trap catches of the tobacco caterpillar (*Spodoptera litura*) on groundnut during winter 2000-01 at 99% probability level and during the rainy season of 2000 at 90% probability level. Total sunshine hours had a strong positive and significant association with the trap catches during both the winter seasons. Relative humidity affected catches positively during the rainy season and negatively during the winter season. Pooled analysis showed significant variation among the different winter seasons in terms of catches (Mahalingam *et al.*, 2003). Monitoring studies of *S. litura* on groundnut carried out using synthetic sex pheromone traps indicated that the peak male moth catch was observed during the 2nd week of February while the highest number of egg masses were observed during the 2nd week of February and the highest number of larvae/20 m2 during the 4th week of February. The moth catches in pheromone traps were found to be positively correlated with number of egg masses and larval counts of *S. litura* on groundnut in the field (Sreenivasulu *et al.*, 2003).

4.10.4. Host plant resistance

Developing a crop variety resistant to pests is the most ideal in solving the pest problem. Virginia bunch varieties (V40 and Ah 629) and Virginia runner varieties (NCA 17840, NFG 79, EC 21989) are known to be resistant. All India Co-ordinated Research Project, on groundnut has identified 21 genotypes to be resistant to *S. litura*. The National Research Centre for groundnut has identified NRCG 384, 1103, 1591, 1686, 4563, 7166 and 7243 to be resistant to this insect. Also the **ICG** numbers 667, 1587, 9934, 3302, 5123 and 620 were found to be showing good amount of resistance to the *S. litura*. Five groundnut (*Arachis hypogaea*) genotypes (ICGV86350, ICGV86590, ICGV87264, ICGV87165 and GBFDS273) showed consistent superiority for resistance to *Spodoptera litura* (Nadaf *et al.*, 1995).

ICGV 86699 peanut germplasm line is high-yielding elite groundnut (*Arachis hypogaea* subsp. *hypogaea* var. *hypogaea*) germplasm line, released in 1994 and developed from a single-plant selection made in CS29 (*A. batizocoi*/ *A. duranensis*/*A. hypogaea* cv. NC2), is noted for its resistance to major diseases and it also suffers less damage due to *Empoasca kerri* and *Spodoptera litura* than control cultivars under field conditions (Reddy *et al.*, 1996).

Sreenivasa *et al.* (1997) showed that larval mortality of *Spodoptera litura* was highest on the groundnut variety ICGV 86031 (82.8%) followed by ICGV

86350 (53.32%) and Dh-3-30 (43.02%). Development was shortest on Dh-3-30 (32.25 days, adults surviving for a maximum of 10.5 days), and longest (37.5 days) on ICGV 86031 with adults surviving a maximum of 9.50 days. ICGV 86350 was intermediate.

4.10.5. Plant Products

The leaves of *Azadirachta indica, Vitex negundo,* rind of *Citrus sinensis* (orange) and rhizome of *Zingiber officinale* (ginger), 10 g of each, were assess for their efficacy against larvae of *Spodoptera litura*. Larval mortality was highest after 96 h on leaves treated with extracts of orange (90%), followed by *V. negundo* (83%), *A. indica* (80%) and ginger (70%) (Sahayaraj and Sekar,1996). *Calotropis gigantea* was found to be the most toxic plant product to *Spodoptera litura*. followed by *Vitex negundo. Azadirachta indica* and *Pongamia glabra* [*P. pinnata*] (Sahayaraj and Paulraj, 1998).

Neem seed kernel extract at 3% protects the crop from foliage feeders including *S. litura*. It acts as an anti-feedant, ovipositional repellant as well as growth disruptor. It is safe to beneficial insects and is considered as ecofriendly. Annona oil concentrate 2.5 EC and neem seed kernel extract 1500 EC were the most effective in reducing foliar damage and numbers of larvae and egg masses, and in increasing the pod yield (Raman *et al.*, 2000). Patil *et al.*(2002) determined the efficacy of botanicals (*Vitex negundo* (5%), *Argemone mexicana* (5%), *Calotropis gigantea* (5%), neem seed kernel extract (NSKE; 5%), and pongamia oil) with entomopathogenic fungi *N. rileyi* against *Spodoptera litura* in groundnut (cv. Dh-3-30) ecosystem. Both 0.05% monocrotophos and untreated control were maintained for comparison. In 1996, *N. rileyi* at 2x1011 conidia per ha combined with either *V. negundo*, NSKE or *A. mexicana* were the most effective treatments, next to monocrotophos spray. NSKE and *V. negundo* combined with *N. rileyi* recorded higher yields than *N. rileyi* alone and the untreated control. In 1997, two sprays of *N. rileyi* combined with NSKE, with *V. negundo* or with *A. mexicana* at 15-day intervals were the next best treatments to monocrotophos. No difference was observed in pod yields between combination treatments and chemical control in both years. NSKE + *N. rileyi* and *V. negundo* + *N. rileyi* were as effective as the monocrotophos spray. Although the highest benefit:cost (B:C) ratio was obtained in N. rileyi spray alone (15:1), it failed to afford complete or desired level of protection against the defoliator. Among the treatments, NSKE + *N. rileyi*, *V. negundo* + *N. rileyi* and *A. mexicana* + *N. rileyi* were the most cost-effective, with B:C ratios of 12.28:1, 14.22:1 and 14.01:1, respectively. NSKE and *V. negundo* enhanced the bioefficacy of *N. rileyi* against *S. litura* in groundnut, resulting in high B:C ratio and yields comparable to that of monocrotophos, aside from being safer to environment.

Among five organic manures, neem cake, pongamia cake, castor cake (all at 1.0 t/ha), farmyard manure (FYM), and vermicompost (at 10.0 t/ha), and 11 nitrogen rates (10, 15, 20, 25, 30, 35, 40, 45, 50, 55 and 60 kg/ha) neem cake was the most effective management of insect pests on groundnut, but was at par with pongamia cake. A significantly higher yield (40.3 q/ha) was obtained from neem cake-treated plot, followed by castor cake- and pongamia cake-treated plots (36.9 and 36.5 q/ha, respectively). Predator population was lower in plots treated with both neem and pongamia cakes. The natural enemy population was directly proportional to the insect population, irrespective of the treatments. Castor cake-, FYM-, vermicompost- and chemical fertilizer-treated plots attracted more coccinellids and spiders than neem and pongamia cake-applied plots. Pest and natural enemy populations increased gradually with increase in N levels. Organic cakes were safer to the natural enemies; however, their population was totally dependent on pest population. Predator population increased with the increase in nitrogen level, but this effect was due to rise in pest population. Yields also increased with N level increase (Venkateswarlu, 2003).

4.10.6. Biological control

- Several parasites and predators including spiders and birds attack and feed on different life stages of *S. litura*. Minimized use of insecticides encourages the natural enemies.
- Release of the egg parasite *Telenomus remus* @ 50,000/ha (3 releases)
- Release of larval parasite *Apanteles africanus* @ 5000/ha (3 releases)

4.10.7. Microbial control

Spray nuclear polyhedrosis virus @ $1.5x10^{12}$ polyhedral occlusion bodies/ ha in the evening hours when the pest is in early instars (I-III). Repeated sprays may be necessary with repeated broods. Combinations of microbial pesticides (NPV, *Bacillus thuringiensis* subsp. *kurstaki* and *Beauveria bassiana*) (along with chemical insecticides (fenvalerate at 0.005% and monocrotophos at 0.025%), i.e. half of the recommended dose) and microbial pesticides themselves were superior to either microbial pesticides or recommended rates of individual chemical insecticides alone. B. *thuringiensis* + fenvalerate at 0.005% was the best treatment, recording the highest larval population reduction, lowest leaf damage (20.15%) and highest pod yield (15.03 g/plant) (Jayanthi and Padmavathamma, 2001).

4.10.8. Chemical control

Out of 7 insecticides tested against *Spodoptera litura* on groundnut, carbofuran at 1 kg a.i./ha was the most effective, resulting in the highest number of pods per 10 plants (226) and pod weight per 10 plants (148 g), followed by phorate and 1 kg a.i./ha (213 pods/10 plants and 140.0 g/10 plants) and quinalphos 0.65% a.i. (207.3 pods/10 plants and 133.0 g/10 plants)

(Mohapatra *et al.*, 1995). Chemical control should be the last option since indiscriminate use can result in development of resistance and destruction of natural enemies. If all other methods of management fail and heavy out break of the pest is seen, one can go for need-based chemical insecticide sprays. When the pest reaches an economic threshold level of 3 egg masses/15 m or 7 larvae/m row apply either quinalphos 0.05% or chlorpyriphos 0.05% or endosulfan 0.07%. Out of seven recommended chemicals quinalphos (0.5%) resulted in the lowest percentage defoliation (25.1), highest pod yield (4756 kg ha-1) and highest shelling percentage (77.50), followed by monocrotophos (27.7% defoliation, 5058 kg yield and 76.83 shelling percentage). All other chemicals tested performed poorly compared to quinalphos and monocrotophos (Patil *et al.*, 1997). The efficacy of diflubenzuron individually and in combination with buprofezin, methomyl, chlorpyrifos, triazophos, profenofos and neem oil was tested by Kumar and Krishnayya (1999) against major lepidopteran pests of groundnut like *Aproaerema modicella, Spodoptera litura* and *Helicoverpa armigera.* Diflurobenzuron (0.0125%) in combination with profenofos (0.025%) and methomyl (0.025%) was proven effective against these pests and resulted in the highest pod yields. The percent larval population reduction in treatments with profenofos and methomyl at 2 days after sowing (DAS) was 85.6 and 81.9, 66.2 and 61.6, and 62.8 and 59.0 of *A. modicella, S. litura* and *H. armigera,* respectively. Diflurobenzuron increased the efficacy of other insecticides and resulted in higher pod yields. Neem oil (1.0%) treatments at 2 DAS resulted in 52.9, 30.7 and 39.9% larval population reduction of *A. modicella, S. litura* and *H. armigera,* respectively.

Dharne *et al.* (2001) evaluated different insecticides against tobacco caterpillar (*Spodoptera litura*) on groundnut cv. JL-24.and concluded that 0.1% profenofos gave maximum protection up to 14 days of spray, recording 7.78% foliage damage. Although yield differences were non-significant, 0.03% phosphamidon recorded the highest return of rupees 9.25 per rupee invested compared to the rest of the insecticidal treatments. Manjanaik *et al.* (2002) tested the efficacy of phosalone 0.05%, methylparathion [parathion-methyl] 0.05%, quinalphos 0.05%, monocrotophos 0.05%, dimethoate 0.03%, chlorpyrifos 1 ml/litre and carbaryl 0.02%) and a plant product (neem [*Azadirachta indica*] oil 5 ml/litre) against *S. litura* on groundnut, and reported that quinalphos produced the lowest rate of defoliation and high crop yield. Monocrotophos produced the highest shelling quality. All chemical treatments resulted in higher crop protection compared to the control and neem oil.

4.10.9. Integrated Pest management

Integrated pest management module consisting of seed treatment with imidacloprid and mancozeb (3 g/kg), use of trap crop (castor), manual picking of larvae and egg masses, use of pheromone trap, spraying of *S. litura* nuclear polyhedrosis virus at 500 LE/ha and use of larval bait with chlorpyrifos

(0.05%) proved significantly effective by managing the populations of *S. litura* and sucking pests on groundnut and gave higher yield and cost-benefit ratio compared with the chemical control schedule, consisting of sprays of monocrotophos (0.05%), chlorpyrifos (0.05%), quinalphos (0.05%) and endosulfan (0.07%), and farmers practice, i.e. spray with cypermethrin (0.02%) at 60 days after sowing (Sreenivasulu *et al.*, 2002).

References

Adair, E.W. (1918). Preliminary list of insects associated with cotton in Egypt. *Agric. J. Egypt.* Cairo. 8: 80-88.

Ahmad, R. and Gangwar, B. (1984). Some observations on the pests of winged bean in Andamans. *Entomon.* 9(4): 295.

Albert, H. (1951). Les insects vivants sur le cacaoyer en afrique occidentale. *Mem. Inst. Franc Afr.* Noire, Dakar. 115: 174.

Anonymous. (1974 to 1991). Progress Report, Annual *Kharif* Oilseeds Work shop, All India Coordinated Research Project on Oilseeds Research (AICORPO), Directorate of Oilseeds Research, Hyderabad.

Anonymous. (1977). Report of the Department of Scientific and Industrial Research for the year ended 31 March 1977. New Zealand, 70pp.

Anonymous. (1982). Annual Report 1982-83. International Crops Research Institute for the Semi-Arid Tropics, Patancheru, Andhra Pradesh, India.

Anonymous. (1984). Annual Progress Report (1983-1984) of AICRP on Rodents Control. (Ludhiana:Punjab Agril.Uni.).

Anonymous. (1985 to 1998). Annual progress report, All India Coordinated Research Project on Groundnut (ICAR), National Research Center for Groundnut, Junagadh-362 001, Gujarat, India.

Anonymous. (1985). Annual Report, All India Coordinated Research Project on Biological Control of Crop Pests and Weeds. Indian Institute of Horticulture Research, Bangalore.

Anonymous. (1988). Annual report *International Crop Research Institute for the Semi-Arid Tropics (ICRISAT).*, Patancheru, Andhra Pradesh, India.

Anonymous. (1988). Annual Report, All India Coordinated Research Project on Biological Control of Crop Pests and Weeds, Biological Control Centre, (ICAR- NCIPM)., Bangalore, India. Tech. Doc. No. 23: 1-260.

Anonymous. (1994). Annual Report. Project Directorate of Biological Control, Bangalore, India, 143 pp.

Anonymous. (1994). Annual Research Report AICRP biological control, Gujarat Agricultural University, Anand, Gujarat.

Ansari, M. A., Pawar A. D. and Ahmed S. N. (1987). A note on pathogenicity of naturally occurring bacterium *Serratia marcescens* Bizio on some lepidopteron pests. *Plant protection Bulletin of India.* 39: 27-28.

Armes, N. J., Wightman, J. A., Jadhav, D. R. and Rao, G. V. R. (1997). Status of insecticide resistance in *Spodoptera litura* in Andhra Pradesh, India. *Pesticide Science.* 50: 240-248.

Armes, N. J., Wightman, J. A., Jadhav, D. R. and Ranga Rao, G. V. (1994). Status of insecticide resistance in *Spodoptera litura* in Andhra Pradesh, India *Proceedings of the National Scientists Forum On Spodoptera litura* held on 2-4 April 1996 at ICRISAT.

Ayyar, T. V. R. (1921). A list of parasitic hymenoptera of economic importance from south India. Rept. *Proceedings of the 4th Entomologists meeting,* Pusa, Calcutta, pp 363-366.

Ayyar, T. V. R. and Margabandhu, V. (1934). Hymenopterous parasites of economic importance in South India. *Madras Agricultural Journal.* 22: 430-446.

Ayyar, T.V.R. (1963). *Hand book of Economic Entomology for South India.* (2nd ed.).Govt. Press, Madras, 526 p.

Balikai, R. A., Bagali, A. N. and Ryagi, Y. H. (1999). Incidence of *Spodoptera litura* (Fabricius) on grapevine, *Vitis vinifera* L. *Insect Environment* 5(1): 32.

Barrion, A. T. and Litsinger, J. A. (1987). A larval parasite of swarming caterpillar and common cutworm in the Philippines. *Int. Rice Res. Newsl.* 12: 34-35.

Battu, G. S. (1977). Occurence of *Parasarcophaga miseara* (Walker) and *Campoletis* sp. as parasites of *Spodoptera litura* (F.) from India. *Current Sci.* 46: 568-569.

Battu, G. S., Bindra, O. S. and Rangarajan, M. (1971). Investigation on microbial infection of Insect pest in Punjab. *Indian Journal of Entomology.* 33: 271-280.

Beeson, C. F. C. and Chatterjee, S. N. (1935). On the biology of the Ichneumonidae (Hymenoptera). *Indian For. Rec.,* 8: 151-168.

Bhanukiran, Y., Rao, P. A. and Rao, K. T. (1998). Construction of life table for tobacco caterpillar, *Spodoptera litura* (Fabricius) on rabi groundnut during post-rainy season in southern peninsular region of India. *Journal of Entomological Research.* 22: 163-167.

Bhatnagar, R. K. and Srivastava, Y. N. (1985). Impact of reptiles on food production and environment. Non insect pest and predators. *Proceedings of the National Symposium on Impact of Non-insect pest and predators on Food production and Environment.* March 12-14, 1984, All India Scientific Writers Society. pp. 116-120, 265-266.

Bhatnagar, V. S., (1981). Arthropod parasitoids of insect pests (excluding *Heliothis* sp.) recorded in Andhra Pradesh, India, *Cropping System (Entomology)* Progress report ICRISAT.

Bhatnagar, V. S. (1981). Are effective parasites of *Heliothis* eggs found on pigeonpea and chickpea? *International Pigeon pea Newsletter.* 1: 32.

Bhatnagar, V. S., Pawar, C. S., Jadhav, D. R. and Devic, J. C. (1985). Mermithid nematodes as parasites of *Heliothis* sp. and other crop pests in Andhra Pradesh. *Proceeding of Indian Academy of Sciences. Animal Science.* 94: 509-515.

Bilapate, G. G., Pawar V. M. and Thombre, V. T. (1980). Life table and intrinsic rate of increase of *Spodoptera litura* (F) on Sunflower. *Indian Journal of Agriculture Research.* 18: 463-466.

Braune, H. J. and Kan, S. K. (1981). Species composition and biology of predator parasite complex of the taro cluster caterpillar *Spodoptera litura* (F). in Western Somoa. *Alfua Agricultural Bulletin*. 6: 46-50.

Braune, H. J. (1982). Effect of structure of host egg mass on the effectiveness of egg parasite of *Spodoptera litura* (F) (Lepidoptera : Noctuidae). Western Samoa, *Drosera* 1: 7-16.

Braune, H. J. and Kan, S. K. and Trevaranus, H. (1981). Studies on effectiveness of *Apanteles sp.* a larval parasite of *Spodoptera litura.* (F). *Alafua Agricultural Bulletin*, Apia: Western Samoa. 6: 12-22.

Butani , D. K. (1984) Spices and pest problems: 3. Coriander. *Pesticides*. 18(9): 15-17.

Chari, M. S. and Rao, R. S. N. (1990). Biological suppression of tobacco pests. In Proceedings of Indo-USSR Joint Workshop On Problems & Potentials of Biocontrol of Pests and Diseases (S. P. Singh. Ed.). June 26-28, 1990, Biological Control Centre, NCIPM, Bangalore, pp 77-87.

Chari, M. S., Bharpoda, T. M. and Patel, S. N. (1985). Studies on integrated management of *Spodoptera litura* Fb. in tobacco nursery. *Tobacco Research* 11 (2): 93-98. Cherian, M. C. and Brahmachari, K. (1941). Notes on three predatory Hemiptera from South India. *Indian Journal of Entomology*. 3: 115-119.

Cherian, M. C. and Brahmachari, K. (1941). Notes on three predatory Hemipterons from South India. *Indian Journal of Entomology*. 3: 115-119.

Chhibber, R.C., Bhattacharya,A.K., and Pathak,A.K. (1978). Utilization of lipids by *Spodoptera litura* (Fab) on nine food plants. Currant Science 47(19):733-735.

Chockalingam, S. and Somasundaram P. (1983). Effect of thermal stress on the consumption and utilization of flower of Zinna elegans in the final instar larva of the moth *Spodoptera litura* (Noctuidae) in insect ecology and resource management Muzaffarnagar, India. Sanatan Dharm College.

Devi, P. S. V. (1995). Soil treatment with *Nomuraea rileyi*: a promising technique for the control of *Spodoptera litura* on groundnut. *Biocontrol Science and Technology*. 5: 361-364.

Dhandapani, N. and Babu, P. C. S. (1995). Efficacy of nuclear polyhedrosis virus against *Spodoptera litura* on groundnut in different locations of Tamil Nadu. *Madras Agricultural Journal*. 82: 531-533.

Dhandapani, N., Janarthanan, R. and Kumaraswami, T. (1982). Epizootic occurrence of nuclear polyhedrosis virus of *Spodoptera litura* F. *Current Science*. 51: 793-794.

Dharne, P. K., Patil, S. K. and Patil, R. B. (2001). Field evaluation of some insecticides against tobacco caterpillar (*Spodoptera litura* Fabricius) on groundnut. *Journal of Maharashtra Agricultural Universities*. 26(2): 164-165.

Dharne, P. K. and Patel, S. K. (2000). Screening of promising groundnut genotypes for their reaction to *Spodoptera litura*. *International Arachis Newsletter*. 20: 67-69.

Duong-Nguyen Hai, Thanh HoNgoc, Doan T., Yen NguyenThi, Tam T. T. M., Dung PhanThanh, Phuong, L. T. T., Duong, N. H., Thanh, H. N., Yen, N. T. and Dung, P. T. (1998). Diseases and pests of *Hevea brasiliensis* in Vietnam. Symposium on natural rubber (*Hevea brasiliensis*) Vol 2 - physiology and exploitation and crop protection and planting methods session, Ho Chi Minh City, China 14-15 Oct 1997. pp 80-91.

Geethalakshmi, L., Muthukrishnan, N., Chandrasekaran, M. and Raghuraman, M. (2000). Chrysopids biology on *Corcyra cephalonica* and feeding potential on different hosts insects. *Annals of Plant Protection Sciences* 8 (2): 132-135.

Gopali, J. B., Nadaf, H. L., Patil, R. K., Lingappa, S., Hanamaratti, N. G. and Sattigi, N. (1996). Larval and ovipositional behaviour of *Spodoptera litura* (F.) in response to resistant and susceptible cultivars of groundnut. *Crop Improvement*. 23: 49-52.

Im, D. J., Aguda, R. M and Shepard, B. M. (1989) Virus diseases of some lepidopterous rice pests in the Philippines. *International Rice Research Newsletter*. 14(2): 35-36.

Intachat, J. and Kirton, L. G. (1997). Observations on insects associated with *Acacia mangium*in Peninsular Malaysia. *Journal of Tropical Forest Science*. 9(4): 561-564.

Jagginavar, S. B., Biradar, A. P., Bagali, A. N. (1999). Menace of *Spodoptera litura* (Fabricius) on grape. *Insect Environment* 5(2): 76-77.

Jalali, S. K., Singh, S. P. and Ballal, C. R. (1987). Role of the host plants of *Spodoptera litura* (Fabricius) on the degree of parasitism by *Cotesia migriniventris* (Cresson) (Hymenoptera: Braconidae). *Indian Journal of Agricultural Sciences*. 57: 676-678.

Janardan Singh and Bardhan, A. K. (1974). Effectiveness of DD-136, an entomophilic nematode against insect pest of agriculture importance. *Current Science*. 43: 622.

Jayanthi, P. D. K and Padmavathamma, K. (2001). Joint action of microbial and chemical insecticides on *Spodoptera litura* (Fab.) (Lepidoptera: Noctuidae). *Journal of Tropical Agriculture*. 39(2): 142-144.

Joshi, B. G., Sitaramaiah S., Satyanarayanan. S. V. V. and. Ramaprasad, G. (1979). Note on natural enemies of *Spodoptera litura* (F). and *Myzus persicae* (Sulz). on flue cured tobacco in Andhra Pradesh. *Sciences and Culture*. 23: 56-57.

Jyoti-Tripathi, Dwijendra-Singh and Shalini-Mathur. (1997). Outbreak of *Spodoptera litura* on brahmi - an important medicinal plant. *Insect Environment*. 2(4): 134.

Kamal, M. (1951). The biological control of the cotton leafworm (*Prodenia litura* F.). in Egypt. *Bull Soc. Fouad ler. Ent.*, 35: 221-270.

Kapoor, K. N., Gujrati, J. P. and Gangrade, G. A. (1975). Chemical control of soybean leafminer, *Stomopteryx subsecivella* Zeller (Lepidoptera: Gelechiidae). *Indian Journal of Entomology*. 37: 286-291.

Karla, V. K., Singh Harvir, Ram, H., and Yadava, T. P. (1989). *Scopula albiflava* Warren (Geometridae: Lepidoptera). damaging groundnut crop in Haryana. Haryana Agricultural University. *Journal of Research* 19 (1): 74-75.

Kore, S. S. and Bhinde, V. P. (1978). Bacterial diseases of tobacco caterpillar *Spodoptera litura* (F). *Journal of Maharashtra Agricultural University*. 3: 37-37.

Krishnaiah, K., Prasad, V. G., and Mohan, N. J. (1975). New record of alternate hosts of tobacco caterpillar, *Spodoptera litura* (Fab). (Noctuidae: Lepidoptera) and Bihar hairy caterpillar, *Diacrisia oblique* Wlk (Arctiidae: Lepidoptera). *Current Science*. 44: 138.

Krishnaiah, K., Ramakrishnan, N. and Reddy P. C. (1985). Control of *Spodoptera litura* (F). on black gram by nuclear polyhedrosis. *Indian Journal of Agricultural Sciences*. 55: 775-776.

Krishnamurti, B. and Usman, S. (1954). Some parasites of economic importance noted in Mysore State. *Indian Journal of Entomology*. 16: 327-344.

Kulkarni, S. A (1989). Bio-ecology and management of *Spodoptera litura* (F). (Lepidoptera: Noctuidae) on groundnut (*Arachis hypogaea* L.). Ph.D.Thesis, UAS, Dharwad, India, pp 270.

Kulkarni, N. S. and Lingappa, S. (2002). Seasonal incidence of entomopathogenic fungus, *Nomuraea rileyi* (Farlow) Samson on *Spodoptera litura* in groundnut, soybean and potato - a comparative study. *Karnataka Journal of Agricultural Sciences*. 15(1): 63-70.

Kumar, B.V. N. and Regupathy, A. (2000).Generating base line data for insecticide resistance monitoring in *Spodoptera litura* (Fabricius). *Pesticide Research Journal*. 12(2): 232-234.

Kumar, D. A. and Krishnayya, P. V. (1999). Effect of diflubenzuron in combination with selected insecticides on major lepidopteran pests of groundnut, *Arachis hypogaea* Linn. *Journal of Applied Zoological Researches*. 10(1): 1-5.

Liu-Tashiu, Liu T. S., Chen YungWu, Huang, Sheng Chung Sheu, ChianShinn Tsai, Suhhuey Yih Meeishiouh. (1998). Insect pests of gladiolus and lily plants. Proceedings of the symposium on research and development of gladiolus, lilies and chrysanthemums, Taichung, Taiwan 12-13 June 1997. Special Publication Taichung District Agricultural Improvement Station No. 40, 139-150.

Macfarlane, R; Abington, J.B. and Walton, P.D. (1987). Cluster caterpillar of taro (*Spodoptera litura*). Solomon Islands,Ministry of Agriculture and Lands Research Department, Agriculture Division, annual-report-1985., 9-10 pp.

Mahalingam, C. A., Saminathan, V. R. and Venkatesan, S. (2003). Impact of weather factors on pheromone trap catches of tobacco caterpillar (*Spodoptera litura*). *Indian Journal of Agricultural Sciences*. 73: 438-440.

Manjanaik, C., Jyothilakshmi, S., Chandrappa, H., Somasundra, K. M. and Herle, P. S. (2002). Efficacy of insecticides and plant products (neem oil) against *Spodoptera litura* (Noctuidae: Lepidoptera) on groundnut in coastal zone. *Environment and Ecology*. 20(1): 105-108.

Manjula, K., Nagalingam, B. and Rao, P. A. (2003). Occurrence of *Nomuraea rileyi* on *Spodoptera litura* and *Helicoverpa armigera* in Guntur District of Andhra Pradesh. *Annals of Plant Protection Sciences*. 11: 224-227.

Manjula, K., Rao, P. A., Nagalingam, B. (2004). Record of *Nomuraea rileyi* (Farlow) Samson on *Helicoverpa armigera* Hubner in kharif groundnut. *Indian Journal of Plant Protection*. 32 (2): 125.

Manoj Kumar and Mukhtar Ahmad. (1998). Noctuid infestation to the introduced species of Paulownia. *Annals of Forestry*. 6(2): 177-185.

Meshram, P. B. and Joshi, K. C. (1994). A new report of *Spodoptera litura* (Fab.). Boursin (Lepidoptera: Noctuidae) as a pest of *Jatropha curcas* Linn. *Indian Forester*. 120(3): 273-274.

Mohapatra, H. K., Dhir, B. C., Senapati, B. and Mishra, B. K. (1995). Evaluation of various insecticides against *Spodoptera litura* (F.) in groundnut seedlings. *Current Agricultural Research*. 8: 79-81.

Muraleedharan, N. and Kandasamy, C. (1981). Proceedings of the Twenty-Fifth Scientific Conference 11th September 1981. *Bulletin of United Planters' Association of Southern India.* 37: 122 pp.

Mustafee, T. P. (1977). *Spodoptera litura* (Fabricius) causing damage to ramie in Assam. *Current Science.* 46(10): 350.

Nadaf, H. L., Patil, R. K., Giriraj, K. and Hanamaratti, N. G. (1995). Field performance of groundnut genotypes with inherent resistance to biotic stresses. *Crop Improvement.* 22: 261-263.

Nagalingam, B., Rao, B. N. and Rosaiah, B. (1979). First record of *Spodoptera litura* Fabri. on certain minor species. *Current Research.* 8: 171-172.

Naik, C. M., Ramaswamy, G. R. (2005). Role of spray adjuvants in improving efficacy of Spodoptera litura NPV on groundnut in coastal zone. *Environment and Ecology.* 23: 17-19.

Narayanan, E. S. and Subba Rao, B. R. (1955). Studies in insect parasitism I-III. The effect of different hosts on the physiology on the development and behaviour and on the sex- ratio of *Microbracon gelechiae* Ashmead (Hymenoptera: Braconidae). *Beitr. Ento.* 5: 36-60.

Narayanan, K. (1985). Susceptibility of *Spodoptera litura*(F) to a granulosis virus. *Current Science.* 54: 1288-1289.

Narayanan, K. and Jayaraj, S. (1979). *Spodoptera litura* (F) as a host for *Nosema* sp. *Current Science.* 48: 276.

Narendran, T. C. and Joseph, K. J. (1977). Studies on some aspect of host specificity with reference to *Brachymeria lasus* Walker, a polyphagous chalcid parasite of Lepidopteron insect. Insects and host specificity. The Macmillan Company of India Ltd., India, 85-89.

Nath, P. and Singh, A. K. (1998). Effect of intercropping of groundnut with millets and pigeonpea on the relative incidence of insect pests. *Annals of Plant Protection Sciences.* 6: 151-154.

Oblisami, G., Ramamoorthy, K. and Rangaswamy, G. (1969). Studies on the pathology of some crop pests of South India. *Mysore Journal of the Agricultural Sciences* 3(1): 86-98.

Obulapathi, K., Rao, P. K., Padmavatamma, K., Reddy, K. S. (2000). Efficacy of certain botanical and biopesticides in the control of *Spodoptera litura* (Fab.) on groundnut. *Indian Journal of Plant Protection.* 28(2): 165-168.

Oikawa, M. Nagaoka, H. Morita, K. and Asano, S. (1999). Relationship between host plants and toxicity of *Bacillus thuringiensis* formulation against common cutworm, *Spodoptera litura.* Annual Report of the Kanto Tosan Plant Protection Society 46: 89-92.

Okada, M. (1993). Report on A.P.O.Seminar, 1991, No.1991, No.91, p.121, Tokyo, Japan.

Pandey, S. N. and Srivastava, R. P. (1967). Growth of larvae of *Prodenia litura* F. in relation to wild food plants 29(3): 229-233.

Pandey, S. N. and Rangarajan, M. (1967). *Prodenia litura* (F), a new host of *Serratia marcescens* Bizio. *Indian Journal Entomology.* 29: 14-17.

Patel , J. S. (1943). Indian Central Jute Committee. Annual Report. Agri. Res. Scheme, 1942-43, Calcutta, 51 pp.

Patel, A. G. and Vyas, H. N. (1984). Studies on predatory capacity of ladybird beetle, *Menochilus sexmaculata* Fabricius against *Aphis craccivora* Koch under laboratory condition. *Pesticides.* 18(11): 8-9.

Patel, B. H. and Pillai, G. K. (1988). Studies on the spider fauna of groundnut fields in Gujarat, India, *Journal of Biological Control.* 2(2): 83-88.

Patel, D. J., Patel, B. A., Chavda, J. C. and Patel, H. V. (1988). Record of *Meloidogyne javanica* on groundnut in Gujarat, India, *International Arachis Newsletter.* 3: 16-17.

Patel, J. S. (1944). Indian Central Jute Committee, Calcutta, India. *Annual Report of Agricultural Research Scheme* for the year 1942-1943. pp. 43-44.

Patel, M. H. (1988). Studies on predation of *Heliothis armigera* Hubner and *Spodoptera exigua* (Hb.) by insectivorus birds with special reference to *Acridotheres tristis*L. *M.Sc. Thesis submitted to Gujarat Agricultural University, Anand,* 77 pp.

Patel, N. G. (1980). The bionomics and control measures of tobacco bug, *Nesidiocoris tenuis* Retutrg (Miridae: Hemiptera), *Gujarat Agric. Univ. Res. J.* 5:60.

Patel, N. G. (1980). The bionomics and the control measures of the tobacco bug *Nesidiocoris tenuis* Retutrg (Miridae: Hemiptera). *Gujarat Agricultural University Research Journal.* 5: 60.

Patel, R. C., Patel, J. C. and Patel, J. K. (1971). New record of parasites of *Heliothis armigera* (Hub.) and *H. peltigera* (Schiff). from Gujarat. *Indian Journal of Entomology* 33: 223-224.

Patil, R. K., Gowda, M. V. C. and Nadaf, H. L. (1991). Screening of Spanish bunch breeding lines of groundnut against *Spodoptera litura* (F.) damage. *International Arachis Newsletter.* 10: 26-27.

Patil, R. K., Lingappa, S., Hiremath, R. V. and Hiremath, I. G. (2003). Seasonal incidence of *Nomuraea rileyi* (Farlow) Samson on *Spodoptera litura* (F.) in groundnut ecosystem. *Journal of Entomological Research.* 27: 109-113.

Patil, R. K., Lingappa, S., Hiremath, I. G. and Hiremath, R. V. (2003). Efficacy of botanicals with *Nomuraea rileyi* (Farlow) Samson in groundnut ecosystem. Biological control of lepidopteran pests Proceedings of the Symposium of Biological Control of Lepidopteran Pests, July,17-18,-2002, Bangalore, India.; 313-317.

Patil, R. K., Mahabaleshwar-Hegde, Kulkarni, K. A. and Hegde, M. (1997). Efficacy of insecticides against *Spodoptera litura* on groundnut. *Karnataka Journal of Agricultural Sciences.* 10: 674-677.

Patil, R. K., Ravishankar, G., Mannikeri, I. M., Giriraj, K. and Rayar, S. G. (1996). Effect of sowing time on production potential and incidence of *Spodoptera litura* on groundnut cultivars. *Journal of Oilseeds Research.* 13: 18-21.

Pawar A. D. (1976). *Andrallus spinidens* (F) (Asopinae: Pentatomidae: Hemiptera) as a predator of the insect pest of rice in Himachal Pradesh, India. *Rice Entomol Newsletter*. 4: 23-24.

Phadke, C. H. and Rao, G. V. (1978). Studies on the entomogenous fungus *Nomuraea rileyi* (Farlow) Samson I. *Current Science*. 47: 511-512.

Prasad, J. and Kushwaha, K. S. (1990). Role of pathogenic microbes in management of the *S. litura* (F) in cabbage and cauliflower crops. *Indian Journal Entomology*. 52: 641-652.

Prasad, S.R.; Krishnayya, P.V.; Vijayalakshmi, K. (1999). Insecticidal efficacy of cryolite against *Spodoptera litura* Fabricius. Annals-of-Plant-Protection-Sciences., 7: 2, 150-153.

Pruthi, H. S. (1969). Text book of Agricultural Entomology pp. 213.

Rai, P. S. (1974). Record of *Chelonus formosanus* (Sonam). (Hymenoptera: Bracinidae) a parasite of *Spodoptera litura* (F) from Mysore state. *Current Science*. 43: 30.

Rajendra, M. K. and Patel, R. C. (1971). Studies on life history of a predatory pentatomid bug *andrallus spinidens* (Fabr.). *Journal of Bombay Natural History Society*. 68: 319-327.

Rajgopal, K., V. Nandagopal and Bhagat, N. R. (1988). Preliminary screening of Virginia groundnut genotypes against tobacco caterpillar *Spodoptera litura* (F). *Journal of Oilseeds Research*. 5: 162-165.

Rama Prasad, G. (1996). Integrated management of *Spodoptera litura* F. *in India Proceedings of the National Scientist Forum on Spodoptera litura* (F). 2-4 Apr 1996, pp 96-103.

Ramamurthy, K., Oblisami G. and Rangaswami, G. (1967). An epizootic fungus disease of Sunhemp pest. *The Mysore Journal of Agricultural Sciences*. 1: 225-226.

Raman, G. V., Reddy, V. R. K., Rao, M. S., Srimannarayana, G. and Venkateswarlu, B. (2000). Effect of annona and neem formulations on Rabi groundnut crop against *Spodoptera litura*. Fab. *Pesticide Research Journal*. 12(2): 174-177.

Ramegowda, G. K. and Basavanagoud, K. (2001). Management of insecticide resistance in *Spodoptera litura* (Fabricius) by microbial agents and newer molecules. *Karnataka Journal of Agricultural Sciences*. 14(4): 1088-1092.

Ranga Rao, G.V. and Wightman, J.A. (1994). First Annual Rabi/Summer Groundnut Research Workers' Group Meeting held at Indian Institute of Technology, Kharagpur, West Bengal from 12- 15 September, 1994.

Rao, B. N. and Patel, R. C. (1976). Bionomics of *Chelonus formosanus* Sonam. An egg-larval parasite of *Spodoptera litura* (F). *Indian Journal of Entomology*. 36: 103-109.

Rao, R. S. N, Satyanaraya S. V. V and Soundrarajan V. (1981). Note on new addition to the natural enemies of *Spodoptera litura* (F.) and *Myzus persicae* Sulz. on flue cured tobacco in Andhra Pradesh. *Science and Culture*. 47: 98-99.

Rao, R. S. N. (1983). Record of *Chrysopa (=Apertochrysa) crassinervis* Esbenpeterson from India. *Current Science*. 52: 438-439.

Rao, R. S. N. and Satyanaraya, S. V. V. (1984). Note on more addition to the natural enemies complex of *Spodoptera litura* (F). and *Myzus persicae* Sulz, on tobacco in Andhra Pradesh. *Current Science*. 53: 201-202.

Rao, R. S. N., Joshi, B. G. and Satyanarayana, S. V. V. (1982). Studies on alternate host plants of *Myzus persicae* Sulz. during non-tobacco season. *Tobacco Research*. 8(1): 77-79.

Rao, V. G. and Phadke, C. H. (1977). A muscardine disease of tobacco leaf eating caterpillar. *Current Science*. 46:648-649.

Rao, V.T. (1956). *Indian Journal of Entomology*. 18: 123-126.

Raychoudhury, N., Bhowmik, A. K., Avinash Jain, Joshi, K. C., Banerjee, S. K. and Jain, A. (1998). Relative resistance in teak clones to *Spodoptera litura* (Fabricius) Boursin (Lepidoptera: Noctuidae), in relation to certain leaf nutrients. *Indian Journal of Forestry* 21(4): 321-326.

Reddy, L. J., Nigam, S. N., Moss, J. P., Singh, A. K., Subrahmanyam, P., McDonald, D. and Reddy, A. G. S. (1996). Registration of ICGV 86699 peanut germplasm line with multiple disease and insect resistance. *Crop Science*. 36: 821.

Roychoudhury, N., Kalia, S., and Joshi, K. C. (1997). *Spodoptera litura* (Fabricius) Boursin (Lepidoptera: Noctuidae), a major pest of Sesbania spp. and its species preferences. *Indian Journal of Forestry*. 20(1): 67-69.

Sahayaraj, K and Paulraj, M. G. (2001). Behaviour of *Rhynocoris marginatus* (Fabricius) (Heteroptera: Reduviidae) to chemical cues from three lepidopteran pests. *Journal of Biological Control*. 15(1): 1-4.

Sahayaraj, K. (1994). Biocontrol potential evaluation of the reduviid predator *Rhynocoris marginatus* (Fabricius) to the serious groundnut pest *Spodoptera litura* (Fabricius) by functional response study. *Fresenius Environmental Bulletin*. 3 (9): 546-550.

Sahayaraj, K. (1999). Field evaluation of the predator, *Rhynocoris marginatus* (Fab.) on two groundnut defoliators. *International Arachis Newsletter*. (19): 41-42.

Sahayaraj, K. (2000). Evaluation of biological control potential of *Rhynocoris marginatus* on four groundnut pests under laboratory conditions. *International Arachis Newsletter*. 20: 72-74.

Sahayaraj, K. (2001). Biopesticidal effect of neem and NPV on production potential and behavior of *Rhynocoris marginatus* to groundnut pest *Spodoptera litura*. *International Arachis Newsletter*. 21: 46-48.

Sahayaraj, K. (2002). Field bioefficacy of a reduviid predator *Rhynocoris marginatus* and plant products against *Aproaerema modicella* and *Spodoptera litura* of groundnut. *Indian Journal of Entomology*. 64(3): 292-300.

Sahayaraj, K. and Paulraj, M. G. (1998). Screening the relative toxicity of some plant extracts to *Spodoptera litura* Fab. (Insecta: Lepidoptera: Noctuidae) of groundnut. *Fresenius Environmental Bulletin*. 7: 557-560.

Sahayaraj, K. and Sekar, R. (1996). Efficacy of plant extracts against tobacco caterpillar larvae in groundnut. *International Arachis Newsletter*. 16: 38.

Sahayaraj, K. and Sivakumar, K. (1995). Groundnut pest and pest stage preference of a reduviid predator *Rhynocoris kumarii* Ambrose and Livingstone (Heteroptera: Reduviidae). *Fresenius Environmental Bulletin*. 4: 263-269.

Sain, M., Bentur, J. S. and Kalode, M. B. (1983). *Spodoptera litura* - a voracious feeder *Marsilea* weed. *Current Science*. 52:436-437.

Sathe T. V. (1987). New record of natural enemies of *Spodoptera litura* (F) in Kolhapur, India. *Current Science*. 56:1083-1084.

Sharma, H. C., Pampapathy, G., Dwivedi, S. L. and Reddy, L. J. (2003). Mechanisms and diversity of resistance to insect pests in wild relatives of groundnut. *Journal of Economic Entomology*. 96: 1886-1897.

Siddaramaiah, A. L., Narendrappa, T. and Panchabhavi, K. S. (1986). Occurrence and severity of *Metarrhizium anisopliae* on *Spodoptera litura* Fb. in groundnut field. *Current Research*, 15 (5-6): 48-49.

Singh, G. and Hoi, V.C. (1972). Effect of host plants on biology of *Spodoptera litura* (Fab.) *Malayasian Agril. Res.* 1(1):14-23.

Singh, H.N. and Byas, (1975). Larval development of *Prodenia litura* F. in relation to certain cultivated food plants. *Indian Journal of Entomology*. 37: 1-6.

Singh, K. J. and Dabral, K. C. (1977). Occurrence of tobacco caterpillar, *Prodenia litura* Fabri. on *Opium poppy*. *Entomology Newsletter*. 7: 21.

Singh, R. N. and Singh, K. M. (1975). Exploration of vertebrate predators for crop pests. II. Feeding potential of bush lizard, *Calotes versicolor* Daudin on tobacco caterpillar, *Prodenia litura* Fabricius. *Indian Journal of Entomology*. 35: 258-260.

Singh, H. and Singh, G. (1975). Biological studies on *Heliothis armigera* (Hubn.). in the Punjab. *Indian Journal of Entomology*. 37: 154-164.

Singh, S. P. and Jalali, S. K. (1997). Management of *Spodoptera litura* (Fabricius). (Lepidoptera: Noctuidae), pp.27-65. In: *Spodoptera litura* in India: *Proceedings of the National Scientists Forum on Spodoptera litura* (F.). (April 2-4, 1996). ICRISAT Asia Centre (J.A.Wightman and G.V. Ranga Rao)., ICRISAT, Patancheru, Andhra Pradesh, India, 146+6 pp.

Singh, S. S. G. (1980). *Spodoptera litura* Fabr. and *Toxoptera aurantii* B.de F. - two new pests on black pepper. *Indian Journal of Entomology*. 42: 330.

Singh, A. K. and Paras-Nath, (2003). Effect of intercropping on the population and groundnut leaf damage by *Spodoptera litura* (Fab.). *Shashpa*. 10(1): 33-38.

Sreenivasa, A. G., Patil, R. K., Rahman, S. M. and Lingappa, S. (1997). Development of *Spodoptera litura* as influenced by groundnut genotypes. *Karnataka Journal of Agricultural Sciences*. 10: 872-876.

Sreenivasa, A. G., Patil, R. K., Rahman, S. M. and Lingappa, S. (1997). Development of *Spodoptera litura* as influenced by groundnut genotypes. *Karnataka Journal of Agricultural Sciences*. 10(3): 872-876.

Sreenivasulu, B., Naidu, V. G. and Prasad, P. R. (2003). Monitoring of *Spodoptera litura* Fabr. through sex pheromones in groundnut. *Journal of Applied Zoological Researches*. 14: 148-150.

Sreenivasulu, B., Naidu, V. G. and Prasad, P. R. (2002). Integrated pest management in groundnut against *Spodoptera litura* (Fabr.) and other sucking pests. Pest *Management and Economic Zoology*. 10: 193-196.

Sridhar, V. and Prasad, V. D. (1996). Life-table studies on natural population of *Spodoptera litura* (Fabricius) on groundnut, *Arachis hypogaea* Linn. *Annals of Plant Protection Sciences*. 4: 142-147.

Srinivas, K. and Rao, P. A. (1999). Management of *Spodoptera litura* (F.) infesting groundnut by mating disruption technique with synthetic sex pheromone. *Journal of Entomological Research*. 23: 115-119.

Srivastava, O. S., Malik, D. S. and Thakur, R. C. (1971). Estimation of losses in yield due to the attack of arthopod pests in Soyabean. *Indian Journal of Entomology*. 33: 224-225.

Stechmann, D. H. and Semisi, S. T. (1984). Insect control in west Samoa with particular reference to present state of biological and integrated control. *Anzeiger fur schadl. Pflan. Umwelt*. 57: 65-70.

Stevenson, P. C. (1993). Biochemical resistance in wild species of groundnut (*Arachis*) to *Spodoptera litura* (Lepidoptera: Noctuidae). *Bulletin-OILB-SROP*. 16(5): 155-162.

Tamaki, Y., Noguchi, H and Yashima, T., (1973). Sex pheromones of the *Spodoptera litura* (Lepidoptera: Noctuidae): Isolation, identification and synthesis. *Applied Entomology and Zoology*. 8: 200-203.

Thakur, J. R., Sharma, P. C. and Bharadwaj, A. K. (1976). Occurrence of *Spodoptera litura* (Fabricius) and *Euproctis fraterna* more on apple. *Indian Journal Entomology*. 36(3): 226.

Thomas, R. T. S., Jacob, A. and Nair, M. R. G. K. (1969). Host biology relation of *Spodoptera litura* F. (Lepidoptera: Noctuidae). *Indian Journal of Agricultural Sciences*. 39:400-402.

Thontadarya, T. S. and Nangia, N. (1983). Natural enemies of tobacco caterpillar infesting Soyabean. *Current Research*. 12: 105-106.

Tiwari, S. N., Rathore, Y. S., Bhattacharya, A. K. and Sachan, G. C. (1988). Classification of groundnut varieties based on growth reaction of *Spodoptera litura* (F) and Mahalanobis D2 statistics. *Candian Journal of Entomology*. 50(2): 202-208.

Vaish, O. P. and Agrawal, S. C. (1978). Food preference and gowth index of *S. litura* F. (Lepidoptera: Noctuidae). *Indian Journal of Agriculture sciences*. 48: 365-367.

Venkatesan, S., Nigam, S. N., Naina-Mohammed, S. E. and Ganesan, K. N. (2000). Natural fungal pathogenicity on groundnut defoliator *Spodoptera litura*. *International Arachis Newsletter*. (20): 74-75.

Venkatesha, M.G., Bhat, P.K. and Gowda,D.K.S.(1993). Record of some lepidopteran pests' outbreak on *Coffea arabica* L. *Journal of Coffee Research*. 23(1):59-61.

Venkateswarlu, P., Sitaramaiah, S., Sreedhar, U., Rao, S. G., Sawant, S. K., and Rao, S. N. (2003). Role of organic and inorganic manures on the incidence of insect pests and their natural enemies in rabi groundnut. Biological control of lepidopteran pests Proceedings of the Symposium of Biological Control of Lepidopteran Pests. July,17-18, 2002, Bangalore, India. 295-299 pp.

Vos, J. G. M. and Frinking, H. D. (1998). Pests and diseases of hot pepper (*Capsicum* spp) in tropical lowlands of Java, Indonesia. *Journal of Plant Protection in the Tropics*. 11(1): 53-71.

Wightman, J. A and Ranga Rao, G. V. (1993). A Groundnut Insect Identification Handbook for India. Information Bulletin No. 39 ICRISAT. Patancheru, Andhra Pradesh, India, pp. 28–37.

Yadav, L. S. and Yadav, P. R. (1983). Pest complex of Cowpea (*Vigna sinensis* Savi) in Haryana. *Bulletin of Entomology*. 24: 57-58.

Yashima, T., Tamaki, T., Kamano, S. and Oyama, M. (1974). Field evaluation of a synthetic sex pheromone "litlure" as an attractant for males of *Spodoptera litura* (F.) (Lepidoptera: Noctuidae). *Applied Entomology and Zoology*. 9: 147-152.

Zaz, G. M. and Kushwaha, K. S. (1983). Quantitative incidence of tobacco caterpillar *Spodoptera litura* (Fb.) and related natural enemies in cole crop. *Indian Journal of Entomology*. 45: 201-202.

Table 4.3 contd.

Table 4.3. List of parasitoids of *Spodoptera litura* (Fabricius)

Parasitoid species	Order: Family	Satge of Pest attacked	Reference
Apanteles sp. (octonarius group)	Hymenoptera Braconidae	Larvae	Joshi *et al.*, 1979;
Apanteles beneficus	"	"	Chari *et al.*, 1985
Apanteles colemani	"	"	Sathe, 1987; Patel *et al.*, 1971
Apanteles prodeniae	"	"	Sathe, 1987
Apanteles nioro	"	"	Sathe, 1987
Bracon brevicornis	"	"	Thontadarya and Nangia,1983
Bracon gelechiae	"	"	Thontadarya and
Chelonus sp.	"	"	Narayanan and Subba Rao, 1955; Nangia, 1983
Chelonus carbonator	"	Egg-Larvae	Ayyar, 1921; Rao and Satyanaraya, 1984;
Chelonus haliopae	"	Larvae	Rao, 1983; Patel *et al.*, 1971
Chelonus formosanus	"	Egg-Larvae	Patel *et al.*, 1971; Rai 1974
Cotesia (=Apanteles) plutellae	"	Larvae	Patel *et al.*, 1971;
Cotesia (=Apanteles) ruficrus	Hymenoptera Braconidae	Larvae	Albert, 1951; Zaz and Kushwaha, 1983;
Cotesia (=Apanteles) marginiventris	"	"	Jalali *et al.*, 1987;
Echthromorpha sp.	"	"	Sathe, 1987
Enicospilus sp.	"	"	Sathe, 1987
Meteorus sp.	"	"	Chari and Rao, 1990
Microplitis sp.	"	"	Zaz and Kushwaha, 1983
Microplitis prodiniae	"	"	Rao and Kuriyan, 1950; Sathe, 1987
Microgaster sp.	"	"	Kalra *et al.*, 1989
Rogas sp.	Hymenoptera Braconidae	Larvae	Bhatnagar, 1981
Zele chlorophthalma	"	Larvae	Kamal, 1951; Bhatnagar, 1981; Rao and Satyanarayana 1984
Zele nigricornis	"	"	Adair, 1981
Brachymeria lasus	Hymenoptera Chalcididae	"	Chari and Rao, 1990; Narendran and Joseph, 1977

Contd...

Table 4.3. Contd...

Brachymeria sp.	Hymenoptera Chalcididae	Pupae	Thontadarya and Nangia, 1983
Hybothoracini sp.	Hymenoptera Chalcididae	Larvae-Pupae	Rao *et al.*, 1981 and Chari and Rao 1990
Lasiochalcidia erythropoda	Hymenoptera Chalcididae	Larvae	Bhatnagar, 1981; Rao and Satyanarayana, 1984
Euplectrus gopimohani	Hymenoptera Eulophidae	"	Patel, 1944 Patel, 1980
Tetrastichus ayyari	Hymenoptera Eulophidae	"	Sathe, 1987
Trichospilus pupivora	Hymenoptera Eulophidae	Larvae	Anantanarayanan 1934; Sathe 1987
Aleiodes (=Paniscus) testaceus	Hymenopter Ichneumonidae	"	Beeson and Chatterjee, 1935
Barylypa rufa	"	"	Hafez, 1951
Campoletes chlorideae	"	"	Battu, 1977; Sathe, 1987
Campoletes sp.	"	"	Battu, 1977
Charops obstusus	"	Egg-Larval	Patel *et al.*, 1971 Patel, 1980
Diadegma argenteopilosa	"	"	Ayyar and Margabandhu, 1934; Sathe, 1987
Enicospilus sp.	"	"	Chari and Rao, 1990
Erioborus sp.	"	"	Bhatnagar, 1981
Ichneumon sp.	"	"	Anon., 1988
Netelia ferruginea	"	"	Krishnamurti and Usman, 1954; Sathe, 1987
Telenomus sp.	Hymenoptera Scelionidae	Egg	Anon., 1985
Telenomus sp.nr.*manolus*	"	"	Anon., 1982
Telenomus remus	"	"	Patel *et al.*, 1976; Rao and Patel, 1976; Joshi *et al.*, 1979; Zaz and Kushwaha, 1983;
Trichogramma chilonis	Hymenoptera Trichogramm-atidae	"	Joshi *et al.*, 1979; Bhatnagar, 1981; Thontadarya and Nangia, 1983; Zaz and Kushwaha, 1983
Trichogramma australicum	"	"	Joshi *et al.*, 1979
Parasarcophaga albiceps	Diptera Sarcophagidae	Larvae	Rao, 1983

Contd...

Table 4.3. Contd...

Sarcophaga albiceps	"	Larvae	Bhatanagar, 1981
Sarcophaga dux	Diptera Sarcophagidae	Larvae-pupae	Joshi *et al.*, 1979; Chari and Rao, 1990
Sarcophaga peregrina (Robineauedesvoidy)	"	Larvae	Zaz and Kushwaha, 1983
Blepharella setigiera	"	Larvae-pupae	Joshi *et al.*, 1979 Chari and Rao, 1990
Peribaea orbata	Diptera Tachinidae	Larvae	Rao and Patel, 1976
Strobilomyia aegyptia	"	"	Patel *et al.*, 1971; Joshi *et al.*, 1979; Bhatnagar, 1981
Strobilomyia orbata	"	"	Patel *et al.*, 1971
Fannia leucostica	Diptera Muscidae		Zaz and Kushwaha, 1983

Table 4.4. List of predators of *Spodoptera litura* (Fabricius)

Predator species	Order/Family	Stage of the pest predated	Reference
Cheiracanthium danieli Clubionidae	Araneae	Larvae	Sitaramaiah *et al.*, 1980
Lyposcelia sp.	Lyposcelia	"	Zaz and Kushwaha, 1983
Marpissa mandali	Araneae Salticidae	"	Sitaramaiah *et al.*, 1980
Marpissa sp.	Araneae Salticidae	"	Sitaramaiah *et al.*, 1980
Olios punctipes	Araneae Heteropodidae	"	Sitaramaiah *et al.*, 1980
Oxyopes sp.	Araneae Oxyopidae	"	Sitaramaiah *et al.*, 1980
Oxyopes wroughtonii	Araneae Oxyopidae	"	Sitaramaiah *et al.*, 1980
Phidippus punjabensis	Araneae	"	Sitaramaiah *et al.*, 1980
Thomisus lobosus	Araneae Thomisidae	"	Sitaramaiah *et al.*, 1980
Thomisus projectus	Araneae Thomisidae	"	Sitaramaiah *et al.*, 1980

Contd...

Table 4.4. Contd...

Cyrtopeltis (=Nesidiocoris) tenuis	Heteroptera Miridae	Eggs, larvae	Sitaramaiah *et al.*, 1980
Andrallus spinidens	Hemiptera Pentatomidae	"	Cherian and Brahmachari, 1941; Rajendra and Patel, 1971; Pawar, 1976
Cantheconidia furcellata	Heteroptera Pentatomidae	"	Kapoor *et al.*, 1975
Cantheconidia melacanthus	Hemiptera Pentatomidae	"	Kapoor *et al*, 1975
Coranus atricapillus	Hemiptera: Reduviidae	"	Sitaramaiah and Ramprasad, 1982
Coranus spinisuitis	"	"	Sitaramaiah and Ramprasad, 1982;
Harpactor costalis	"	"	Sitaramaiah *et al.*, 1975; Sitaramaiah and Ramprasad, 1982
Rhynocoris squalis	"	"	Rao *et al.*, 1982; Rao and Satyanaraya, 1984
Rhynocoris kumarii	Heteroptera Reduviidae	"	Sahayaraj, 1994
Rhynocoris marginatus	Heteroptera Reduviidae	"	Sahayaraj, 1994
Rhynocoris fuscipes	Heteroptera Reduviidae	"	Cherrian and Brahmachari, 1942; Sahayaraj, 2000
Sphedanolestes sp.	Heteroptera Reduviidae	"	Rao *et al*,1981
Dieuches sp.	Heteroptera Lygaeidae	"	Rao *et al*,1981
Coccinella sp.	Coleoptera Coccinellidae	"	Sitaramaiah and Ramprasad, 1982
Epilachna sp.	Coleoptera Coccinellidae	"	Cherrian and Brahmachari, 1942
Icaria marginata	Hymenoptera Vespidae	Larvae	Rao *et al*, 1981
Polistes stigma	Hymenoptera Vespidae	"	Rao *et al.*, 1982
Ropalidia sp.	Hymenoptera Vespidae	"	Joshi *et al*, 1979

Contd...

Table 4.4. Contd...

Apertochrysa crassineruis	Neuroptera Chrysopidae	"	Rao, 1983
Chrysopa sp.	Neuroptera Chrysopidae	Eggs	Rao *et al.*, 1982; Sitaramaiah and Ramprasad, 1982
Chrysopa crassinervis	Neuroptera Chrysopidae	Eggs	Rao and Satyanarayana, 1984
Chrysoperla carnea	Neuroptera Chrysopidae	Eggs/ larvae	Kamal, 1951; Geethalakshmi *et al.*, 2000
Calotes versicolor	Reptilia Agamidae	"	Singh and Singh, 1975
Uromastix hardovickii	Reptilia Agamidae	"	Bhatnagar and Srivastava, 1985
Liposcelis sp.	Pscoptera: Lipuscelidae	Egg	Zaz and Kushwaha, 1983

Table 4.5. List of entomopathogens of *Spodoptera litura* (Fabricius)

Entomopathogen species	Order/Family	Stage of the pest attacked	Reference
Aspergillus flavus	Moniliales Moniliaceae	Larvae	Battu *et al.*, 1971
Beauveria bassiana	"	"	Zaz and Kishwaha, 1983
Metarrhizium anisopliae	"	"	Siddaramaiah *et al.*, 1986
Nomuraea rileyi	"	"	Rao and Phadke, 1977; Phadke and Rao, 1978;
Entomophthora sp.	Entomophthorales Entomophthoraceae	"	Prasad and Kushwaha, 1990
Bacillus cereus	Eubacteriales Bacillaceae	Larvae	Kore and Bhinde, 1978; Oblisami *et al.* 1969
Bacillus thuringiensis	"	"	Zaz and Kushwaha, 1983
Micrococcus sp.	"	"	Zaz and Kushwaha, 1983
Streptococcus sp.	"	"	Zaz and Kushwaha, 1983
Streptococcus faecalis	"	"	Battu *et al.*, 1972; Zaz and Kushwaha, 1983
Pseudomonas aeruginosa	Eubacteriales Pseudomonadeceae	"	Prasad and Kushwaha, 1990

Contd...

Table 4.5. Contd...

Serratia marcescens	Eubacteriales Enterobacteriaceae	"	Pandey and Rangarajan, 1967; Zaz and Kushwaha,1983; Ansari *et al*, 1987
Mattesia sp.	Neogregarinida Ophrycoystidae	"	Anonymous, 1994
Nosema sp.	Microspora Nosematidae	"	Narayanan and Jayaraj, 1979;
Granulosis virus	"	"	Battu *et al.*, 1971; Narayanan, 1985;
Nuclear polyhedrosis	Virales	Larvae	Ramakrishnan and Tiwari, 1969; Battu *et al.*, 1972; Ramakrishnan, *et al.*, 1975; Zaz and Kushwaha, 1983; Chári *et al*, 1985; Krishnaiah *et al.*, 1985,
Cytoplasmic polyhedrosis virus	"	"	Narayanan, 1985;
Hexamermis spp.	Nematoda Mermithidae	"	Bhatnagar *et al.*, 1985
Neoplectana carpocapsae	Nematoda Steinernematidae	"	Janardan singh and Bardhan, 1974
Ovomermis albicans	Nematoda Mermithidae	"	Bhatnagar *et al.*, 1985
Pentatomimermis spp.	Nematoda Mermithidae	"	Bhatnagar *et al.*, 1985

Table 4.6. List of bird predators of *Spodoptera litura* (Fabricius)

Bird species	Order/Family	Stage of the pest attacked	Reference
Acridotheres tristis	Passeriformes Sturnidae	Larvae-adults	Patel, 1988
A.ginginianus	Passeriformes Sturnidae	Larvae-adults	"
Bubulcus ibis	Ciconiiformes Ardeidae	Larvae-adults	"
Corvus macrorhynchos	Passeriformes Ploceidae	Larvae	"
C. splendens	Passeriformes Dicruridae	Larvae-adults	"

Table 4.6. Contd...

Dicrurus adsimilis	Passeriformes Dicruridae	Larvae	"
Passer domesticus	Passeriformes Ploceidae	Larvae-adults	"
Pseudibis papilosa	Ciconiiformes Threshiornithidae	"	"
Sturnus roseus	Passeriformes Sturnidae	"	"

□□□

Helicoverpa armigera (Hubner)

J. G. BAPODRA, K. GUNATHILAGARAJ
AND V. NANDAGOPAL

5.1 Introduction

Helicoverpa armigera Hubner is a polyphagous pest and widely distributed through out the world. To mention a brief nomenclature of *Heliothis*, it was during 1806, J. Hubner of Augsburg, Germany described the genus as *Heliothis* (Hemming, 1937). Later Ochsenheiner (1808) included species *armigera* among the 14 nominal species he placed in his concept of genus *Heliothis*. Subsequently in the same year Hubner, 1808 published the figure of *Heliothis jucundus*, a north American species (Hemming, 1935). Hubner (1808) placed this in a new a generic name as *Melipotis jucunda*. Hampson (1903) used genus chloridae for *armigera* and it was followed by (Warren, 1914). Hardwick (1965b) established the genus *Helicoverpa* in comparison with *Noctua armigera* Hubner (1808) as its type species. Again Boursin (1965) placed *Helicoverpa* as synonym of *Chloridea*. Hardwick (1970) reiterated with reason that *Helicoverpa* and *Heliothis* were generically distinct. Nye (1980) proposed to designate gender and stem on *Heliothis* Ochsenteimer (Insecta- Lepidoptera) and in the International workshop majority of *Heliothis* Scientists recommended to retain the established nomenclature for *H. armigera* (Nye, 1982) to the International Commission of the Zoological nomenclature. Bhatnagar and Davies (1978) recorded 50 species of crop plants and more than 150 alternative host plants in Andhra Pradesh.

5.2 Nature of damage

Being polyphagous, it feeds on variety of plant species. The neonate feed by scrapping **(Plate 1)**. The grown up caterpillars are voracious feeder and do extensive damage even at low population in groundnut. The caterpillars feed on the flowers and foliage and defoliate the plant. In most cases , the larvae feeds characteristically by thrusting its head into the pod and leaving the rest of its body outside in pigeonpea. It thus attacks groundnut (flowers, leaves), gram (pods), cotton (buds, flowers, bolls), tobacco (capsules, seeds), sunflower (flower, seeds), tomato (fruits), sorghum, maize and wheat (milky grains).

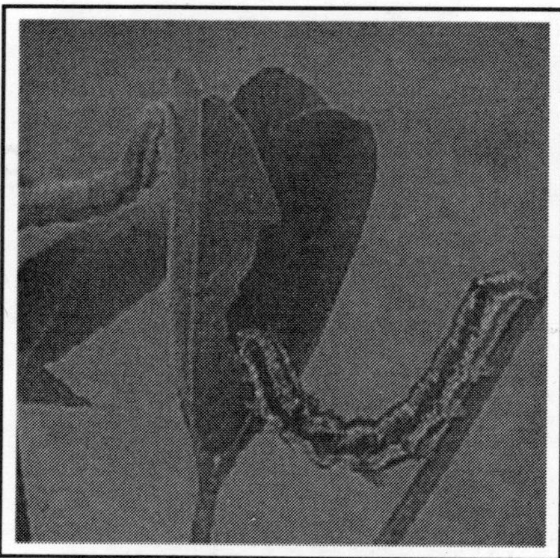

Plate 1 Feeding of larva

5.3 Economic importance

Helicoverpa causes serious damage to cotton, chickpea, pigeon pea, sorghum, maize, wheat, groundnut, sunflower, tobacco, tomato and lucerne crops. According to Srivastava (1959), it causes 40 to 50 per cent damage to tomato fruits in Tamil Nadu. Cotton suffered a loss of 41 to 56 per cent occurred in Madhya Pradesh (Kausik *et al.*, 1969). Losses estimated in other crops include chickpea (84%) (Sithannatham *et al.*, 1984). Chari *et al.*(1983) worked out loss in seed yield to a tune of 88.5 kg/ha in Gujarat in sorghum (18- 26%) Rawat *et al.*(1970) and sunflower 41.6 per cent (Margal, 1990).

5.4. Biology of *H. armigera*

The biology of the insect has been reported by a few workers (Singh and Singh, 1975; Chaudhari and Sharma, 1981; Bapodra, 1988).

Egg: The females lay eggs one at a time and distribute on tender parts of host plants. The eggs are shining and yellowish white at first and changing to dark brown before hatching. The incubation period is longer in cold weather and shorter in hot weather, being 2 to 5 days in India (Singh and Singh, 1975). The average incubation period was 3.11 days when larvae were reared on groundnut (Bapodra, 1988). The viability of eggs varies from 63 to 90 per cent **(Plate 2).**

Larva: The first instar larva of *H. armigera* is translucent, yellowish white in colour. The body of full grown larva is pale green in colour with reddish brown head dark broken grey lines along the sides of the body and short hairs scattered all over the body. Bapodra (1988) observed six larval instars on groundnut. The larval duration varied from 21 to 40 days in California,

18 to 51 days in Ohio (Wilcox *et al.*, 1956) and 8 to 12 days in the Punjab, India (Singh and Singh, 1975) on tomato. The average larval period was recorded 21.40 days on groundnut (Bapodra, 1988).

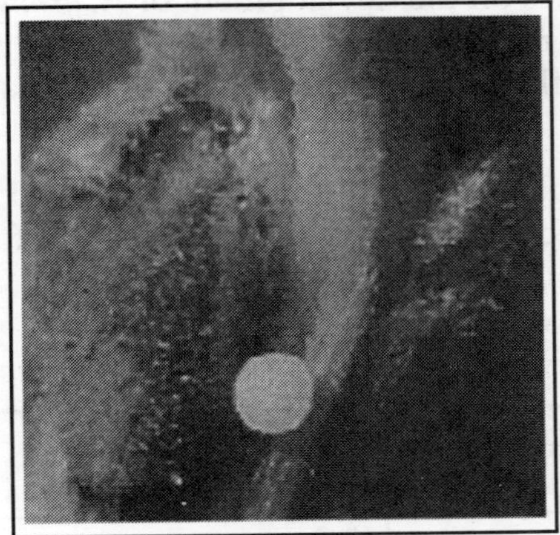

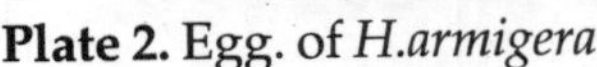

Plate 2. Egg. of *H.armigera*

Plate 3 : Pupae of *H. armigera*

Pupa: The pupa is brown in colour and becomes dark **(Plate 3)** prior to emergence of adults. The pupal period of *H. armigera* ranged from as 14 to 40 days in Southern Rhodesia, 12 to 25 days in Tanganika, 14 to 37 days in Uganda, 15 to 25 days in Nyasaland and 5 to 8 days in India (Pearson and Darling 1958). The average pupal period was 11.41 day on groundnut (Bapodra, 1988).

Adult: The female moth is stout bodied **(Plate 4)** with yellowish brown in colour. Forewings are marked with greyish wavy lines and black spots of varying size on the upper side and black kidney shaped marking and a round spots on the under side. The hind wings are whitish and lighter in colour with broad blackish band along outer margin. Tuft of hairs are present on the tip of abdomen of females only. The average pre-oviposition and

Plate 4 : *H. armigera* adult

oviposition periods were 3.7 days and 6.3 days respectively on groundnut (Bapodra, 1988) and fecundity was 4735 eggs/female, Bapodra (1988). Larval growth varies with host plants. The mean egg, larval and pupal periods of *Helicoverpa armigera* ranged from 3.0 to 3.7, 9.7 to 20.7 and 8.3 to 14.0 days, resp., during September-February. Studies on the development of *H. armigera* and growth indices on different hosts revealed that chickpea, pigeon pea, cotton, safflower, groundnut and sunflower were preferred. (Venkataiah *et al.*, 1994).

The average life span was 46.7 days, when larvae were reared on groundnut (Bapodra, 1988). The detailed duration of different stages of the *H. armigera* in Uttar Pradesh (Paras Nath, 1993) is given in Table 5.1.

Table 5.1. Biology of *H. armigera*

Stages **Mean ± SEM**	**Duration (Days)** **Range (day)**	
Eggs	3.68±0.15	2-5
LARVAE		
1st Instar	3.20±0.14	2-4
2nd Instar	3.28±0.15	2-4
3rd Instar	3.56±0.15	2-5
4th Instar	3.88±0.15	3-5
5th Instar	4.32±0.20	3-6
6th Instar	4.36±0.16	3-6
Total larval period	22.60±0.41	19-28
Pre-pupa	2.60±0.15	1-4
Pupa	6.24±0.23	5-8
Pre-oviposition	2.40±0.16	1-4
Oviposition	3.32±0.21	2-5
Post-oviposition	1.76±0.16	1-3
ADULT		
Male	4.20±0.30	2-7
Female	7.48±0.18	6-9
Total life	40.88±0.93	34-50
Fecundity	723.16±42.29	230-1095

5.5 Seasonal activity

Annual cycle of *Helicoverpa* is completed through four generations in Punjab (Singh and Singh, 1975) whereas Bhatnagar (1980) reported 7 to 8 generations in an year in Andhra Pradesh.

Light trap studies carried out at Anand, Gujarat revealed that the pest population was low from June-August which slowly increased up to December. The growth of population was rapid from December to February and reached a peak during March and rapidly declined during May (Anon., 1987). Similarly the pheromone study carried out at Junagadh, Gujarat indicated that the pest remained very active during January-March and September-December. Butani and Kapadia (1995) reported that the maximum moth catches was observed during January, March and September. Bapodra (1988) reported that the infestation of *Heliothis* commenced in first week of December on chickpea crop and it attained the peak during third week of December. He further stated that *Helicoverpa* population had negative correlation with minimum temperature and sunshine hours.

The patterns of emergence and mating, fecundity and bio-ecology of *H. armigera* were studied by Rosaiah and Reddy (1999). Peak emergence at 20.00-22.00 h during summer, 22.00-24.00 h during rainy season, and 24.00-02.00 h during winter. Most insect pairs mated at 01.00 h in summer, 03.00-05.00 h in rainy season and 0400-06.00 h in winter, and the mating period lasted for 44-62 minutes. The pre-oviposition, oviposition and post-oviposition periods varied from 2.6 to 3.5, 3.8 to 7.2 and 2.1 to 2.5 days in summer and winter seasons, respectively. Weather parameters like temperature, wind and rain affected pest build up. Studies of *H. armigera* in the different seasons showed that mung bean in August, groundnut in September, cotton in October, pigeon pea in November, castor in December, chickpea in January and tomato in February supported its population.

5.6 Alternate Hosts

This species has been reported to damage about 96 crops and 61 weeds (Mule and Chari, 1978) and the main host are chickpea, pigeon pea, field bean, tomato and cotton (Patel and Koshiya, 1999). Amin (1988) reported this species damaging groundnut considerably. Larvae of *H. armigera* were found to infest the weed *Physalis minima* in groundnut fields (Men and Kandalkar, 2001) and grape cv. Thompson (Balikai *et al.*, 2001).

5.7 Mechanisms of resistance

5.7.1 Host Preference

A study on seasonal abundance of *Helicoverpa armigera* showed that during kharif season the pest started its activity in groundnut from the first week of July 1993. Thereafter, the pest moved to cotton from the last week of July and started to build up its population during the month of August to mid-September. Simultaneously, the pest infestation was also noticed in sunflower and pearl millet during this period but the population was very low in sunflower. However, in pearl millet it was at a peak during September

1993. In the rabi season, pest activity was observed in chickpea during November to January-February, and the population was at its peak during December. During the summer season, the pest started its activity on groundnut in February and was active up to June (Patel and Koshiya, 1997). Accessions belonging to *Arachis cardenasii, Arachis duranensis, Arachis kempff-mercadoi, Arachis monticola, Arachis stenosperma, Arachis paraguariensis, Arachis pusilla* and *Arachis triseminata* showed multiple resistance to the *Aproaerema modicella, Helicoverpa armigera, Empoasca kerri. Arachis cardenasii* (ICG 8216), *Arachis ipaensis* (ICG 8206), *A. paraguariensis* (ICG 8130) and *Arachis appressipila* (ICG 8946) showed resistance to leaf feeding and antibiosis to *Spodoptera litura* under no-choice conditions. Six lines, derived from wild relatives, showed resistance to *H. armigera* and *S. litura*, and/or leaf miner. Plant morphological characteristics such as main stem thickness, hypanthium length, leaflet shape and length, leaf hairiness, standard petal length and petal markings, basal leaflet width, main stem thickness and hairiness, stipule adnation length and width, and peg length showed significant correlation and/or regression coefficients with damage by *H. armigera, S. litura*, and leafhoppers, and these traits can possibly be used as markers to select for resistance to these insect pests (Sharma *et al.*, 2003). In groundnut, host preference studies were very much limited probably owing to the difficulty of rearing the larvae all through instars, as this species is highly cannibalistic. Twelve genotypes were field screened against the early instars larvae of *H. armigera*. There was no difference among the entries in the number of the eggs and larvae recorded. However, significant difference existed between genotype in the % damaged plant. ICBSB-6143 was the least damaged (50% foliage damaged {TG-27(E), ICGS-76, ICBSB-6143, ICGS-1 and SGB-4}. Fifteen genotypes were screened against *H. armigera* of the 3rd instars following choice test in a diallel set, by taking two genotype at a time in 105 possible combinations. The genotype such as IR-13A, BG-2 and IR-17 registered moderately resistance against this insect (26 to 50% damage) (our unpublished data).

5.8 Natural enemies

List of parasitoids (Table 5.2), predators (Table 5.3), entomopathogens (Table 5.4) and bird predators of *H. armigera,* recorded in India are given.

5.9. Management Strategies

5.9.1 Biological control

Ibrahim (1980) recorded 103 species of parasites of *Helicoverpa armigera* belonging to 10 different families of Hymenoptera and Diptera. Whereas Pawar *et al.* (1986) listed 76 parasitoids and 33 predators feeding on this insect. Use of *Trichogramma chilonis* @ 1 lakh/ha and *Chrysoperla carnea* (Stephens) @ 50000/ha at 40 and 50 days after sowing of groundnut can effectively check the *Helicoverpa* population in Tamil Nadu (Sundram *et al.*, 1994).

Prey preference was studied by Sahayaraj and Sivakumar (1995) in 5th-instar larvae of *Rhynocoris kumarii* on 3 lepidopteran pests of groundnuts, *Helicoverpa armigera, S. litura* and *Stomopteryx subsecivella* [*Bilobata subsecivella*]. The predator preferred *S. litura* (48%), followed by *H. armigera* (34%) and *B. subsecivella* (18%). The predation efficiency of *Rhynocoris marginatus* against *Helicoverpa armigera* and *Spodoptera litura* was studied on groundnuts by Sahayaraj (1999). Populations of *H. armigera* and *S. litura* were significantly reduced following predator release. Pod yield was also highest (1867.77 kg/ha) in plots where the predator was released.

In the laboratory, maximum predation of *R. marginatus* was observed on *Aproaerema modicella*, followed closely by *Helicoverpa armigera* and *Spodoptera litura*. Predation on *Amsacta albistriga* was very low, presumably due to the small size of the prey (Sahayaraj, 2000). *Rhynocoris marginatus* fourth and fifth instar nymphs preferred *H. armigera*, followed by *S. litura*. Adult predators preferred *S. litura* the most, followed by *H. armigera*. Predators at all life stages showed the least preference for *A. modicella*. Sahayaraj and Paulraj (2001).

Urmila Sharma *et al.* (1997) identified *Geocoris ochropterus* and *G. jucundus* as predators of *Aphis craccivora, Empoasca kerri, Helicoverpa armigera* and *Spodoptera exigua*.

Jayanthi and Padmavathamma (1996) showed that the nuclear polyhedrosis virus (NPV) of *Spodoptera litura* was not cross infective to other groundnut pests such as *Helicoverpa armigera, Aproaerema modicella, Empoasca kerri* and *Aphis craccivora* and was safe to their predators (*Coccinella septempunctata, Menochilus sexmaculatus* [*Cheilomenes sexmaculata*] and *Chrysoperla carnea*) and parasitoids (*Bracon gelechiae* and *Tetrastichus* sp). *Bacillus thuringiensis* subsp. kurstaki was highly effective against the larvae of lepidopterous pests but not against homopteran insects. *B.t.* was also safe to coccinellid predators and larval parasitoids of *A. modicella*, except adults of *C. carnea*. *Beauveria bassiana* was pathogenic to groundnut pests and coccinellid predators, however, it was safe to use in the presence of larval parasitoids of *A. modicella*. *Beauveria bassiana* was pathogenic to all stages of the pest on groundnut (2-72% mortality) at 1.18x1010. This susceptibility decreased with age (Rajiv-Rathod, 2002).

Four sprays of NPV suspension @ 250 LE/ha at five days interval after appearance of pest are recommended for effective control of *Helicoverpa* attacking groundnut in Gujarat (Anon., 1996). A mixture of NPV @ 250 LE/ha and chlorpyriphos @ 125 g a.i./ha control successfully the population of *Helicoverpa* in groundnut in Tamil Nadu (Dhandapani *et al.*, 1993). HaNPV application at 250 LE/ha was found to be the most effective in reducing the larval population of *Helicoverpa* and increased yields of groundnut, with the highest ICBR of 1:3.82 in comparison to high and low doses of HaNPV when four sprays of the virus were applied at 5 days interval starting from the pest appearance (Butani *et al.*, 1997). Sonalkar and Deshmukh (1999) tested seven different adjuvants (sucrose, jaggery, urea, chickpea flour, cotton seed oil cake, groundnut oil cake and egg white) for their effects on the efficacy of nuclear polyhedrosis virus against third instar larvae of *Helicoverpa armigera.* Chickpea flour, egg white, sucrose and jaggery at 1, 3, 0.5 and 0.5% concentration, respectively, had significant impact on virus efficacy.

The persistence of a virus can be increased by the addition of robin blue which acts as UV protectant for nucleopolyhedrosis virus (NPV). Addition of robin blue increased the mortality of third-instar larvae to 74.7%, indicating that robin blue increased the persistence of the virus by protecting it against UV rays (Reddy and Divakar, 2001).

Conservation of natural enemies of *Helicoverpa* by using selective chemicals viz., endosulfan and phosalone are recommended by Odak (1982) and Sukhoruchenko *et al.* (1977) as compared to other insecticides. II and III instars of *Helicoverpa armigera* were more susceptible to the pathogen, *B. bassiana* (1.18x104, 106, 108 and 1010) than other larval stages.

Manjula *et al.* (2003) studied the occurrence of *N. rileyi* in *Spodoptera litura* and *H. armigera* in groundnut fields. The numbers of cadavers and live larvae were recorded at fortnightly intervals from September 2001 until the end of the cropping season. *N. rileyi* was initially observed during the first fortnight of September, and the level of infection gradually increased to 5-10, 25-50 and 50-100% during the second fortnight of September, first fortnight of October and second fortnight of October, respectively. These periods were characterized by a mean maximum temperature of 31-32 degrees C, minimum temperature of 22-23 degrees C, morning relative humidity (RH) of 79-84%, and evening RH of 29-42%.

Manjula *et al.* (2004) observed the occurrence of *N. rileyi* on *H. armigera* and *S. litura* on cotton, chilli, tomatoes, red gram, black gram and groundnut. Cotton and groundnut recorded 100% mycosed larvae of *H. armigera* and *S. litura* due to infection by *N. rileyi*. Mycosed larvae were low in sweet pepper. The incidence of *N. rileyi* showed positive association with morning relative humidity. *H. armigera* recorded higher infection compared to *S. litura* in groundnuts. However, higher infection was observed in *S. litura* compared to *H. armigera* in black gram.

5.9.2 Cultural control

Early planting, destruction of crop residues and use of traps has provided highly effective components of the pest management systems for soybean and Cotton (*Gossypium* spp.).

5.9.3 *Cropping pattern*

(1) Intercropping: Intercropping of Cotton with Cowpea, onion, coriander, soybean, blackgram and sorghum is advocated against *Helicoverpa* by ICRISAT. Inter-crop of Bajra and pigeonpea 2:1 reduced the incidence of *Helicoverpa* and gave higher yield than sole crop. Similarly intercropping of maize with pigeonpea is recommended for *Helicoverpa* attacking pigeon pea crop.

(2) Trap-cropping and mixed cropping: The tomato crop when boardered with rustica tobacco as trap crop and marigold when mixed with tomato crop reduced the infestation of *Helicoverpa.*

5.9.4 *Light and pheromone trap*

Use of light and pheromone trap helps in population reduction of this pest. The peak activity of the moths were recorded during May and July at Hebbal and at Raichur during September, November and December months. The peak activity was preceded by rains indicating the possible existence of the relationship between the rain and the emergence of adult moths (Jayaramaiah and Jagadeesh Babu, 1992). The low catches of moths at Hebbal from August to December due to reduction in the temperature from 19°C to 15°C indicated that positive relation exists between minimum temperature and adult activity (Srinivas, 1984). Nandagopal *et al.* (2003) evaluated the efficacy of different types of pheromone traps and their modification against *H. armigera* . They concluded that there was no significant difference in the performance of funnel traps and the trap with funnel and flap. The trap without funnel and flap was efficient in capturing the moths. When the distance between the lid and funnel of these traps was reduced from 4 (normal) to 2 cm, the efficiency significantly increased (21.3 males/trap/alternate day) compared with any other type of trap.

5.9.5 *Chemo-sterilants*

Mule and Chari (1978) reported that the tepa and thiotepa cent per cent sterilized the males and caused reduction in eggs viability and fecundity when treated on virgin females.

5.9.6 *Anti-feedants*

Chari and Patel (1973) reported the anti-feeding property of tryphenyl tin hydroxide (TTH) and triphenil tin acetate (TTA) against *H. armigera*

5.9.7 *Chemicals*

Application of recommended insecticides *viz.,* endosulfan 0.07%, phosalone 0.07% DDVP 0.05%, monocrotophos 0.05%, fenvalerate 0.015%, cypermethrin 0.009% and quinolphos 0.05% are effective for the control of this pest (Bapodara,1988). Diflurobenzuron (0.0125%) in combination with profenofos (0.025%) and methomyl (0.025%) was effective against major Lepidopteran pests of groundnut like *Aproaerema modicella, Spodoptera litura* and *Helicoverpa armigera* and resulted in the highest pod yields. The percent larval population reduction in treatments with profenofos and methomyl at 2 days after sowing (DAS) was 85.6 and 81.9, 66.2 and 61.6, and 62.8 and 59.0 of *A. modicella, S. litura* and *H. armigera,* respectively. Diflurobenzuron increased the efficacy of other insecticides and resulted in higher pod yields. Neem oil (1.0%) treatments at 2 DAS resulted in 52.9, 30.7 and 39.9% larval population reduction of *A. modicella, S. litura* and *H. armigera,* respectively. (Kumar and Krishnayya, 1999). Loganathan *et al.* (2001) carried out bioassay experiments to determine the LC_{50} and LT_{50} values of *B. thuringiensis* var. *galleriae* against the larval instars of *H. armigera* on groundnut. The LC_{50} values obtained against the first, second, third, fourth, fifth and sixth instars of *H. armigera* were 0.38, 1.82, 2.03, 3.75, 4.15 and 5.09 ml/l for 48 h, respectively. The LT50 values against these instars were 12.42, 19.22, 28.47, 34.58, 42.45 and 48.77 h, respectively. They concluded that early instars of *H. armigera* are more susceptible to *B. thuringiensis* var. *galleriae* than the late instars.

References

Achan, P. D., Mathur, K. C., Dharmadhikari, P. R. and Manjunath, T. M. (1968). Parasite of *Heliothis* spp. in India. Pages 129-149 in *Commonwealth Institute of Biological Control Technical Bulletin* 10, Ascot, Berks, UK.

Amin, P. W. (1988). Insect and mite pests and their control. In *Groundnut* (Ed. P.S. Reddy). pp. 393-452. Indian Council of Agricultural Research, New Delhi.

Anonymous. (1987). Annual Report All India Coordinated Research Project on Biological Control of Crop Pests and Weeds, Indian Council of Agricultural Research, IIHR, Bangalore, India. Tech. Doc. No. 20: 1-236.

Anonymous. (1987). Technical Bulletin, Plant Protection in Groundnut, National Research Centre for Groundnut, Junagadh, Gujarat.

Anonymous. (1989). Annual Progress Report (1987-1988 and 1988-1989). of AICRP on Rodents Control. Junagadh, Gujarat Agricultural University, Junagadh.

Anonymous. (1989). Twenty-seventh meeting of the Research Council held on Dec 22, 1989 at TNAU, Coimbatore. Agenda notes. Directorate of Research, TNAU, Coimbatore. 137 p.

Anonymous. (1996). Annual Research Report, Department of Entomology, College of Agriculture, Junagadh.

Balikai, R. A., Ravikumar, M. R. and Ryagi, Y. H. (2001). *Helicoverpa armigera* (Hubner) as berry borer in grapevines of Northern Karnataka. *Insect Environment*. 6(4): 175-176.

Bapodra, J. G. (1988). Studies on biometrical analysis, biology, population dynamics and chemical control of *Heliothis armigera* (Hubner) on different hosts. Ph.D thesis submitted to G.A.U., Sardar Krishinagar, Gujarat.

Bhatnagar V. S., Sithanantham, S., Pawar, C. S., Jadhav, D. R., Rao, V. K. and Reed, W. (1983). Conservation and augmentation of natural enemies with reference to IPM in chickpea and pigeonpea. pp 157-180 in Proceeding of International Workshop on IPM in grains legumes. 4-9 April 1983, Goiana, Brazil.

Bhatnagar, V.S and Davies, J.G. (1978). Factors affecting population of gram pod borer, *Heliothis armigera* (Hubner). (Lepidoptera: Noctuidae). in the period 1974-77 of Patancheru (A.P). Oriental Entomology, *Mini workshop on population ecology in relation to insect of economic importance*, Bangalore.

Bhatnagar, V. S., Lateef, S. S., Sithanantham, S., Pawar, C. S. and Reed, W. (1983). Research on *Heliothis* at ICRISAT. *In Proceedings of the International Workshop on Heliothis Management*, 15-20 Nov, 1980, ICRISAT Center, Patancheru, A.P., India pp.385-396.

Bhatnagar, V.S. (1980). Report on research on *Heliothis* complex at ICRISAT (India). 1974-79 presented at the *All India Workshop on Consolidation of Pest Management recommendation and guidelines of Research* 24-26 April, 1990, Udaipur.

Bottrell, D. G., Young. J. H., Price, R. G. and Adams, R. H. (1968). Parasites reared from *Heliothis* spp. in Oklahoma in 1965 and 1966. *Annals of the Entomological Society of America* 61: 1053-1055.

Boursin, C. (1965). Errata et addenda a mon travail " Les Noctuidae Trifinae de France et de Belgique" paru dans ce meme Bulletin, 1964, No.6. p.204. Bulletin Mensule de la Societe Linneenne de Lyon 34: 182-187.

Butani, P. G. and Kapadia, M. N. (1995). Population studies of *Helicoverpa armigera* (Hubn.). through pheromone trap catches and its correl;ation with weather parameters under South Saurashtra (Gujarat). paper presented in *National Seminar on IPM in Agriculture* JKV, Akola (India)., 29-30th, December, 1995.

Butani, P. G., Kapadia, M. N. and Parsana, G. J. (1997). Comparative efficacy and economics of nuclear polyhedrosis virus (NPV) for the control of *Helicoverpa armigera* (Hubner) on groundnut. Journal of Oilseeds Research. 14: 85-87.

Carl, K. P. (1978). *Heliothis armigera* ; parasite survey and introduction of *Apanteles kazak* to New Zealand Commonwealth Institute of biological control Report, European Station, Delemont, Switzerland, 8 pp.

Chari, M. S and Patel, N. G (1973). Control of *Heliothis armigera* damaging tobacco capsules. *Madras Agricultural Journal*. 60: 631.

Chari, M. S., Patel, N. M. and Patel, A. R. (1983). Paper presented at "All India Workshop on Crop losses due to insect pest", Hyderabad.

Cherian, M. C. and Subramaniam, C. K. (1940). *Tetrastichus ayyari* Rhow., a pupal parasite of some moth borers in South India. *Indian Journal of Entomology*. 2: 75-77.

CIBC. (1968a). Annual report, 1968. Commonwealth Institute of Biological control Commonwealth Agricultural Bureau, Farnham Royal, UK.

CIBC. (1974). Biology and breeding techniques for parasites and predators of *Ostrinia* spp. and *Heliothis* spp. U. S. PL- 480 Project Final Technical Report, 21 June 1967-20 June 1973, Bangalore, Karnataka, India.

CIBC. (1978). Possibilities of biological control of *Heliothis armigera* and *Heliothis zea* CIBC status paper 13, Ascot, Berks, UK.

Danks, H. V., Rabb, R. L. and Southern, P. S. (1979). Biology of insect parasite of *Heliothis* larvae in North Carolina, *Journal of Georgia Entomological society*. 14: 36-64.

Donohoe, M. C. and Pitre, H. N. (1977). *Reduviolus roseipennis* behavior and effectiveness in reducing number of *Heliothis zea* on cotton. *Environmental Entomology*. 6: 872-876.

Elsey, K. D. (1972). Predation of eggs of *Heliothis* spp. on tobacco. *Environmental Entomology* 1: 433-438.

Graham, H. M. (1970). Parasitism of bollworms, tobacco cutworm and loopers by *Trichogramma semifumatum* in the Lower Rio Grande Valley, Texas. *Journal of Economic Entomology*. 63: 686-688.

Habib, R. (1973). Memorandum on possibilities of the biological control for *Heliothis armigera*. Commonwealth Institute of Biological control, Ravalpindi, Pakistan, 18 pp.

Hafez, M. (1951). Notes on introduction and biology of *Microplietis demolitor* Wilk (Hymenoptera : Braconidae). *Bull. Soc. Fouad. Icr. Entomol*. 35: 107-121.

Hampson, G.F. (1903). Catalogue of lepidoptera Phalaenae in the British Museum4. London: British Museum, 689 pp.

Hanumanna, M., Thippeswamy, C. and Thimaiah, G. (1980). Insecticidal control of *Heliothis armigera* (Hubner). on Bengal gram. *Current Research*. 9: 11-13.

Hardwick, D. F. (1965b).The corn earworm complex. *Memories of the Entomological Society of Canada*. 40: 1-247.

Hardwick, D. F. (1970). A generic revision of the North American Heliothidinae (Lepidoptera: Noctuidae). *Memories of the Entomological Society of Canada*. 73: 1-59.

Hemming, F. (1935). A note on Jacob Hubner's Erste Zutrage zur Sammlung exotischer Schmetterlinge od 1908.Stylops 4: 38-48.

Hemming, F. (1937). Hubner 1. London : Royal Entomological Society of London. 605 pp.

Hubner, J. (1808). Jacob Hubner's Erste Zutrage zur Sammlung exotischer Schmetterlinge od 1908.Stylops 4:38-48.

Hussain, M. A. and Mathur, U. B. (1924). A preliminary list of parasite economic importance bred in the Punjab. pp 119-121 in Rep. *Proc. 5th Entomol. Meeting*, Pusa, Bihar, India.

Huston, J. C. (1939). Report on the work of the Entomological division. Adm. Rep. Dir. Agric. Ceylon, 1937. 37-42.

Ibramhim, A. E. (1980). Biotic factors effecting different species in the genera *Heliothis* and *Spodoptera* in Egypt. Inst. Pl. Prot. Report, pp 2-5.

ICRISAT. (1974- 85). Annual Report,International Crops Research Institute for the Semi-Arid Tropics, Patancheru, A.P., 502 324, India.

Jayanthi, P.D.K. and Padmavathamma, K. (1996). Cross infectivity and safety of nuclear polyhedrosis virus, *Bacillus thuringiensis* subsp. *kurstaki* Berliner and *Beauveria bassiana* (Balsamo) Vuille to pests of groundnut (*Arachis hypogaea* Linn.) and their natural enemies. Journal of Entomological Research. 20: 211-215.

Jayaramaiah, M. and Jagadeesh Babu, C. S. (1992). Present status of *Helicoverpa armigera* in Pulses and strategies for its management in Karnataka. (eds, Sachan J. N.). *Helicoverpa management : Current status and future strategies*, Proc. of first National Workshop held at Directorate of pulses Research Kanpur 208 024, India 30-31 Aug., 1990 75-83 pp.

Kaushik, U. K., Rathore, V. S., Sood, N. K. (1969). Incidence of bollworms and losses caused to cotton in Madhya Pradesh (India). *Indian Journal of Entomology*. 31: 175-177.

Khan, R. M. and Sharma, S. K. (1972). *Cataglypsis bicolor* Fab. (Hymenoptera: Formicidae). as a predator of few Noctuid larvae. *Madras Agricultural Journal*. 59: 192.

King, E. G., Powel, J. E. and Smith, J. W. (1981). Prospect for utilization of parasites and predators for management of *Heliothis* spp. paper presented in *International workshop on Heliothis management* held at ICRISAT, Patancheru, India from 15-20, November, 1981.

Koshiya, D.J. and Patel, H.K., (1987). Life table and population dynamics of *Heliothis armigera* Hubn. on groundnut. *G.A.U. Research Journal*. 12, 11-14.

Krishnamurti, B. and Usman, S. (1954). Some parasites of economic importance noted in Mysore State. *Indian Journal of Entomology*. 16: 327-344.

Kumar, D. V and Krishnayya, P. V. (1999). Effect of diflubenzuron in combination with selected insecticides on major lepidopteran pests of groundnut, *Arachis hypogaea* Linn. *Journal of Applied Zoological Researches*. 10(1): 1-5.

Lewis, W. J. and Brazzel, J. R. (1968). A three years study of parasite of bollworm and the tobacco buworm in Mississippi. *Journal of Economic Entomology*. 61: 673-676.

Lincoln, C. Phillips, J. R., Whitcomb, W. H., Powell, G. C., Boyer, W. P., Bell, K. O. Jr., Dean, G. L., Matthew, E. J., Graves, J. B., Newsom, L. D., Clower, D. F., Bradley, J. R. Jr. and Bagent, J. L. (1967). The bollworm- tobacco budworm problems in Arkansas and Louisiana. Arkansas Agriculture Experiment Station Bulletin 720. 66 pp.

Lingren, P. D., Guerra, R. J., Nickelsen, J. W. and White, C. (1970). Host and host age preferences of *Campoletis perdistinctus*. *Journal of Economic Entomology* 63: 613-618.

Loganathan, M., Babu, P. C. S. and Balasubramanian, G. (2001). The dose and time-mortality responses of *Helicoverpa armigera* (Hub.) to *Bacillus thuringiensis var. galleriae*. *Madras Agricultural Journal*. 88(7/9): 493-495.

Manjula, K., Rao, P. A. and Nagalingam, B. (2004). Record of *Nomuraea rileyi* (Farlow) Samson on *Helicoverpa armigera* Hubner in kharif groundnut. *Indian Journal of Plant Protection*. 32(2): 125.

Manjula, K., Nagalingam, B., and Rao, P. A. (2003). Occurrence of *Nomuraea rileyi* on *Spodoptera litura* and *Helicoverpa armigera* in Guntur District of Andhra Pradesh. *Annals of Plant Protection Sciences*. 11 (2): 224-227.

Manjunath, T. M. (1972). Biological Studies on *Trichogrammatoidae armigera* Nagaraja, a new dimorphic eggs parasite of *Heliothis armigera* Hubner in India. *Entomophaga*. 17: 131-147.

Manjunath, T. M., Patel, R. C. and Yadav. D. N. (1976). Observation in *Heliothis peltigera* (Schiff). (Lep: Noctuidae). and its natural enemies in Anand (Gujarat State in India). *Proceeding Indian Academy of Science*, 83 (2): 55-65.

Manjunath, T. M., Phalak, V. R. and Subramaniyam, S. (1970). First record of egg parasite of *H. armigera* Hb. (Lepidoptera: Noctuidae). In India. Technical bulletin of CIBC, Bangalore, India, 13: 111-115.

Margal, S.M. (1990). Studies on estimation of losses in sunflower cultivars due to *Heliothis armigera* Hb. (Lepidoptera : Noctuidae) and possible occurrence of biotypes. Ph.D. thesis submitted University of Agricultural Sciences, Bangalore 146 pp.

Mathur, K. C. (1967). Notes on *Banchopsis ruficornis* Cameron (Hym: Ichneumonidae). an internal larval parasite of *Heliothis armigera* (Hubner). in India. *Current Science*. 39: 167.

Mathur, K. C. (1970). Four new records of parasites of attacking *Heliothis armigera* Hb. In Himachal Pradesh. *Current Science*. 39:167.

Men, U. B. and Kandalkar, H. G. (2001). *Physalis minima* L. an alternate weed host of *Helicoverpa armigera* Hub. *Insect Environment*. 7(1): 27.

Mistry, A., Yadav, D. N., Patel, R. C. and Parmar, B. S. (1984). Field evaluation of nuclear polyhedrosis virus against *Heliothis armigera* Hubner (Lepidoptera: Noctuidae) in Gujarat. *Indian Journal of Plant Protection*. 12: 31-33.

Mule, N. M. and Chari, M. S. (1978). Chemo sterilization of male moths of tobacco capsule borer (*Heliothis armigera* Hubn.). *Tobacco Research*. 4: 16-23.

Dhandapani, N., Sundara Babu, P. C., Sathiah, N. and Jayaraj, S. (1993). Combined use of nuclear polyhedrosis viruses of *Spodoptera litura* (Fb.) and *Heliothis armigera* (Hb.) on groundnut (*Arachis hypogaea* L.). *Journal of Applied Entomology*. 116 (5) : 523-526.

Nandagopal, V., Pradip-Fulmali, Gedia, M. V. and Rajiv-Rathod. (2003). Development of an efficient pheromone trap for catching *Helicoverpa armigera* in a pigeon pea ecosystem in the field. International Pest Control. 45 (1) : 32-36.

Nye, I. W. B. (1980). *Heliothis* Ochsenheimer, 1816 (Insecta, Lepidoptera): proposal to designate gender and stem. Bulletin of Zoological Nomenclature 37: 186-189.

Nye, I. W. B. (1982). The nomenclature of *Heliothis* and associated Taxa (Lepidoptera : Noctuidae): Past and Present. *In* Proceeding of the International Workshop on *Heliothis* Management, 15-20 November,1981, International Crops Research Institute for the Semi-Arid Tropics, Patancheru, A.P., India.

Ochsenheiner, F. (1808). Die Schmetterlinge Von Europa 4. Leipzig. 226pp.

Odak, S. C. (1982). Biological control of gram pod borer *Heliothis armigera* (Hubn.) through encouragement and augmentation of parasite *C. chloridae*. Paper presented at *National Workshop on Methodology of Biological Control* held at Ahmedabad. 70-75 pp.

Paras Nath. (1993). .Studies on the groundnut insect pest and their management in east Utter Pradesh. Final technical report of ICAR AD-HOC Project, Dept. of Entomology and Agril. Zoology, Institute of Agril. Sciences, Banaras Hindu University, Varanasi. 114 pp.

Patel, C. C and Koshiya, D. J. (1997). Seasonal abundance of American bollworm, *Helicoverpa armigera* on different crop hosts at Junagadh (Gujarat). *Indian Journal of Entomology*. 59: 396-401.

Patel, C. C. and Koshiya, D. J. (1999). Population dynamics of gram pod borer, *Helicoverpa armigera* (Hubner) Hardwick on cotton, pigeonpea and chickpea. *Gujarat Agricultural University Research Journal*. 24 (2) : 62-67.

Patel, M. H. (1988). Studies on predation of *Heliothis armigera* Hubner and *Spodoptera exigua* (Hb.) by insectivorus birds with special reference to *Acridotheres tristis* L. M.Sc. Thesis submitted to Gujarat Agricultural University, Anand, 77 pp.

Patel, R. C., Patel, J. C. and Patel, J. K. (1971). New record of parasites of *Heliothis armigera* (Hub.) and *H. peltigera* (Schiff). from Gujrat. *Indian Journal of Entomology* 33: 223-224.

Patel, R. K. and Patitundu, A. (1981). Perennial weed *Aeschynomene indica* (L). an important alternate host plant of pod borer, *Heliothis armigera* (Hub.). *Food Farming and Agriculture*. 11: 54.

Pawar, C. S., Bhatnagar, V. S. and Jadav, D.R. (1986). Host parasite and natural enemies of *Heliothis* spp. in India a campodium progress report 13, International Crops Research Institute for Semi Arid Tropics, Patancheru, A.P.

Pawar, A. D., Divakar, B. J. and Singh, S. N. (1981). Field recovery of *Eucelatoria* sp. Near *E. armigera* (Coq). (Diptera: Tachnidae) from *Heliothis armigera* (Hubner). (Lepidoptera: Noctuidae). in Karnataka, India, *Entomon*. 6: 175-177.

Pawar, C. S. and Jadhav, D. R. (1983). Wasp-predators of *Heliothis* on pigeonpea. *International Pigeonpea Newsletter*. 2: 65-66.

Pawar, C. S., Bhatnagar, V. S. and Jadav, D. R. (1985). Some predatory insects and their parasites. *Science & Culture*. 51(3): 101-102.

Pearson, E.O. and Darling, R.C.M.(1958). The insect pests of cotton in tropical Africa, London, U.K., Empire. Cotton growers and common wealth Institute of Entomology, pp 355.

Rajiv-Rathod (2002). Antagonistic effect of *Beauveria bassiana* fungus against *Helicoverpa armigera*. *Current Agriculture*. 26(1/2): 123-125.

Rangadhamaiah, K., Thontadarya, T. S., Rao, K. J., Jai-Rao, K. (1985). Bionomics of the egg-larval parasite, *Chelonus blackburni* Cameron (Hymenoptera: Braconidae). on the 'tur' pod borer, *Heliothis armigera* (Hubner). (Lepidoptera: Noctuidae). *Mysore Journal of Agricultural Sciences*. 18(3): 203-207.

Rao, V. P. (1968). *Heliothis* spp. and their parasite in India. *Pest Articles and News Summaries* (A). 3: 367-75.

Rao, V. P. (1974). Biology and breeding technique for parasite and predators of *Ostrinia* spp. CIBC Final Technical Report, U.S. PL-480 Project, Bangalore, India 86 pp.

Rawat, S. R., Jakhmola, S. S and Sohuch, R. H. (1970). Assessment of losses of hybrid sorghum (S.H. 1). to earhead caterpillar and comparison of insecticidal control. *PANS*. 16: 367-379.

Reddy, A. R. and Divakar, B. J. (2001). Effect of Robin blue on the efficacy of nuclear polyhedrosis for the management of *Helicoverpa armigera* (Hub.) in groundnut crop.Perspectives-in-biotechnology-Proceedings-of-a-national-symposium, Warangal, India,-26-27-February, 1999. 41-43 pp.

Ridgway, R. L. and Lingren, P. D. (1972). Predaceous and parasitic arthropods as regulators of *Heliothis* population. *Southern Cooperative Series Bulletin*. 169: 48-56.

Rosaiah, B. and Reddy, A. S. (1999). Bioecology of *Helicoverpa armigera* (Hubner) in Nagarjuna Sagar Project area of Andhra Pradesh. *Journal of Insect Science*. 12 (1): 27-33.

Rossler, Y., Bar, D., Gerling, D. and Gonzalez, D. (1975). Integrated control in cotton: Early season pest and Natural enemies complexes. in Proc. Sixth Scientific Conference of the Israel Entomological Society, Tel-Aviv, Israel.

Sahayaraj, K. (1999). Field evaluation of the predator, *Rhynocoris marginatus* (Fab.) on two groundnut defoliators. *International Arachis Newsletter*. 19: 41-42.

Sahayaraj, K. (2000). Evaluation of biological control potential of *Rhynocoris marginatus* on four groundnut pests under laboratory conditions. *International Arachis Newsletter*. 20: 72-74.

Sahayaraj, K. and Paulraj, M.G. (2001). Behaviour of *Rhynocoris marginatus* (Fabricius) (Heteroptera: Reduviidae) to chemical cues from three lepidopteran pests. Journal of Biological Control. 15 (1) : 1-4.

Sahayaraj, K. and Sivakumar, K. (1995). Groundnut pest and pest stage preference of a reduviid predator *Rhynocoris kumarii* Ambrose and Livingstone (Heteroptera: Reduviidae). *Fresenius Environmental Bulletin*. 4: 263-269.

Seshu Reddy, K. V. (1973). Studies on the gram caterpillar, *Heliothis armigera*. (Lep: Noctuidae). with special to its host preference and estimation of loss in red gram. Ph.D. thesis. University of Agriculture Sciences, Bangalore, 132 pp.

Sharma, H. C., Pampapathy, G. Dwivedi, S. L. and Reddy, L. J. (2003).Mechanisms and diversity of resistance to insect pests in wild relatives of groundnut. *Journal of Economic Entomology*. 96 (6): 1886-1897.

Shepard, M. and Sterling, W. (1972). Incidence of parasitism of *Heliothis* spp. (Lep: Noctuidae) in some cotton field of Texas. *Annals of Entomological Society of America*. 65: 759-760.

Singh, H. and Singh, G. (1975). Biological studies on *Heliothis armigera* (Hubn.) in the Punjab. *Indian Journal of Entomology*. 37: 154-164.

Sithanantham, S., Rao, V. R., and Ghaffar, M. A.(1984). International review of crop loss caused by insects on chickpea pages 269-283. In Proceedings of the national seminar on crop losses due to insect pests. 7-9 January 1983, APAU, Hyderabad, India.

Sivagami, V., Mohansundaram, M, and Subba Rao, P. V. (1975). Record of parasites and predators of some south Indian crop pests. *Indian Journal of Entomology.* 37: 100-101.

Snow, J. W., Hamm, J. J. and Brazzel, J. R. (1966). *Geranium carolinianum* as an early host of *Heliothis zea and H. virescens* (Lepidoptera: Noctuidae). in the southeastern United State, with notes on associates parasites. *Annals of Entomological Society of America.,* 59: 506-509.

Sonalkar, V. U. and Deshmukh, S. D. (1999). Evaluation of certain adjuvants in increasing the efficacy of nuclear polyhedrosis virus against *Helicoverpa armigera* (Hub.). *Journal of Experimental Zoology.* 2(1): 89-91.

Srinivas, N. (1984). The influence of moon light and some weather factors on selected species of insect caught in light trap. M. Sc. Thesis, University of Agricultural Sciences, Bangalore, 224 pp.

Srivastava, P. M. (1959). Control of fruit borer *Heliothis armigera* (Hubn) of tomato. *Indian Journal of Horticulture.* 16: 187-188.

Subba Rao, B. R. (1955). A new species of *Chilonis* on *Heliothis armigera* (Fab.). *Indian Journal of Entomology* 17: 63-64.

Sukhoruchenko, G. I., Niyazoo, O. D., and Alekseev, Yu. A. (1977). The effect of modern pesticides on beneficial b and injurious insect fauna of cotton. *Entomologichekoe Obozrenie* 56 (1) : 3-14.

Sundram, M. K., Dhandhapani, N., Swamiapan, M., Sundara Babu, P. L. and Jayaraj, S. (1994). A study on the management of some pests of groundnut with biocontrol agents. *Journal of Biological Control.* 8(1): 1-4.

Urmila-Sharma, Yadav, D. N. and Sharma, U. (1997). A new record of arthropod predators in groundnut at Anand (India). Gujarat Agricultural University Research Journal. 22: 157.

Urs, N. V. R. R. and Govindu, H. C. (1971). *Metarrhizium anisopliae* (Metchinkoff) Sorokin. and its host range. *Mycopathologia Mycologia Applicata.* 44 (4): 317-320.

Van den Bosch, R. and Hagen, K. S. (1966). Predaceous and parasitic arthropods in California cotton fields. *California Agriculture Experiment station Bulletin.* 820: 32.

Venkataiah, M., Sekhar, P. R., Jhansilakshmi, V., Rajasri, M and Rao, N. V. (1994). Studies on the interaction of the host plant species with the growth and development of *Helicoverpa armigera* Hubner. *Journal of Research-APAU.* 22 (3-4): 102-105.

Verma, A. N., Verma, N. D. and Singh, R. (1971). Chemical control of *Prodenia litura* Fabr. (Lep: Noctuidae). on Cauliflower. *Indian Journal of Horticulture.* 28: 240-243.

Warren, W. (1914). *In* Die Gross-Schmetterlinge der Erde, ed. A. Seitz, Stutttgart. 511 pp.

Whitcomb, W. H. and Eason, R. (1967). Life history and predatory importance of the striped lynx spider (Araneida: Oxyopidae) *Arkansasa Academy of Science Proceedings*. 21: 54-58.

Wilcox, J. Howland, A. F. and Cambell, R. E. (1956). Investigations on the tomato fruitworm, its seasonal history and method of control. U.S. Department of Agriculture, Bulletin 1147. Washington, D.C., U.S.A. pp 47.

Table 5.2 contd.

Table 5.2. List of parasitoids of *Helicoverpa armigera* (Hubner)

Species	Order: Family	Host stage	Reference
Trichogramma chilonis	Hymenoptera Trichogrammatidae	Egg *et al.*, 1970;	ICRISAT 1974-85; Achan *et al.*, 1968; Manjunath, 1972
Trichogramma chilotraeae	"	"	Manjunath *et al.* 1970,
Trichogramma pretiosum	"	"	Graham, 1970
Trichogramma exiguum	"	"	King *et al*, 1981
Trichogramma evanescens	"	"	Habib, 1973
Trichogramma sp.	"	"	Bhatnagar *et al.* 1982, Rao 1974, Ridgway and Lingren, 1972; Danks *et al.*, 1979
Trichogramma achaeae	"	"	Manjunath, 1972; Manjunath *et al.* 1970
Trichogramma brasiliensis	Hymenoptera: Trichogrammatidae	Egg	CIBC, 1968
Trichogrammatoidea sp.	"	"	Bhatnagar *et al.*, 1982
Trichogrammatoidea armigera	"	"	Achan *et al.*, 1968; Manjunath, 1972
Trichogrammatoidea bactrae sp. *fumata*	"	"	Bhatnagar *et al.*, 1982
Eriborus argentiopilosus	Hymenoptera Ichneumonidae	Larvae	Achan *et al.*, 1968; ICRISAT, 1974-85; Rao, 1968; Bhatanagar *et al.*, 1982; Rao, 1974; Manjunath, 1972; Carl, 1978; Rossler *et al.*, 1975
Agrypon nox	"	"	Mathur, 1970
Attractodes sp.	"	"	Mathur, 1970
Banchopsis ruficornis	"	"	Hussain and Mathur,1924; Achan *et al.*, 1968; Manjunath , 1972
Barichneumon sp.	"	"	Mathur 1967, Seshu Reddy, 1973
Campoletis (=*Ecphoropsis*) sp.	"	"	Manjunath, 1972

Contd...

Table 5.2. Contd...

Campoletis chlorideae Uchida	"	"	ICRISAT 1974-85; Rao, 1974; Achan *et al.*, 1968; Manjunath, 1972
Campoletis maculipes (Tsch.)	Hymenoptera Ichneumonidae	Larvae	Patel and Patitundu, 1998; Manjunath *et al.*, 1976
Disophrys sp.	"	"	ICRISAT, 1974-85
Enicospilus sp.	"	"	Achan *et al.* 1968; Seshu Reddy, 1973; Manjunath, 1972
Enicospilus sp. nr. *shinkanus* Uchida	"	"	Pawar *et al.*, 1985
Pristomerus sp.	"	"	Patel *et al.*, 1971
Temelucha sp.	"	"	Bhatanagar *et al.*, 1982
Xanthopimpla stemmator Thunberg	"	Larvae Pupal	Bhatanagar *et al.*, 1982
Campoletis flavicincta (Ashm.)	"	Larvae	Ridgway and Lingren, 1972; Danks *et al*, 1979
Campoletis sonorensis (Cameron)	"	"	Lingren *et al.*, 1970 Ridgway and Lingren, 1972; Danks *et al*, 1979
Charops sp.	"	"	Habib, 1973
Telenomus sp.	Hymenoptera Scelionidae	Larvae	Manjunath, 1972
Goniozus sp.	Hymenoptera Bethylidae	"	Sivagami *et al.* 1975
Bracon lafroii	Hymenoptera Braconidae	"	Seshu Reddy, 1973
Apenteles sp. (*glomeratus* group)	"	"	Achan *et al.*, 1968
Apenteles sp. nr. *glomeratus* L.	"	"	Achan *et al.*, 1968
Apenteles marginiventris (Cresson)	"	"	Ridgway and Lingren, 1972; Danks *et al.*, 1979
Apenteles kazak Telenga	"	"	Carl, 1978; Habib, 1973
Apenteles maculitarsis (Cam)	"	"	Carl 1978; Habib 1973;

Contd...

Table 5.2. Contd...

Apenteles sp.	"	"	Achan *et al.*, 1968; Ibrahim, 1980; Habib, 1973
Apenteles ruficrus Haliday	"	"	Achan *et al.*, 1968
Bracon sp.	"	"	Achan *et al.*, 1968
Bracon brevicornis Wesmael	"	"	Achan *et al.*, 1968; Manjunath, 1972; Ibrahim 1980; Habib 1973; CIBC 1978
Bracon cushmani Muesbeck	"	"	CIBC, 1974
Bracon gelechiae Ashmead	"	"	Achan *et al.*, 1968, Koshiya and Patel, 1987
Bracon greeni Ashmead	"	"	Achan *et al.*, 1968
Bracon hebetor Say	"	"	CIBC, 1974
Barylypa humeralis Brauns.	"	"	Ibrahim, 1980
Cardiochiles nigriceps (Vierick)	Hymenoptera Braconidae	Larvae	Snow *et al.*, 1966; Lewis and Brazzel 1968 Ridgway and Lingren, 1972; Danks *et al.*, 1979
Cardiochiles nigricollis (Cam.)	"	"	Habib, 1973
Chelonus sp.	"	Egg-Larvae	ICRISAT, 1974-85
Chelonus haliopae Gupta	"	"	Achan *et al.*, 1968,
Chelonus narayani Subba Rao	" "	" Larvae	Subba Rao, 1955 Rangadhamaiah *et al.*, 1985
Chelonus texanus (Cress)	"	"	Ridgway and Lingren, 1972; Danks *et al*, 1979
Microchelonus curvimaculatus Cameron	"	Egg- Larvae	ICRISAT, 1974-85
Microplitis maculipennis (Szepligeti)	"	Larvae	Krishnamurthy and Usman, 1954
Microplitis sp. nr.*pallidipes* Szepligeti	"	"	Achan *et al.*, 1968
Microplitis croceipes (Cresson)	"	"	Snow *et al.*, 1966; Lewis and Brazzel 1968; Ridgway and Lingren, 1972; Danks *et al.*, 1979
Microplitis demolitor Wilkinson	"	"	Hafez, 1951

Contd...

Table 5.2. Contd...

Microplitis rufiventris Wesm.	"	"	Ibrahim, 1980
Odentepyris sp.	"	"	Achan *et al.*, 1968
Paraphylax sp.	"	"	CIBC, 1974
Rogas sp.	Hymenoptera Braconidae	Larvae	Achan *et al.*, 1968
Rogas sp. ?*testaceus* Spinola	Hymenoptera Braconidae	Larvae	Achan *et al.*, 1968
Brachymeria responsator Walker	Hymenoptera Chalcididae	Pupa	Achan *et al.*, 1968
Tetrastichus ayyari Rohwer	Hymenoptera Chalcididae	Pupa	Cherian and Subramaniam, 1940
Euplectrus spp.	Hymenoptera Eulophidae	Pupa	Mathur, 1970
Sarcophaga orientoides S.W.	Diptera Sarcophagidae	Egg	CIBC, 1974
Winthemia sp. nr? *diversoides* Bar.	Diptera Tachinidae	Larvae	Achan *et al.*, 1968
Carcelia (Stenometopia) illota Curran	"	Larvae / pupa	Achan *et al.*, 1968; Rao, 1974; Manjunath, 1972
Carcelia kolkiana Townsend	"	"	Achan *et al.*, 1968
Carcelia peraequalis Mensil	"	"	Achan *et al.* 1968
Carcelia raoi Mesnil	"	"	Rao, 1968
Carcelia sp.	"	"	Bhatanagar *et al*, 1982
Compsilura concinnata Meigen	"	Larvae/ pupa	CIBC, 1974
Drino (Prosturmia) imberbis Wiedemann	"	Larvae	Achan *et al.*, 1968, Seshu Reddy, 1973
Eucelatoria bryani Sabrosky	"	"	CIBC, 1968; Bottrell *et al.*, 1968
Exorista fallax Meigen	"	"	Achan *et al.*, 1968
Exorista japonica Townsend	"	"	Achan *et al.*, 1968
Exorista xanthaspis Wiedemann	"	"	Rao, 1968
Goniophthalmus halli Mesnil	Diptera Tachinidae	Larvae/ pupa	Rao, 1968; Achan *et al.* 1968; Seshu Reddy, 1973; Rao 1974; Habib 1973;

Contd...

Table 5.2. Contd...

			Manjunath, 1972
Isomera cinerascens Rondani	"	Larvae	Achan *et al.* 1968
Pales coeruleo-nigra Mensil	"	"	CIBC, 1974
Palexorista laxa (Curran)	"	"	Manjunath *et al.*, 1976; Patel and Patitundu, 1981; Rao, 1974 Habib, 1973; Achan *et al.*, 1968; Manjunath, 1972
Palexorista solennis Walker	"	"	Patel and Patitundu, 1981
Sisyropa apicata Bar.	"	"	Achan *et al.*, 1968
Spallanzania sp.	"	"	Achan *et al.*, 1968
Strobliomyia aegyptia Vill.	"	"	Achan *et al.*, 1968
Strumiopsis inferens Townsend	"	"	Patel and Patitundu, 1981
Suensonomyia n. sp.	"	"	Achan *et al.*, 1968
Thecocarcelia incedens Rondani	"	"	Achan *et al.*, 1968
Voria edentata Bar.	"	"	Achan *et al.*, 1968
Voria ruralis Fallen	"	"	Achan *et al.*, 1968
Archytas marmoratus (Townsend)	"	"	Shepard and Sterling, 1972 Ridgway and Lingren, 1972; Danks *et al*, 1979
Eucarcelia illota	"	"	Seshu Reddy, 1973
Lespesia aletiae (Ril.)	"	"	Ridgway and Lingren, 1972;Danks *et al*, 1979
Microplitis maculipennis	"	"	Krishnamurthy and Usman, 1954
Senometopia illota (Curran)	"	"	Koshiya and Patel, 1987
Winthemia rufopicta (Bigot)	"	"	Ridgway and Lingren, 1972; Danks *et al.*,1979
Mepachymerus ensifer Thompson	Diptera Chloropidae	"	Verma *et al.*, 1971
Hexamermis sp.	Nematoda Mermithidae	Pupal	Achan *et al.*, 1968; Seshu Reddy, 1973
Ovomermis albicans (Seibold)	Nematoda Mermithidae	Pupal	Bhatanagar and Devis, 1978

Table 5.3. List of predators of *Helicoverpa armigera* (Hubner)

Species	Order: Family	Host stage	Reference
Euborellia annulipes (Lucas)	Dermaptera Carcinophoridae	Eggs	Bhatanagar *et al.*, 1983
Euborellia stali Dohrn	"	"	Bhatanagar *et al.*, 1983
Nala lividipes (Dufour)	Dermaptera Labiduridae	"	Bhatanagar *et al.*, 1983
Humbertiella sp.	Dictyoptera Mantidae	"	Bhatanagar *et al.*, 1983
Orius insidiosus (Say)	Hemiptera Anthocoridae	Eggs	Lincoln *et al.*, 1967
Orius maxidentex Ghauri	"	"	Bhatanagar *et al.*, 1983
Jalysus spinosus (Say)	Hemiptera Berytidae	"	Elsey, 1972
Geocoris punctipes (Say)	Hemiptera Lygaeidae	Eggs, first instar larvae	Lopez *et al.*, 1976; Lincoln *et al.*, 1967; Ridgway and Lingren, 1972
Paromius gracilis (Rambur)	"	"	Bhatanagar *et al.*, 1983
Nabis sp.	Hemiptera Nabidae	"	Lincoln *et al.*, 1967; Donohoe and Pitre, 1977
Nabis ferus (L.)	"	"	Ridgway and Lingren, 1972
Nabis alternatus Parshley	"	"	Ridgway and Lingren, 1972
Tropiconabis capsiformis (Germar)	"		Bhatanagar *et al.*, 1983
Podisus maculiventris (Say)	Hemiptera Pentatomidae	Third instar larvae	Lopez *et al.*, 1976
Cantheconidea furcellata (Wolff.)	"	larvae	Bhatanagar *et al.*, 1983
Catamiarus brevipennis (Serv.)	Hemiptera Reduviidae	"	Bhatanagar *et al.*, 1983
Coranus spinicutis Reut.	"	"	CIBC, 1974
Ectrychotes dispar Reut.	"	"	Bhatanagar *et al.*, 1983
Onchocephalus annulipes Stal.	"	"	CIBC, 1974
Rhinocoris fuscipes Stal.	"	"	CIBC,1974
Rhinocoris marginatus (F.)	"	"	Bhatanagar *et al.*, 1983
Sycanus indagator Stal.	"	"	CIBC, 1974

Contd...

Table 5.3. Contd...

Delta companiforme esuriens F.	Hymenoptera Eumenidae	"	Pawar and Jadhav, 1983
Delta conoideum (Gmelin)	"	"	Pawar and Jadhav, 1983
Delta pyriforme (F.)	"	"	Pawar and Jadhav, 1983
Cataglypsis bicolor F. Formicidae	Hymenoptera	"	Khan and Sharma, 1972
Camponotus sericeus F. Formicidae	Hymenoptera	larvae	Manjunath *et al.*, 1970
Sphex argentatus F.	Hymenoptera Sphecidae	"	Bhatanagar *et al*, 1983
Polistes olivaceus Degeer	Hymenoptera Vespidae	"	"
Repalidia marginata Lepeltier	"	"	"
Vespa orientalis F.	"	"	"
Vespa tropica haemotodes Bequaert	"	"	"
Chrysopa carnea Stephens	Neuroptera Chrysopidae	Eggs, first instar larvae	Manjunath *et al*,1970, Lopez *et al*, 1976 Ridgway and Lingren 1972
Chrysopa oculata Say	"	"	Ridgway and Lingren 1972
Chrysopa rufilobris Burm.	"	"	"
Chrysopa sp.	"		Bhatanagar *et al*, 1983
Calosoma sp.	Coleoptera Carabidae	Large larvae, pupae	Lincoln *et al.* 1967, Van den Bosch and Hagen 1966
Calleida decora (F.)	"	Eggs, Small larvae	Shepard and Sterling, 1972
Scymnus sp.	Coleoptera Coccinellidae	Eggs	Lincoln *et al.*, 1967; Ridgway and Lingren 1972
Hippodamia convergens (Guerin – Meneville)	"	Eggs, first instar larvae	Lincoln *et al.* 1967; Ridgway and Lingren, 1972
Coleomegilla maculata (De Geer)	"	Eggs, first and second instar larvae	Lincoln *et al.* 1967; Ridgway and Lingren, 1972
Menochilus sexmaculatus F.	"		Bhatanagar *et al.*, 1983
Oxyopes salticus Hentz	Araneida Oxyopidae	second instar larvae	Whitcomb and Eason, 1967

Contd...

Table 5.3. Contd...

Oxyptila reenae (Basu)	Araneida Thomosidae		Bhatanagar *et al.*, 1983
Thomisus sp.	"		"
Clubiona sp.	Araneida Clubionidae		"
Neoseona theis (Walck.)	Araneida Aranidae		"
Leucauge tessellata (Thorb.)	"		"

Table 5.4. List of entomopathogens of *Helicoverpa armigera* (Hubner)

Pathogen	Order : Family	Host stage	Reference
Bacillus thuringiensis	Eubacteriales Bacillaceae	Larvae	Majumdar *et al.*, 1956
Serratia marcescens	Eubacteriales Enterobacteriaceae	"	Seshu Reddy, 1973
Beauveria bassiana	Moniliales Moniliaceae	"	Jayaramaiah, 1981
B. brogniartii	Moniliales Moniliaceae	"	Jayaramaiah, 1981
Metarhizium anisopliae	Deuteromycotina Hyphomycetes	"	Urs and Govindu, 1971; Gopalakrishnan and Narayanan, 1988
Nomuraea rileyi	Moniliales Moniliaceae	"	Gopalkrishnan and Narayanan, 1988
Vairimorpha sp.	Protozoa Microsporidia Burenellidae	"	Chinnamade Gowda, 1990
Nuclear Polyhedrosis Virus	Virales Baculoviridae	"	Anon., 1987, 1989

Table 5.5. List of bird predators of *Helicoverpa armigera* (Hubner)

Bird species	Order: Family	Stage of the pest attacked	Reference
Acridotheres tristis Linnaeus	Passeiformes Sturnidae	Larvae-adults	Patel, 1988
A. ginginianus Lathum	Passeiformes Sturnidae	Larvae-adults	Patel, 1988
Bubulcus ibis Boddart	Ciconiiformes Ardeidae	Larvae-adults	Hanumanna *et al.*, 1980

Contd...

Table 5.5. Contd...

Corvus macrorhynchos Sykes	Passeriformes Ploceidae	Larvae	Patel, 1988
C. splendens Vieilot	Passeriformes Ploceidae	Larvae-adults	"
Dicrurus adsimilis Hodgson	Passeriformes Dicruridae	Larvae	"
Passer domesticus Jardin and Selby	Passeriformes Ploceidae	Larvae-adults	"
Pseudibis papilosa (Temminck)	Ciconiiformes Threshiornithidae	"	Hanumanna *et al.*, 1980
Sturnus roseus Linnaeus	Passeiformes Sturnidae	"	Patel, 1988

□□□

Hairy Caterpillars

T. MURALIKRISHNA AND T.V.PRASAD

6.1 Introduction

Among various hairy caterpillars which are polyphagous pests of crops, red hairy caterpillars, *viz., Amsacta albistriga* Walk and *Amsacta moorei* Butler and Bihar hairy caterpillar, *Spilosoma obliqua* (Walk.) are important insect pests of groundnut. *A. albistriga* occurs in South India (Nagrajan *et al.*, 1957; Nagrajan and Ramachandran, 1958; Venkataraman *et al.*, 1970; Saroja *et al.*, 1971; Paramsivam *et al.*, 1973; Logiswaran *et al.*, 1982; and Gunathilagaraj and Babu, 1987), while, *S. obliqua* occurs in north Indian whereas, *A. moorei* occurs both in south and north India (Gupta *et al.*, 1966; Patel and Patel,1965; Bindra and Kittur,1961; Srivastava and Goel, 1962; Trehan and Talgeri, 1947; Singh and Pandey 1995; Sinha *et al.*, 1975; Nath and Singh, 1996; Ramesh Arora *et al.*, 1996). Though these pests are sporadic in nature, in certain years they become serious and cause heavy losses (Mukundan, 1964; Siva Rao *et al.*, 1977; Gunathilagaraj and Babu, 1987; Nath and Singh, 1996).

6.2 Distribution

Red hairy caterpillars, *A. albistriga* and *A. moorei* were observed to be intermating at Pollachi tract in Tamil Nadu (Mukundan,1964; Ramaswamy and Kuppaswamy,1973). *A. albistriga* occurs in Coimbtore, Madurai and Pollachi districts (Kuppaswamy *et al.*, 1965) and adjoining areas of Tamil Nadu (Nagrajan and Ramachandran, 1958). It also occurred in central and northern India (Amin, 1988) and recorded along with *A. albistriga* in Tamil Nadu (Mukundan, 1964; Ramaswamy and Kuppuswamy, 1973). In certain years epidemics of this pest also occurs in Srikakulum, Visakhapatanam, Cuddapah, Kurnool, Ananthpur and Chitoor districts of Andhra Pradesh (Siva Rao *et al.*, 1977). An outbreak was recorded in Chittoor district in 1975 (Siva Rao *et al.*, 1977; Qayyum *et al.*, 2000). The Bihar hairy caterpillar occurs in Punjab, Uttar Pradesh, Delhi, Bihar, Bengal, Assam, Gujarat, Madhya Pradesh and Maharastra (Lefroy, 1906; Shaw *et al.*, 1988; Nath and Singh , 1996; Ramesh Arora *et al.*, 1996 and Anon.,1993).

6.3 Economic importance

Whenever there is severe outbreak of red hairy caterpillars, a total loss over large area is common in south India. In Andhra Pradesh a massive programme totally with non-synthetic pesticides approaches was launched by voluntary organizations in collaboration with Government Agencies and the Directorate of Oilseeds Research (ICAR) by setting up of bonfires in an area of over 2000 ha during 1989-90 and light traps covering over 5000 ha in the subsequent years from 1991-94 (Lakshminarayan *et al.*, 1995). Studies revealed that there were 1-6 peak emergence of moths in different locations and the duration of each peak was 2 to 4 days. The pest damage was contrastingly low in light trap operated villages. Out of 1696 ha treated , only 3 per cent area was damaged as against 56 per cent in untreated villages. Similarly 6% area was damaged in the light trap operated villages compared to 40% in the untreated during 1993 (Qayyum *et al.*, 2000).

6.4. Nature of injury and extent of loss

Young larvae of *A. albistriga* and *A. moorei* feed gregariously by scrapping **(Plate 1)** on the under surface of the leaves. The grown up larvae, which are voracious feeders, defoliate the crop presenting a cattle grazed field. Full-grown larvae devour not only the entire foliage but flowers and growing points. When large number of larvae feed on the crop, the field with only bare stems of crop leading to total loss of yield (Nagrajan *et al.*, 1957). Young larvae of *S. obliqua* feed by scrapping leaves, which become papery thin and dry, while full grown larvae **(Plate 2)** devour the foliage completely. During

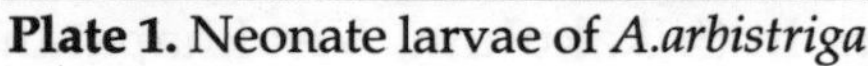

Plate 1. Neonate larvae of *A.arbistriga*

Plate 2. Full grown larva of *A.albistriga*

the outbreaks, of *A. albistriga*, a total loss over a large area is very common (Nagrajan *et al.*, 1957). There was an estimation of saving of groundnut to the extent of 75 per cent by controlling the pest through large scale campaign over large area (Mukundan, 1964). Avoidable loss of 26 per cent in the controlled plots were achieved (Ramaswamy *et al.*, 1968). There was no clear cut relationship between larval numbers and yield loss has been worked out and the most of the pest control experiments indicated the reduction in larval number over 24 hour after treatment without reporting yield loss (Narayanan and Ranga Rao, 1959; Kuppaswamy *et al.*, 1965; Ramaswamy *et al.*, 1968; Venkataraman *et al.*, 1970; Saroja *et al.*, 1971; Paramsivam *et al.*, 1973). Pachori *et al.* (1980) reported economic threshold level for 'Jyoti 2' as 2 larvae/m row, which caused 27 per cent yield loss and 43 per cent foliage loss. A density of 15 or more larvae of *Spilosoma obliqua* Wlk. (*Diacrisia obliqua*) per 1 m row length caused significantly more damage than 5 (or fewer) larvae/m. With an increase in every 1 larva/m, there was an increase of 2.80% in the leaf area consumed and a decrease of 0.17 g in the weight of pods and 0.12 g in the weight of kernels per plant. The economic injury level was 2 larvae/m, causing an economic loss of about 43% in foliage and 27% in yield (Pachori *et al.*, 1980).

6.5. Biology

The field biology of *A. albistriga* has been reported by many workers (Nagrajan and Ramachandran, 1958; Mukundan, 1964; Ramaswamy and Kuppuswamy,1973). Similarly the biology of *A. moorei* has been worked out by many workers (Trehan and Talgeri, 1947; Bindra and Kittur, 1961; Patel and Patel,1965; Gupta *et al.*, 1966 and Mathur,1966). For *S. obliqua,* several workers have reported the biology (Trehan and Bagal, 1957; Mathur,1962, Singh and Gangrade,1974; Pachori *et al.*, 1980). Both *A. albistriga* and *A. moorei* have one generation in a year. However, a second generation was reported from Pollachi region of Tamil Nadu in the month of August (Ramaswamy *et al.*, 1968) and in *A. moorei* in Rajasthan (Yadav *et al.*, 1966) and 2 or 3 generations in Madhya Pradesh (Bindra and Kittur, 1961). While *S. obliqua* has 8 generations in a year (Lefroy, 1906). All these three species of hairy caterpillars were considered to be major pests of groundnut for a decade ago, however, now, the status has changed to an extent that they are unpredictably important in isolated pockets such as Tumkur of Karnataka, Chittor and Anantapur districts of Andhra Pradesh and Pollachi area of Coimbatore district in Tamil Nadu. The biology of *A. moorei* was studied on different groundnut varieties by Singh and Pandey, (1995) in the laboratory at 26-31 degrees C. Kaushal was the most preferred variety for growth and development, while NSC 343 was the least preferred. The incubation, larval and pupal periods varied from 4.58 (PR 312) to 5 (IOGS 1) days, 19.85 (Kaushal) to 26.27 (NSC 343) days, and 11.64 (Kaushal) to 16.75 (NSC 343) days, resp.

6.5.1. Adults

Adults of *A. albistriga* (**Plate 3**), *A. moorei* and *S. obliqua* (**Plate 4**) emerge from soil at the on set of south west monsoon (usually June). The former two species are brownish white moths with a 40-50 mm wing open. The fore

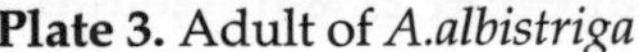

Plate 3. Adult of *A.albistriga*

Plate 4. Adult of *S. obliqua*

wings are completely white in *A. albistriga* and *A. moorei* while they are brown in *S. obliqua* with 40-50 mm wing open and a red abdomen. The forewing have black spots. *A. albistriga* can adopt and synchronizes with the weather to its best advantage. The mass emergence of moths is well synchronized with rains, occurring in June to July in Arcot and Salem, in May in Pollachi, and in August to October in Madurai and Ramnathpuram districts of Tamil Nadu (Nagrajan and Ramchandran, 1958). The adult mate immediately after emergence and oviposit in the same night in the case of *A. albistriga* and 2-9 days after mating in *S. obliqua.*

6.5.2. Egg

Cream coloured eggs are laid in clusters (**Plate 5**) and each cluster contains over 40 eggs and occasionally up to 830 in *A. albistriga*. The eggs are laid on the available host plants, hedges, weeds or even clods of earth. The incubation period is about 3 – 4 days. In the case of *A. moorei*, the eggs are laid in clusters of 30-400 and a individual female can lay eggs up to 1500 eggs (Trehan and

Plate 5. Egg mass of *A.albistriga*

Tulgeri, 1947). The incubation period is similar to *A. albistriga, S. obliqua* female lays eggs from 533 to 1287 with an average of 875. The eggs are scattered all over the plant (Mathur, 1962) or in large groups on the lower surface of the leaves in a single layer. They are greenish and spherical (Patel, 1944). The egg period lasts for 3 to 9 days.

6.5.3. Larvae

The neonate larvae of *A. albistriga,* which are initially light brown but turn reddish yellow as they grow up. The larvae come out of eggs and feed gregariously by scrapping the green portion of leaves giving them papery appearance. The older larvae are ash brown coloured, densely hairy and feed individually, but when fully grown after 20-31 days, they attain reddish colour. Larvae has 6 instars and mean larval period on groundnut is 33.45 days (Sivasankara Reddy *et al.,* 2003).

The larva of *A. moorei* completes 6 instars in 23-40 days. The newly hatched larvae feeds gregariously and full grown larva is about 5 cm long, black at both the ends and dirty yellow in between. The larval period ranges from 20-40 days depending on hosts (Patel, 1944), but usually it takes 35 – 45 days after 5-6 moults. The larval period of *A. moorei* was worked out to be 18.4 days (Ahlawat *et al.,* 1993). The hairiness of all the species of hairy caterpillar makes them conspicuous, especially the larger ones. Percentage survival and weight gain of larvae of *Spilosoma obliqua* on 9 varieties groundnut indicated that after 8 days of feeding, larval survival was highest on M-145, followed by Pol-2 and C-501. By 24 days after feeding began, it was still highest on M-145, followed by Dwarf Mutant and JH-62. Weight gain by larvae after 4 days of feeding was significantly higher on M-145, C-501, Dwarf Mutant and JH-62 than on any of the other varieties. After 20 and 24 days of feeding, maximum larval weight was recorded on M-145, followed by Dwarf Mutant and C-501. No differences in weight between male and female larvae were recorded (Tiwari *et al.,* 1989). Out of 9 groundnut varieties tested in the laboratory for resistance to *Spilosoma obliqua,* M-13 was the most resistant and M-145 and Dwarf Mutant the most susceptible. On the basis of growth and development parameters, the varieties could be grouped in 6 clusters.

6.5.4. Pupa

Pupation takes place in soil at a depth of 10-20 cm, under trees, hedges, shady corners or bunds in the case of *A. albistriga* and at 2.5 to 16 cm soil depth in the case of *A. moorei.* The pupation of *S. obliqua* takes place on ground under dried foliage and debris in a silken cocoon formed by hairs of larvae. The pupa of *A. albistriga* remains in soil in diapause till the receipt of next monsoon. However, about 23 per cent emergence of adults was recorded during September as second generation (Sivasankara Reddy *et al.,*2004). The pupal period of *S. obliqua* was reported to be 16-22 days depending on the host sp. It varies from 9 days (on Jute) (Patel, 1944) to 11.5 days (on sweet potato) (Treha and Bagal, 1957) **(Plate 6).** The biology of *S. obliqua* at Uttar Pradesh has been studied by Paras Nath (1993) is given in Table 6.1. Laboratory studies with *Spilosoma obliqua* [*Spilarctia obliqua*] on groundnuts showed the average incubation, larval, prepupal and pupal periods as 5.60, 24.72, 1.76 and 11.46 days, resp. The average pre-oviposition, oviposition and post-oviposition periods were 2.40, 2.88 and 4.00 days, resp. The life cycle was completed in 53.06 days (Nath and Singh, 1996). The biology of the arctiid *Diacrisia obliqua* [*Spilosoma obliqua*] was studied by Shaw et al., (1988) on *Dolichos lablab* [*Lablab purpureus*], *Sesamum indicum* [sesame], *Phaseolus mungo* [*Vigna mungo*], *Cynodon dactylon*, neem (*Azadirachta indica*), sweet potatoes, groundnuts, okra, maize and rice in the laboratory. Sweet potato and *P. mungo* were the most favourable food plants, resulting in the shortest larval period, and the greatest larval survival and size. Groundnut was the poorest food plant with only 33.3% larval survival. First- to 3rd-instar larvae did not survive on leaves of okra, maize, rice, *C. dactylon* or neem.

Plate 6. Pupae of *S.obliqua*

The growth, development and survival of larvae of *Spilosoma obliqua* were studied on groundnut, sesame, sunflower, lucerne and lablab (*Dolichos lablab* [*Lablab purpureus*]) in the laboratory at 26°C. Larvae had 6, 7 or 8 instars when reared on groundnut, 6 or 7 instars on lucerne, sunflower and *L. purpureus* and 6 instars on sesame. Fecundity was greatest on lucerne (745.22 eggs/female), followed by sunflower (704.58 eggs/female), *L. purpureus* (397.15 eggs/female), sesame (333.30 eggs/female) and groundnut (308.13 eggs/female) (Dahale *et al.*, 1988).

6.6. Alternate hosts

The hairy caterpillars, in general polyphagous in nature causing extensive damage particularly to groundnut. The other plants recorded as host are

Life stages of RHC, A. albistriga

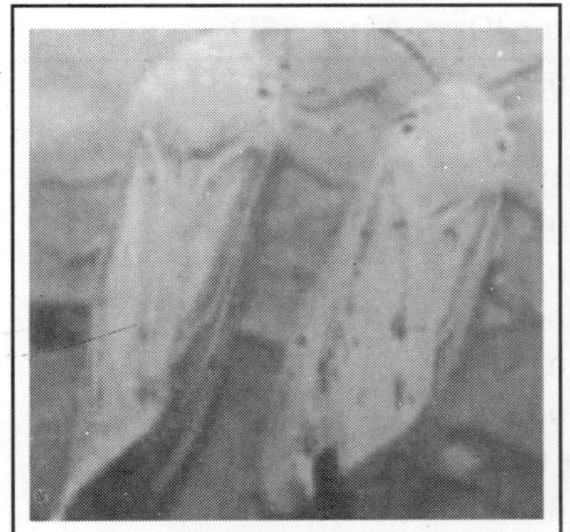
a. Moths after emergence

b. Moth laying eggs

c. Freshly laid eggs

d. Eggs before batching

e. Neomate larvate

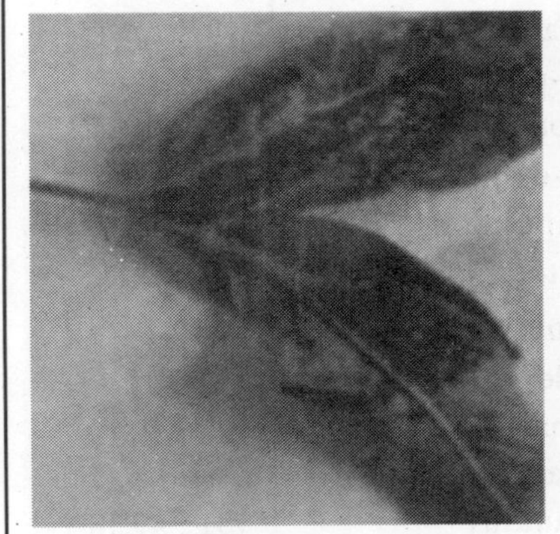
f. Second Instar larvae

Table 6.1a. Biology of Red hairy caterpillar, *Amsacta albistriga* (Sivasankara Reddy *et al*,2003)

Stages	Duration (Days) Mean ± SD
Eggs	2.50± 0.07
Larvae	
1st Instar	2.5± 0.05
2nd Instar	3.25± 0.12
3rd Instar	6.00± 0.41
4th Instar	6.90± 0.72
5th Instar	7.30± 0.35
6th Instar	7.50± 0.15
Total larval period	33.45± 0.39
Pre-pupa	3.96± 0.09
Adult (Longevity)	
Male	2.42± 0.03
Female	5.10± 0.21

Table 6.1 b. Biology of Bihar hairy caterpillar, *S. obliqua*

Stages	Duration (Days) Mean ± SEM	Range (day)
Eggs	5.60± 0.33	3-9
Larvae		
1st Instar	3.72± 0.18	3-5
2nd Instar	3.84± 0.15	3-5
3rd Instar	3.96± 0.20	3-6
4th Instar	4.00± 0.21	3-6
5th Instar	4.48± 0.10	4-5
6th Instar	4.80± 0.10	4-6
Total larval period	24.72± 0.45	19-29
Pre-pupa	1.76± 0.13	1-3
Pupa	11.44± 0.72	10-13
Pre-oviposition	2.40± 0.13	1-3
Oviposition	2.88± 0.18	2-4
Post-oviposition	4.00± 0.15	3-5
Adult		
Male	10.96± 0.45	8-16
Female	9.24± 0.26	6-11
Total life	53.06± 0.75	46-62
Fecundity	951± 49.38	632-1145

finger millet, cowpea, castor, sorghum, cotton (Nagrajan *et al.*, 1957); sorghum, pearl millet, maize, soybean, horse gram, green gram, black gram, cluster bean, pigeon pea, sesame, jute, sunhemp (Rao, 1973) and several other weeds such as *Commelina benghalensis, Rhynchosia minima* and *Portulaca* sp. (Nagrajan and Ramachandran, 1958).

The host plants of *A. moorei* are groundnut, sunflower, castor, maize, sorghum, pearl millet, lab lab, green gram, black gram, sunnhemp, soybean, cluster bean, rice, jute, sweet potato, lucern, kodo millet (Gupta *et al.*, 1966); tobacco, pigeon pea, cotton (Patel and Patel, 1965); cowpea (Bindra and Kittur, 1961) and several fruit trees (Srivatsava and Gohel, 1962).

Food-plants of larvae of *A. moorei* (Btlr.) include groundnut, moong bean (*Vigna radiata*), moth bean (*V. aconitifolia* (*Phaseolus aconitifolius*)), cowpea (*V. unguiculata*), guar (*Cyamopsis tetragonoloba*), bajra (*Pennisetum americanum* (typhoides)), sorghum and various grasses. The wild food-plants include *Gisekia pharnacioides, Tribulus terrestris, Achyranthes aspera, Aerva persica, Cenchrus prieurii, Brachiaria ramosa, Heliotropium subulatum, Boerhavia diffusa, Ipomoea pestigridis* and *Cyperus rotundus* (Verma, 1983). This species also shows preference to weeds such as *Andropovan lalopensis, Euphorbia nariifolia* (Trehan and Tulgari, 1947), *Ziziphus mammularia* (Bindra and Kittur, 1961) *Eclipta alba* and *Euphorbia hirta* (Udayraj Singh *et al.*, 1989).

Among the cultivated plants *A. moorei* feeds on soybean, cowpea , black gram, cotton, kidney bean and water melon. Among the weeds, carpet weed (*Trianthima portulachstum)* and puncture vine (*Tribulus terrestries*). Ahlawat *et al.*, (1993) reported the alternate hosts of *A. moorei* such as cotton, green gram, pearl millet, cow pea, sunnhemp and snap melon.

The host plants of *S. obliqua* are black gram, green gram, soybean, cowpea, sweet potato, sesame, sunflower, and lablab (Dahale *et al.*, 1988).

6.7. Factors influencing population

The species *A. albistriga* can adopt and synchronize moth emergence with rain in June - July in Arcot and Salem, in May in Polachi, in August - October in Madurai and Ramnathpuram area of Tamil Nadu (Nagraj and Ramchandran, 1958). In the case of *S. obliqua* high ambient temperature (32-34°C) favoured population increase. It completes 8 generations in a year with 5 weeks per generation and up to 75 days in cold weather as an exceptional (Lefroy, 1906).

6.8. Natural enemies

Several natural enemies such as parasites, predators and pathogen of the hairy caterpillars have been reported (Table 6.2). *Exortia civiloides* Baranoff, *Sturmia* nr.*inconspicuella* Baranoff, (Krishnamurthy and Usman, 1954; Usman and Puttarudriah, 1955), and braconid *Apanteles creatonoti* Viereck (Bhatnagar, 1948; Krishnamurthy and Usman, 1954) parasitise larvae. *Trichogramma evanescens minutum* Riley (Krishnamurthy and Usman, 1954) parasitises the eggs. However, none of the parasites could reduce the number of hairy caterpillars below ETL. Recently *Telenomus manolus* Nixon caused 94% egg-parasitism and was recorded from 5 localities in Tamil Nadu. However, the emergence of adult parasites ranged from 4 to 59% apparently due to multi-parasitism. This parasite species appears to have some potential for the biological control of hairy caterpillars (Sundaramurthy *et al.*, 1976). The stink bug *Cantheconidea (=Canthecona) furcellata* (Wolff.) is predacious on larvae of red hairy caterpillar (Cherian & Brahmachari, 1941; David and Basheer, 1961).

Plate 7. Parasites emergence

In case of *A. moorei*, several parasitoids have been reported, however, their role in controlling the pest has not been determined. *Apanteles creatonoti* Viereck and *A. bosei* parasitised the caterpillar (Bhatanagar, 1948). Bindra and Kittur (1961) recorded 3 tachinid species *Tachina (=Exorista) fallax* Meigen, *Sturmia inconspicuella* Baranoff and *Carcelia evolens* Townsend parasitising the older larvae. They also reported larval parasitism by a nematode *Mermes indica* V. Linstow and infection by nuclear polyhydrosis virus. Mathur and Mathur (1968) reported infection by *Aspergillus* in older larvae.

Severe outbreaks of the arctiid *Amsacta albistriga* are reported from Tamil Nadu, India, where the pest caused great damage to rain-fed groundnut. The scelionid *Telenomus manolus* attacked eggs of the pest, the braconids *Apanteles colemani* and A. creatonoti attacked young larvae, the tachinids *Sturmia incospicuella* and *Exorista civiloides* attacked mature larvae and the phorid *Megaselia scalaris* attacked pupae. High parasitism by larval parasitoids, especially the tachinids, was found (Gunathilagaraj and Babu, 1987).

Battu and Dhaliwal (1976) reported *a* tachinid, *Carcelia* sp., as parasite. *Rhynocoris fuscipes* is predatory on young larvae. In case of *S. obliqua,* three Hymenopteran and four Dipteran parasites have been recorded which caused up to 30% parasitism, of which 24% was accounted for by Hymenoptera and the rest by Diptera. *Apanteles obliqua* and another species of *Apanteles*

(= *glomeratus* group) were the major parasite. Sahayraj (1999) reported a predator, *Rhinocoris marginatus,* feeding on *A. albistriga* as a least preferred host for the predator. Baskaran *et al.* (1993) reported an entomopathogen nematode on *A. albistriga.* The fourth instar larvae of *A. albistriga* was evaluated using *Steinernema carpocapse* (strain DD 136), *Heteroheboitis bacteriophora* (strain Burliar) and *Heteroheboitis spp.* (strain Chekkanurnai and Melur). *S. carpocapse* was the most effective, followed by *Heteroheboitis spp.* strain Chekkanurnai. The 3 *Heteroheboitis spp.* did not develop in the larvae. The nuclear polyhedrosis virus Baculovirus amsacta was also found to be effective against *Amsacta albistriga* and did not have any adverse effects on the larval parasitoids (Gunathilagaraj and Babu, 1987). Some work has been done on the use of Baculoviruses against *A. albistriga of the* different viruses such as *H. armigera virus,* the NPV of teak defoliator *Hybiaea puera.* The NPV of *A. albistriga* did not show any infectivity on *H. armigera* (Rabindra *et al.,* 1997).

The effect of nine host plants *viz.,* pearl millet, sorghum, castor, groundnut, calotropis, soybean, cotton, sesame and cowpea on the virulence and in vivo production of nuclear polyhedrosis virus (NPV) on groundnut caterpillar (*A. albistriga*) reveled that larvae infected through calotropis leaves had the highest mortality (91.1%), followed by that infected through castor and sesame leaves. Mortality was lowest in pearl millet leaves (57.8%). Larvae on calotropis leaves produced the highest number of NPV. LT50 was lowest when larvae were infected through NPV-treated calotropis leaves (Baskaran *et al.,* 2001).

6.9. Management

The hairy caterpillars were considered to be key pest in about three decades back in the parts of Tamil Nadu, Andhra Pradesh and north Indian states like Gujarat. Now, they are sporadic in nature unpredictably occurring in isolated pockets. Since these larvae having hairs all through the body a different strategy has to be followed in managing these insects. The early instars need to be tackled with dust formulation and older larvae with EC sprays.

6.9.1. Cultural and Mechanical

The use of barrier crop such as barnyard millet (*Echiniclova frumentacea* Link) act as a strong barrier crop against the hairy caterpillars when planted surrounding the groundnut fields against *A. albistriga.* To manage *A. moorei,* Trehan and Talgeri (1947) recommended clean cultivation, collection and destruction of egg masses and hand picking of larvae were found effective in reducing the larval population and associated damage to maize and pearl millet. Nagrajan *et al.* (1957) have advocated destruction of moths on their emergence from soil and later on collection of egg masses and young larvae and destroying them. To avoid the dispersal of larvae, digging of drenches

Table 6.2. List of natural enemies

Species	Parasite	Order/Family	Stage	Reference
Amsacta albistriga	*Telenomus manolus* Nixon	Hymenoptera: Scelionidae	egg	Gunathilagaraj and Babu, 1987
	Apanteles colemani Viereck	Hymenoptera: Braconidae	young larvae	”
	A. creatonoti Viereck	”	”	”
	Palexorista solennis (Wlk.)	Diptera: Tachinidae	”	”
	Exorista xanthaspis Wiedemann	”	”	”
	Megaselia scalaris (Loew)	”	”	”
	Eocanthecona furcellata (Wolff)	Heteroptera : Pentatomidae	larvae	David and Basheer, 1961
	Nuclear polyhedrosis virus	Virales Baculoviridae	”	Jayaraj *et al.*, 1976
	Steinernema carpocapse	Nematode	”	Baskaran *et al.*, 1993
	Heterorhabditis bacteriophora	”	”	Baskaran, 1993
	Heterorhabditis spp.	”	”	Baskaran, 1993
	Baculovirus	virus	”	Rabindra *et al.*, 1997
Amsacta moorei	*Trichogramma brasiliensis* (Ashmead)	Hymenoptera Trichogrammatidae	egg	Tuhan *et al.*, 1987
	Exorista xanthaspis (Wiedemann)	Diptera:Tachinidae	larvae	Verma, 1983
	Apanteles obliquae (Wlkn.)	Hymenoptera Braconidae	”	Siddiqi, 1981
	Meteorus sp.	”	”	Siddiqi, 1981
	Exorista fallax Mg.	Diptera:Tachinidae	larvae/ pupae	Patel and Patel, 1970
	Nuclear polyhedrosis virus	Virales Baculoviridae	larva	Rao and Amonkar, 1980
Spilosoma obliqua	*Apanteles* sp.	Hymenoptera Braconidae	larva	Hameed and Yazdani, 1986
	Carcelia sp.	Diptera:Tachinidae	larva	”
	Blepharella lateralis Macquart	Diptera:Tachinidae	larva	”
	Rhynocoris fuscipes Fabr.	Heteroptera Reduviidae	larva	Murthy, 1989

all-round the fields can check the mobility. Setting up of light traps were found effective in minimizing the population in the field. Setting up of bonfires for a few days after showers are found to be effective (Patel *et al.*, 1981; Kumaraswamy, 1985). In Gujarat, the effectiveness of light traps in controlling *Amsacta sp.* was investigated by recording the larval population after 20 days of traps used. It was ascertained that light trapping was quite effective in reducing the population compared to areas in which the traps were not set up (Patel *et al.*, 1981). Qayyum *et al.*(2000) have consolidated the use of light traps in about 34 villages in Andhra Pradesh including endemic district from 1989-1993. One light trap with 200 W bulb was used to cover about 10 ha area. The observations revealed that out of 1696 ha treated in Nalagonda and Warangal districts of Andhra Pradesh, only 3% area was damaged when compared to 56% damage in untreated villages. The whole programme was implemented in 5000 ha.

6.9.2. Biological control

Though several natural enemies have been reported against the hairy caterpillars, none of them have been used to put them under field conditions. However, the use of viruses and plant products have found to be encouraging. *Amsacta* species infected by NPV which extended to about 19 % on groundnut plants, 52% crawling in between plants and 94 % migrated to fence were found diseased. The infected 4th and 5th instar larvae usually migrate to fence plants which may even 500 m away from the site of treatment and died in large numbers hanging upside down (Jayaraj and Ravindra, 1986). In another attempts, Jayaraj and Rabindra (1986) reported that the population of *A. albistriga* can be suppressed with NPV and *Bacillus thuringiensis* (*Bt*). The *Bt* preparation was BIOTROL at 4 kg/ ha was as good as synthetic pesticide causing mortality in 2-3 days. The NPV at low dose of 2 x 10^5 PIB / lit was on par with BIOTROL, however, the mortality was delayed to 7-9 days. The earlier instars were found to be more susceptible than later instars. Interestingly, NPV applied to groundnut in September of 1978 causes very high mortality in next year in the same period (Ravindra and Balasubhramaniam, 1980). The authors have the clarification that these could have been due to transmission from generation to generation either through infected eggs, accumulation and persistence in soil or through the previous season. The gregarious attitude of the larvae might have also influenced spreading of the disease. The NPV of *A. albistriga* was found to be safe to white mice, poultry birds and common carp fish (Narayanan *et al.*, 1977). In the case of common carp fish, *Cyprius carpis* L. all fishes were found healthy showing normal weight gain and swimming and feeding behaviour were also unaltered (Narayanan *et al.*, 1977). In an another study after about 5 days of spray of NPV 10^7 resulted 39% and 44% larval mortality when sprayed using hand operated and ULV disc applicator, respectively (Logiswaran *et al.*, 1982). To increase the efficiency of NPV by protecting than from U.V. rays, Congo Red found to be more efficient resulting 62% and 72% in dry and wet viruses respectively (Baskaran *et al.*, 1998).

Amsacta albistriga was observed to be attacked by *Enicospilus* sp. from larvae collected on sesamum and groundnut. This is the first record of *Enicospilus* parasitizing *A. albistriga* in India. Twenty-seven of 1500 larvae (1.8%) collected were parasitized by this wasp. These endophagous wasps are nocturnal or crepuscular in habit, with the females laying their eggs within the larvae of the host insects. Mature larvae emerge from host pupae to spin characteristic brown cocoons while pupating in the soil. They may be useful in biological control programmes to control crop pests. The potential of these wasps as biological control agents is however yet to be realized (Veenakumari *et al.*, 2003).

6.9.3. Plant product

The anti-feedant effect of Neem Seed Kernel Extract at 0.1, 0.5 and 1% against *A. moorei* was found to be above 95% under field conditions (Sharma *et al.*, 1992). Similarly Neem Oil and Karanj Oil at 0.1% had greater anti-feedant effect than leaf extract of *Ocium sanctum* or bold seeded mustard oil emulsion (Verma and Singh, 1985). Patel *et al.* (1968) have used extracts of Neem, *Anona squamosa, Mamardica charancherentia, Lantana camara, Pongamia glabra, Diospyrum cardifolia* and garlic against *A. moorei* and found Neem Seed Extract effective against second instar larvae with 92% mortality.

6.9.4. Inter crop

The barrier or inter crops such as barnyard millet, growing of cowpea as an inter crop helps to attract adult moths to lay more eggs for early detection of the pest as a trap crop (Jayaraj and Ravindra, 1986).

6.9.5. Pheromone

Since the hairy caterpillars are the major pest of dry land agriculture in India, though sporadic in nature, its appearance could not be predicted in a routine way during the crop season. If it occurs in a large scale to devastating proportion, it would be worth to monitor its appearance using pheromones. Under this direction Persoon *et al.* (1993) have reported identification of both *A. albistriga* and *S. obliqua*. The pheromone of *A. albistriga* contain Octadecanal, Z9,Z12- octadecadienal, Z9,Z12,Z15- octadecadienal and Z3,Z6,Z9-heneicosatriene in a ratio of 1.0: 1.0: 11.5: 24 and the pheromone of *S. obliqua* was identified as a mixture of (Z3,Z6)-cis-9, 10-epoxy-3,6-heneicosadienal, (Z3,Z6)-cis-9, 10-epoxy-1,3,6- heneicosatriene, Z9,Z12-octadecadienal, Z9,Z12,Z15-octadecatrienal and Z3,Z6,Z9-heneicosatriene in a ratio of 1.0 : 0.5: 1.0 :3.0 :2.6.

6.9.6. Light trap

Patel *et al.* (1987) tested an ultraviolet-light trap for controlling *Amsacta moorei* on groundnut in Gujarat, India, in 1981-83. The traps were installed at a rate of 11/10 ha, 7/8 ha and 8/10 ha in 1981, 1982 and 1983, respectively. An area about 30 ft in diameter around each trap was dusted with insecticide to kill hatching larvae. The first adults were trapped 5, 3 and 7 days after the monsoon showers in 1981, 1982 and 1983, respectively. In 1981 and 1982, 98.7 and 84.1%, respectively, of the total catch over 12 days was made in the first 5 days, while in 1983 73.6% of the total catch was made in the first 6 days. The mean larval population was 20.42-72.25 and 144.5-396.6 individuals/300 ft^2 in areas with and without light traps, respectively, in 1981. The larval population in 1982 was drastically lower at 8.49-11.9 and 18.99-24.66 individuals/300 ft^2; this was a result of the large catch in 1981.

6.9.7. Chemical Control

Gupta and Sharma (1980) evaluated the effectiveness of fentin acetate and fentin hydroxide as anti-feedants against larvae of *Spilosoma obliqua* Wlk. (=*Diacrisia obliqua*). On the basis of larval weight and leaf area consumed, fentin hydroxide was found to be superior to fentin acetate, and was most effective at a concentration of 0.05%. It is recommended that dust formulations of insecticides are used against young larvae and spray formulations against mature larvae. The recommended insecticides (carbaryl, chlorpyrifos, dichlorvos, endosulfan, ethion, fenitrothion, phosalone and quinalphos) were highly toxic to adults of the parasitoids in the laboratory (Gunathilagaraj and Babu, 1987). Out of 17 insecticides tested, phosphamidon, endosulfan and malathion were most effective against *S. obliqua*. A combination of preventive + curative applications was significantly superior to preventive or curative applications alone (Nath and Singh, 1996)

The effectiveness of the insect growth regulator, diflubenzuron and 7 other insecticides for the control of larvae of the groundnut pest Amsacta albistriga was determined by Rabindra and Balasubramanian (1985) in the laboratory, LC50s of diflubenzuron 48 and 60 h after larvae in different instars had fed on treated leaves showed that older larvae were less susceptible than young ones and that efficacy was higher at 60 h than at 48 h. In field-cage tests, sprays of diflubenzuron applied at 30 g a.i./ha produced 94.7% mortality of last-instar larvae in 44 h; fenvalerate at 50 g a.i./ha, methomyl at 200 g a.i./ha and monocrotophos at 200 g a.i./ha were also effective. Spraying quinalphos at 0.05% recorded 95.45% larval mortality of the red hairy caterpillar (RHC, *Amsacta albistriga*) on groundnut, while dusting with methyl parathion at 25 kg/ha recorded 85.75% larval mortality. Both insecticides were effective against the early stages of RHC (Manjula, 2003). The LC50 values of spinosad and thiodicarb were 0.0106 and 0.018%, respectively.

Spinosad was more toxic to *A. albistriga* by 72-fold than thiodicarb. (Srinivasan and Jaydeep-Halder, 2004). Laboratory studies with *S. obliqua* on groundnuts showed the average incubation, larval, prepupal and pupal periods as 5.60, 24.72, 1.76 and 11.46 days, resp. The average pre-oviposition, oviposition and post-oviposition periods were 2.40, 2.88 and 4.00 days, resp. The life cycle was completed in 53.06 days (Nath and Singh, 1996).

6.10. Integrated management

Jayaraj and Ravindra (1986) gave detailed steps to be followed in the management of hairy caterpillars are as under:

- Organize campaigns to hand pick and destroy the pupae after summer ploughing on receipt of showers and from the bunds and shady areas near groundnut fields.
- Grow Cowpea as an intercrop which helps to attract the adult moths to lay more eggs on it and for early detection of occurrence and as a trap crop.
- Set up light traps or bon fires (with old cycle tyres or trash) immediately after receipt of rains in the rainfed kharif season to attract and kill the moths and detect the brood emergence.
- Collect the egg masses and destroy.
- Monitor for the lace-like leaves of Cowpea and groundnut which harbor gregarious early instar larvae, collect them and destroy.
- Spray the specific nuclear polyhedral virus of red hairy caterpillar at 100 larval equivalents / 0.4 ha or 1 acre. Early instars and medium sized larvae are more susceptible than grown up caterpillars.
- Add a wetting agent like Sandovit or Teepol or Triton to potable water and mix NPV suspension and spray as high volume in the evening hours. Avoid using brackish water for spraying and avoid spraying in sunlight.
- Hand pick and destroy grown up caterpillars to the extent possible.
- If possible, dig a trench around the field to avoid migration of the caterpillars to neighboring fields and to trap the larvae in trenches and kill them.
- Organize mass ground spraying under the Panchayat Act 1958 in endemic areas, if necessary, in case of outbreak of the pest.

If necessary, apply one of the insecticides such Quinalphos 1.5% D @ 25kg; Phosalone 4% D @ 25kg; Endosulfan 4% D @ 25kg; Endosulfan 35 EC @ 750 ml; Quinalphos 25 EC @ 750 ml; Phosalone 35 EC @ 750 ml ; Chlorpyriphos 20 EC @ 625 ml. Sreenivasulu *et al.* (2002) worked out the cost benefit ratio conducted a field study in Andhra Pradesh, India to compare the performance of an integrated pest management (IPM) module developed for the management of foliage-feeding insects, especially *Amsacta albistriga*, with respect to an all-insecticide control package and the farmer's practice of applying insecticide spray only once during the entire crop season. There was lower incidence of aphids, jassids, thrips and eggs and larvae of *A. albistriga* in IPM plots. IPM treatments also resulted in maximum net return and higher cost-benefit ratio of 1:2.1 and 1:2.0 in comparison to chemical control and farmer's practice during kharif seasons of 2000 and 2001, respectively.

References

Ahlawat, D. S., Saini, R.K., and Singh, H. (1993). Biology of the red hairy caterpillar, *Amsacta moorei* Butler (Arctiidae: Lepidoptera). *Annals of Biology*. 9(1): 118-122.

Amin, P.W. (1988). Insect and mite pests and their control. In Groundnut (ed. P. S. Reddy) pp.393-452.

Anonymous. (1993). Annual Report (1st April, 1992 to 31st March, 1993). All India Co-ordinated Research Project on Biological Control of Crop Pests and Weeds. Biological Control Centre (ICAR-NCIPM)., Bangalore, India. Tech. Doc. No. 40, 1-313.

Baskaran, R. K. M., Sivakumar, C. V. and Venugopal, M. S. (1993). Biocontrol potential of native entomopathogenic nematodes in controlling red hairy caterpillar, *Amsacta albistriga* (Lepidoptera:Arctiidae) of groundnut (*Arachis hypogaea*). *Indian Journal of Agricultual Science*. 64(9): 655-657.

Baskaran, R. K. M., Venugopal, M. S. and Janarthanan, R. (1998). Ultraviolet protection of the nuclear polyhedrosis virus of the groundnut red hairy caterpillar, *Amsacta albistriga* Wlk. (Lepidoptera: Arctiidae) through Congo red. *Journal of Entomological Research*. 22(4): 343-347.

Baskaran, V. and Narayanasamy, P. (1995). Traditional pest control of Caterpillars. *Publication*, Tamil Nadu. pp. 90.

Bass, M.H. and Arant, F. S. (1973). Insect pests of peanuts. In *Peanuts culture and uses*. Stone Printing Company, Roanoke, Virginia. 393-396.

Basu, M. S. (1995). Peanut Bud Necrotic Disease: Activities in the Indian national program. Pp 61-63 In Recent studies on peanut bud necrotic disease: Proceedings of a meeting, 20 March, 1995, ICRISAT Asia Center, India (Buiel, A. A. M., Parleviliet, J. E. and Lenne, J. M. eds.) Patancheru 502 0324, Andhra Predesh, India: International Crop Research Institute for Semi- Arid Tropics; and P. O. Box 386, 6700 A. J. Wasgeningen, The Netherlands: Department of Plant Breeding, Agriculture University of Wageningen.

Battu, G. S. and Dhaliwal, (1976). *Carcelia* sp. (Tachinidae: Diptera) as a parasite of *Diacrisia obliqua* (Walker) in Punjab. *Current research*. 5(7): 122-3.

Bhatnagar, S. P. (1948). Studies on *Apanteles* Forester (Vivionidae: Parasitic Hymenoptera). from India. *Indian Journal of Entomology*. 10: 133-203.

Bindra, O. S and Kittur, S. U. (1961). Biology and control of the red hairy caterpillar (*Amsacta moorei* B.) in Madhya Pradesh. *Vikram Science.* 5: 62-71.

Cherian, M. C. and Brahmachari, K. (1941). Notes on three predatory Hemipterons from South India. *Indian Journal of Entomology.* 3: 115-119.

Dahale, P. S., Puri, S. N. and Bilapate, G. G. (1988). Biometrics, growth and development of *Diacrisia obliqua* (Walker) on different hosts. *Journal of Maharashtra Agricultural Universities.* 13: 164-168.

David, B. V. and Basheer, M. (1961). Mass occurrence of the predatory stink bug *Cantheconidia (Canthecona) furcellata* (Wolff) on *Amsacta albistriga* Walk. in south India. *Journal of Bombay Natural History Society.* 58(3): 817-819.

Gunathilagaraj, K. and Babu, P. C. S. (1987). *Amsacta albistriga* Wlk. on groundnut and other crops. *FAO Plant Prot. Bull.* 35(2): 63-64.

Gupta, R. L., Pandey, N. D., Ram, S. and Sukhani, T. R. (1966). Bionomics of *Amsacta moorei* Buller (Lepidoptera: Arctiidae). *Allahabad Farmer.* 40(2): 65-70.

Hameed, H. F. and Yazdani, S. S. (1986). Final Report- Integrated control of major hairy caterpillar pests in Bihar, ICAR Adhoc Research Scheme, Department of Entomology, Rajendra Agricultural University, Bihar, Pusa (Samastipur), pp 96.

Jayaraj, S. and Ravindra, J. (1986). Management of Red hairy caterpillar, *Amsacta* spp. ((Lepidoptera: Arctiidae) on groundnut, pp 38-43 in Pest and Diseases Management in Oilseeds, Pulses, Millets and Cotton (eds, S. Jayaraj), Tamil Nadu Agricultural University, Coimbatore, Tamil Nadu.

Jayaraj, S., Sundaramurthy, V. T. and Mahadevan, N. R. (1976). Laboratory studies on the nuclear polyhedrosis of red hairy caterpillar of groundnut *Amsacta albistriga* (Walker). Symposium on Plant Protection Research and Development. Part B- Herbicides, Pesticide degradation and Microbiology, and Integrated Pest Control. *Madras.* 63: 567-569.

Krishnamurti, B. and Usman, S. (1954). Some parasites of economic importance noted in Mysore State. *Indian Journal of Entomology.* 16: 327-344.

Kumaraswamy, (1985). Pest management in Groundnut. pp 268 –269 in Microbial Control and Pest Management (ed., S.Jayaraj), Tamil Nadu Agricultural University, Coimbatore, Tamil Nadu.

Kuppusamy, S., Ramasamy, K. A. and Basheer, M. (1965). Recent insecticidal trials to control red hairy caterpillar on groundnut in Madras State. *Indian Oilseeds Journal.* 9: 195-200.

Laxminarayanan, M., Quyyam, M. A. and Sanghi, N. K. (1995). Innovative approaches for the management of red hairy caterpillar *Amsacta* spp. *Proceedings of the National Sminar on Integrated Pest Management in Agriculture,* December, 29-30,1995, Nagpur, 12 p.

Lefroy, H. M. (1906). Hairy caterpillar pest of crops. *Agricultural Journal of India.* 1:187-91.

Logiswaran, G., Madhava Rao, G., Vasudevan and V. Kannan. (1982). Influence of time of sowing and weather factors on the infestation of leaf miner *Aproaerema modicella* Deventer and yield in rainfed groundnut. *Madras Agricultural Journal.* 69: 359-363.

Mathur, A. C. (1962). Food Plant spectrum of *Diacrisia obliqua* Walk (Arctiidae: Lepidoptera). *Indian Journal of Entomology*. 24: 286-287.

Mathur, Y. K. (1966). Preliminary observation on the behaviour and colour variation in the larvae of *Amsacta mooeri* Butler. (Lepidoptera: Arctiidae). *Journal of Entomology*. 28: 567-570.

Mathur, Y. K. and Mathur, B. L. (1968). Note on parasitic *Aspergillus* on larvae of *Amsacta mooeri* Butler. *Plant protection Bulletin*, FAO, Rome 16:161-162.

Mukundan, M. (1964). Achievements of campaign for the control of the groundnut red hairy caterpillar, *Amsacta albistriga* M. in Madurai division (Madras State). *Madras Agricultural Journal*. 51:46-49.

Murthy, S. N. (1989). Biological suppression of pulse and oilseed pests, pp. 84-106. *Proceedings of the Seminar-cum-Seventh Workshop on Biological Control of Crop Pests & Weeds*. AICRP on BCCP & W, Biological Control Centre (ICAR-NCIPM). Bangalore. Tech. Doc. No. 26, pp 191.

Nagraj, K. R. and Ramchandran, N. (1958). Some adaptation in the habits of the red hairy caterpillar, *Amsacta albistriga* M. *Madras Agricultural Journal*. 45(12): 451-4.

Nagrajan, K. P., Perumal, K. and Shanmugam, N. (1957). The red hairy caterpillar, *Amsacta albristiga* M. and its field scale control. *Madras Agrìcultural Journal*. 44(4): 150-3.

Narayanan, K. L. and RangaRao, D. V. (1959). Control of red hairy caterpillar of Groundnut (*Amsacta albistriga* W.). *Andhra Agricultural Journal*. 46:165-6.

Narayanan, K., Easwaramoorthy, S., Santharam, G., Jayaraj, S. and Muthu, M. (1977). Exposure of white mice to *Baculovirus amsacta* of groundnut red-hairy caterpillar, *Amsacta albistriga* (Walker). *Current Science*. 46(12): 417-419.

Nath, P. and Singh, A. K. (1996). Biology and insecticidal management of Bihar hairy caterpillar, *Spilosoma obliqua* infesting groundnut. *Annals of Plant Protection Sciences*. 4(1): 42-46.

Pachori, R., Rathore, V. S., Rawat, R. R., Jakhmola, S. S. and Khatri, A. K. (1980). Economic threshold of *Diacrisia obliqua* Wlk. on groundnut. *Indian Journal of Agriculture Sciences*. 50: 69-72.

Paramasivan, V., Letchoumanae, S. and Sreeramulu, O. S. (1973). Preliminary studies on the insecticidal control of the groundnut red hairy caterpillar (*Amsacta* spp.). *Madras Agricultural Journal*. 7: 36-38.

Paras Nath. (1993). Studies on the groundnut insect pest and their management in east Uttar Pradesh. Final technical report of ICAR AD-HOC Project, Dept. of Entomology and Agril. Zoology, Institute of Agril.

Patel, B. S., Patel, R. C. and Patel M. S. (1981). A note on effectiveness of light traps for the control of Gujarat Hairy Caterpillar *Amsacta* spp. *GAU. Research Journal*. 7:47-48.

Patel, H. K. and Patel, V. C. (1965). Life history, epidemiology and seasonal history of Gujarat hairy caterpillar (*Amsacta moorei* Butler). *Bansilal and Amritlala Coll. Mag*. 16 and 17: 44-56.

Patel, H.K. (1968). Neem seed suspension-A sure deterrent to hairy caterpillar *Amsacta moorei* B. *Madras Agriculture Journal*. 55: 509-510.

Patel, J. S. (1943). Indian Central Jute Committee. Annual report. Agri. Res. Scheme, 1942-43, Calcutta, pp. 51.

Patel, J. S. (1944). Indian central Jute Committee, Calcutta, India. Annual report of the Agriculture research scheme for the year 1942-43 pp 43-44.

Patel, V. C and Patel, H. K. (1970). New record of *Exorista fallax* Mg. (Tachinidae: Diptera). as a larval-pupal parasite of *Amsacta moorei* B. in Gujarat. *Indian Journal of Entomology*. 32(1): 106-107.

Persoons, C. J., Vos, J. D., Yadav, J. S. and Prasad, A. R., Sighomony, S., Jyothi, K. N. and Prasuna, A. L. (1993). Indo-Dutch Co-operation on pheromone of Indian Agricultural insect pests. In pheromone technology in Europe and the developing countries. Working group: Use of pheromone and other semiochemicals in integrated control held on 10-14,May 1993,Natural Resources Institute, Chathan, England.

Qayyum, M. A., Lakshminarayana, M. and Sanghi, N. K. (2000). Management of red hairy caterpillar, *Amsacta* spp. through light traps. National Seminar on Oil seeds and oils- Research and development needs in the Millennium, Feb, 2-4, 2000, Directorate of Oilseeds Research, Hyderabad.

Rabindra, R. J., Geetha, N., Jayaraj, S., Brown, M., Grazywacz, D. (1997). Evaluation of cross infectivity of some baculoviruses to *Helicoverpa armigera* (Hbn.). *Pest Management in Horticultural Ecosystems*. 3(2): 61-65

Rabindra, R. J. and Balasubramanian, M. (1980). Observations on an induced epizootic of nuclear polyhedrosis virus of *Amsacta albistriga* Wlk. *Current Science*. 49: 279.

Ramaswamy, K. A. and Kuppuswamy, S. (1973). Some observation on the biology of the red hairy caterpillar (*Amsacta* spp.) on groundnut. *Madras Agricultural Journal*. 60(7): 637-9.

Ramaswamy, K. A., Kuppuswamy, S. and Shanthanaraman, T. (1968). Ridding groundnut for red hairy caterpillar. *Indian farming*. 18: 21-22.

Ramesh Arora, Battu, G. S., Bath, D. S., Mohinder Singh, Arora, R and Singh, M. (1996). Management of Bihar hairy caterpillar, *Spilosoma obliqua* (Walker) with a nuclear polyhedrosis virus. *Indian Journal of Ecology*. 23(2): 138-140.

Rao, A. S. and Amonkar, S. V. (1980). Nuclear polyhedrosis of *Amsacta moorei* Buttler (Lepidoptera: Arctiidae). *Current Science*. 49:709-710.

Rabindra, R. J. and Balasubramanian, M. (1980). Observations on an induced epizootic of nuclear polyhedrosis virus of *Amsacta albistriga* Wlk. *Current Science*. 49: 279.

Sahayaraj, K. (1999). Field evaluation of the predator, *Rhynocoris marginatus* (Fab.) on two groundnut defoliators. *International Arachis Newsletter*. 19: 41-42.

Saroja, R., David, L. A. and Lewin, H. D. (1971). Efficacy of some new chemicals for the control of red hairy caterpillar. *Madras Agricultural Journal*. 58: 824-6.

Sharma, I. N. S., Singh, A. K., Singh, S. P., Tauro, P. Narwal, S. S. (1992). Allelopathic potential of some plant substances as anti-feedant against insect pests of jute. Proceedings. First National Symposium. Allelopathy in agroecosystems (Agriculture & Forestry), February 12-14, 1992, held at CCS Haryana Agricultural University, Hisar-125 004, India. 166-168.

Shaw, S. P., Singh, B. B. P. and Parsad, D. (1988). Influence of different host plants on ovipositional preference, larval survival, growth potential and fecundity of Bihar hairy caterpillar, *Diacrisia obliqua* Walker. *Annals of Agricultural Research.* 9(2): 215-223.

Siddiqi, J. I. (1981). Record of *Apanteles obliquae* Wlkn. and *Meteorus* sp. (Hymenoptera: Braconidae) as parasites of the hairy caterpillar *Amsacta moorei* Butl. (Lepidoptera: Arctiidae). *Journal of Bombay Natural History Society.* 78(1): 181-182.

Singh, C. P. and Pandey, M. C. (1995). Studies on the biology of *Amsacta moorei* Butler on groundnut. *Journal of Insect Science* 8(2): 207-208.

Singh, O. P. and Gangrade, G. A.(1974). Biology of *Diacrosia obliqua* Walker (Lepidoptera: Arctiidae) on soybean and effect of loss of chlorophyll on pod and gram. *Jawaharlal Nehru Krishi Viswa Vidyalaya Research Journal.* 8(2): 86-91.

Sinha, M. M., Yadav, R. P. and Kumar, A. (1975). Outbreak of the Bihar hairy caterpillar, *Diacrisia obliqua* Walker in North Bihar. *Entomologist Newsletter.* 5: 47.

Siva Rao, D.V., Swamy, M. T. and Raju, A. P.(1977). Outbreak of *Amsacta albistriga* Walker on groundnut in Andhra Pradesh. *Entomologists' Newsletter.* 7: 3.

Sivasankara Reddy, Y., Rajendra Prasad, P., Srinivasulu B., Govardhana Naidu, V. (2003) Bioecology and management of Redhairy caterpillar, *Amsacta albistriga* Walker in groundnut pub. Acharya N.G.Ranga Agril. University, S.VAg.College, Tirupati p.10.

Sivasankara Reddy, Y., Rajendra Prasad, P. and Uma Maheswari, T. (2004) Studies on pupation behaviour of red hairy caterpiller *Amsacta albistriga* Walker *Indian Journal of Entomology. 66*: 181-190.

Srivastava, A. S. and Goel, B. K. (1962). Bionomics and control of red hairy caterpillar *Amsacta moorei* Butl. *Proceedings of the national Academy of Sciences*, India B 32, 97-100.

Sundaramurthy, V. T., Jayaraj, S. and Swamiappan, M. (1976). *Telenomus manolus* Nixon, a potential egg parasitoid of groundnut red hairy caterpillar *Amsacta albistriga* Walker. *Current Sciences.* 45: 21.

Trehan, K. N. and Talgeri, G. M. (1947). Combating the hairy caterpillar, *Amsacta moorie* in Bombay province. *Indian Farming.* 17: 560-564.

Trehan, K. N. and Bagal, S. R. (1957). Life history, bionomics and control of sweet potato weevil (*Cylas formicarius* N.) with short notes on the pests of sweet potato in Bombay State. *Indian Journal of Entomology.* 19: 245-52.

Tuhan, N. C., Pawar, A. D. and Singh, B. (1987). Preliminary studies on *Trichogramma brasiliensis* Ashmead against the red hairy caterpillar, *Amsacta moorei* Butler. *Journal of Biological control.* 1(1): 72-73.

Udayraj Singh, Agarwal, D. P., Pandey, P. N. and Singh, U. R. (1989). *Eclipta alba* and *Euphorbia hirta* as new food plant and foci of infestation of *Amsacta moorei* Bult. (Lepidoptera: Arctiidae). *Bulletin of Entomology.* 30(2): 238-239.

Usman, S. and Puttarudraiah, M. (1955). A list of insects of Mysore including mites. Department of Agriculture Mysore State- Entomology Series Bulletin 16. pp 194 .

Venkataraman, A., Abraham, E. V. and Sreenivasan, D. (1970). A note on trial of two insecticide for the control of red hairy caterpillar *Amsacta* spp. on groundnut in Tanjavur District. *Madras Agricultural Journal*. 57: 478-9.

Verma, S. K. (1983). *Exorista xanthaspis* (Wiedemann) (Tachinidae: Diptera) a new record on *Amsacta moorei* Butler. *Annals Arid Zone*. 22(1): 41-42.

Verma, S. K. and Singh, M. P. (1985). Antifeedant effects of some plant extracts on *Amsacta moorei* Butler. *Indian Journal of Agricultural Sciences*. 55(4): 298-299.

Yadav, C. P. S., Bhatnagar, K. N. and Srivastava, B. P. (1966). Incidence of *Amsacta moorei* Butler in semi-arid zones of Rajasthan. *Indian Journal of Entomology*. 28:274.

Rabindra, R. J. and Balasubramanian, M. (1985). Diflubenzuron and some new insecticides for the groundnut red hairy caterpillar, Amsacta albistriga Walker. In (ed. Regupathy, A and Jayaraj, S.) Behavioural and physiological approaches in pest management. 175-178 pp.

Patel, R. C., Dodia, J. F., Patel, A. G. and Yadav, D. N. (1987). Management of Gujarat hairy caterpillar, *Amsacta moorei* But. by ultra violet light trap. *Gujarat Agricultural University Research Journal*. 12: 59-61.

Srinivasan, S. and Jaydeep-Halder (2004). Determination of LC50 of spinosad and thiodicarb against red hairy caterpillar, *Amsacta albistriga* (Walker). Insect Environment. 10(2): 56-58.

Verma, S. K. (1983). Host plants of *Amsacta moorei* Butler in the Rajasthan Desert. Bulletin of Entomology. 24: 49-50.

Gunathilagaraj, K. and Babu, P. C. S. (1987) *Amsacta albistriga* Wlk. on groundnut and other crops. *FAO Plant Protection Bulletin*. 35: 63-64.

Baskaran, R. K. M., Srinivasan, T. R., Muthumeena, K., Muthulakshmi, S., Mahadevan, N. R. (2001). Influence of host plants on infectivity and in-vivo production of nuclear polyhedrosis virus on red hairy caterpillar (*Amsacta albistriga*) (Lepidoptera: Arctiidae). *Indian Journal of Agricultural Sciences*. 71(11): 742-744.

Singh, C. P. and Pandey, M. C. (1995). Studies on the biology of *Amsacta moorei* Butler on groundnut. Journal-of-Insect-Science. 8(2): 207-208.

Veenakumari, K., Rabindra, R. J., Naik, C. D. S. and Shubha, M. R. (2003). First record of *Enicospilus* sp. (Hymenoptera: Ichneumonidae) as a parasitoid of *Amsacta albistriga* (Walker) (Lepidoptera: Arctiidae) from Karnataka, India. *Insect Environment*. 9(4): 151.

Manjula, K. (2003). Efficacy of quinalphos and methyl parathion against early stages of red hairy caterpillar. *Insect Environment*. 9(1): 17-18.

Sreenivasulu, B. Naidu, V. G. and Prasad, P. R. (2002). Development of an integrated pest management module in groundnut against the major foliage-feeding pests and the red hairy caterpillar, *Amsacta albistriga* (Walker). *Pest Management and Economic Zoology*. 10(1): 35-39.

Tiwari, S. N., Rathore, Y. S., Bhattacharya, A. K. and Sachan, G. C. (1989). A note on the larval survival and weight gain by *Spilosoma obliqua* Walker on some promising varieties of groundnut. *Indian Journal of Entomology*. 51(1): 116-119.

Gupta, B. M. and Sharma, S. K. (1980). Evaluation of triphenyl tin acetate and triphenyl tin hydroxide as anti-feedants against the larvae of *Diacrisia obliqua* Walker. Journal of Entomological Research. 4: 107-109.

Tiwari, S. N., Rathore, Y. S. Bhattacharya, A. K., Sachan, G. C. (1989). Susceptibility of several varieties of groundnut to *Spilosoma obliqua* walker (Lepidoptera: Arctiidae). *Indian Journal of Entomology*. 50: 179-184.

Nath, P and Singh, A. K. (1996). Biology and insecticidal management of Bihar hairy caterpillar, *Spilosoma obliqua* infesting groundnut. *Annals of Plant Protection Sciences*. 4(1): 42-46.

Pachori, R., Rathore, V. S., Rawat, R. R., Jakhmola, S. S. and Khatri, A. K. (1980) Economic threshold of *Diacrisia obliqua* Wlk. on groundnut. *Indian Journal of Agricultural Sciences*. 50: 69-72.

Shaw, S. P., Singh, B. B. P. and Parsad, D. (1988). Influence of different host plants on ovipositional preference, larval survival, growth potential and fecundity of Bihar hairy caterpillar, *Diacrisia obliqua* Walker. *Annals of Agricultural Research*. 9: 215-223.

□□□

White Grubs

C. P. S. YADAVA AND V. NANDAGOPAL

7.1. Introduction

The term whitegrub or root grub is applied to the larval stage of scarab beetle popularly known as cock chafers, leaf chafers, chafer beetles, May beetles or June beetles. It belongs to the family Scarabaeidae of order Coleoptera. This family is divided into 14 sub-families, *viz.* Scarabaeinae, (coprinae), Aegialiinae, Aphodiinae, Ochodaeinae, Hybosorinae, Geotrupinae, Pleocominae, Glaphyrinae, Acanthocerinae, Troginae, Melolonthinae, Rutelinae, Dynastinae and Cetoniinae. The phytophagous chafers are more highly evolved as compared to scavengers. This is the group of beetles which is causing concern to agriculturists throughout the world, as their grubs are root feeders. The economically important grubs belong to sub-families Melolonthinae and Rutelinae. Some grubs of Dynastinae and Cetoniinae are also root feeders. The beetles have lamellate antennae and the grubs have an arched ('C' shaped) body with large and smooth apical segments which facilitate their active movement in the soil but make them unfit for active movement above the soil surface.

7.2. Distribution

The white grub is a pest of country wide menace, concentrating in certain localities as major problem. Its attack as pest of crop was first reported from Dalmiyanagar Sugar Factory area, Bihar in 1956 (Gupta and Avasthy, 1957).It has been a problem causing severe damage to groundnut and other crops in Rajasthan. It was also reported from Karnataka and Maharashtra around 1970. Now it has spread through out the length and breadth of the country causing damage to different crops. There are a few major species of the white grub. Of the different species of *Holotrichia*, *H. consenguinea* is a major one dominating in Rajasthan, Gujarat, Haryana, Punjab, Uttar Pradesh, Andhra Pradesh and Bihar. *H. serrata* is a key pest in the southern parts of the country occurring in Karnataka, Tamil Nadu and Maharastra. *H. seticollis* occurs in plantation crops of south Karnataka and Kerala. The three species of genus *Lencopholis viz., L.coneophora, L. burmeiatri* and *L. lepidophora* have been reported to cause damage to plantation crops of Kerala and coastal Karnataka. *Aranals dimidiata* has been reported as a serious pest in the hills and some parts of *tarai* area of Uttar Pradesh. The distribution of the white grubs belonging to Melolonthinae and Rutelinae is given in Table 7.1. *Leucopholis lepidophora* was observed causing damage to 25-100% of groundnut apart from sugarcane, paddy [rice], maize, and vegetables in Maharashtra. In summer groundnut, plant mortality at different locations was 56-89% with an average of 80.56% in 1989. In the laboratory, the egg, larval and pupal periods were 21-33, 290-350 and 28-35 days, resp., and adults lived for 17-32 days (Adsule and Patil, 1990). Outbreaks of larvae of *Lachnosterna fissa* (Brenske) on groundnut and bajra (*Pennisetum typhoides*) was observed in the Mohindergarh district of Haryana (Verma, 1975).

7.3. Economic importance

The grubs are subterranean and actively feed on the living roots. Being polyphagous, they feed on the roots of a wide variety of cultivated as well as uncultivated plants. The young grubs after hatching orient towards the roots and start feeding them. Consequent to feeding, the plants show varying degree of yellowing, some get wilted and ultimately die. Such plants can be easily pulled out. The crops with tap root system suffer more as compared to those with adventitious root system. Almost all field crops grown during the rainy season (kharif) in India are damaged, viz. groundnut, sugarcane, pearlmillet, sorghum, ginger *(Zingiber officinale)*, Cowpea, pigeonpea, green gram, cluster bean, chilies, upland paddy etc. The plantation crops like tea and Coffee (*Coffea arabica)* suffer similar damage in their seedling and early growth stages.

The beetles, like the grubs, are also polyphagous and feed on about 250 different types of host trees. Sometimes the adults also cause economic damage to certain fruit, ornamental and other economically important trees.

7.3.1. Nature of injury

White grubs are a serious constraint to groundnut production. In endemic areas, the damage to groundnut ranges from 20- 100 percent. The presence of one grub m^{-2} may cause mortality of 80-100 per cent plants. Because of the tap root system and smaller amount of roots, the damage to groundnut is more pronounced as compared to fibrous rooted crops like pearlmillet, sugarcane and sorghum. Although the damage to roots starts in the early first instar stage but manifests only when grubs reach into late second instar i.e. around middle of August. The maximum damage occurs when the grubs are in the third instar. The attacked plants show varying degrees of wilting, and ultimately die. The grubs also damage the pods by scratching the nut shell, such damaged pods fetch low market value.

Table 7.1. Species of white grubs

Species	Status	Distribution
Melolonthinae		
Holotrichia consanguinea	Major	Uttar Pradesh, Rajasthan, Bihar, Madhya Pradesh, Haryana, Punjab, Gujarat, Andhra Pradesh, Delhi, Tamil Nadu
H. serrata	"	Maharashtra, Karnataka, Tamil Nadu, Kerala, Rajasthan, Bihar, Orrisa, Andhra Pradesh, Delhi
H. seticollis	"	Uttar Pradesh
H. (insularis) reynaudi	Minor	Rajasthan, Karnataka
H. nilgiria	"	Karnataka
H. sculpticollis	"	Karnataka
H. rufoflava	"	Karnataka, Kerala
H. excisa	"	Karnataka,
H. *heterotincta*	"	Karnataka,
H. longipennis	"	Uttar Pradesh, Himachal Pradesh, Eastern India
H. fissa	"	Karnataka, Kerala, Haryana
H. problematica	"	Uttar Pradesh, Haryana,
Leucopholis coneophora	Major	Kerala, Karnataka, Maharashtra
L. urmeistri	Minor	Kerala, Karnataka
L. lepidophora	"	Kerala,
Schizonycha ruficollis	"	Rajasthan, Karnataka ,Punjab, Maharashtra, Uttar Pradesh, Haryana
S.hizonycha fuscescens	"	Rajasthan, Karnataka, Punjab
Autoserica nathlani	"	Uttar Pradesh, Karnataka
A. nilgirensis	"	Karnataka
Maladera sp.	"	Rajasthan, Karnataka ,Punjab ,Haryana
Apogonia ferruginea	"	Uttar Pradesh, Karnataka , Maharashtra
A. cribricollis	"	Assam
A. rauca	"	Uttar Pradesh, Karnataka
Rutelidae		
Anomala dimidiata	Major	Uttar Pradesh, Himachal Pradesh
A. bengalensis	Minor	Rajasthan, Uttar Pradesh, Maharashtra, Karnataka, Tamil Nadu
A. dorsalis	"	Rajasthan, Tamil Nadu
A. ruficapilla	"	Rajasthan
A. bicolor	"	Rajasthan
A. polita	"	Uttar Pradesh (Hills)
A. elata	"	Rajasthan
A. marginipennis	"	Kerala
A. versicolor	"	Tamil Nadu
A. varians	"	Kerala
Rhinyptia laeviceps	Major	Rajasthan, Karnataka, Maharastra,
R. meridionalis	Minor	Rajasthan
Pachynhinadoretus frontatus	Major	"
Adoretus punjabensis	Minor	"
A. lasiopygus	"	"
A. nitidus	"	Jammu and Kashmir

Plate 1. Gaps due to grub feeding

The damaged plants are easily pulled out. They feed on soil organic matters for a few weeks and then to the roots. They also damage pods. The roots show a sharp cut which can be distinguished from the damage by termite where the main root becomes hollow and is filled with soil. Patches of dead plants **(Plate 1)** are seen throughout the field, which later coalesce and produce intensive areas of damage. This damage continues till middle of October when the grubs become full fed and move downward in the soil for pupation.

7.4. Biology

The beetles emerge from the soil during dusky hours after good rain in Mid-May or later. Once active, the beetles follow the daily rhythm of emergence and congregation on host trees at dusk and return to soil at dawn. The beetles feed on the foliage of variety of host trees and bushes found in the close vicinity. However, they do exercise some preference for certain hosts. In some situations large number of beetles settle on a tree which is not a host, only for mating and just after mating shift for feeding to nearby preferred hosts. The host preference depends on the combination of hosts available in a particular locality. The beetles, usually after emergence, do not fly long distances in the evenings but try to avail of short twilight hour for mating and selecting suitable host as they are unable to see clearly in dark. It is the female which first comes out of soil, settles on a host and secretes a pheromone which attracts the male for mating. If a female fails to mate in first 15 to 20 minutes of settling on the host trees, she tries mating next evening. The process of congregation of beetles on host continues every evening for a period of about four weeks to ensure mating of all females. The mated females returns to soil in early morning and starts egg laying and have no preference for farmyard manure or organic matter. Newly hatched grubs may feed or organic matter for some time till they come in contact with living roots.

For feeding, the grub makes a chamber by compressing the surrounding soil particles and then eats the rootlets exposed into the chamber; thereafter it moved vertically a little, to eat more of the same root. After that, the grub moves horizontally making chambers and feeding on the exposed roots.

The grubs continue active feeding from July to Mid-October, some species may continue up to April, and thereafter enter into pupation. Before pupation the grub ceases to feed and descends down in the soil to the depth of 40 –70 cm or more in search of suitable moisture zone. A short prepupal period is spent before entering into pupation during which the grub prepares an earthen chamber for pupation. The beetles remain in the soil in inactive state up to the middle of May at a depth of about one meter. They emerge from their hiding abode after receiving first good rain. Thus, there is only one generation in a year.

In each locality a number of different species of grubs belonging to Melolonthinae, Rutelinae, Cetoniiae or other related sub-families can be found. Their active period varies and some of these, particularly rutelinids in general, pass the winter in their larval form. The full-grown grubs of species of smaller size look like the newly hatched grubs of the large species. This creates confusion in the mind of a casual observer and some workers feel that there are several over-lapping generations of the predominant species.

The beetles are positively photo tactic and some beetles can be caught in light trap. Among the ordinary light sources, fluorescent tube-light attracts the maximum number. But the number of beetles attracted to ultraviolet light is higher than that to fluorescent tube.

7.4.1 *H. serrata*

The detailed biology of the beetle *H. serrata* has been reported by David and Kalra (1966), Majumdar and Teotia (1965) and Veeresh (1977a, b) and as presented by David *et al.* (1986) is given below.

Egg: Egg is creamy white (**Plate 2**), oval when freshly laid and measures 3.3 mm long and 1.7 mm in width. It becomes spherical in 2 or 3 days measuring 2 mm in diameter. The chorion becomes slightly translucent on one end and milky white on the other, with the cephalic appendages of the developing embryo visible. The egg swells up to 4 mm in diameter, one or two days before hatching. The incubation period lasts for 10-12 days at room temp. (27-34°C). Hatching is delayed at lower temp. About 82 per cent of eggs hatch under laboratory conditions.

Larva: The neonate grub is creamy white in colour, measuring 10.8x2.7 mm in size, with the head capsule measuring 2.1 mm across. The grub becomes active in about 4 hours. The first instar larva comes out of the earthen cell when there is sufficient moisture, but remains inside the cell for over 100 days under drought conditions until favorable conditions resume. This stage lasts for 23-36 days with an average of 30.2 days. The second instar grub measures

29x 7 mm with the size of the head capsule being 5.6 mm. The grub feeds on the developing pod (**Plate 3**). Its shape and colour resemble the first instar, but the last abdominal segment is more swollen and dark. The mean duration of this stage is 35.8 days, the minimum and maximum being 26 and 62 days, respectively.

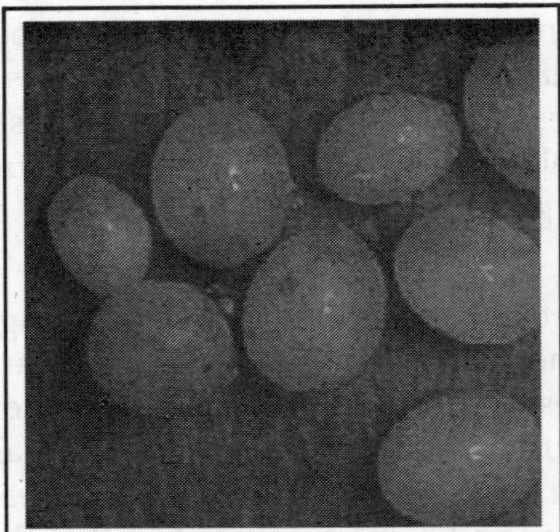

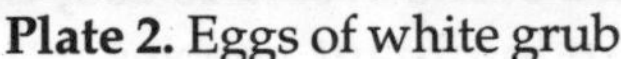

Plate 2. Eggs of white grub

Plate 3. The grub feeding

The third instar is creamy white in colour, 47 x 12 mm in size, with the head capsule measuring 7- 8 mm in width. This instar is an active feeder with powerful mandibles and has five segmented antennae with sensory spots (**Plate 4**). Thoracic segments are distinct, forelegs short, while hind legs are distinctly long. The average duration of the third instar is 102.9 days with a range of 72-188 days. The tenth abdominal segment has on its vertical side the raster with a longitudinal group of pelidia on either side of a median bare region- septula. The septula is long, narrow anteriorly and broad posteriorly. The tegillae are well developed, covering the venter between the barbulae and the pelidia (Veeresh and Reddy, 1978).

Pupa: Pupation occurs in an earthen cell at 20-30 cm depth in the soil. The pupa measures 32x 15 mm in size, creamy white in colour initially and turning brown later. Pupal period lasts for 11.8 days with a range of 10-14 days.

Adult: The adult is reddish brown, often with a dark tinge and measures 22- 25 mm in length; head oblique, transverse, punctuation prominent (**Plate 5**). Antennae ten segmented and apical three segments form a lamellate club. Hind tibia with spiny ridge. Male slightly smaller than female.

7.4.2 H. consanguinea

This grub was first recorded damaging sugarcane in india by Gupta and Avesthy (1957). The detailed biology of the beetle *H. consanguinea* has been reported by Kalra and Kulshrestha (1961) and as presented by David *et al.* (1986) is given below.

Plate 4. Grownup grub

Plate 5. Adults of white grub

Egg: The egg is milky white in colour and elongated in shape, measuring 3.0 mm in length and 1.4 mm in width. With the development of the embryo, the width increases to 2.3 mm and the egg becomes almost spherical in about six days. By about the seventh day, the egg turns dirty white in colour with a comma-shaped mark on the side of it, indicating the position of mandibles which become distinct on the eighth day. The incubation period id 8-10 days at 28-38°C, with a mean relative humidity of 60-68 per cent under laboratory conditions in the month of July. The larvae emerge by rupturing the egg shell in a zigzag manner.

Larva: The neonate larva measures 5.5 mm in length and is milky white in colour. The tip of mandibles is shiny brown. It remains inactive for three hours after hatching, during which period its body length increases to 9 mm. The larva feeds on roots of grasses for a few days, later on moving to crops. The growth of larva depends on the host on which it feeds; those feeding on groundnut roots grow faster than those that feed on grass roots. They also feed on pods of groundnut. Full grown larva is 'C' shaped and measures about 53 x7 mm with three pairs of prominent thoracic legs. The abdominal segments being translucent, intestinal contents are clearly visible. Two parallel rows of spines (the raster) on the ventral surface of the terminal abdominal segment are useful in distinguishing it from other species of white grubs. The larval period varies from 8-10 weeks between July and September when the maximum temperature is 28-29°C and relative humidity is 65-96 per cent.

Pupa: Pupa is exarate, creamy white when freshly formed, gradually changing to straw colour and ultimately to brown. It measures 20 –23 mm long and 8-10 mm across the thorax. As the pupal stage advances, the head, antennae and legs turn brownish, eyes dark brown and the elytra yellowish brown in colour. Pupation takes place in an earthen cell about 30-150 cm below the soil surface during late September/early October. Pupal period varies from 12-16 days during October-November, when the maximum temperature ranges from 27-32°C and relative humidity from 59-93 per cent.

Adult: On emergence from the pupal case, the adult has a soft, creamy-yellow abdomen and thin, crumpled, pinkish, papery elytra which later become hard and stiff. The colour changes gradually to dark brown. Males measure 16-22 mm and 8.5-12 mm and the females measure 16.5 –22.0 mm and 9-12 mm, respectively, in length and across the thorax.

7.4.3. *H. insularis*

The detailed biology of the beetle *H. insularis* has been reported by Srivastava and Khan (1963) and as presented by David *et al.* (1986) is given below.

Egg: Freshly laid egg is shiny white in colour, oval in shape and on an average measures 3.5 mm in length and 1.5 mm in width. As the development proceeds, the egg turns dirty white in colour and becomes spherical in shape. When ready to hatch, a depression appears in the center of the egg, around which the curved embryo is visible.

Larva: Neonate grub is milky white in colour with a dirty white head and light brown mandibles. It measures 5x 2.2 mm. Second instar grub is more curved, active and cannibalistic and measures 21 x 5.5 mm. The full grown grub is 41.0 x8.0 mm in size with a dark brown head. The abdominal segments have distinct integumental folds which are translucent, with the intestinal contents being clearly visible. Ventral surface of the anal segment bears a triangular patch of hairs, with two parallel rows of thick hairs in the middle.

Pupa: Pupation takes place in an earthen cell, which measures 32 and 14 mm in length and width, respectively. The pre-pupal larva is yellowish in color. Fresh pupa is exarate and yellowish, turning brown and gradually darkening at the time of adult emergence. It measures 18 - 21 mm and 8 -9 mm in length and width, respectively.

Adult: Freshly emerged adult has thin, crumpled and yellowish papery elytra, with a soft yellow abdomen. Later, the elytra harden and the colour gradually changes to dark brown. It measures 18 –20 mm 7 –9 mm in length and width, respectively. Females are much larger than the males and distinctly oval in shape, while the males are oblong. In both the sexes, posterior end of the abdomen is not covered by the elytra.

7.4.4.*H. robustus*

The detailed biology of the beetle *H.robustus* has been reported by Kalra and Sharma (1965) and Hukkam Singh and Rizvi (1968) and as presented by David *et al.* (1986) is given below.

The adult is shiny black in colour and measures 1.5 –2.0 cm in length and about 1.0 cm in width. The duration of the different immature stages and the period of egg laying for the common species of cockchafer beetles which damage groundnut. The life history given in table 7.1.

Table 7.2. Life history of white grubs

Species	Period of egg laying	Duration in days			
		Egg period	Larval period	Pupal period	Total life cycle (Days)
H. consanguinea	June- July	8-10	56-70	12-16	76-96
H. serrata	March-May	10-12	121-202	10-14	141-228
H.insularis	May-June	8-12	54-78	15-22	77-112
A.impressicolle	Sept.-Dec.	14-17	-	14-17	-
A.bengalensis	Feb.-May	12-16	22-102	105-206	160-286
L.mansueta	June- July	10-13	About 600	-	About 730

7.5.Seasonal History

All the white grub species occurring in India have a life cycle of about one year, with the known exeption of *L. mansueta* which has a two year life cycle (Gupta *et al.*, 1968). Grubs of *Holotrichia* spp., feed voraciously on groundnut roots during July and August. Pupation commences from middle of September and continues up to February in the tropics. The Dynastid grubs are active during February to March.

7.6.Host Range

7.6.1.Adult

The adult beetles of *H. serrata* feed on the foliage of *Azadirachta indica* a. Juss., *Swietenia mahagoni* Linn., *Syzigium jambolonum* DC, *Musa sapientum* Linn., *Rosa* sp., *Psidium guajava* Linn., *Achras sapota* Linn., *Vitis vinifera* Linn., *Pyrus malus* Lin.., *Prunus persica* benth. and HK., *Ficus religiosa* Linn., *Cassia fistula* Linn., *Pithecolobium dulce* Benth., *Acacia Arabica* Wild., *Morus alba* Linn., (Veeresh, 1977a), *Ailanthus* sp., *Delonix regia* Rafin., *Terminalia arjuna* W. and A., (David and Ananthanarayana, 1974) and *Butea monosperma* (Lam.) Kuntz., (Majumdar and Teotia, 1965). Neem (*A. indica*) and *Acacia* sp., are the preferred hosts. Severe defoliation of grapevine (*V. vinifera*) and *M.alba* has also been reported (Veeresh, 1977a).

The food plants for adult beetles of *H.consanguinea* include *Mangifera indica* Linn., *T. arjuna, Rosa damascena* Mill., *Zizyphus jujuba* Lam., *A.indica, Ficus glomerata* Roxb., *F. religiosa* (Gupta and Avasthy, 1957b; Kalra and Kulshreshta, 1961) and *B.monosperma* (Avasthy, 1967).

H.insularis beetles have been recorded feeding on *Moringa oleifera* Lam., *Carissa carandas* Linn., *Eugenia jambolana* Lam., *Lawsonia inermis* Linn., *A. arabica, Punica granatum* Linn., *Grewia asiatica* Linn., and *Cratoeva religiosa* Hook and Thomas (Avasthy, 1967). Beetles of *Schizonycha ruficollis* feed on *Rosa* sp., *V.vinifera, Tamarindus indicus* Linn., *Citrus* sp., and *P.guajava* (Paras Nath and Janardhan Singh, 1980).

7.6.2.Grub

The grubs of *H. serrata* feed on roots of *P. guajava, P. persica,* vegetables, pulses, oil seeds, millets, creals and tobacco, *Nicotiana tabacum* Linn., (Veeresh, 1977a) besides roots of grasses. These also damage groundnut (*A. hypogaea*) and tomato (*Lycopersicum esculentum* Mill) (David and Kalra, 1966). Grubs of *H.consanguinea* feed on roots of castor (Hari deva Vasu, 1970). *A. hypogaea* (Anon., 1975), pearl millet (*Pennisetum typhoideum* Linn.,) and guinea grass (*Panicum maximum* Jacq.) (Dhaliwal, 1980). *A.impressicolle* feeds on *Typha* sp. (Cyperaceae) (Banerjee and Mookherjee, 1960) apart from sugarcane. *S. ruficollis* feeds on *A.hypogaea* and *Citrus sp.*, while grubs of *Anomala dorsalis* Fab., *A. bengalensis* and *A. alata* feed on *Capsicum annum* Linn., *P.typhoideum* Linn., and *Sorghum vulgare* Pers (Sachan and Pal, 1974). The average distance traveled by larvae of *Holotrichia serrata* (F.) in soil at depths of 12-19 cm in India during their lifespan was 151.2 cm in a lawn and 408.3 cm in cultivated fields. Migration of the larvae depended on the type of crop, distance between plants and the type of soil. It was least in lawns, with plenty of roots, and highest in soil with no roots. Red soil was more favourable for movement and survival than black or gravelly soils. Onion, French bean (*Phaseolus vulgaris*), groundnut and ridge gourd (*Luffa acutangula*) were preferred (in that order) by larvae migrating in search of food-plants. Crop roots did not influence the movement of 1st-instar larvae, but the movement of 2nd- and 3rd-instar larvae was guided by the frequency and availability of roots (Shivayogeswara and Veeresh, 1983).

7.7. Management strategy

For managing the population of white grubs in groundnut crop, efforts can be directed against beetles or grubs. There can not be one uniform practice for all the situations. The groundnut is cultivated in rainfed as well as irrigated conditions. In rainfed condition the crops sown with the onset of monsoon in June-July. This coincide with the emergence of beetles, but where irrigation is available, the farmers usually go for early sowing of the crop. Therefore, some of the practices which are applicable in rainfed condition may not be suitable for advance planted crop, as this does not coincide with the emergence of beetles. Looking to the variations in the conditions and suiting to these requirements, the management strategies have been devised separately for rainfed/irrigated conditions. Neverthless, the beetle control strategy is suitable for both the conditions. The farmers may select any of the practices mentioned below suiting to their condition and requirement. The various strategies have been mentioned below:

7.7.1. Management through beetle control

Most of scarab beetles emerge from soil with first good shower (pre-monsoon or monsoon) and congregate on some preferred hosts and can be killed conveniently by spraying of insecticides on the host trees. This technology is also successful where the number for host trees are more. Select one host tree from a group of host trees within the radius of 20m and spray 0.05% monocrotophos during day time and pheromone dispensers be placed (2-3/ tree) on this host tree in the evening continuously for three evenings after beetle emergence. No time gap should be given between onset of rain, spraying and installing pheromone dispensers. This should be done for 3-4 days of first emergence of beetles.

The beetles which are attracted by pheromone, settle and feed on the sprayed host trees and get killed soon by contact as well as stomach action of the insecticide and are thus not able to return to the soil next morning for egg-laying. This way number of trees to be sprayed is drastically reduced which make economical and less time consuming. Generally, it is argued that the beetle control will be effective only when the entire village/region is sprayed through community effort. But this is not true. As the beetles do not fly long distances, individual effort by a farmer shall also provide protection to his crops.

For beetle control, pheromone dispenser, insecticide, sprayer, bamboo pole etc. should be kept ready with the farmer prior to the onset of monsoon. This will enable the farmer to start the spraying operation immediately after the first shower, without waiting for outside agencies and thereby not loosing any time, as the first 3-4 days are crucial for the operation.

Local farmers be associated in selecting the preferred hosts of the area/ village as they know the trees defoliated by the beetles in previous years. Beetle control reduces the pest problem in the operational area, and thereby provides protection to all crops, including the advance sown ones.

The beetle control can be done without pheromone also but in this case number of trees to be treated be more. Some trees can be lopped completely to minimize the spraying cost and to cover more area. The lopping should be done before onset of monsoon. While lopping, some trees should be left uplopped, to attract all the beetles of the adjoining area to these unlopped trees, which should be sprayed; the unlopped trees should be uniformally distributed in the area. The branches of neem trees with foliage, when freshly cut and planted before dusk in the infested fields, attract large number of adults, which can be easily destroyed (Gupta, 1973). Mass collection of adults were undertaken with great success in the case of *H. consanguinea* (Kalra and Kulshreshtra,1961). Mass collection and destruction of *H. serrata* have been undertaken with spectacular success in Belgaum district of Karnataka, where about three lakhs beetles were destroyed in one day (Veeresh,1977). Similarly in a season about nine million beetles were collected and destroyed in Maharashtra (Rao Deo,1974; Rao Deo *et al.*,1976). This resulted a reduction of beetles by 80 to 85% during subsequent three years from 1970 to 1972 compared to the 1969 year when the adult beetles were collected on community basis.

Fluoresvent light traps are highly efficient to attract *H.consanguinea* and using this technique, thousands of adult beetles have been collected and killed (Gupta,1958; Avesthy,1967).

7.7.2. *Management of grubs*

7.7.2.1. Monsoon sown crops (rainfed/irrigated)

a) *Seed treatment (only for groundnut):* To protect groundnut crop from grubs, seed treatment with chlorphyriphos 20 EC or quinalphos 25 EC at 25 ml/kg seed is quite effective. For this, a known quantity of seed should be placed in a seed treatment drum or polyethylene bag and the required quantity of insecticide added to it. The drum/bag should then be swirled gently to provide uniform coat of the insecticide on the seed. Vigorous shaking should be avoided. The seed thus coated should be spread on cement floor or plastic sheet under shade so that the insecticidal coat gets dried. This seed should be sown in furrows. The interval between coating and sowing should be short (4-6 hours). Pre-monsoon or early sown crops may not get effective protection by the seed treatment.

b) *Soil treatment:* Pre-sowing soil treatment with Phorate 10 G at 25 kg/ha or quinalphos 5 G at 30 kg/ha effectively protects the crop. The insecticide be applied in furrows and seeds be placed in the same furrow. Since seed treatment is equally effective but cheaper, soil treatment may not be done in groundnut crop.

7.7.2.2. Advance sown crop and standing crop treatment

For advance sown, seed treatment or soil treatment may not be done as this may become ineffective by the time beetles emerge, lay eggs and grubs develop. For such crop situations, application of quinalphos 25 EC or chlorpyriphos 20 EC at 4 1/ha with irrigation be done three weeks after the first good monsoon shower. Insecticidal application in standing crop could be done in the following ways:

i. Required amount of insecticide be applied at the point from which the water enters into the plot. This can be accomplished by putting the insecticide in an earthen pitcher having a hole in the bottom. A cotton wick should be placed in the hole for controlling the drops. The drops should be so calibrated by the movement of the cotton wick that the required amount of insecticide is applied uniformly with irrigation. This pitcher can be placed on a tripod.

ii. Soil collected from the field (approx. 80-100 kg/ha) be treated with required amount of insecticide (4 1/ha) and the insecticide mixed soil be uniformly applied in standing crop near root zone. The application should be followed by light irrigation so that the insecticide percolates downwards. Insecticide mixed soil can also be used in rainfed crops provided it rains soon after the application. Farmers can use this method between mid -July to mid-August, the recommended period, as in areas where rainfed groundnut crop is taken, regular showers are expected during this period.

7.7.2.3.Cultural practices to manage grubs

i. In the infested areas, summer fodder crops provide good ground for the survival of a large number of grubs. These fields should be ploughed from the end of July to middle of August after taking the last cutting to expose the grubs to predatory birds. Two ploughings at this stage reduce the grub population by about 70 per cent. If predatory birds are absent, a man with bucket in hand may follow the plough and pick the exposed grubs. Tractor fitted with disc is usually used in ploughing the fields in sandy soil areas. The disc does not expose the grubs to the predatory birds. Therefore, two initial ploughings should be done with a soil-turning plough.

ii. Early sowing (pre-monsoon, irrigated) of crops should be done to allow roots to get established so that a certain level of grub population may be tolerated by the crops without any noticeable damage. This will be useful in areas where grub population is quite low. This may, however, be taken as a very short term strategy.

iii. Trash trapping in fallow land 2-3m apart help attracting the grubs to a limited area (Veeresh,1977).

7.7.2.4.Biocontrol

Larvae of *Holotrichia consanguinea* were found to be infected with *Bacillus popilliae* var. *holotrichiae* and showed milky-disease symptoms. The bacterium was also found to infect the other scarabaeids H. serrata, *Anomala bengalensis* and *Autoserica nathani* (=*Maladera nathani*). Some larvae were also found to be infected with green-muscardine disease caused by the fungus *Metarhizium anisopliae* var. *anisopliae*. Mixed infection varied from 4.0 to 7.69% (Vyas *et al.*, 1986). Vyas *et al.*(1991) investigated on the use of spore suspensions of *Bacillus popilliae* var. *holotrichiae* using healthy larvae of the scarabaeid *Holotrichia consanguinea* which were released on microplots of groundnut plants in Vatrak, Gujarat. With treatments of 5000, 500 and 50 million spores/m2, mean larval mortality was 62.86, 44.98 and 39.23% compared with 7.65% in control plots treated with talc. Maximum mortality (72.2%) was observed 40 days after treatment.

The biocontrol potential of the nematodes *S. glaseri* (strain NC-34) and *S. carpocapsae* (strain DD-136) against *H. consanguinea* was evaluated in a laboratory test using the filter paper exposure method at 25°C and at dosage levels of 0, 5, 10, 20 and 40 infective juveniles/final instar grub (Shanti and Sivakumar, 1991.). The LC_{50} values were considerably lower for *S. glaseri* (13.38 infective juvenile/grub) than for *S. carpocapsae* (33.84). LT_{50} values showed the same trend 152.0-85.0 h for *S. glaseri* and 223.8-119.1 h for *S. carpocapsae*. The effectiveness of *Steinernema glaseri* was evaluated against *Holotrichia consanguinea* under laboratory and field conditions (Vyas and Yadav, 1993). The results of trials showed the nematode to infect 40 and 100% of the grubs, after 10 and 20 days of treatment respectively at an initial dose of 100 000 infective juveniles/m2. Shashi-Sharma *et al.* (1998) studied in the laboratory using *Metarhizium anisopliae, Beauveria brongniartii* and *B. bassiana* and indicated that they are potential candidates for the control of *Holotrichia consanguinea*. A species of *Holotrichia* near *H. consanguinea* (Blanch.) is one of the most serious pests of groundnut, bajra, sorghum and sugar-cane in Gujarat and other parts of India. Diseased larvae collected from Kapadvanj were found to be infected with a local milky disease. This is the first record of such a disease from India and the causal organism appeared to be different from others causing similar diseases elsewhere. In subsequent tests in 1969, healthy larvae were successfully artificially infected with the disease (Patel *et al.*, 1978). Crow (*Corvus splendens*) and common mynah (*Acridotheres tristis*) were important predators of larvae of *Lachnosterna* (*Holotrichia*) spp. exposed during the pre-sowing ploughing of groundnut fields. A simple equation for estimating the size of larval populations in fields is presented (Yadava *et al.*, 1975).

Table 7.3. List of parasitoids of white grub

Species	Order: Family	Reference
Adapsilia flaviseta Aldrich	Diptera: Pyrgotidae	Rao *et al.*, 1971
Anthia sexguttata Fb.	Coleoptera: Carabidae	Anon., 1988
Autocrates aeneus Parry	-	Anon., 1991; Singh *et al.*, 2003
Camposmeris collaris	Hymenoptera:Scoliidae	Channa Basavanna, 1954
Hypopi sp.	Acari: Histiostomatidae	Singh *et al.*, 2003
Lebridura reparia Pallas	Labriduridae	Anon.; 1988
Scolia sp.	Hymenoptera:Scoliidae	Channa Basavanna, 1954
Scolia auripennis	Hymenoptera:Scoliidae	Anon., 1991
Scarites indus Oliver	Coleoptera: Carabidae	Anon., 1988
Scarites sp.	Coleoptera: Carabidae	Anon., 1988
Tiphia rufo-femorata	Hymenoptera : Tiphiidae	Singh *et al.*, 2003

Table 7.4. List of entomopathogens of white grub

Species	Order: Family	Reference
A. Bacteria		
Bacillus popilliae Dutky	Bacillaceae: Firmicutes	David *et al*., 1973, Patel *et al*., 1977; Shinde and Sharma, 1971; Vyas *et al*. 1986; Vyas *et al*.,1991; Nath and Singh, 1988
Bacillus cereus	Bacillaceae	Nath and Singh, 1988
Bacillus thuringiensis	Bacillaceae	Nath and Singh, 1988
Clostridium sp.	Bacillaceae	Nath and Singh, 1988
Diplococcus sp.	Bacillaceae	Nath and Singh, 1988
B. Fungi		
Beauveria brongniartii (Sacc.) Petch	Moniliales : Moniliaceae	Jayaramaiah 1981, Jayaramaiah and Veeresh 1983; Vereesh, 1977; Vyas *et al*. 1997
Beauveria bassiana (Bab.) Vuili	"	Gupta and Avasthy, 1960; Jayaramaiah and Veeresh, 1983; Rao and Vijayalakshmi, 1995
Metarhizium anisopliae (Metscinkoff) Sorokin	Deuteromycotina Hypomycetes	Jayaramaiah, 1981; Jayaramaiah and Veeresh, 1983; Vyas *et al*. 1997; Rao and Vijayalakshmi, 1995
Aspergillus parasiticus		Rao and Vijayalakshmi, 1995
C. Nematode		
Steinernema feltiae	Nematoda Steinernematidae	Easwaramoorthy, 1986
Steinernema glaseri (Steiner)	"	Vyas and Yadav, 1993
Rhabditis sp.		Veeresh, 1980
Rickettsiella popillae	Rickettsiae	Nath and Singh, 1988

Table 7.5. List of predators of white grub

Bird species	Order : Family	Reference
Bank myna, *Acridotheres ginginianus*	Passeriformes : Sturnidae	Anon., 1988; Parasharya *et al*., 1994
Black drongo, *Dicrurus adsimilis*	Passeiformes: Dicruridae	Anon., 1988; Parasharya *et al*., 1994

Contd...

Table 7.5. Contd..

Brahminy myna,*Sturnus pogodarum*	Passeriformes : Sturnidae	Anon., 1988
Cattle egret, *Bubulcus ibis*	Ciconiformes: Ardeidae	Parasharya *et al.*, 1994
Common myna, *Acridotheres tristis*	Passeiformes: Sturnidae	Anon., 1988; Singh *et al.*, 2003; Parasharya *et al.*, 1994; Nath and Singh, 1988
Green bee eater, *Merops orientalis* Latham	-	Anon., 1988
House crow, *Corvus splenens*	Passeriformes: Ploceidae	Anon., 1988; Parasharya *et al.*, 1994; Nath and Singh, 1988
House sparrow, *Passer domesticus*	Passeriformes: Ploceidae	Anon., 1988
Indian roller, *Coracias bengalensis*	-	Anon., 1988
Jungle crow, *Corvus ma crorhynchos*	Passeriformes: Ploceidae	Singh *et al.*, 2003; Parasharya *et al.*, 1994
Rosy pastor, *Sturnus roseus*	Passeriformes : Sturnidae	Anon., 1988
Spotted owlet, *Athene brama*	Strigiformes : Strigidae	Anon., 1988
White breasted king fisher, *Ceryle rudis*		Anon., 1988
Other Biocontrol Agents		
Spiders		
Oxyopes pandae	Arachnida	Nath and Singh, 1988
Stregodyphus sp.	Araneae	Anon., 1988
Thomisus sp.	Arachnida	Nath and Singh, 1988

Species	**Common name**	**Reference**
Non - arthropod Predators		
Bufo melanosticus	Indian Toad	Nath and Singh, 1988
Calotes versicolor Daudin	Lizard	"
Hemidactylus flavivirdis	Lizard	"
Gecho gecho	Lizard	"
Mammals		
Cuon alpinus Pallas	Stray dog	"
Felis sonstantina	Stray cat	"

Chemical Control

Holotrichia serrata (F.) causes severe damage to the nodules and roots of groundnut. The most effective chemical in reducing infestation and increasing the yield in both seasons was a dust containing 5% aldrin applied at 2.5 kg toxicant/ha at sowing, followed by dusts containing 0.65% lindane or 5% dieldrin (Gandhale *et al.*, 1982), now most of them were banned.

The efficacy of phorate 10G and Ekalux [quinalphos] 5G against the white grub complex [scarabaeids] other than Holotrichia on groundnut and sugarcane was evaluated by Janardan-Singh and Nath (1985).. Phorate was applied at 20.0 and 25.0 kg a.i./ha and quinalphos at 20.0, 25.0 and 30.0 kg a.i./ha. Both insecticides and their different doses were significantly superior to the control. Phorate at 25 kg and quinalphos at 30 kg/ha reduced the white grub population on sugarcane by 86.13 and 79.18%, respectively, one month after treatment, and by 86.78 and 79.43%, respectively, on groundnut. Both doses of phorate were significantly superior to the 3 doses of quinalphos. There was an increase in the population of white grubs at the time of harvest which is thought to be due to the migration of *Autoserica nathani* [*Maladera nathani*], *Apogonia ferruginea, A. uniformis* and *A. rauca* from *Acacia arabica* [*A. nilotica*] and *Azadirachta indica* trees surrounding the fields. The yield of groundnut was significantly higher from the treated plots compared with the control.

Insecticides like chlorpyriphos [chlorpyrifos] 20 EC, quinalphos 25 EC (at 5, 10, 15, 20 and 25 ml\kg seed), cypermethrin 25 EC (5, 10 and 15 ml) and imidacloprid (1, 2 and 3 g) were effective in controlling white grubs. Higher rates of chlorpyriphos and quinalphos reduced germination (19.2 and 21.5 plants/2 m row) compared to the control (24.7 plants). The lower and medium doses of insecticides and all the doses of cypermethrin and imidacloprid were safe. In addition, phorate 10 G at 1.88 and 2.5 kg a.i.\ha was applied as pre-sowing soil application, mixed with seed and applied in standing crops between the rows at 30, 45 and 60 days after sowing. Phorate applied by mixing with seed significantly reduced seed germination at both the rates of 18.75 and 25 kg/ha (16.7 and 14.8 plants/m row, respectively), whereas the germination was not affected in the other two methods. M-145 was most tolerant with no adverse effects observed, followed by M-13 (7.8% germination reduction) whereas PG-1 was the most susceptible with 16.0% reduction in germination over control. These studies advocate cultivar specific seed treatment instead of blanket recommendation (Chaudhary and Dashad, 2002).

To evaluate newer insecticides for controlling white grubs and their deleterious effects on seed germination of different groundnut cultivars (M-13, M-145 and PG-1), insecticides chlorpyrifos 20 EC, quinalphos 25 EC at 5, 10, 15, 20 and (25 ml/kg seed), cypermethrin 25 EC (5, 10 and 15 ml) and imidacloprid (1, 2 and 3 g) were applied as seed treatment. In addition, phorate 10 G at 18.75 and 25 kg/ha as presowing soil application, mixing with seed and application in standing crop between the rows at 30, 45 and 60 days after sowing was also tested. All insecticides were effective in controlling white grubs. Higher doses of chlorpyrifos and quinalphos reduced germination (19.2 and 21.5 plants/12 m row) compared to control (24.7 plants). The lower and medium doses of insecticides and all the doses of cypermethrin and imidacloprid were safe. Phorate applied by mixing with seed significantly reduced seed germination at both the doses of 18.75 and 25 kg/ha (16.7 and 14.8 plants/m row, respectively), whereas the germination was not affected in other two methods. Variety M-145 was most tolerant where germination was not hampered, followed by M-13 (7.8% reduction), whereas PG-1 was most susceptible with 16.0% reduction in germination over control. These studies advocate cultivar specific seed treatment instead of blanket recommendation (Chaudhary and Dashad, 2002).

References

Adsule, V. M. and Patil, S. M. (1990). *Leucopholis lepidophora*: a new white grub pest of groundnut in Western Maharashtra. *Groundnut News*., 2: 2, 7.

Anonymous, (1991). Annual progress report Groundnut 1990. 38th annual *Kharif* Oilseed research worker's group meeting. All India Co-ordinated Research Project on Oilseeds. pp 16-18.

Anonymous, (1988). Annual Report, International Crop Research Institute for the Semi-Arid Tropics (ICRISAT)., Patancheru, Andhra Pradesh, India.

Avasthy, P.N. (1967). The problems of white grubs of sugarcane in India. Proceedings of the Interantional Society of Sugarcane Technology. 12:1321-1333.

Banerjee, S. N. and Mookherjee, A. L. (1960). Note on *Alissonotum impressicolle* Arr., (Scarabeidae: Coleoptera)., a new pest of sugarcane in West Bengal. *Proceedings of the Biennial Conference of Sugarcane Research Developement Wkrkers*, 4, 513-520.

Channabasavanna, G. P. (1954). Insect pest of some cash crops- nature of damage and control. Mysore Agricultural College Year Book. 89 pp.

Chaudhary, O. P. and Dashad, S. S. (2002) Deleterious effects of various insecticides used to control white grubs in groundnut cultivars. *Research on Crops*. 3(1): 164-170.

Chaudhary, O. P. and Dashad, S. S. (2002). Deleterious effects of various insecticides used to control white grubs in groundnut cultivars. *Journal of Applied Zoological Researches*. 13(2/3): 240-243.

David, H and Ananthanarayana, K. (1974). Some aspects of the biology of white grub, *Holotrichia serrata* F., in sugarcane in South India. *Whitegrub Newsletter*. 1:14-15.

David, H. and Kalra, A. N. (1966). Some observation on *Holotrichia (Lachnosterna) serrata* F., a beetal pest of sugarcane in Hospet area of Mysore state. *Indian Sug*., 16, 195-198.

David, H., Alexander, K. C. and Ananthanarayan, K. (1973). Milky disease bacterium of the sugarcane white grub. *Curr. Sci.*, 42:695-696.

David. H, V. Nandagopal and K. Ananthanarayanan.(1986). Recent studies on the white grub *Holotrichia serrata* F. infesting sugarcane. *Journal of Soil Biology*. 6:117-127.

Dhaliwal, J. S.(1980). White grubs as a pest of forage in Punjab. *Whitegrubs Newsletter*. 3, 1-2.

Easwaramoorthy, S. (1986). New approaches in Sugarcane pest control, In Sugarcane Entomology in India Eds David H., Easwaramoorthy, S., and Jayanthi R, Coimbatore, 437-458 pp.

Gandhale, D. N., Naik, L. M. and Khandge, S. V. (1982). Soil application of various insecticidal dusts for the control of white grubs (*Holotrichia serrata* F.) on groundnut (*Arachis hypogaea* L.). *Pesticides*. 16: 21-22.

Gupta, B. D. and Avasthy, P. N. (1960). Insect pests of sugarcane and their control. Indian Cent. Sug. Cane Comm., New Delhi. 34 pp.

Gupta, B.D. (1958). A new floresscent light trap. *Indian Institute of Sugarcane Research Newsletter.* 4:17.

Gupta, B.D. and Avesthy, P.N. (1957). First record of *H. consanguinea* Blanch in sugarcane in India. *Current Science.* 26:114-115.

Gupta, K.M.(1973). Neem leaves attract white grub beetles. *Indian Journal of Entomology.* 35:276.

Gupta, K.M., K.M. Rai and Ramasudan Dyal. (1968). First record of *Lepidiota mansueta* Burm., a new white grub beetle damaging sugarcane in U.P. *Indian Sugar.* 18:757.

Hari deva Vasu.(1970). *Ricinis communis* Linn. As a new host record for the white grub,*Holotrichia consanguinea* Blanch. Indian Jouranla of Entomology, 32: 182.

Hukkam Singh and Rizvi,S. M. A. (1968). Preliminary observation on *Heteronychus robustus* Arr., a beetal pest of Sugarcane. *Indian Sug*. 18, 251.

Janardan-Singh and Nath, P. (1985). Effects of some granular insecticides on white grubs population in groundnut and sugarcane. *Pesticides*.19: 66-68.

Jayaramaiah, M and Veeresh, G. K. (1983). Studies on the symptoms of infection caused by the new silkworm white muscardine fungus, *Beauveria brongniartii* (Sacc.) Petch to different stages of white grub, *Holotrichia serrata* Fab. (Coleoptera: Scarabaeidae). *Journal of Soil Biology and Ecology.* 3: 7-12.

Jayaramaiah, M. (1981). Studies on the entomogenous fungus, *Beauveria brongniartii* Sacc. Petch. in relation to white grub, *Holotrichia serrata* Fab. and silkworm, *Bombyx mori* L. and possibilities of its use in the management of the white grub. Ph.D. thesis (unpublished)., University of Agricultural Sciences, Bangalore.

Kalra, A. N. and Sharma,N.C.(1965) . *Heteronychus robustus* Arr., a new beetle pest of sugarcane. Indian Sugars, 15, 613.

Kalra,A.N and Kulshreshta,J.P. 1961. Studies on the biology and control of *lachnosterna consanguinea*. Blanch, a pest of sugarcane in Bihar (India). Bulletin of Entomological Research 52: 577-87.

Majumdar, N. and Teotia, T. P. S. (1965). Biology of *Holotrichia serrata* F. a serious pest of Palas, *Butae monosperma* Lam. *Indian Journal of Entomology.* 27:153-159.

Nath, P. and Singh, J (1988). White grub management in sugarcane crop. Pesticides. 22 (12): 23-27.

Paras Nath and Janardhan Singh (1980). *Schizonycha ruficolis* F. as a beetle pest of rose. *Whitegrubs Newsletter.* 3:4-5.

Parasharya, B. M., Dodia, J. F., Mathew, K. L. and Yadav, D. N. (1994). Natural regulation of white grub (*Holotrichia* sp,: Scarabaeidae) by birds in agroecosystem. *Journal of Biosciences.* 19 (4): 381-389.

Patel, R. M., Patel, C. B. and Vyas, H. N. (1978). Record and some observations on local milky disease in white grubs (*Holotrichia* sp. near *consanguinea* Blanch) in India.*Indian Journal of Entomology*. 39: 181-182.

Patel, R. M., Patel, C. B. and Vyas, H. N. (1977). Record and some observations on local milky disease in white grubs (*Holotrichia* sp. near *consanguinea* Blanch.) in India. *Indian Journal of Entomology,* 39 (2): 81-182.

Rao Deo, A. K. (1974). A note on white grub control campaign in Maharashtra State. *Whitegrubs Newsletter*. 1: 19-20.

Rao Deo, A.K. and Deshpande, S.V., Pur, S.N. and Bilapate, G.G. (1976). A large scale campaign for the control of white grubs (*H.serrata*) in Maharashtra State. *PANS.* 22: 223-228.

Rao, G. N. and Vijayalakshmi, U. (1995). A note on the occurrence of certain parasitic fungi on insect pests of sugarcane. *Curr. Sci.*, 28:295.

Rao,V.P.,Ghani, M.A.,Sankaran,T. and mathur, K.C. (1971). A review of the biological control of insects and other pests in South East Asia and the Pacific Region, 14 pp.

Sachan, J. N. and Pal, S. K. (1974). White grub and their control at Centra Arid Zone Research Institute, Jodhpur. *Whitegrub Newsletter*., 1, 20-24.

Shanti, A. N. and Sivakumar, C. V. (1991). Comparative virulence of *Steinernema glaseri* Steiner and *Steinernema carpocapsae* (Weiser) to the chafer *Holotrichia consanguinea* (Coleoptera: Scarabaeidae). *Indian Journal of Nematology.* (publ. 1993). 21 (2) :149-152.

Shashi Sharma, Gupta, R. B. L., Yadava, C. P. S., Sharma, S., Reddy, P.P. (ed.), Kumar, N. K. K. (ed.) and Verghese, A. (1998). Preliminary studies on activity of three entomofungi against whitegrub, *Holotrichia consanguinea*. Advances in IPM for horticultural crops. *Proceedings of the First National Symposium on Pest Management in Horticultural Crops: environmental implications and thrusts.* Bangalore, India, 15-17 October 1997. 1998, 230-232.

Shinde, V. K. R. and Sharma, K. (1971). *Bacillus popillae* Dutky pathogenic to *Lachnosterna consanguinea* Blanch. *J. Econ. Ent.*, 64: 1301-1302.

Shivayogeswara, B. and Veeresh, G. K. (1983). Movement and migration of white grubs (*Holotrichia serrata* F.) in soil with varying cropping pattern and soil types. *Journal of Soil Biology and Ecology.* 3: 136-146.

Singh, M. P., Bisht, R. S. and Mishra P. N. (2003). Survey to explore the natural enemies of white grub (*Holotrichia* spp.) in Garhwal Himalayas. *Indian Journal of Entomology*. 65 (2): 202-210.

Srivatsava, A.S and Khan, R.M. (1963). Bionomics and control of *Holotrichia insularis* (Melolonthidae:Coleoptera). *Indian Journal of Entomology.* 25:347-354.

Veeresh, G. K. (1977). Root grub controlling campaign in Karnataka. *Whitegrub Newsletter*. 1: 28-33.

Veeresh, G. K. and N. V. M. Reddy. (1978). Some aspect of larval taxonomy of white grubs of Karnataka – India. In "Soil Biology and Ecology in India" (Edwards C. A. and G. K. Veeresh, eds.). Univ. agric. Sci. Banglore. p216- 222.

Veeresh, G. K. (1977). Root grub controlling campaign in Karnataka. *White grub Newsletter.* 1: 28-33.

Veeresh, G. K. (1977a). Studies on the Root grubs of Karnataka. *U. S. A. Monograph*. 2:87.

Veeresh, G. K. (1977b). Bionomics and control of white grubs in India. *Sugarcane Newsletter.* 9:44-56.

Veeresh, G. K. (1980). Recent findings on bio-ecology of white grubs. *White grubs News Letter*, 3: 5-6.

Veeresh, G.K. (1977). Bionomics and control of white grubs in India. *Sugarcane Newsletter.* 9: 44 -56.

Verma, A. N. (1975). Occurrence of white grubs in Haryana. *Current Science*. 44: 32-33.

Vyas, H. G., Joshi, D. P., Patel, K. C., Yadav, D. N. and Dodia, J. F. (1986). Natural incidence of milky disease of white grubs in Gujarat. *Indian Journal of Agricultural Sciences.,* 56(3): 213-214.

Vyas, H. G., Joshi, D. P., Patel, K. C., Yadav, D. N. and Dodia, J. F. (1986). Natural incidence of milky disease of white-grubs in Gujarat. *Indian Journal of Agricultural Sciences.* 56: 213-214.

Vyas, R. V., and Yadav, D. N. (1993). *Steinernema glaseri* (Steiner). Travassos, a biological control agent of root grub *Holotrichia consanguinea* Blanchard. *Pakistan Journal of Nematology*. 1 (11): 41-44.

Vyas, R. V., Patel, H. R., Patel, D. J. and Patel, N. B. (1997). Biological suppression of root-knot nematode - white grub pest complex attacking groundnut. *International Arachis Newsletter*. 17: 40-41.

Vyas, R. V., Yadav, D. N. and Patel, R. J. (1991). Efficacy of *Bacillus popilliae* var. *holotrichiae* against white grub (*Holotrichia consanguinea*). *Indian Journal of Agricultural Sciences*, 61(1): 80-81.

Vyas, R.V., and Yadav, D.N. (1993). *Steinernema glaseri* (Steiner). Travassos, a biological control agent of root grub *Holotrichia consanguinea* Blanchard. *Pakistan Journal of Nematology*. 11(1):41-44.

Vyas, R.V., Yadav, D.N. and Patel, R.J. (1991a). Efficacy of *Bacillus popilliae* var. *holotrichiae* against white grub (*Holotrichia consanguinea*). *Indian Journal of agricultural Sciences.* 61(1):80-81.

Vyas, R.V., Yadav, D.N. and Patel, R.J. (1991b). Studies on mass production of *Bacillus popilliae* var. *holotrichiae,* a promising pathogen of white grub in Gujarat. *Gujarat Agriculture University Research Journal.* 17(1):30-37.

Yadava, C. P. S., Pandey, S. N., Bhardwaj, S. C. and Yadava, S. R. S. (1975). Predation of white grubs (*Holotrichia* spp.) by birds. *Indian Journal of Entomology*. 35: 169.

□□□

Termites

T. V. PRASAD AND RAJIV RATHOD

8.1. Introduction

Twenty six species of termites belonging to the families Termitidae and Hodotermitidae of the order Isoptera damage groundnut crop through out the world. In India five species of termites *viz., Macrotermes thoracalis* Sjostedt, (Rambur), *Odontotermes* sp. *Odontotermes brunneus* (Hagen), *Trinervitermes biformis* (Wasmann), *Microtermus sp.* damage not only the groundnut pods, but they also feed on roots, stems and some times even the foliage. Groundnut crop is infested with *Odontotermes obesus, O. brunneus, O. wallonensis* and *Microtermes obesi* (Rao *et al.*, 1991).

8.2. Nature of damage

Termites penetrate into the tap root, hollow out the contents and replace with mud, leading to wilting and death of plants (**Plate 1**). They live in colony. The feed on roots by replacing with mud. This type of damage is very often to be seen in sandy and red soils of Northern and Central India caused by

Plate 1. Tap root damaged by temites

Odontotermes obesus (Rawat *et al.*, 1970). On mature pods, they feed on the shell and scarify the pods externally and occasionally bore into them, paving way for invasion by soil fungi that can attack kernels. Kernels thus infected rot due to attack by saprophytic fungi and aflatoxin producing fungi (Mc Donald and Harkness, 1967).

8.3. Economic importance

O. obesus has been found to cause 5 to 15 per cent plant mortality in Madhya Pradesh (Kaushal and Deshpandy, 1967; Rawat *et al.*, 1970) 10-45 per cent in western Rajasthan (Roonwal, 1979) and 46 per cent pods bored by termites in Madhya Pradesh (Kaushal and Deshpandy, 1967). Pod damage is more in sandy and light red soil, and if there is delay in harvesting of the groundnut.

Singh and Singh (1992) tried to determine the damage caused by termites (*Microtermes* and *Odontotermes*) and the ant *Dorylus orientalis* to groundnut in monoculture and in intercrop systems. Damage was greater in the monoculture and in sorghum-intercropped plots than in other intercrops in all seasons.

8.4. Bionomics

The social organization in termites comprises four different castes which are structurally and functionally distinct. The distinguishing features of the different caste have been studied by Pawar and Sood (1964) which is given below.

O. *obesus*

This species build large mounds in India. The mounds are tall, unilocular with buttresses, or dome-shaped and multilocular with or without turrets or in the form of lumps of earth within the main nest, underground. The mounds may reach a height of three meters and are sub-cylindrical. Generally, only one queen and a king are present, though more than one pair is also reported. In such a case, each pair is present in a separate royal chamber. Swarming starts with the beginning of monsoon at dusky hours and may continue during night. The ratio of soldiers: workers: nymphs in the mound is 7.7: 49 : 43.3 in the non-mound building mounds and 5.5 : 66.5 : 28 in the mound building.

T. biformis

It is a common foraging termite in India. The workers go out on foraging expeditions in thousands, a little before sunset in regular columns, with the soldiers guarding against ants and other predators. The nests of this species are generally subterranean and consist of ramifications of galleries and small chambers which are coated with black excretory matter. The small dome-shaped mounds are covered with grass.

8.5. Caste Systems in Termites

Termites are the social insects exhibiting polymorphism living as communities. It comprises four castes which are distinct structurally and functionally. The features distinguishes between castes are described below based on detailed information given by earlier workers (Pawar and Sood, 1964; Imms, 1977).

8.5.1. Primary reproductives

Primary reproductives represents the royal pairs constituting – the queen and king of the colony. They clearly exhibits two pairs of large membranous wings, which are shed during and after the process of swarming of founding new colonies. The wings are narrow at the base and broad and round at the apex. The well sclerotized body is oily or smoky brown in colour with well developed compound eyes and paired ocelli. The winged termites, in large number are attracted to light and are predated. Those termites which survive after shedding their wings, always move in pairs, enter the soil or crevices in the wood, where they mate and reproduce to start a new colony.

The female which is referred to as the queen measured approximately 6cm X 1.3 cm in size when mature. With advance in age, the abdomen of the gravid female gradually gets distended, due to large number of eggs within. The queen is confined to the royal chamber with the male and their function is to reproduce solely and continue to multiply the colony to larger size. The fecundity of a female varies from 2,000 to 22,000 eggs per day. The life span of a queen varies from 5 – 15 years, sometimes, which may extend up to 50 years.

8.5.2. Supplementary reproductives

When compare to the body of the primaries the supplementary reproduce are less sclerotized and pigmented. The speciality of these forms are not found generally in the colonies headed by the royal pairs, however, they appear several weeks after the death of any one or both the royal members to continue the development of the colony.

8.5.3. Workers

The pale, slightly sclerotized looking more of a nymphs are the sterile forms of termites which are called workers. The head is bent downwards, wider than the reproductives, however they are not larger than that of soldiers. They possess reasonable well developed compound eyes in the workers of Hodotermitidae, but are generally vestigial or absent. These are the forms which exhibit care for the eggs and young ones and in times of danger, they carry them to safer places. They tend the queen, forage for food and cultivate the fungus gardens. They nurse the young ones till they are getting independent. More importantly they are the actual enemies of crop plant and roots.

8.5.4. Soldiers

Like workers, the soldiers are also sterile forms, which are easily distinguished by their large sclerotized heads, apart form well chitinized head and thorax. The abdomen is dirty white in colour with delicacy. There are four well defined types of soldiers.

i) *Mandibulate soldiers :* These are the most commonly occurring forms with large powerful mandibles that may assume a strikingly grotesque shape,

ii) *Nasute soldiers :* They possess vestigial mandibles, but there is a long median frontal nasus or rostrum at the end of which frontal gland opens and these soldiers occur only in Nasutitermitinae (Termitidae),

iii) *Nasutoid soldiers :* They are found among Rhinotermitidae in which the mandibles vary in size, but the frons, clypeus and labrum are produced into a snout like structure bearing a dorsal channel for the flow of frontal gland secretions,

iv) *Phragmotic soldiers :* They are found in Kalotermitidae in which the head is strongly sclerotized, more or less plug like in structure that can be used to block the nest openings. Their mandibles are reduced.

The soldiers defend the colony mainly against other insects like ants and vertebrate predators. Typically soldiers attack with their mandibles, which can pierce or slice and can be used defensively. Other soldiers attack insect intruders with adhesive secretions of the salivary or frontal glands.

The other forms include,

i) Pseudergates, which replace the workers in their absence,

ii) The Achrestogonimes, which are alates that do not swarm but remain in the nest. Wings are lost and having atrophied gonads. Generally they do not take part in the maintenance of the colony, and

iii) The inter-castes, which are morphologically intermediate between the different castes.

8.6. Morphology of Imago and Soldier

The distinguishing features of imago and soldiers of the different species of termites infesting groundnut are given in detail by Chhotani (1980).

***Odontotermes obesus* (Imago)**

Head capsule dark brown, post clypeus pale brown, pronotum and abdomen somewhat paler than head. The former with 'T' shaped pale mark, densely pilose. Head capsule sub circular (width 2.43-2.80 mm); fontanelle minute; eyes large, sub circular, projecting out, maximum diameter 0.73 - 0.90 mm; ocelli large, oval, situated very close to eyes; antennae 19 segmented; post clypeus swollen, about half as long as wide; pronotum sub-trapezoidal, somewhat emarginate both anteriorly and posteriorly; wings brownish. Length of the body 23-30 mm with wings and 10-14 mm without wings.

***Odontotermes obesus* (Soldier)**

Head capsule yellowish to dark brown or dark reddish brown, antennae pale basally and dark distally, body yellowish to yellowish brown; head sparsely and body moderately pilose. Head capsule oval, converging in front (1.201.67 x 1.07-1.37 mm); antennae with 16-17 segments, third segment shortest; labrum tongue-shaped; mandibles thin, sabre-shaped, left mandible with a prominent tooth at the base of distal third, postmentum arched, convex laterally; pronotum saddle-shaped, deeply notched both anteriorly and posteriorly. Body length about 4-6 mm.

***T. biformis* (Imago)**

Head capsule and abdomen brown with reddish tinge above, pronotum pale brown, sternite brownish, pale medially; head and body densely hairy. Head capsule sub circular, broader than long to base of mandibles (length 1.2-1.4 X 1.52-1.82 mm), medially depressed a little; epicranial suture either present or absent; fontanelle pale; Y-shaped or triangular; eyes sub-oval, moderately large (0.5-0.55 mm dia.); ocelli oval, translucent, length 0.180.25 mm, from eyes situated at a distance of one fifth to one third of length; antennae 15 or 16 segmented, segment three longer than two; postclypeus swollen, length a little less than half its width; labrum dome shaped; pronotum trapezoidal, anterior margin straight, not incised, posterior margin either straight or weakly convex, with a weak to strong medial emargination; wings hairy and punctate. Body length with wings 18-19 mm and without wings 9-12 mm.

***T. biformis* (Soldier)**

Soldier major : Head capsule brown, rostrum reddish brown and pale apically, abdomen pale brown to dark smoky brown dorsally. Head capsule . without rostrum, almost round or pear-shaped, not constricted behind antennae (head length with rostrum 1.35-1.60 mm, width 1.25-1.50 mm) in profile with a weak concavity at base of rostrum; occiput greatly arched in profile; rostrum conical, uplifted apically, tip with a few short hairs; antennae 12-15 segmented, segment three variable, dependil1g on number of segments, slightly longer by nearly 2.5 times that of second; mandibles vestigeal without spines; pronotum, saddle shaped, anterior lobe greatly upturned and broadly invaginated at anterior margin, posterior margin weakly convex; abdomen elongate; tergites without any long hairs; sternites with long and short hairs.

Body length 4.5-5.0 mm.

Soldier minor: Head capsule yellowish to dark brown, rostrum dark, abdominal tergites fuscous. Head capsule sub-rectangularly oval, length about one and a half times that of width (length with rostrum 1.4-1.7 mm, without rostrum 0.85 – 1.10 mm; width 0.55-0.77 mm) in profile, concave; rostrum long, cylindrical, awl-shaped; antennae long, 12-14 segmented, segment three about twice as long as second or a little longer; mandibles vestigeal, without any spines; pronotum small, saddle-shaped, not notched anteriorly and convex posteriorly. Abdomen as in soldier major. Body length 3 - 4 mm.

8.7. Seasonal activity

In general the termites activities are enhanced under drought condition (Roonwal, 1981). In sandy and sandy loam soil, even in red soil termites assume economically important (Agarwala, 1955). Those fields Where the wheat straw, groundnut shells and paddy straw are used as mulch, the termite activity increase tremendously.

8.8. Alternate Host

O. obesus

Apart from groundnut, this species has been reported to damage fruit trees, grasses, cereals, oil trees and weeds which include, *Abies pindrow* Royle, *Acacia* sp., *Artocarpus integrifolia* Linn., *Cassia fistula* Linn., *Calophyllum* sp., *Cynodon dactylon* Pers., *Dalbergia sissoo* Roxb., *Eucalyptus* sp., *Mangifera indica* Linn., *Morus* sp., *Pyrus sinensis* Lindl., *Shorea robusta* Gaertn., *Tamarindus* sp., *Zyzyphus* sp., *Bambusa arundinacea* Willd., *Psidium guajava* Linn., *Triticum aestivum* Linn., *Cedrus deodara* (Roxb.), *Gossypium* sp., *Saccharum* spp., *Cocus nucifera, Ipomoea batatus, Citrus aurantifolia , betel vine, Areca catechu, Zea mays* and *Camellia sinensis* Linn, etc.

8.9. Mechanisms of resistance

Pod shell flour from 21 cultivars with different reactions to *Microtermes* and *Odontotermes* was analysed biochemically. Crude fibre content was positively correlated with percentage termite damage, but contents of fat and starch, although variable, were not correlated with termite damage. Of 8 minerals estimated, Zn and Mn contents were negatively correlated with termite damage (Rasheedunisa, 1986). Termite damage was assessed at harvest from varietal screening trials conducted at Bhubaneswar over 2 seasons, the 1987 rainy season and 1988-89 post rainy season. In 1987, 23 groundnut varieties were grown and of these, 20 varieties were grown in 1988-89. Pod damage by termites differed significantly among the varieties in both seasons. In 1987, pod damage was 6% or less in all varieties except VG77 (11.1%) and CGC9 (9.9%). In 1988-89, pod damage ranged from 2.5 to 28.1% and was higher than in the earlier season in all varieties except one. Four varieties were relatively less susceptible to termite attack (viz. JL24, J11, TMV2 and CO1). Among these less susceptible varieties, JL24 gave the highest pod yield in both seasons (1.32 and 2.01 t/ha in 1987 and 1988-89, respectively) (Mohapatra *et al.*, 1995).

8.10. Management Strategies

8.10.1. Biological control

Adults and larvae of the predator beetle, *Omphra pilosa* Klug (Coleoptera: Carabidae) feed exclusively on all castes of Isoptera, with each adult consuming on average 15 workers and 2.1 alates per day. The consumption rate increased with increasing prey density, both larvae and adults were most abundant in millet, groundnut and soybean and lived under weed stubble that was infested with termites.

Other natural enemies recorded on various species of termites include Sycanus collaris Fabr. (Reduviidae: Hemiptera), *Cydnocoris gilvus* Burm. (Reduviidae: Heteroptera) Venkatesan *et al.* (1999), *Tetraponera rufonigra* (Hymenoptera: Formicidae) (Mehrotra, 1988); *Rhynocoris marginatus* (Reduviidae: Hemiptera) (Ambrose *et al.*, 1990); *Coranus soosai* (Reduviidae: Hemiptera) (Ambrose and Alexander, 1989) and grey langur, *Presbytis entellus* (Moore, 1985). The occurrence and activity of the predator, *Omphra pilosa*, were studied in 1987-1990 in Karnataka, India. Adults and larvae of the predator fed exclusively on all castes of Isoptera. The adults each consumed on average 15 workers and 2.1 alates per day. The consumption rate increased with increasing prey density. Larvae stored termites in their tunnels. The relative frequencies of the predator in millet, groundnut, soybean, paddy [rice], mangoes, mulberries and a Eucalyptus plantation were investigated. Both larvae and adults were most abundant in millet, groundnut and soyabean ecosystems and lived under weed stubble that was infested with termites (Kumar and Rajagopal, 1990).

In addition several fungal pathogens were also recorded infecting *Odontotermes obesus*. Of them *Beauveria bassiana* was the most effective followed by *Metarhizium anisopliae, M. flavoviride, Paecilomyces lilacinus* and *P. fumosoroseus* (Khan *et al.*, 1993; Gurusubramanian *et al.*, 1999). Laboratory studies clearly demonstrated the potential of the reduviid predator *Acanthaspis pedestris* as a biological control agent of the groundnut pest *Odontotermes assmuthi* (Sahayaraj, 1995).

8.10.2. Cultural control

The effects of intercropping groundnut with sunnhemp on the incidence of termites (*Microtermes obesi* and *Odontotermes* spp.) and their damage were studied in the rainy and post-rainy seasons of 1987-88 in Andhra Pradesh,

India. Termite incidence in different cropping systems was determined by baiting with bamboo stakes.

In a preliminary trial with 4 mulches to protect drying groundnut pods against scarification by *Odontotermes* spp. and *Microtermes obesi*, sunn hemp mulches increased termite damage, while neem cake and *Ipomoea fistulosa* mulches seemed to act as repellent barriers between the soil and the pods; *Celosia argentea* mulches afforded intermediate protection. Damage was less in pods placed on top of mulches than mixed in them (Gold *et al.*, 1989).

Damage to groundnut plants and pods was evaluated at harvest. Intercropping groundnut with sunnhemp did not affect termite abundance or damage to groundnut (Gold and Wightman,1991).

The use of mulches to reduce *Microtermes obesi* and *Odontotermes* spp. damage was studied Gold *et al.* (1991). Scarification of groundnut was 80-90% lower for pods dried in neem cake *or Ipomoea fistulosa* [*I. carnea*] mulches than for pods dried directly on the soil. Similarly, subterranean attack by termites on bamboo baits was 20-40% lower in mulched plots than in controls. However, mulches were ineffective in protecting bullock manure from removal by termites. Termite location of concentrated food sources, such as bullock manure, was followed by recruitment of workers and construction of protected runways. In contrast, termites foraging on limited food sources, such as groundnut pods, did not construct runways and had greater exposure to mulches. Groundnut was known to be infested with *Odontotermes obesus, O. brunneus, O. wallonensis* and *Microtermes obesi*. The field was treated with neem (*Azadirachta indica*) cake at sowing and 40 d after crop emergence so that the cake initially formed a layer about 2 cm thick over the soil. Treatment did not reduce the numbers of Isoptera in the field but gave a small (14.3%) decrease in the number of scarified pods and increased yield (Rao *et al.*, 1991).

8.10.3. Chemical control

Mishra (1999) studied the efficacy of chlorpyrifos 10 G as a soil treatment at 0.50, 0.75 and 1.00 g a.i./ha and chlorpyrifos 20 EC as a seed dressing at 12.5 ml/kg seed against termites with phorate 10 G at 1.00 kg a.i./ha applied to the soil in the furrows at sowing during the *kharif* seasons of 1996-97 at Mahispat, India. The results showed that plant damage was significantly low with chlorpyrifos 20 EC seed dressing and phorate 10 G soil applications. The percentage pod infestation at harvest was lowest with chlorpyrifos 20 EC seed dressing and chlorpyrifos 10 G soil applications at 1.00 kg a.i./ha. However, pod yield was significantly higher with chlorpyrifos 10 G at 1.00 kg a.i./ha, chlorpyrifos 20 EC as a seed dressing and phorate 10 G at 1.00 kg a.i./ha. As seed dressing with chlorpyrifos 20 EC at 12.5 ml/kg seed had the lowest plant damage, percentage pod infestation and highest pod yield, it was concluded to be the best treatment. Soil application of chlorpyrifos 10 G and phorate 10 G were equal to chlorpyrifos seed dressing in terms of termite control and pod yield.

References

Agarwala,. (1955). Control of Sugarcane termites. *J. Econ Ent.,* 48(5): 535-537.

Ambrose, D. P. and Alexander, P. M. (1989). Impact of mating on the post-embryonic development of *Coranus soosaii* Ambrose and Vennison (Insecta-Heteroptera-Reduviidae).Proceedings of the Indian Academy of Sciences, Animal Sciences., 98 (5):305-311.

Ambrose, D. P., Saju, T. and Sahayaraj, K. (1990). Prey influence on the development, reproduction and size of assassin bug *Rhynocoris marginatus. Environment and Ecology.* 8(1B): 280-287.

Chhotani, O. B. (1980). Termites pests of agriculture in the Indian Region and their control. Tech. Monograph, Zoo, Survey, India, 4:87.

Giradii, R. S. (2000). Termite and ant manace to earthworm in field vermi composting beds. *Insect Environment.* 6(1):26-27.

Gold, C. S. and Wightman, J. A. (1991). Effects of intercropping groundnut with sunnhemp on termite incidence and damage in India. Insect Science and its Application. 12 (1-3):177-182. (In Proceedings of the Second International Conference on Tropical Entomology, held in Nairobi, Kenya, on 31 July-4 August 1989).

Gold,C.S., Wightman,J.A., Pimbert,M. (1989). Mulching effects on termite scarification of drying groundnut pods International-Arachis-Newsletter., 6, 22-23.

Gold,C.S., Wightman,J.A., Pimbert,M.P.(1991). Effects of mulches on foraging behaviour of *Microtermes obesi* and *Odontotermes* spp. in India. Insect-Science-and-its-Application. 1991; 12(1-3): 297-303.

Gurusubramanian, G., Tamuli, A. K. and Ghosh, R. K. R. (1999). Susceptibility of *Odontotermes obesus* (Rambur) to *Beauveria bassiana* (Bals.) Vuill. *Insect Science and its Application.* 19 (2-3):157-162.

Imms, A. D. (1977). A general text book of Entomology (O.W. Richards and R.G. Davies, eds.). John wiley and sons, New York, 1354pp.

Kaushal, P. K. and Deshpandy, R. R. (1967). Losses to groundnut by termites. *J.N. Krishi Vishwavidyalaya Res J.* Jabalpur, 1:92-93.

Khan, H. K., Jayaraj, S. and Gopalan, M. (1993). Muscardine fungi for the biological control of agroforestry termite *Odontotermes obesus* (Rambur). *Insect Science and its Application.* 14(4):529-535.

Mc Donald, D. and Harkness, C. (1967). Aflatoxin in the groundnut crop at harvest in northern Nigeria. *Tropical Science.* 9:148-61.

Mehrotra, A. K. (1988). Activity and feeding habit of *Tetraponera* (=*Sima*) *rufonigra* (Jerdon) (Formicidae: Pseudomyrmecinae: Hymenoptera) in outer Siwalik hills. *Annals of Entomology.* 6(1): 57-58.

Mishra, H. P. (1999). Efficacy of chlorpyrifos against termites in groundnut. *Indian Journal of Entomology*. 61(4):326-329.

Moore, J. (1985). Insectivory by grey langurs. *Journal of the Bombay Natural History Society*. 82(1):38.

Pawar, J. G. and Sood, N. K. (1964). When forearmed against them, termites need hold no terror for the farmer. *Indian Farming*. 14: 17-19.

Rao, D. V. R., Singh, K. N., Wightman, J. A. and Rao, G. V. R. (1991). Economic status of neem cake mulch for termite control in groundnut. *International Arachis Newsletter*. 9: 12-13.

Rasheedunisa, (1986). Preliminary investigations on some chemical constituents of pod shell in relation to resistance of groundnut cultivars against termites. *Journal of Oilseeds Research*. 3(2): 195-200.

Rawat, R. R., Deshpande, R. R. and Kaushal, P. K.(1970). Comparative efficacy of different modern insecticides and their methods of application for the control of termite, *Odontotermes obesus* in groundnut, *Arachis hypogaea, Madras agric J*. 57(2):83-87.

Roonwal, M. L. (1979). Termites life and termite control in Tropical South Asia. Scientific Publishers, Jodhpur, Rajasthan.

Roonwal, M. L. (1981). Termites of agricultural importance in India and their control. In Progress in soil biology and ecology in India" (Vereesh, G.K. ed). *Tech Ser. 37. Univ agric Sci.*, Bangalore p. 253-256.

Venkatesan, S., Seenivasagan, R. and Karuppasamy, G. (1999). Nutritional influence of prey on the predatory potential and reproduction of *Sycanus collaris* Fabricius (Reduviidae: Heteroptera). *Entomon*. 24(3): 241-252.

Singh,T.V.K., Singh,K.M.(1992). Effect of different intercrops on termites and oriental army ant, Dorylus orientalis Westwood damage to groundnut. Indian-Journal-of-Plant-Protection. 20(2): 129-132.

Sahayaraj,K.(1995). Developmental stages and biocontrol potential of a reduviid predator, *Acanthaspis pedestris*, against termites on groundnut. International-Arachis-Newsletter.; (15): 57-59.

Kumar,P., and Rajagopal,D. (1990). Carabid beetle, *Omphra pilosa* Klug (Coleoptera: Carabidae) a potential predator on termites. Journal-of-Biological-Control. 1990; 4(2): 105-108.

Rao,D.V.R., Singh,K.N., Wightman,J.A., Rao, G.V.R.(1991). Economic status of neem cake mulch for termite control in groundnut. International-Arachis-Newsletter., No. 9, 12-13.

Mohapatra,H.K.,Padhi,J., Samalo,A.P., Patra,G.J.(1995). Screening promising groundnut varieties against termite damage at Bhubaneswar, Orissa, India. International-Arachis-Newsletter.; (15): 59-60.

□□□

9

Aphis craccivora Koch

C. A. MAHALINGAM AND M. V. GEDIA

9.1.Introduction

Aphids in general were considered to be a sporadic insect pest of groundnut till late 70s (Basheer, 1958; Sachan, 1959; Narayanan, 1962; Dorge *et al.*, 1966) and assumed a regular and serious pest status from early 80s (Brar and Sandhu, 1975; Brar, 1981; Nandagopal, 1992; Jena and Kurla, 1996). *Aphis craccivora* Koch is one of the most serious pests of groundnut in India, especially during the dry season (with a population peak in July), and outbreaks occur in the absence of regular monsoons Brar and Sandhu (1975). There are seven species of aphids recorded on groundnut as vectors of Peanut Stripe Viral diseases. These are *Aphis craccivora* Koch., *Aphis gossypii* Glover, *Hysteroneura setariae* (Thomas), *Aphis citricola*, *Rhopalosiphum maidis* (Fitch) and *Myzus persicae* (Sulzer) (Singh *et al.*, 1993).

9.2. Economic importance

A. craccivora is commonly called as groundnut aphid or cowpea aphid or black bean aphid. The greatest damage results from virus diseases which are transmitted by *A. craccivora*. It can act as vector for about 30 phytopathogenic viruses which can also do much damage in non-leguminous crops. Among the seven aphid species that were recorded mainly as vector for peanut stripe virus (PStV), *A. craccivora* is the major pest species occurring on groundnut causing pod loss up to 40% (Jena and Kuila, 1997). *A. craccivora* has been found to transmit about 23 viral diseases which included Bunchy top, Chlorosis, Chlorotic spot, Mosaic, Ring mosaic, Ring mottle and Peanut stripe virus (Srineevasalu *et al.*, 1991) occurring in India (Table 9.1).

The other species *H. setariae* is reported to a vector of Peanut Stripe Virus (Singh *et al.*, 1993) apart from transmitting four viruses in other crops *viz.*, maize, black gram, pepper and sugarcane. Similarly *A. glycine* is a vector of PStV and soybean stunt and mosaic virus. *A. citricola* transmits 19 viruses apart from PStV and Peanut green mosaic virus in groundnut. *A. gossypii* has been found to transmit 26 viruses in different crops apart from PStV and Yellow mottle mosaic virus in groundnut. *R. maidis* transmits 13 plant viruses apart from PStV in groundnut. *M. persicae* has been recorded transmitting 66 virus diseases in different crops in addition to transmitting 4 viral diseases in groundnut which include PStV, groundnut mottle virus, Peanut stunt virus and peanut green mosaic virus.

Table 9.1. Virus diseases of groundnut transmitted by the *A. craccivora* and their symptoms (Sreenivatulu *et al.*, 1991)

S. No.	Viruses	Symptoms
1.	Bean yellow mosaic	Chlorotic ring spotting
2.	Groundnut sclerotic spotting	chlorotic spotting
3.	Groundnut eyespot	chlorotic spotting, chlorotic ring spotting, line pattern
4.	Groundnut Rosette assistor	symptomless infection
5.	Peanut Mottle	Mottle, leaf rolling, interveinal mosaic, Systemic necrosis, line pattern stunting
6.	Peanut stripes	Mottle, stripes, stunting, blotches
7.	Peanut stunt	chlorosis, mottle, green vein banding, leaf rolling, systematic leaf deformation.

Aphids congregate on young leaves and young leaf buds **(Plate 1)**, sometimes even on flowers (**Plate 2)** and aerial pegs. They desap the phloem vessels. Under heavy infestation, plant become chloratic and leaves curl. Rangaswamy and Rao (1964) reported that due to heavy infestation of pegs, the pod formation is adversely affected. Young plants are stunted by early attack and severe attack at the time of flowering and see formation affects the

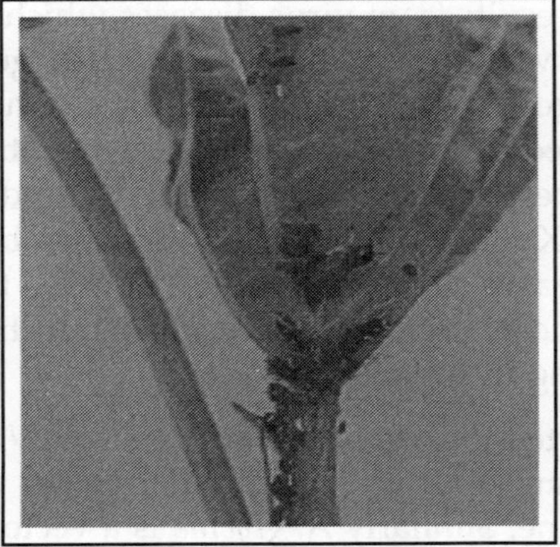

Plate 1. Aphids on the young leaf bud

Plate 2. Aphids on the flower

yield and can produce wilt symptoms. In addition abnormalities due to virus disease (resetting, stunting, mosaic, mottle etc.,) can be observed. More detailed information on the aphids role as vector in spreading virus diseases has been given elsewhere

9.3.Biology

Aphid passes through four moults before becoming adult with each instar lasting for 1 or 2 days. It completes 31 overlapping generations during one year. Low rate of fecundity was associated with low nitrogen content of the host (Waghery and Singh, 1965). The adult lives up to 16-68 days during winter and 9-20 days during the rest of the year. The reproduction is mostly by parthenogenesis and rarely by sexual. This being a rainy season pest, the peak infestation was during July to August in south India and August in Gujarat, August to September in Madhya Pradesh (Behura, 1956; Khan And Hussain, 1965) and July to September in Punjab (Bakhetia and Siddhu, 1977). The aphid biology is considerably affected by the prevailing weather and the host plant species. The reproductive period of *A. craccivora* ranged from 5-30 days in different season producing 15-124 nymphs by a single female (Bakhetia and Siddhu, 1977). *Aphis craccivora* Koch. remains active throughout the year and has 31 overlapping generations. The durations of the pre-reproductive period, the reproductive period and the total life-span and the number of young per female are discussed in detail by Bakhetia and Sidhu (1977). Rainfall wipes out the aphid population and heavy population build up occurs during the intermittent drought periods.

Reproductive rate of *Aphis craccivora* Koch on groundnut was studied in the laboratory by Talati and Butani (1980). Only parthenogenetic viviparity was observed. The rate of reproduction per day varied from 1 to 13 for different adults, and the total number of nymphs produced varied from 38 to 99 at the prevailing temperatures of 22-28 deg C. The reproductive capacity tended to be lower at temperatures outside this range. The offspring from a single gravid adult aphid averaged 1743.2 in 15 days. The build-up was arithmetic in the first 7 days, and geometric thereafter as the older offspring themselves began to reproduce. On *Dolichos lablab* in the laboratory in India at 31 ± 1°C and 91± 1% RH, the durations of the nymphal and adult stages of *Aphis craccivora* averaged 4.4 and 7.6 days, respectively, and the numbers of nymphs deposited averaged 45.7/adult female (Verma *et al.*, 1983).

The biology of *A. craccivora* was studied in the laboratory at 22.7-27.2°C and 89.1% RH, on several hosts including groundnut (Srikanth and Lakkundi, 1988). On *Vigna unguiculata*, nymphs survived for 4.84 days and the adult longevity was 10.24 days. Fecundity and rate of reproduction were 47.50 and 6.09, respectively on this plant. Longer development, shorter adult longevity and lower fecundity were observed on *L. purpureus, V. mungo, V. radiata* and groundnut. The aphid did not develop beyond the 1st instar on peas and soybean. *Aphis craccivora* underwent four nymphal instars on cowpea in the laboratory, the nymphal period averaging 5.6, 5.1, 5.15 and 4.86 days, resp., in May-June, August-September, October-November and March-April. The corresponding durations of the total life cycle were 11.07, 11.15, 10.79 and 10.42 days (Patel and Srivastava, 1989).

9.4. Host plants

A. craccivora was observed to feed on 14 different plant species (mostly Leguminosae) in the field at different times of year (Bakhetia and Sidhu, 1977). The host plants of *A. craccivora* have considerable influence on their population build up apart from its life cycle. Longer development, shorter adult longevity and lower fecundity were observed on *Lablab purpurious, Vigna mungo, V. radiata* and groundnut. The aphid could not develop beyond first instar on peas and soybean (Srikhanta and Lakhanali, 1988). Host plants include mostly Leguminosae, also some Malvaseae, Caesalpinia, Caricaceae, Convolvulaceae, Chenopodiaceae, Cucurbitaceae, Compositae, Malpighiaceae, Mimosaceae, Oleaceae, Scrophulariaceae, Solanaceae, Tamaricaceae and ornamentals. An extensive survey of the Terai area of northeastern Uttar Pradesh, revealed that 25 species of aphids were recorded on more than 78 species of food plants. Among these, 21 species belonged to Aphididae, one to Drepanosiphinae, two to Greenidinae and one to Pemphigidae. *A. craccivora, A. gossypii, Melanaphis sacchari, R.maidis, Schizaphis graminum, Lipaphis erysimi, Macrosiphum miscanthi* [*Sitobion miscanthi*] and *Myzus persicae* were categorised as pests of several crops such as *Brassica* spp., *Dolichos lablab, Cajanus cajan, Cucurbita maxima, Lagenaria vulgaris, Raphanus sativus, Pennisetum glaucum* and *Zea mays* (Ahmad *et al.*, 1995).

9.5. Alternate hosts

The alternate hosts of *A. craccivora* include, *Cajanus cajan, Capcicum annum, Cicer arientinum, Arachis hypogaea*, peas, sorghum, *Vicia faba*, Vigna unguiculata, *Dolichos lablab, Vigna mungo*, bean, *Colocasia esculenta*, cucumber, Lucern, soybean, Lupines, *B. juncea, Chrysanthemum coronarium, Vigna mungo, Meghamia macrophylla, Astragallus sinicus*, citrus, Limabean, pepper, blackgram, water melon, *Lens esculentus*, potato, *Vigna radiata*, cotton, *Gliricidia sepium*, Holly hocks, pegionpea, *Inula viscose*, sugarbeat, *Acer saccharium*, Brasica, Brocoli, Bussels, cabbage(*Brassica oleracea* var. *capitata*), celery, lattuce, peach (*Prunus persica*), *Pilea cedierei, Caspella bursa, Medicago minima, M.truncatula, Phaseolus vulgaris, Robinia pseudaelia* etc. Various host plants were recorded by Patel and Patel (1971) which include *Dolichos lablab, Vigna unguiculata, Gliricidia sepium, Pisum stivum, Medicago sativa, Carthamus tinctorius, Coccinia indic, Psidium guajava, Cyamopsis tetragonoloba, Chrysanthemum indicum, G. sepium, Portulaca oleracea, Spermacoce hispida, Cassia auriculata,* and *C. tora*.

9.6. Ecology

Studies were carried out in Gujarat, India, in 1969-70 to determine the seasonal history of *A. craccivora* in the rainy season. Collections in sticky traps placed in fields indicated that alate migrated from alternate food-plants during August. Populations of alate and apterae on the kharif crop reached peak numbers 22 days after alate were first collected in traps (on 4th August). Examination of various plants (wild and cultivated) found near groundnut fields revealed that alternate food-plants included val (*Dolichos lablab*), cowpea (*Vigna unguiculata* (catjang)) and the weeds aval (*Cassia auriculata*) and kuvadio (*C. tora*) during the rainy season (from July to the end of September), Cowpea (*Vigna unguiculata*), *Gliricidia sepium* (maculata), val, pea and lucerne during the cold season (from October to mid-March), and val, lucerne, cowpea and the summer groundnut crop during the hot season (about March to June). Aphids collected from valod (*Dolichos* sp.), safflower (*Carthamus tinctorius*), tindoli (*Coccinia indica*), lucerne, guavar (*Cyamopsis tetragonoloba*), guldaudi (*Chrysanthemum indicum*), *G. sepium*, and the weeds kuvadio, luni (*Portulaca oleracea*) and sheshmuli (*Spermacoce hispida*) were successfully reared on groundnut in the greenhouse. The aphid can complete its development entirely on groundnut when it is grown throughout the year (Patel and Patel, 1972).

The effect of water stress on *Aphis craccivora* on groundnut was examined by Rao *et al.* (1991). *A. craccivora* was more dense on plants that had sufficient water from irrigation, with population density decreasing in a linear manner with increasing distance from the irrigation source. Rain reduced the population by 90%, probably due to the physical effect of the rain falling on the plants. Nandagopal *et al.* (2004) studied the population dynamics of aphids (*Aphis craccivora* Koch and *Hysteroneura setariae* Thomes) in relation to weather parameters in groundnut. The peak population of *A. craccivora* was recorded between December and January. *H. setariae* population was low in the experimental fields. The population of *A. craccivora* was correlated with air temperature, soil temperature and vapour pressure. Correlation coefficients between weather parameters and *A. craccivora* populations were 0.59 and 0.53 during the cropping season and fallow period.

9.7.Management

9.7.1. Cultural

In field tests in 1964 at two places in Tamil Nadu, India, the effects of spacing, rouging and weeding on the incidence of rosette disease of groundnut and its vector *A. craccivora* were investigated. Among eight spacings (varying from 6 X 6 in. to 12 X 12 in.) and four cultural treatments investigated, it was found that closer spacing and late weeding (weeding twice with an interval of 30 days) reduced aphid infestation and the incidence of rosette disease and increased the yield. Rouging also reduced disease incidence (Kousalya *et al.*,

1971). In another study Singh *et al.* (1991) conducted field experiments on the effect of intercropping in groundnut (to control insect pests) on the abundance of their natural enemies. The organically manured treatments, i.e. farmyard manure, neem cake, and vermi-compost recorded the lowest pest population of jassid *Empoasca kerri* and aphid *Aphis craccivora* compared to straight fertilized treatments. The groundnut plants that received organic manures recorded low nitrogen and higher level of phenols and tannins whereas the straight fertilized groundnut leaves recorded higher nitrogen and lower phenol and tannin levels. Pest incidence was positively correlated with nitrogen content and negatively correlated with phenols and tannins. The organic manures induced the production of phenols and tannins in groundnut indicating the importance of induced resistance in groundnut insect pest management. Venkateswarlu *et al.* (2003) evaluated five organic manures, neem cake, pongamia cake, castor cake (all at 1.0 t/ha), farmyard manure (FYM), and vermi-compost (at 10.0 t/ha), and 11 nitrogen rates (10, 15, 20, 25, 30, 35, 40, 45, 50, 55 and 60 kg/ha) on insect pests and their natural enemies in groundnut (cv. TMV-2) in Andhra Pradesh, India. The insect pests recorded were *Aphis craccivora,* and other insects such as *Spodoptera litura,* and *Empoasca kerri*. The predators recorded were *Coccinella septempunctata, Coccinella transversalis, Micraspis vincta, Micraspis discolor, Cheilomenes sexmaculata, Brumoides suturalis, Oxyopes wroughtoni* and *Clubiona* sp. Among the cakes, neem cake was the most effective, but was at par with pongamia cake. A significantly higher yield (40.3 q/ha) was obtained from neem cake-treated plot, followed by castor cake- and pongamia cake-treated plots (36.9 and 36.5 q/ha, respectively). Predator population was lower in plots treated with both neem and pongamia cakes. The natural enemy population was directly proportional to the insect population, irrespective of the treatments. Castor cake-, FYM-, vermicompost- and chemical fertilizer-treated plots attracted more coccinellids and spiders than neem and pongamia cake-applied plots. Pest and natural enemy populations increased gradually with increase in N levels. Organic cakes were safer to the natural enemies; however, their population was totally dependent on pest population. Predator population increased with the increase in nitrogen level, but this effect was due to rise in pest population.

9.7.2. Biological control

The incidence of *A. craccivora* increased from 10% to 97% within a weeks time in the month of August virtually disappeared after the natural activity of predators *viz; M. sexmaculatus, C. septumpunctata,* the syrphid *Xenthogramma scutellara* (F.) and the chrysopid *Chrysopa carnea* (Steph) with an average of 5 larvae , 20 pupae and 10 adults per plant until early September (Patel and Patel, 1971). Predacious insects recorded on aphids include coccinellids, *Menochilus sexmaculatus* (F.) and *Coccinella septempunctata* L., the syrphid, *Xanthogramma scutellare* (F.) and the chrysopid *Chrysopa carnea* Steph. of these *M. sexmaculatus* was the most numerous, averaging 5 larvae, 20 pupae and 10 adults/plant until early September, when a sharp decline in adult density indicated migration to other areas owing to lack of food on the originally infested crop (Patel *et al.*, 1976). The predators *Menochilus sexmaculatus* (F.), *Brumoides suturalis* (F.) (*Brumus suturalis*) and *Coccinella septempunctata* L. were found breeding and feeding on the aphid in the field (Bakhetia and Sidhu, 1977). Talati and Butani (1979) observed that a single larva of *Coccinella septempunctata* consumed as many as 467 nymphs of *A. craccivora* during its development time of 7 days, while a single adult consumed up to 143 nymphs per day.

Each *Menochilus sexmaculatus*, predator consumed an average of 474 aphids during the 6 days of larval development and the 16-49 days of adult life, which indicates that *M. sexmaculatus* has good potential as a biological control agent. Attention is also drawn to *Coccinella septempunctata* as a voracious predator that might usefully be released for biological control of A. craccivora, and *Xanthogramma scutellare* and *Chrysopa carnea* are also cited as predators of this pest on various food-plants (Patel and Vyas, 1984). Butani and Bharodia (1984) identified *Hippodamia variegata* (Goeze), *Coccinella septempunctata* L. and *Menochilus sexmaculatus* (F.) as predators of *Aphis craccivora* Koch on groundnut. It was clearly demonstrated that the aphid population was more (347 aphids /10 plants) when the predatory coccinellids both larvae and adults were in the beginning (4.3 predators/ 10 plants) of their build up. When the predators reached one per one plant, the population of aphid were considerably reduced to 33 per 10 plants (Upadhyay and Vyas, 1987). The maximum incidence of the aphid population on groundnut was recorded between July and September. Increase in the numbers of aphids were controlled by the coccinellid predators *Coccinella* sp. and *Menochilus sexmaculatus* [*Cheilomenes sexmaculata*]. 5-10 larvae of *C. sexmaculata* per plant successfully controlled aphid populations during mild weather (Kenchaiahand Porte, 1989). An extensive survey of the Terai area of northeastern Uttar Pradesh, revealed that 25 species of aphids were recorded on more than 78 species of food plants. The majority of parasitoids (which included *Trioxys indicus*, *Lipolexis scutellaris*, *Lysiphlebia mirzai*, *Lysiphlebus delhiensis* and *Aphidius uzbekistanicus*) were polyphagous and varied greatly in the rate of parasitization of one or more aphid hosts. The biological control potential of *T. indicus*, *L. mirzai* and *L. delhiensis* is known, whereas *L. scutellaris* is a little-known species. Six species of aphid hyper parasitoids were recorded, including *Alloxysta pleuralis* and *Phaenoglyphis* sp. (Ahmad *et al.*, 1995). *Geocoris ochropterus* and *G. jucundus* were identified as predators of *Aphis craccivora* (Urmila-Sharma and Yadav, 1997).

The efficacy of *R. marginatus* in controlling *Aphis craccivora* infesting groundnut cv. TMV 7 was studied by Sahayaraj *et al.*(2003). The biological control potential of *R. marginatus* increased with increasing prey density, although it decreased for prey at the late instar and adult stages. The first and second instars of *R. marginatus* had higher predatory potential compared to the late instars and the adults.

9.8.Natural enemies

Several species of parasites and predators have been reported on *A. craccivora* in India (Table 9.2).

Table 9.2. Natural enemies on *A. craccivora*

Natural enemy	Order: Family	Reference
Alloxysta pleuralis (Cameron)	Hymenoptera, Alloxystidae	Singh and Srivastava, 1987
Aphelinus abdominalis Dalman	Hymenoptera, Aphelinidae	Murthy, 1989
Aphidius absinthii Marshall	Hymenoptera, Aphidiidae	Murthy, 1989
Brumoides suturalis (F.)	Coleoptera, Coccinellidae	Bakhetia and Sidhu, 1977
Cheilomenes sexmaculata (F.)	Coleoptera, Coccinellidaee	Patel *et al.*, 1976 Kenchaiah and Porte, 1989 Bakhetia and Sidhu, 1977
Chilocorus nigritus (F.)	Coleoptera, Coccinellidae	Murthy, 1989
Chrysopa sp.	Neuroptera, Chrysopidae	Patel and Vyas, 1984
Chyrsopera carnea (Steph.)	Neuroptera, Chrysopidae	Pawar *et al.*, 1985 Patel *et al.*, 1976
Coccinella septempunctata (Linnaeus)	Coleoptera, Coccinellidae	Patel *et al.*, 1976; Bakhetia and Sidhu, 1977
Episyrphus balteatus	Diptera Syrphidae	Ben Halima Kamel *et al.*, 1994
Fusarium pallidoroseum Cooke (Sacc)	Deuteromycotina, Hyphomycetes	Faizal *et al.*, 1996; Rani *et al.*, 1999
Geocoris ochropterus	Hemiptera, Lygaeidae	Urmila Sharma *et al.*, 1997
Geocoris jucundus	Hemiptera, Lygaeidae	Urmila Sharma *et al.*, 1997
Ischiodon scutellaris (Fabr.)	Diptera, Syrphidae	Pawar *et al.* 1985 Patel *et al.*, 1976

Contd...

Table 9.2. Contd...

Leucopis fumidilarva Tanas	Diptera, Chamaemyiidae	Saxena, 1981
Lysiphlebus fabarum Marshall	Hymenoptera, Braconidae	Murthy, 1989
Menochilus sexmaculatus (F.)	Coleoptera, Coccinellidae	Patel and Vyas, 1984
Motacilla caspica	*Motacillidae*	*Chakravarthy and Lingappa, 1979*
Motacilla flava	Motacillidae	Chakravarthy and Lingappa, 1979
Orius sp.	Heteroptera, Anthocoridae	Murthy, 1989
Orius albidipennis (Reut.)	Heteroptera, Anthocoridae	Ghosh *et al.*, 1981
Pantala sp.	Odonata, Libellulidae	Mehto *et al.*, 1986
Paragus spp.	Diptera, Syrphidae	Murthy, 1989
Praon sp.	Hymenoptera, Braconidae	Stary and Ghosh, 1978; Rajendra-Singh *et al.*, 1987
Spilocaria bissellata (Muls.)	Coleoptera, Coccinellidae	Saharia, 1980
Syrphophagus aphidivorus (Mayr)	Hymenoptera, Encyrtidae	Murthy, 1989
Trioxys indicus Subba Rao and Sharma	Hymenoptera, Aphidiidae	Srivastava and Singh, 1988; Singh and Srivastava, 1989; Rajendra Singh and Agarwala, 1992

9.8.1. Host plant resistance

Brar and Sandhu (1975) reported that, varieties with a bunchy or semi-spreading growth habit were found to be more favourable to *Aphis craccivora* Koch, because of the good shelter afforded, than were spreading varieties. *Dolichos lablab* was found to be the most suitable for development of *Aphis craccivora* arhar and groundnut were unsuitable as food-plants. (Reddy *et al.*, 1983). The cultivars Ah 7983 and Faizpur 1-5 in bunch type and Ah 8048 in spreading were found to be resistance supporting low rate of multiplication of aphids. Amin (1980 and 1987) screened over 6000 cultivars in screen house and found EC 36892 to be resistant. Also, a wild species *Arachis checoense* showed high degree of resistance against this species. The possible factors in resistance are the length and density of leaf hairs and stiffness of leaves. The chemical analysis undertaken in the cultivar EC 36892 suggested the presence of isoflavinoides like substances confirms limited resistance at the probing stage itself. While, a high concentration of tannin known as procyanidin in the phloem sap might be a major constrain to prolong ingestion of the sap by the aphid (Grayer *et al.*, 1992). Within 24 hours of release of alates and apterus left the resistant genotype and continue to feed on the susceptible lines. The interference with phloem locations for feeding on resistant genotype could explain the low rate of development and reproduction, this probably may account for greater field mortality of nymphs (85%) on the resistant line EC 36892. Groundnut genotypes PBS 24005, PBS 24006, PBS24030, Girnar 1, PBS 21031 and PBS 21020 were categorized as moderately resistant to aphid infestation Samdur *et al.* (1999).

9.8.2.Chemical control

Upadhyay and Vyas (1986) reported that a spray of thiometon applied against *Aphis craccivora* and *Empoasca kerri* at 0.03 and 0.05% was found to be the least harmful to the coccinellids, followed by demeton-O-methyl. Malathion at 0.05% was highly toxic to both coccinellids, *Menochilus sexmaculatus* and *Coccinella septempunctata*. The comparative toxicity to coccinellids in ascending order was thiometon < demeton-O-methyl < phosphamidon < monocrotophos < dimethoate < malathion. Chauhan *et al.* (1988) determined the efficacy and persistence of 0.03% dimethoate, 0.04% monocrotophos, 0.03% phosphamidon, 0.03% oxydemeton-methyl and 0.05% malathion against *A. craccivora* on chickpea. They concluded that all the insecticides were effective, with phosphamidon, monocrotophos and oxydemeton-methyl causing >70% mortality for up to 10 days after application. It is recommended that applications of phosphamidon, monocrotophos, oxydemeton-methyl and dimethoate should be repeated at 13- to 15-day intervals, those of malathion at 7-day intervals.

Dimethoate gave the most effective level of control against *Aphis craccivora* on groundnut. Other promising pesticides included endosulfan and carbaryl, and using quinalphos dust resulted in the highest pod yield (Jena *et al.*, 1997). The efficacy of insecticides for controlling *Aphis craccivora* was dimethoate > endosulfan > carbaryl > quinalphos > dichlorvos > chlorpyrifos > malathion. Chlorpyrifos was the least toxic to *C. transversalis* followed by parathion-methyl, quinalphos, dichlorvos, carbaryl, malathion, HCH, dimethoate, and endosulfan (Jena and Kuila, 1997a). At present only insecticidal control is recommended in the management of this pest. Phorate 1.1 kg a.i./ha; carbofuran, isophenphos and fensulfothion @ 1 kg a.i./ha, 5% diazinon @ 25 kg/ha were found effective in checking this pest. Spraying of host trees for controlling adults with carbaryl 2%, Quinolphos 0.025% or Monocrotophos 0.05% was found effective. The median lethal concentration of imidacloprid was 0.327 ppm against *A. craccivora* (Babu and Santharam, 2001). Similarly they have also reported imidacloprid at 100 ml/ha was effective in controlling the population of aphids on groundnut and was superior to the standard chemicals (Babu and Santharam, 2001a).

References

Amin, P. W. (1988). Insect and mite pests and their control. In *Groundnut* (Ed. P.S. Reddy). pp. 393-452. Indian Council of Agricultural Research, New Delhi.

Ahmad,M.E., Rajendra Singh, and Singh,R. (1995). Survey of parasitoids of aphids in northeastern Uttar Pradesh for possible use in biological control. *Annals of Entomology.*, 13(2), 87-96.

Amin, P.W. (1980). 'Gaps in groundnut pest research in India'. Paper presented at the All India Co-ordinated Research Project on Oilseeds Group Meeting, April 81, Ludhiana, Punjab.

Babu, K. R. and Santharam, G. (2001). Acute toxicity of imidacloprid to leaf miner and aphids on groundnut. *Madras Agricultural Journal.* 87(10/12): 677-678.

Babu, K. R. and Santharam, G. (2001a). Bioefficacy of imidacloprid against thrips and aphids on groundnut, *Arachis hypogaea* L. *Madras Agricultural Journal.* 87(10/12): 605-608.

Bakhetia, D. R. C. and Sidhu, A. S. (1977). Biology and seasonal activity of the groundnut aphid *Aphis craccivora* Koch. *Journal of Research of Punjab Agricultural University* 14: 299-303.

Basheer, M. (1958). Observations on *Aphis craccivora* infesting *Gliricidis maculata* and its early method of control. *Indian Journal of Entomology.* 20: 66-7.

Behura, B. K. (1956). Observation on biology of *Aphis craccivora* Koch. (Aphididae: Homoptera). *Proceedings of the 43rd Indian Science Congress.* 3: 295.

Ben Halima Kamel, M., Rabasse, J. M., Ben Hamouda, M. H. (1994). The aphids of citrus fruits and their natural enemies in Tunisia. *Tropicultura.* 12(4) :145-147.

Brar, K. S. and Sandhu, B. K. (1975). A note on the evaluation of resistance among different cultivars against aphid and grey weevil. *Science and culture* 41: 445-448.

Brar, K. S. (1981). Note on the comparative susceptibility of groundnut varieties to *Aphis craccivora* Koch. *Madras Agriculture Journal.* 68: 207-208.

Brar, K. S. and Sandhu, G. S. (1975) A note on the evaluation of resistance among different cultivars of groundnut against aphid and grey weevil. *Science and Culture.* 41: 445-448.

Butani, P. G. and Bharodia, R. K. (1984). Relation of groundnut aphid population with its natural predator, lady bird beetles. *Gujarat Agricultural University Research Journal.* 9: 72-74.

Chakravarthy, A. K. and Lingappa, S. (1979). Wagtails as predators of field bean aphids. *Journal of the Bombay Natural History Society.* 76: 367.

Chauhan, R., Sucheta, K., Ombir Singh, Khokhar, K. S. and Singh, O. (1988). Bioefficacy and persistence of some insecticides against bean aphid, *Aphis craccivora* Koch infesting chickpea. *Haryana Journal of Horticultural Sciences.* 17(1-2): 102-105.

Dixon, A.F. (1985). Structure of aphid population. *Annual Review of Entomology.* 30:155-174.

Dorge, S. K., Dalaya, V. P. and Kaul, O. B. (1966). Studies on bionomics and control of groundnut aphid (*Aphis craccivora* Koch). *Labdev Journal Science Technology.* 4: 165-167.

Faizal, M. H., Susamma Mathai and Mathai, S. (1996). Evaluation of different substrates for mass production of *Fusarium pallidoroseum* an entomogenous fungus. *Indian Journal of Entomology.* 58(2): 99-102.

Ghosh, D. Poddar, S. and Raychaudhuri, D. N. (1981). Natural enemy complex of *Aphis craccivora* Koch and *Lipaphis erysimi* (Kalt.) in and around Calcutta, West Bengal. *Science and Culture.* 47(2) : 58-60.

Grayer, R. J., Kimmins, F. M., Padgham, D. E., Harborne, J. B., Ranga-Rao, D. V.(1992). Condensed tannin levels and resistance of groundnuts (*Arachis hypogaea*). against *Aphis craccivora. Phytochemistry.* 31(11): 3795-3800.

Jena, B. C. and Kuila, B. (1997a). Effect of insecticides on population of aphids and their natural enemies in groundnut. *Madras Agricultural Journal.* 84(2): 101-102.

Jena, B. C., Kuila, B. and Mishra, B. K. (1997). Aphid menace in groundnut and its control. *Indian Journal of Entomology.* 59(1): 24-26.

Jena, B.C. and Kuila, B. (1996). Jassid infestation in groundnut and its chemical control. *Indian Journal of Entomology.* 58(4): 378–381.

Kenchaiah, R. P. M. and Porte, B. S. (1989). Some observation on population fluctuation of insect pests of groundnut in Karnataka. *Plant Protection Bulletin.* 417-10.

Kenchaiah, R. P. M. and Porte, B. S. (1989). Some observations on population fluctuation of insect pests of groundnut in Karnataka. *Plant Protection Bulletin Faridabad.* 41: 3-4.

Khan, M. K. and Hussain, M. (1965). Role of coccinellid and syrphid predators in biological control of groundnut aphids. *Indian oilseed Journal.* 9: 67-70.

Mehto, D. N., Singh, K. M. and Singh, R. N. (1986). Natural enemy complex on insect pest complex in chickpea *Cicer arietinum* Linn. *Bulletin of Entomology* 27: 1-12.

Modawal, C.N. (1941). Biological notes on *Chilomenes sexmaculata* Fab. *Indiann Journal of Entomology.* 3: 139-40.

Murthy, S. N. (1989). Biological suppression of pulse and oilseed pests, pp. 84-106. Proceedings of the Seminar-cum-Seventh Workshop on Biological Control of Crop Pests & Weeds. AICRP on BCCP & W, Biological Control Centre (ICAR-NCIPM). Bangalore. Tech. Doc. No. 26, pp 191.

Nandagopal, V.(1992). Studies on the integrated pest management in groundnut in Saurashtra. Ph.D. thesis submitted to the Saurashtra University, Rajkot, pp. 247.

Nandagopal, V., Gedia, M. V. and Makwana, A. D. (2004). Population dynamics of aphids (*Aphis craccivora* Koch and *Hysteroneura setariae* Thomes) in relation to weather parameters in groundnut, *Arachis hypogaea* L. *Journal of Oilseeds Research.* 21(1): 98-103.

Narayanan, E.S.(1962). *Insect Pests of Groundnut and methods of their control in groundnut - A Monograph*, pp.222-8. Seshadri, C.R. (Ed.) Indian Central Oilseeds Committee, Hyderabad.

Patel, A. G. and Vyas, H. N. (1984). Studies on predatory capacity of lady bird beetle, *Menochilus sexmaculata* Fabricius against *Aphis craccivora* Koch under laboratory condition. *Pesticides*, 18(11): 8-9.

Patel, M. B. and Srivastava, K. P. (1989). Biology of groundnut aphid, *Aphis craccivora* Koch on cowpea, *Vigna unguiculata* (Linnaeus) Walpers. *Bulletin of Entomology*. 30: 65-73.

Patel, R. C., Yadav, D. N. and Patel, J. R. (1976). Natural control of groundnut aphid, *Aphis craccivora* Koch in Central Gujarat. *Current Science*. 45(1): 34-35.

Patel, R. M. and Patel, C. B. (1972). Factors contributing to the carry over of groundnut aphid (*Aphis craccivora* Koch.) through the off season in Gujarat. *Indian Journal of Entomology*. 33: 404-410.

Patel, R. M. and Patel, C.B. (1971). Factors contributing to the carry over of groundnut aphid (*Aphis craccivora* Koch.) through the off season in Gujarat. *Indian Journal of Entomology*. 33(4): 404-410.

Pawar, C. S., Bhatnagar, V. S. and Jadav, D. R. (1985). Some predatory insects and their parasites. *Science & Culture*, 51(3): 101-102.

Puttarudraiah, M. and Channabasavanna, G.P. (1955). Beneficial coccinellids of Mysore.II. *Indian Journal of Entomology*. 17:1-7.

Rajendra Singh and Agarwala, B. K. (1992). Biology, ecology and control efficiency of the aphid parasitoid *Trioxys indicus*: a review and bibliography. *Biological Agriculture and Horticulture*. 8(4): 271-298.

Rajendra-Singh, Tripathi, R. N. and Singh, R. (1987). Records of aphid parasitoids from Terai belt of eastern Uttar Pradesh - I. *Journal of Aphidology*. 1: 89-92.

Rangaswamy, S. V. and Rao, D. (1964). Classified list of minor pests of important crops in India. *Aphids Lal Baugh*. 9(3): 9-12.

Rani, O. P. R., Susamma Mathai and Mathai, S (1999). Mass production of entomogenous fungus *Fusarium pallidoroseum* (Cooke) Sacc in naturally available solid and liquid media. *Insect Environment*. 5(2): 59-60.

Rao, G. V. R., Wightman, J. A., Wadia, K. D. R., Rao, D. V. R. and Rao, R. C. N. (1991). Influence of water stress on groundnut aphids. *International Arachis Newsletter*. 9: 14.

Rao, K. R. (2002). Induced host plant resistance in the management of sucking insect pests of groundnut. *Annals of Plant Protection Sciences*. 10(1): 45-50.

Reddy, K. D., Gargav, V. P. and Misra, D. S. (1983). Studies on host preference of *Aphis craccivora* Koch. *Entomon*. 8: 75-78.

Sachan, J. N. (1959). Common insect pests of crops and vegetables of Jobner (Rajasthan). *Rajsthan Agriculture*. 7: 39-44.

Saharia, D. (1980). Some aspects of the biology of coccinellid predators associated with *Aphis craccivora* Koch on cowpea. *Journal of Research, Assam Agricultural University*. 1: 82-89.

Sahayaraj, K., Delma, J. C. R. and Martin, P. (2003). Biological control potential of aphidophagous reduviid predator *Rhynocoris marginatus*. *International Arachis Newsletter*.23: 29-30.

Samdur, M. Y., Nandagopal, V., Mathur, R. K. and Manivel, P. (1999). Evaluation of advanced breeding lines of groundnut for resistance to thrips and aphids. *International Arachis Newsletter*. 19: 19-21.

Saxena, A. P. (1981). *Leucopis fumidilarva* Tanas (Diptera: Chamaemyiidae), a new predator of aphids and leaf hoppers on potato crop. Note. *Journal of the Indian Potato Association*. 8(2): 96-98.

Singh, R. and Srivastava, M. (1989). Bionomics of *Trioxys indicus* (Hymenoptera: Aphidiidae) a parasitoid of *Aphis craccivora* (Hemiptera: Aphididae): Effect of host haemolymph on the numerical response of the parasitoid. *Entomophaga*. 34(4): 581-586.

Singh, R. and Srivastava, P. N. (1987). Potential host-habitat location by *Alloxysta pleuralis* (Cameron) (Alloxystidae: Hymenoptera). *Zeitschrift fuer Angewandte Zoologie*. 74 (3): 337-341.

Singh, T.V.K., Singh, K.M. and Singh, R.N. (1991). Influence of intercropping. III. Natural enemy complex in groundnut. *Indian Journal of Entomology*. 53(3): 333-368.

Singh, V., Ghewande, M. P. and Reddy, P. S. (1993). Peanut stripe virus disease - present status and its management. *International Journal of Pest Management*. 39(4): 422-430.

Sreenivasulu, P., Kuhn, C.W., Naidu, R.A., Demski, J.W., Reddy, D.V.R. and Nayudu, M.V. (1991). Viruses infecting peanuts: Taxonomy, identification, and disease management. Research Bulletin 406, The Georgia Agricultural Experiment Stations, The University of Georgia College of Agricultural and Environmental Sciences, p. 34.

Srikanth, J. and Lakkundi, N. H. (1988). Instar period, fecundity and longevity of cowpea aphid, *Aphis craccivora* Koch seven leguminous hosts. *Journal of Aphidology*. 2 (1-2) : 18-21.

Srivastava, M. and Singh, R. (1988). Bionomics of *Trioxys indicus* Subba Rao and Sharma, an aphidiid parasitoid of *Aphis craccivora* Koch. 29. Impact of host-extract on reproductive behaviour of the parasitoid at different host densities. *Journal of Applied Entomology*. 106 (4): 408-412.

Stary, P. and Ghosh, A. K. (1978). Further records of aphid parasitoids (Hymenoptera: Aphidiidae) from Meghalaya, India. *Oriental Insects*. 12: 77-80.

Talati, G. M. and Butani, P. G. (1979). Predatory capacity of *Coccinella septempunctata* on groundnut aphid. *Indian Journal of Plant Protection*. 7: 107.

Talati, G. M. and Butani, P. G. (1980). Reproduction and population dynamics of groundnut aphid. *Gujarat Agricultural University Research Journal*. 5: 54-56.

Upadhyay, V. R. and Vyas, H. N. (1986). Comparative toxicity of certain insecticides to predatory coccinellids associated with sucking pests of groundnut in Saurashtra. *Indian Journal of Plant Protection.* 13: 91-93.

Upadhyay, V. R. and Vyas, H. N. (1987). Integrated pest control of major insect pests of groundnut in Gujarat in relation to pest population levels and coccinellid predatory populations. *National Symposium on Integrated Pest Control-Progress and Perspectives* (Oct. 15 - 17,1987) (abstracts). *Association for Advancement of Entomology,* Trivandrum, Kerala, India, 6 pp.

Urmila-Sharma, Yadav, D. N. (1997). A new record of arthropod predators in groundnut at Anand (India). *Gujarat Agricultural University Research Journal.* 22(2): 157.

Venkateswarlu, P., Sitaramaiah, S., Sreedhar, U., Rao, S. G., Sawant, S. K. and Rao, S. N. (2003). Role of organic and inorganic manures on the incidence of insect pests and their natural enemies in rabi groundnut.Biological control of lepidopteran pests Proceedings of the Symposium of Biological Control of Lepidopteran Pests,-July-17-18,-2002,-Bangalore,-India; 295-299.

Verma, S. N., Gargav, V. P. and Menon, M. V. S. (1983). On the life history and host preference of the bean aphid, *Aphis craccivora* Koch. *Pranikee.* 4: 389-395.

Waghery, R. N. and Singh S. R. (1965). Effect of N, P and K on the fecundity of groundnut aphid, *Aphis craccivora* Koch. *Indian Journal of Entomology.* 127: 331-4.

□□□

Leaf Hoppers

V.NANDAGOPAL AND A. D. MAKWANA

10.1.Introduction

Some of the leaf hoppers (Homoptera: Cicadellidae) are also the insect pests of groundnut in India and in different parts of the world. The importance of leaf hoppers as pest of groundnut has not generally been recognized by farmers particularly in the semi - arid tropics as they do not cause spectacular damage, however, they can be important in reducing pod yields. Among the various species of leaf hoppers that infest groundnut in India (Sandhu and Brar,1977; Nandagopal and Reddy,1987) *Empoasca kerri* Pruthi and *Balclutha hortensis* Lindb. are dominant. Seasonal influence on the life cycle of groundnut jassid, *Empoasca kerri* Pruthi. was studied by Singh *et al.* (1989). The larval period was 9.37 days in August, when the mean maximum temperature was 34.05°C. The length of the life cycle decreased with decreasing temperature. The relationships between meteorological conditions and the incidence of *Empoasca kerri* on groundnuts was studied by Dubey and Thorat (1994). The Cloudy weather indicates the possibility of attack by the cicadellid on groundnuts after 1-2 weeks, and increased hours of sunshine in the week could be responsible for increased activity in July-August. Jayanthi *et al.* (1993) studied the influence of abiotic factors on the build-up of insect pests in groundnuts cv. MH 4. Populations of the cicadellid *Empoasca kerri* showed a positive correlation with maximum daily temperature, sunshine, rainfall and evening relative humidity and a negative correlation with minimum daily temperature, morning relative humidity and wind speed.

10.2.Distribution

Leaf hoppers have become major pest of groundnut in recent years in parts of Gujarat, Maharashtra, Karnataka and Tamil Nadu (Sandhu and Brar,1977; Patel and Vora, 1981; Nandagopal and Reddy,1987; Amin,1988).

10.3. Economic importance and nature of injury

The nymphs as well as adults suck sap from the leaves mainly from under surface. The initial injury to groundnuts appears on young leaves and veins which become white. The terminal leaves may wilt if plants attacked are young. In older plants wilting of leaflets is rare. Prolonged exposure to leaf hoppers results in a 'V' shaped yellowing (**Plate 1**) on the tips of leaflets which may spread and the entire leaflet may become yellow. Groundnuts when infested with leaf hoppers look yellow when compared to unaffected crop. Groundnut leaf hoppers has been mentioned as a serious pest in parts of India (Patel and Vora, 1981) but no precise information on yield loss was reported. However, Saboo and Puri (1979) obtained an yield increase of over 40 per cent by controlling leaf hoppers and thrips. Amin (unpublished data) estimated 9 per cent loss in pod yield and 17 per cent haulm yield at a moderate level of infestation. Vyas (1984) estimated 22 per cent yield loss at Junagadh in Gujarat. *B. hortensis* caused a yield loss of about 584 kg/ha (Anon, 1988). The difficulty in excluding damage by other sucking insect pests such as thrips and aphids that infest the crop simultaneously is the major hurdle in the estimation of yield losses from leaf hoppers.

Empoasca kerri is a small yellowish green hopper measuring about 3 mm in length; males are slightly smaller and greener than females (**Plate 2**) (Pruthi, 1940). Eggs are broad and rounded at one end and gradually tapering at the other end. These are inserted into the leaf tissues mainly near the mid rib and veins on the under surface of the leaves. The eggs and nymphal period range from 7 to10 and 7 to14 days respectively. The maximum number of nymphs obtained from a single female ranged between 30-40. Generally two peaks are common during early September and middle of December in the rainy and post rainy season crops respectively.

A sequential sampling plan for *Empoasca kerri* based on negative binomial distribution using common 'K' of 5.7567 and EIL of 4. 3090 was developed.

Plate 1. 'V'Shaped Yellowing

Plate 2. Adult leaf hopper

Three infestation classes as light, moderate and severe were used for the plan. In addition, the OC-curve and ASN were studied for predicting the probable number of plants required for different population densities. The plan was highly efficient, particularly at very high and low levels of infestation. The probability of rating the population correctly is 90%. A sequential sampling plan for groundnut jassid, *Empoasca kerri* (Singh *et al.*, 1996).

10.4. Economic threshold level

The economic threshold level for the jassid *B. hortensis* has been worked out for different stages of the crop in three varieties of groundnut (Nandagopal, Un-Published data) (Table 10.1).

Table 10.1. Economic thresholds level for *B. hortensis*

Stage of the crop	Variety	ETL (No. of jassid/5m x 7*)
Vegitative		
Kharif	GG2, Girnar 1, GAUG 10	25
	JL 24,	20
Rabi	GG2, Girnar 1	18
Pod filling		
Kharif	GG2, GAUG 10	25
	JL 24,	20
	Girnar 1	38
Rabi	GG2,	23
	Girnar 1	20
Maturity		
Kharif	JL 24,	38
	GG2,	55
	GAUG 10	68
	Girnar 1	45

* Factor calculated based on the sweepnet month diameter.

10.5.Alternate hosts

E. kerri has been recorded as a pest of castor *(Ricinus communis)*, cluster bean, maize *(Zea mays)* (Pruthi,1940); groundnut (Saboo and Puri, 1979; Amin and Mohammad, 1980); soybean (Handa *et al.*,1978); green gram and black gram (*Phaseolus mungo)* (Reghupathi *et al.*,1975; Singh and Singh,1978; Choudhary *et al.*, 1980); Cowpea *(Vigna unguiculata)* (Verma and Lal,1976); and pigeon pea (Verma and Pant,1979). In addition to groundnut *B.hortensis* has been recorded as a pest of Cowpea *(Vigna unguiculata)*, soybean, pigeon pea, green and black gram (*Phaseolus mungo)*, cluster beans, Castor *(Ricinus communis)* and maize *(Zea mays)* (Nandagopal, 1992, Nandagopal and Reddy, 1987).

A survey of the food plants of *Empoasca kerri* was carried out in crops and weeds in India (Singh *et al.*, 1991). Groundnuts were the preferred food plant of the pests in the kharif season, followed by redgram [pigeon peas], castor [*(Ricinus communis)*], cowpea (*Vigna unguiculata*), sem [*Lablab purpureus*] and *Sesbania bispinosa*. In winter, the pest occurred on many plants, especially *R. communis* and faba beans. In summer, although groundnuts were present, Fenugreek *(Trigonella foenumgraceum)*, *Trigonella corniculata* and *Chappan kaddu* (from the Cucurbitaceae) were preferred. Plants of the Leguminosae made up 56% of the food plants of the pest. *Empoasca kerri* preferred groundnuts in the kharif season, followed by red gram, castor, cowpeas, sem and *Sesbania bispinosa*. In winter, the pest occurred on many plants, especially *R. communis* and *faba* beans. In summer, although groundnuts were present, fenugreek, *Trigonella corniculata* and *Chappan kaddu* (from the Cucurbitaceae) were preferred. Plants of the Leguminosae made up 56% of the food plants of the pest (Singh *et al.*, 1991).

10.6. Biology

10.6.1.E. kerri

It is a yellowish-green insect measuring about 3 mm in length; males are slightly smaller and greener than female (Pruthi, 1940). The eggs are inserted into the leaf tissue mainly near midrib, veins from the under surface of the leaves and rarely in the tender stem. They are broad and rounded at one end and gradually tapering towards the other end. The maximum number of nymphs obtained from the eggs laid by a single female was about 40 on cv. TMV 2.

10.5.2.Balclutha hortensis

A detailed study on the biology of this insect was reported by Nandagopal *et al.*(1996).

Egg :It measured mean of 0.44 mm in length and 0.17 mm in width (Plate 3). The incubation period was from 6-12 days with a mean of 8.07 days (Table 10.2).

First instar: The neonate nymph was yellowish green in colour and was having developed antennae. The male nymphs measured a maximum of 0.45 mm in length 0.23 mm in width, while female was with 0.66 mm length and 0.29 mm width. Both male and female moulted within the 12 hr of hatching from the eggs.

Second instar: The female instar took 1.16 days while male took 1.13 days to complete the instar. The males measured 0.57 mm length and 0.264 mm width.

Third instar: There was slight increase in the size of the male nymphs compared to female. This stage took 2.04 and 2.16 days respectively for female and males. Two small wing pad are clearly visible.

Fourth instar: The size of the insect increased to double to that of preceding instar. In the hind legs, spurs are seen from femur to tarsus. The wing pad are expanded in size. The genital organ are visible. The male took 1.18 days while the female took 2.12 days. It measured 1.58 and 0.26 mm for male and 2.43 and 0.49 mm for female, length and width respectively.

Adult: The female measured mean of 2.82 mm and the male only 1.91 mm. The female continue to be in pre-oviposition for an average period 6.06 days with a maximum of 9 days. Oviposition started on the 7th day after mating and continues oviposition for 10 days. It laid average of 30.16 eggs in its life span. The maximum number of eggs laid was during 3rd to 6th days of oviposition. A maximum of 6 eggs were laid in a single day. Under the field condition the sex ratio was always high toward males (1.21 : 1, Male : Female). The total life span of the female was 40.03 days and it was 23 days for male.

Table10.2. Biology of the jassid *B. hortensis*

Instar	Numb-ered observed	Duration (Days±sd)	Range (Days)
I Female	30	12h	8.0– 2.0
Male	30	12h	8.0– 2.0
II Female	24	1.16 ± 1.07	0.5 – 4.0
Male	30	1.13 ± 1.02	0.5 – 4.0
III Female	24	2.04 ± 0.14	1.0 - 0.4
Male	28	2.16 ± 0.17	2.0 – 5.0
IV Female	22	1.18 ± 0.11	1.0 – 2.0
Male	26	2.12 ± 0.12	2.0 – 4.0
Female			
Pre-oviposition period (days)	15	6.06 ± 1.15	5.0-9.0
Oviposition period (days)	12	10.03 ± 0.18	9.0-11.0
Egg period (days)	40	8.07 ± 1.16	6.06-12.0
Post oviposition period(days)	12	3.0 ± 1.18	1.0-6.0
Adult longevity: (days)			
Female	11	6.12 ± 4.19	9.0-22.0
Male	10	7.05 ± 3.11	5.0-8.0
Total life cycle (days)			
Female		40.03	
Male		23.00	

10.7. Natural enemies

The natural enemies which are recorded against the leaf hoppers are given in table 10.3. Among them the Dipteran predator *Crossopalpus* sp. found to play a dominant role in the suppression of the leaf hoppers (Nandagopal, 1988). No field trials using any of the natural enemies are attempted in groundnut against the leaf hoppers. The predator took 221 minutes to pray on 10 leaf hoppers, when the space was stimulated to field condition, it took 27 leaf hoppers in 10 hr period.

Table 10.3. Natural enemies of the leaf hoppers

Species	Order/Family	Reference
Orius maxidentex Ghauri	Hemiptera: Anthocoridae	Amin, 1987
Crossopalpus sp.	Diptera :Empididae	Nandagopal, 1988
Brumoides suturalis (Fabricius)	Coleoptera: Coccinellidae	"
Cheilomenes sexmaculata (Fabricius)	"	"
Coccinella septempunctata Linnaeus	"	"
Hippodamia variegata (Goeze)	"	Singh *et al.*, 1991
Stenus pictus Motschulsky	Coleoptera: Staphylinidae	"
Chrysoperla carnea (Steph.)	Neuroptera : Chrysopidae	"
Monomorium indicum Forel	Formicidae: Hymenoptera	"
Laius malleifer Champion	Coleoptera : Melyridae	"
Geocoris sp.	Hemiptera : Lygaeidae	"
Polistes olivaceus(De Geer)	Hymenoptera : Vespidae	"
Orius sp.	Hemiptera : Anthocoridae	"
Cryptocephalus sp	Coleoptera: Chrysomelidae	"
Campylomma sp.	Heteroptera: Miridae	"

10.8. Host plant resistance

The gelechiid leaf-miner *Aproaerema modicella* was most abundant on the most stressed plants. The cicadellid, *Empoasca kerri* had the reverse distribution. NC Ac 343 had the lowest leaf-miner and cicadellid populations. M 13 and ICGS 11 had intermediate characteristics (Wheatley *et al.*, 1989). F1 hybrids from crosses between resistant NcAc2230 and susceptible ALR1 (*Virginia* bunch type) and ALG33 (Valencia type) were sown in the field and damage assessed by visual ratings on a scale of 1-9 (1 = resistant) based on yellowing of the third leaf from the top 52 days after sowing. The F1s had resistance scores (4.5 and 5.0) intermediate between the parents (NcAc2230 = 2, ALR1 = 7.5 and ALG33 = 8.0). Resistance was positively related to the number of trichomes present on the petiole. F2 segregation data indicated that resistance was controlled by 3 recessive genes (Varman *et al.*, 1989). Breeding lines were developed by the bulk pedigree method from crosses made in 1982 between high-yielding genotypes and jassid [cicadellid] resistant germplasm sources NC Ac2214 and NC Ac2240 (Anon,1994). ICGV86252, ICGV86455 and ICGV86462 are sequentially branching (subsp. fastigiata) while ICGV86393 is alternate branching (subsp. hypogaea). In trials at several locations in India during 1988-90, the 4 lines gave similar pod yields to the controls (ICGV87123, Kadiri 3 and NC Ac343) in the rainy season (1.76-1.98 t/ha), but yielded about 1 t/ha more in the post-rainy season (3.04-3.5 t/ha). Seed oil contents (46-49%) were similar to those of the controls. The lines exceeded the controls for jassid resistance (scores of 1.6-2.2 vs. 3.7-8.0 on a 1-9 scale of increasing damage). Leaf toughness was measured in 3 released varieties, 4 germplasm accessions and 11 advanced intra- and interspecific derivatives using an improvised top pan balance based on leaf toughness (Nandagopal *et al.*,1993). The single top pan of an electric balance was removed and a plastic tube, fitted at one end with a 1.5 cm long steel needle, was fixed on the axis of the pan assembly. The leaflet was held perpendicular to the needle with the abaxial surface facing it and pressed at 4 predetermined sites. The readings on the vernier were noted at the time of piercing and were expressed as toughness in mg mm-2. There were large variations in the intensity of yellowing, due to feeding by the cicadellid, and leaf toughness among the genotypes tested. The intensity of yellowing was lowest in the interspecific derivatives CS17 (M13 X Arachis villosa), and highest in V6. There was a negative relationship between percentage intensity of yellowing and cumulative leaf toughness.

Fifteen groundnut varieties collected from all over India were sown in Junagadh, and the incidence of *E. kerri* was studied. Groundnut variety GG 3 had the lowest jassid population, whereas variety Somnath harboured the maximum number of leaf hoppers (Sherasva and Butani, 1997). A list of resistant lines including released varieties is being given in the table 10.4.

There were large variations in the intensity of yellowing, due to feeding by the cicadellid, and leaf toughness among the genotypes tested. The intensity of yellowing was lowest in the interspecific derivatives CS17 (M13 x *Arachis villosa*), and highest in V6. There was a negative relationship between percentage intensity of yellowing and cumulative leaf toughness. (Nandagopal *et al.*, 1993). Groundnut variety GG 3 had the lowest jassid population, whereas variety Somnath harboured the maximum number of jassids. (Sherasva and Butani, 1997). *Arachis cardenasii* and *A. diogoi* recorded resistance for the (thrips (*Scirtothrips dorsalis*) and leaf hopper (*Empoasca kerri*)) (Varman *et al.*, 2000). The organically manured treatments, *i.e.* farmyard manure, neem cake, and vermicompost recorded the lowest sucking insect pests, namely jassid *Empoasca kerri* and aphid *Aphis craccivora* on the groundnut compared to straight fertilized treatments. The groundnut plants that received organic manures recorded low nitrogen and higher level of phenols and tannins whereas the straight fertilized groundnut leaves recorded higher nitrogen and lower phenol and tannin levels. Pest incidence was positively correlated with nitrogen content and negatively correlated with phenols and tannins. The organic manures induced the production of phenols and tannins in groundnut indicating the importance of induced resistance in groundnut insect pest management (Rao, 2002). Twenty-eight advanced groundnut breeding lines were evaluated for resistance to leafhopper and yield performance under field conditions by John *et al.*, (2004). TCGS 750 (Tirupati 3 x ICGV 86031) was resistant to jassids, recording the lowest level of damage on leaflets (6.0% on average). This line had a pod yield of 1549 kg/ha and shelling outturn of 68%. TCGS 742 (TCGS 645 x TCGS 635) and TCGS 750 G (TCGS 645 x TCGS 635) were moderately resistant to jassids, with 13.5 and 17.5% jassid incidence and pod yields of 1.82 and 1.15 t/ha, respectively. TCGS 737, which had the highest pod yield (1970 kg/ha), showed moderate resistance to jassids. TCGS 750 E registered 39% damage, but had a pod yield of 1681 kg/ha; thus, this line showed tolerance of jassids. Intercropping with pigeon peas or *V. radiata* increased the incidence of *Empoasca kerri* and *Caliothrips indicus* and reduced the incidence of *Myllocerus dentifer* and *Spilosoma obliqua*. Intercropping with sorghum reduced the incidence of *E. kerri, C. indicus, S. obliqua* and *Spodoptera litura* and increased the incidence of *M. dentifer*. Intercropping with groundnut or sesame had little effect on pest incidence. The effects of intercropping on the timing of infestations are noted. It is concluded that intercropping with sorghum reduced the numbers of insect pests the most. (Singh *et al.*, 1991). ICGV 86031 peanut germplasm bred by the bulk pedigree method from a cross between F334A-B14 and NC Ac2214, this Spanish type groundnut germplasm (PI561917) was source of resistance to *Empoasca kerri, Thrips palmi, Spodoptera litura, Aproaerema modicella* and bud necrosis virus. (Dwivedi *et al.*, 1993). ICGV 86699 peanut germplasm line is high-yielding elite groundnut (*Arachis hypogaea* subsp. *hypogaea* var. *hypogaea*) germplasm line, released in 1994 and developed from a single-plant selection made in CS29 (*A. batizocoi/ A. duranensis//A. hypogaea* cv. NC2), suffers less damage due to *Empoasca kerri* and *Spodoptera litura* than control cultivars under field conditions (Reddy *et al.*, 1996).

Accessions belonging to *Arachis cardenasii, Arachis duranensis, Arachis kempff-mercadoi, Arachis monticola, Arachis stenosperma, Arachis paraguariensis, Arachis pusilla* and *Arachis triseminata* showed multiple resistance to the *Empoasca kerri* apart from resistant to leaf miner *Aproaerema modicella, Helicoverpa armigera* (Sharma *et al.*, 2003).

Table 10.4. Resistant lines including released varieties against leaf hoppers in groundnut

s. no	nrcg	icg	variety	synonym	org	hbt	other traits
1.	4631	7439	2635-76 S		NGA	HYB	
2.	–	–	4-1		UNK	–	
3.	4140	–	45-28		IND	FST	
4.	135	1217	Ah 3276		IND	VUL	
5.	4826	7840	Ah 8142		IND	VUL	
6.	9773	9934	CGC 4007		IND	VUL	34

Contd...

Table 10.4. Contd...

7.	9773	9935	CGC 4018		IND	VUL	6,13
8.	–	–	CGC 5-1		IND	VUL	
9.	–	–	CGC 5-2		IND	VUL	
10.	9767	–	CGC 6		UNK	HYB	
11.	3729	1108	G-Local bunch		UNK	FST	
12.	5807	760	K 4-11-II(R)		UNK	HYR	
13.	4786	7728	M 63-74		NGA	HYB	
14.	4155	779	Marlab-1		MEX	HYR	15
15.	4782	7724	NC 2		USA	HYR	
16.	4993	8118	NCAC 10060		USA	HYR	
17.	4812	7807	NCAC 1092		USA	HYB	
18.	4912	8020	NCAC 17634		USA	HYB	
19.	4745	7658	NCAC 180		USA	FST	
20.	6688	5044	NCAC 2242		USA	HYB	
21.	4178	2249	NCAC 23	NFG 47	USA	HYB	
22.	4176	411	NCAC 2700	NFG 43	BRA	FST	
23.	872	2288	NCAC 841		USA	HYB	15
24.	3818	–	No.420		UNK	VUL	
25.	4822	7831	OG 3-36		IND	VUL	
26.	4615	7418	PI 162534		URY	VUL	
27.	1634	6293	PI 261942		PRY	FST	
28.	4529	7305	PI 262013		ZWE	FST	
29.	4522	7288	PI 262093	139/66	BOL	FST	
30.	1637	7660	PI 268551		UGA	FST	
31.	4620	7425	PI 268814		ZWE	VUL	
32.	4862	7898	PI 407454	TIFRUST 4	ECU	FST	
33.	6417	9511	RPM 007		MOZ	HYR	
34.	6421	9516	RPM 016		MOZ	HYR	
35.	6425	9531	RPM 039		MOZ	HYR	
36.	6427	9534	RPM 048		MOZ	HYR	
37.	4612	7415	SP 1		MWI	HYB	
38.	4161	802	Sel. 11 (PC)		IND	HYR	
39.	4561	7362	Sel. from PI 262085		ZWE	FST	
40.	4556	7357	Sel.From PI 261964		ZWE	FST	
41.	6927	3505	US 2	PI 119922	PRY	VUL	25,27
42.	4701	7544	VRR 356		IND	VUL	
43.	4588	7389	Valencia type	E 2	ZWE	FST	

10.9. Management

10.9.1. Cultural

Intercropping of groundnut with per millet reduced the leaf hoppers on groundnut at ICRISAT (Amin, 1988). The leaf hopper numbers were highest in sole groundnut (244 nymphs/10m rows) compared with 3 rows of groundnut and 1 rows of pearl millet recording 177 nymph. Similarly the jassid infestation was more on groundnut when intercropped with castor (3:1) than when intercropped with pearl millet. This trend has been confirmed by Nandagopal (1992). Sorghum intercropped with pigeon pea gave the largest reduction in the green leafhopper, *Empoasca kerri* population (25.6-62.3%) followed by green gram [*Vigna radiata*] (29.3-49.1%), groundnut (22.9-47.1%) and rice bean [*Vigna umbellata*] (6.0-52.8%) compared to pigeon pea alone. (Sekhar *et al.*, 1997). Nath and Singh (1998) observed the lowest incidence of *E. kerri* in the groundnut + sorghum intercrop.

10.9.2. Plant Products

The effects of neem (*Azadirachta indica*) products, i.e. neem oil, neem leaf extract (500 g leaves extracted with 500 ml of distilled water) and neem decoction (500 g of leaves boiled in 500 ml of distilled water), on *Balclutha hortensis* were tested on dried groundnut leaves by Nandagopal *et al.* (1990). Highest mortality (97.5%) of the cicadellid resulted from neem oil at 25%. The most cost-effective treatment was neem leaf extract at 5% with a resulting mortality of 83.4%. The residual efficacy of monocrotophos, dimethoate, cypermethrin, phosphamidon and 2% concentration of neem oil to *Balclutha hortensis* infesting groundnut was determined by Nandagopal and Soni (1992). Cypermethrin was the most persistent and toxic insecticide over 9 days causing >50% mortality, followed by dimethoate and phosphamidon. Neem oil and monocrotophos were the least persistent insecticides and caused >50% mortality only up to 24 h after application.

A laboratory study with groundnut seedlings indicated that *Balclutha hortensis* [*B. incisa*] depended mainly on phloem (61.9%) for its sap requirement, while *Creontiades pallidus* and Aphis craccivora were dependent entirely on phloem and xylem for its requirement, respectively. In another study, neem [*Azadirachta indica*] extracts had antifeedant effect on *B.hortensis* (Nandagopal and Rajgopal, 1996).

Patil *et al.* (2003) reported that the populations of leaf miner larvae, jassids and thrips were significantly reduced by the application of neem seed kernel extract (NSKE) at 5%, which was at par with the control insecticide, monocrotophos, at 0.05%. Monocrotophos at 0.05% recorded the highest pod yield (1271 kg/ha), followed by NSKE at 5% (1162 kg/ha), but NSKE at 5% recorded the highest cost:benefit ratio (1:10.92) compared with the other treatments.

10.9.3.Biocontrol

Adults of *Crossopalpus* sp. were observed preying on cicadellids. *Empoasca kerri, Balclutha hortensis* and *Exitianus taeniaticeps* were attacked in the laboratory (Nandagopal, 1988). *Geocoris ochropterus* and *G. jucundus* were identified as predators of *Aphis craccivora, Empoasca kerri, Helicoverpa armigera* and *Spodoptera exigua* (Urmila-Sharma and Yadav, 1997).

10.9.4. Chemical control

Several insecticide have proved effective against jassid. The EC and dust formulation such as quinalphos, phosalone, endosulphan, Carbaryl and Monocrotophos were found effective against leaf hoppers (Saboo and Puri, 1979). Monocrotophos, dimethoate and demeton-O-methyl (methyl-O-demeton), all at 0.03%, gave good protection to groundnut against *E. kerri* for up to 21 days (Patel and Vora, 1981).

Similarly Patel and Vora (1981) reported dimethoate, phosphomidon, paramethrin were found to be effective against jassid giving 87 to 91% mortality three days after application. They have also reported that dimethaote and monocrotophos were effective for over 21 days. Soil application of aldicarb (Temik) @ 2.25 kg ai/ha at 15 and 30 days after sowing gave maximum protection for 60 to 80 days. Pod yields were significantly higher in plots sprayed with 0.075% monocrotophos than in untreated plots, and incidence of *Aproaerema modicella, Empoasca kerri, Caliothrips dorsalis* and *Frankliniella schultzei* was significantly lower. (Sivasubramanian and Palanisamy, 1986). The effect of imidacloprid foliar spray persisted for three weeks. Increased pod yield in the treatment was recorded at the time of harvest. Spray of thiometon applied against sucking pests (*Aphis craccivora* and *Empoasca kerri*) on groundnut at 0.03 and 0.05% was found to be the least harmful to the coccinellids, followed by demeton-O-methyl. Malathion at 0.05% was highly toxic to both coccinellids. The comparative toxicity to coccinellids in ascending order was thiometon < demeton-O-methyl < phosphamidon < monocrotophos < dimethoate < malathion. (Upadhyay and Vyas, 1986). Application of aldicarb at 2.25 kg a.i./ha 15 and 30 days after sowing gave the maximum protection from all pests of groundnut, including the cicadellid *Empoasca kerri*, for 60-80 days (Bhushan *et al.*, 1988). Shetgar *et al.* (1989) recommended that 2 applications of monocrotophos (sprayed in solution with 0.05% a.i.) resulted in control leafminers, jassids and thrips on groundnut. Phosphamidon (0.05 or 0.02%) also gave good cost: benefit ratios.

Spraying dimethoate at 200 to 250 ml ai/ha was effective (Wightman and Ranga Rao, 1993). Peter and Sundararajan (1991) reported that monocrotophos 250 g ai/ha , pyraclofos 500 g ai/ha and synthetic pyrethroid deltamethrin 15 g ai/ha were effective against leaf hoppers. Treatment schedule of quinalphos dust followed by endosulfan and dimethoate was found promising (Jena and Kuila, 1996).

E. kerri attained its peak activity in the early part of September. The most suitable pesticide was quinalphos dust, which drastically reduced pest damage and increased yield substantially. Endosulfan and dimethoate were also promising insecticides (Jena and Kuila, 1996).

Insecticides like deltamethrin > dimethoate > monocrotophos were also effective (Kennedy *et al.,* 1992). Jyothirmai *et al.* (2002) reported that the application of imidacloprid 70WS at 5 g/kg seed as seed treatment + 0.01% imidacloprid (Confidor 200SL) as foliar spray recorded the most drastic reduction in damage caused by jassids (*Empoasca kerri*) in groundnut and increase in pod yield. Babu and Santharam (2002) reported that imidacloprid 200 SL treatments were significantly superior to the standard chemicals in reducing the leaf hopper, *Empoasca kerri* population compared to the untreated control.

Table 10.5. List of natural enemies of leaf hoppers (*Empoasca kerri* Pruthi, *Balclutha hortensis* Linb.) (Homoptera: Cicadellidae)

Natural enemy	Order : Family	Reference
Leucopis fumidilarva Tanas	Diptera: Chamaemyiidae	Saxena, 1981
Orius maxidentex Ghauri	Hemiptera: Anthocoridae	Amin, 1987
Brumoides suturalis (Fabricius)	Coleoptera: Coccinellidae	Nandagopal, 1988
Campylomma sp.	Heteroptera: Miridae	Singh *et al.*, 1991
Cheilomenes sexmaculata (Fabricius)	Coleoptera: Coccinellidae	Nandagopal, 1988
Chrysoperla carnea (Steph.)	Neuroptera :Chrysopidae	Singh *et al.*, 1991
Coccinella septempunctata Linnaeus	Coleoptera: Coccinellidae	Nandagopal, 1988
Crossopalpus sp.	Diptera : Empididae	Nandagopal, 1988
Cryptocephalus sp	Coleoptera: Chrysomelidae	Singh *et al.*, 1991
Geocoris sp.	Hemiptera :Lygaeidae	"
Hippodamia variegata (Goeze)	Coleoptera: Coccinellidae	"
Laius malleifer Champion	Coleoptera :Melyridae	"
Monomorium indicum Forel	Formicidae: Hymenoptera	"
Orius sp.	Hemiptera :Anthocoridae	"
Polistes olivaceus(De Geer)	Hymenoptera :Vespidae	"
Stenus pictus Motschulsky	Coleoptera: Staphylinidae	"

References

Amin, P. W. (1987). Insect pests of groundnut in India and their management. In: *Plant Protection in Field Crops,* pp. 219-233. Veerabhadra Rao, M. and Sithanantham, S. (Eds). Plant Protection Association of India, Rajendranagar, Hyderabad, India.

Amin, P. W. (1988). Insect and mite pests and their control. In *Groundnut* (Ed. P.S. Reddy). pp. 393-452. Indian Council of Agricultural Research, New Delhi.

Amin, P. W. and Mohammad, A. B. (1980). Groundnut pest research at ICRISAT. *In proceedings of the International Workshop on Groundnuts,* pp 158-166, ICRISAT, Patacheru, A.P. 502 324, India.

Anonymous, (1994). Annual Research Report AICRP biological control, Gujarat Agricultural University, Anand, Gujarat.

Anonymous, (1988). Annual report *International Crop Research Institute for the Semi-Arid Tropics (ICRISAT).*, Patancheru, Andhra Pradesh, India.

Babu, K. R. and Santharam, G. (2002). Bioefficacy of imidacloprid against leaf hopper, *Empoasca kerri* on groundnut. *Annals of Plant Protection Sciences.* 10(1) : 69-71.

Bhushan, V. S., Ramasubbiah, K., Nath, V. V. N. and. Azam, K. M. (1988). Bio-efficiency of different doses of aldicarb (Temik) against certain groundnut pests. *Pesticides.* 22 (4) : 23-26.

Choudhary, J. P., Yadav, L. S., Poonia, R. S. and Rastogi, K. B. (1980). Some observations on field population of *Empoasca kerri* Pruthi, a jassid pest on mung bean crop. *Haryana Agriculture University Research Journal.* 10 (2) : 250-252.

Dubey, R. C. and Thorat, P. G. (1994). The infestation of jassid on groundnut, sunflower and aphid on safflower at Akola (Maharashtra) in relation with meteorological factors. *Plant Protection Bulletin Faridabad.* 46 (1) : 43-47.

Dwivedi, S. L., Reddy, D. V. R., Nigam, S. N., Rao, G. V. R., Wightman, J. A., Amin, P. W., Nagabhushanam, G. V. S., Reddy, A. S., Scholberg, E. and Ramraj, V. M. (1993). Registration of ICGV 86031 peanut germplasm. *Crop Science.* 33 (1) : 220.

Handa, S. K., Diksit, A. K., Awasthi, M. D. and Bhattacharjee, N. S. (1978). Residues of phorate, disulfoton and carbofuran in soybean crop. *Indian Journal of Entomology.* 39 (4) : 354-356.

Jayanthi, M., Singh, K. M. and Singh, R. N. (1993). Population build up of insect pests on MH 4 variety of groundnut influenced by abiotic factors. *Indian Journal of Entomology.* 55 (2) : 109-123.

Jena, B. C. and Kuila, B. (1996). Jassid infestation in groundnut and its chemical control. *Indian Journal of Entomology.* 58(4): 378–381.

John, K., Vasanthi, R. P., Muralikrishna, T. and Venkateswarlu, O. (2004). Evaluation of advanced groundnut breeding lines for resistance to jassids. *International Arachis Newsletter.* (24) : 13-14.

Jyothirmai, T., Krishna, T. M., Ramaiah, M., Padmavathamma, K. and Rao, A. R. (2002). Efficacy of different insecticides against jassids on groundnut. *Journal of Entomological Research.* 26(4): 291-295.

Kennedy, F. J. S., Rajamanickam, K., Balagurunathan, R. and Lourduraj, A. C. (1992). Yield loss due to insect pests and their control in groundnut. *Groundnut News.* 4 (2): 7.

Nandagopal, V. (1988). *Crossopalpus* sp. (Empididae: Diptera) - an efficient predator of leaf hoppers of groundnut. *International Arachis Newsletter.* (4): 19-20.

Nandagopal, V. (1992). Studies on Integrated pest management in Groundnut in Saurashtra, Ph. D. thesis, Saurashtra University, Rajkot, 246 pp.

Nandagopal, V., Murthy, T. G. K., Radhakrishnan, T. and Sen, P. (1993). Leaf toughness in relation to resistance to jassid *Balclutha hortensis* Lindb. in groundnut. *International Arachis Newsletter.* 13: 20-21.

Nandagopal,V. (1992). First record of insect pests and predators of thrips and leaf hoppers on groundnut in India. *International Arachis News Letter*. 11: 26.

Nandagopal, V. and Rajgopal, K. (1996). Feeding site preference for sucking pests on groundnut and the effect of neem products on the *Balclutha hortensis* Lindb.Karnataka Journal of Agricultural Sciences. 9 (2) : 345-348.

Nandagopal, V. and Soni, V. C. (1992) . Residual effect of some insecticides and neem-oil against jassid on groundnut. Journal of Maharashtra Agricultural Universities. 17 (3) : 420-422.

Nandagopal, V., Soni, V. C. and Gedia, M. V. (1996). Biology of the jassid *Balclutha hortensis* Lindb. (Cicadellidae: Hemiptera) a major pest of groundnut in India. *Entomon*. 21: 221-224.

Nandagopal, V and Reddy, P. S. (1987). Out breaks of *Balclutha hortensis* Lindb. a new pest species of groundnut. *Journal of Oilseeds Research*.4: 300.

Nandagopal, V., Jaroli, R. K., Kumar, R. and Reddy, P. S. (1990). Neem products: possible insecticides on the groundnut jassid *Balclutha hortensis*. *International Arachis Newsletter*. 8: 22.

Nath, P. and Singh, A. K. (1998). Effect of intercropping of groundnut with millets and pigeonpea on the relative incidence of insect pests. *Annals of Plant Protection Sciences*. 6 (2) : 151-154.

Patel, B.R. and Vora, V. J. (1981). Efficacy of different insecticides for the control of groundnut jassid, *Empoasca kerri* Pruthi. *Pesticide*. 15: 33-34.

Patil, B. V., Kalegore, N. K., Kamble, S. K., Shinde, S. V., More, S. N. and Dalvi, S. T. (2003). Efficacy of different neem products against major pests of kharif groundnut. *Journal of Soils and Crops*. 13 (2) : 267-270.

Peter, C. and Sundararajan, R. (1991). Chemical control of major insect pests of groundnut in Tamil Nadu. *Journal of Insect Science*. 4 (1) : 64–66.

Pruthi, H. S. (1940). Description of some new species of *Empoasca walsch* (Eupterygidae: Jaasoidea) from North India. *Indian journal of Entomology*. 11: 1-10.

Rao, K. R. (2002). Induced host plant resistance in the management of sucking insect pests of groundnut. *Annals of Plant Protection Sciences*. 10(1): 45-50.

Reddy, L. J., Nigam, S. N., Moss, J. P., Singh, A. K., Subrahmanyam, P., McDonald, D. and Reddy, A. G. S. (1996). Registration of ICGV 86699 peanut germplasm line with multiple disease and insect resistance. *Crop Science*. 36(3): 821.

Reghupati, A., Rathnaswamy, R., Venkatanarayana, D. and Subramanian, T. R. (1975). Physiology of yellow mosaic virus in green gram, *Phaseolus aureus* Roxb. with reference to its preference by *Empoasca kerri* Pruthi. *Current Science*. 44: 577-8.

Saboo, K. C and Puri, S. N. (1979). Effect of insecticides on incidence of sucking pests and yield of groundnut, *Arachis hypogaea* Linn. *Indian Journal of Entomology*. 40: 311–315.

Sandhu, G. S and Brar, K. S. (1977). Insect pests of groundnut and their control. *Agriculture and Agro Industries Journal*. 10: 7-10.

Saxena, A. P. (1981). *Leucopis fumidilarva* Tanas (Diptera: Chamaemyiidae), a new predator of aphids and leaf hoppers on potato crop. Note. *Journal of the Indian Potato Association.* 8(2): 96-98.

Sekhar, J. C., Singh, K. M., Singh, R. N., Yeshbir-Singh and Malik, K. S. (1997). Impact of ntercropping on the incidence of green jassid, *Empoasca kerri* on pigeonpea: *Indian Journal of Entomology.* 59(1): 119-123.

Sharma, H. C., Pampapathy, G., Dwivedi, S. L. and Reddy, L. J. (2003). Mechanisms and diversity of resistance to insect pests in wild relatives of groundnut. *Journal of Economic Entomology.* 96(6): 1886-1897.

Sherasva, R. A. and Butani, P. G. (1997). Population build up of jassid *Empoasca kerrri* Pruthi (Ciccadellidae: Hemiptera) in popular varieties of groundnut. *Gujarat Agriculture University Research Journal.* 23(1): 49-51.

Shetgar, S. S., Bilapate, G. G., Puri, S. N., Patil, V. V. and Londhe, G. M. (1989). Bioefficacy of different insecticides against foliage pests of groundnut. *Pestology.* 13(11): 12-17.

Singh, R. N. and Singh, K. M. (1978). Incidence of insect pests in early varieties of red gram *Cajanus cajan* L. *Indian Journal of Entomology.* 40(3): 229-244.

Singh, T. V. K., Singh, K. M. and Singh, R. N. (1996). A sequential sampling plan for groundnut jassid, *Empoasca kerri. Indian Journal of Entomology.* 58(1): 36-42.

Singh, T. V. K., Singh, K. M. and Singh, R. N. (1991). Influence of intercropping. I. Incidence of major pests in groundnut (*Arachis hypogaea*) Linn. *Indian Journal of Entomology.* 53(1): 18-44.

Singh, T. V. K., Singh, K. M. and Singh, R. N. (1991). Host range of groundnut jassid, *Empoasca kerri* Pruthi. *Indian Journal of Entomology.* 53(1): 1-17.

Singh, T. V. K., Singh, K. M. and Singh, R. N. (1989). Seasonal influence on the life cycle of groundnut jassid, *Empoasca kerri* Pruthi. *Bulletin of Entomology.* 30: 161-163.

Sivasubramanian, P. and Palanisamy, G. A. (1986). Loss in yield due to leaf hopper and thrips in groundnut. *Madras Agricultural Journal.* 73: 530-531.

Upadhyay, V. R. and Vyas, H. N. (1986). Comparative toxicity of certain insecticides to predatory coccinellids associated with sucking pests of groundnut in Saurashtra. *Indian Journal of Plant Protection.* 13: 91-93.

Urmila-Sharma and Yadav, D. N. (1997). A new record of arthropod predators in groundnut at Anand (India).*Gujarat Agricultural University Research Journal.* 22(2): 157.

Varman, P. V., Ganesan, K. N. and Mothilal, A. (2000). Wild germplasm: potential source for resistance breeding in groundnut. *Journal of Ecobiology.* 12(3): 223-228.

Varman, P. V., Kennedy, F. J. S. and Raveendran, T. S. (1989). Inheritance of leaf hopper (*Empoasca kerri* Pruthi) resistance in groundnut (*Arachis hypogaea* L.). *Journal of Oilseeds Research.* 6: 364-368.

Verma, S. and Lal, R. (1976). Evalution of phorate, monocrotophos and carbaryl against the pests of cowpea crop. *Entomologists' Newsletter.* 6: 52-53.

Verma, S. and Pant, N. C. (1979). Movement of phorate in soil and its residues and residual toxicity. *Indian Journal of Entomology.* 41: 110-116.

Vyas, H. N. (1984). Losses due to insect pests in groundnut (*Arachis hypogaea*). Proceedings of the All India work shop on crop losses on insect pests. *Indian Journal of Entomology*. 2 : 312-317.

Wheatley, A. R. D., Wightman, J. A., Williams, J. H. and Wheatley, S. J. (1989). The influence of drought stress on the distribution of insects on four groundnut genotypes grown near Hyderabad, India. *Bulletin of Entomological Research*. 79 : 567-577.

Wightman, J. A and Ranga Rao, G. V. (1993). A Groundnut Insect Identification Handbook for India. Information Bulletin No. 39, ICRISAT. Patancheru, Andhra Pradesh, India, pp. 28–37.

□□□

11

Thrips

K, GUNATHILAGARAJ AND V. NANDAGOPAL

11.1.Introduction

Thrips species attack groundnut as a complex, starting from vegetative stage to harvest. Four species are considered to be economically important. All the species of the thrips are polyphagous in nature occurring in almost all the groundnut growing countries in Asia and causes extensive damage to the crop. The thrips *Frankliniella schultzei* was previously reported as the main vector of tomato spotted wilt virus (TSWV) on groundnut in India. However, it was discovered that specimens believed to be *F. schultzei* from the ICRISAT slide collection were in fact *Thrips palmi*, a vector of the disease in Japan. An examination of thrips in the ICRISAT slide collection revealed that *Scirtothrips dorsalis, S. oligochaetus, F. schultzei* and *T. palmi* had been present on the ICRISAT farm since 1980. Field studies conducted near Hyderabad, Andhra Pradesh, and Raichur, Karnataka, during February 1989 revealed that adults of *S. dorsalis* and *S. oligochaetus* were present in the leaf terminals, on old leaflets and in the flowers of groundnuts, while larvae were concentrated between the folded terminal leaflets. *F. schultzei* was present in the flowers and *T. palmi* in the terminal leaf buds. *S. dorsalis* was the most abundant species while *S. oligochaetus* was comparatively rare; *F. schultzei* and *T. palmi* had an irregular distribution. These species (and *Megalurothrips usitatus*) were used to test individuals for the presence of TSWV using dot immunobinding assay. *T. palmi* was positive for the virus, but viral antigens were not detected in *M. usitatus* and *S. oligochaetus*. Results with *S. dorsalis* and *F. schultzei* were inconclusive because of nonspecific reactions (Palmer *et al.*, 1990). Tobacco streak virus (TSV) recently caused an epidemic in groundnut crops in Andhra Pradesh, India. In the epidemic area TSV occurred in many widely distributed weeds of which *Parthenium hysterophorus* probably plays a major role in its spread by thrips. Three thrips species, *Megalurothrips usitatus, Frankliniella schultzei,* and *Scirtothrips dorsalis* were vectors in the presence of infected pollen. Sunflower and marigold could act as sources of the inoculum. In limited tests, the virus was not seed-transmitted in the groundnut cv. JL-24 or in the sunflower hybrids KBSH-41, -42, -44, and -50, MSFH-17, and ZSH-976 (Rao *et al.,* 2003).

However, since last few years these species caused severe damage to the groundnut crop in India both damaging the standing crop during the early vegetative stage and as a vectors of viral diseases (Vijayalakshmi, 1994).

11.2.Distribution

The major thrips species attacking groundnut are *Scirtothrips dorsalis* Hood, *Frankliniella schultzei* Trybom, *Thrips palmi* Karny and *Caliothrips indicus* Bagnell and *Megalurothrips usitatus* (Bagnall). The first three species are dominant in southern parts of India while the fourth one is predominantly occurring in western India (Nandagopal and Vasanta, 1991; Ranga Rao and Wightman, 1993). At Hyderabad, Ranga Rao and Wightman (1993) determined the abundance and distribution of the thrips species complex inhibiting terminals and open flowers. *S. dorsalis* was the dominant species accounting for 72 per cent of the foliar thrips. Before flowering *T. palmi* was the second most abundant species (23%) followed by *F. schulzei* (5%). After flower production the foliar samples comprised *S. dorsalis* (93%), *T. palmi* (6%) and *F. schulzei* (1%). In flowers *F. schulzei* was the predominant species (68%) followed by *S. dorsalis* and *T. palmi* (16%).

11.3. Economic importance

The pest appears at early stage of the crop and sometimes even 15 days after sowing. It builds up its population during dry spell of the *kharif* season, though it also appears during summer season. Upadhyay and Vyas (1983) recorded 28 and 22.5 per cent avoidable loss in *kharif* and summer seasons, respectively. The losses in yield due to damage of this pest was estimated 2.3 per cent (*Kharif*) and 5 per cent (summer). Ghewande (1987) recorded the yield loss to the tune of 17 to 40 per cent in Gujarat. The species *S. dorsalis* is an important yield reducer in Karnataka (Thimmaiah and Panchabhavi,1973), Orissa (Senapati and Patnaik,1973) and in Maharashtra (Saboo and Puri,1978). *C. indicus* is reported to be a "menace" to groundnut, particularly in dry years (Ananthakrishnan,1973), and *T. palmi is* the major vector of Peanut bud necrosis virus disease (PBND) (Vijayalakshmi, 1994).

11.4. Nature of damage

Thrips being polyphagous in habit has a wide host range. Thirty crops belonging to 9 families were recorded as hosts in India by various workers, majority of them belong to Leguminosae and groundnut is one of them. Earlier

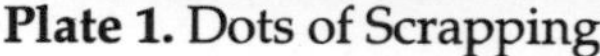

Plate 1. Dots of Scrapping

Plate 2. Adult of Thrips

Poos (1941) described the nature of damage by thrips. He stated that thrips feed by rasping the upper surface of the leaves (**Plate 1**), when they were in bud stage, the unfolded leaves thus distorted, a condition known as "pouts". There is frequent discoloration which may be tinged with red or even black colour. Ananthakrishnan (1973) reported that if there was a heavy incidence of the thrips, the leaves were totally bleached and gave white silvery appearance. There were numerous dark spots of excreta. The leaves later on dried up. In severe cases, the plant showed stunted growth with stripped leaves and became pale. The feeding symptoms are described based on (Amin and Palmer, 1985; Acharya and Koshiya, 1987; Vijayalkshmi, 1994). Bud necrosis disease is the most devastating disease of groundnut (*Arachis hypogaea* L.), caused by a tospovirus. The virus is transmitted by *Thrips palmi* Karny (Thripidae: Thysanoptera). Estimated losses caused by this virus are more than 89 million US dollars per year. A study was carried to determine the occurrence of *Thrips palmi* Karny in the Marathwada region of Maharashtra state on 85 different plants in the year 1992-93. The levels of *Thrips palmi* were graded as high, moderate and absent on the basis of their population. Thirty plant species were found to be heavily infested by the thrips (Kendre *et al.,* 2000).

11.4.1. C. indicus

C. indicus was originally referred as *Selenothrips indicus* Begnall, and *Hercothrips indicus* and *Heliothrips indicus.* According to Ayyar (1929), it was a serious pest of groundnut in Egypt. Both adults and nymphs inhibit older leaves and feed on the upper surface. Feeding results in white spots or streaks. In severe infestations the plants are stunted with stippled leaves and ultimately such leaves dry up. High population of this species occurs during hot dry periods in India.

11.4.2. F. schultzei

This species also has several synonyms such as *F. dampfi, F. sulphurea, Physopus schulzei* and others. Adults and nymphs inhabit the terminal leaves of the groundnut plants and remain hidden in the crevices around the leaf bud. The feeding injury to young unopened leaves or leaf buds appears as light green patches. As the leaves grow, this injury develops into white scars. In severe infestations the leaves remain small, become puckered and their margins become distorted. The affected plants remain stunted and lose their normal green colour. The nymphs also inhabit flowers and feed on pollen. The periods from August to September and January to February are the peak periods of activity.

11.4.3. M. usitatus

This species is also reported in literature as *Frankliniella nigricornis, F. obscuricornis, Physothrips usitatus, P.u. cictipennis, Taeniothrips distalis* and *T. longisstylus.* This species does not survive on groundnut leaves though they feed on the leaves by scraping.

11.4.4. Scirtothrips dorşalis Hood

This species of thrips is referred as *Anaphothrips dorsalis, Caliothrips minutissimus, Scirtothrips andraeae,* and *S. padmae.* Feeding results in the formation of dull yellowish green patches on the upper surface and brown necrotic areas and silvery sheen on the lower surface of the leaves. The leaves become thickened and some curling also occurs. In severe infestations the plants get stunted and the leaves become blighted. High population occurs from February to April, but breeding continues throughout the year.

11.4.5. Thrips tabaci Lindeman

Even though *T. tabaci* occurs on groundnut it may not survive solely on groundnut.

11.4.6. Thrips palmi Karny

This species is found in the folded leaves and flowers (Wightman and Ranga Rao,1994).

11.5. Alternate hosts

The thrips species that invade the groundnut crops, being a passive fliers, are found on many other crops and weeds. Alternate hosts plants for *Frankliniella schultzei* (11.1), *Thrips palmi* (Table 11.2), *C. indicus* (Table 11.3) *Scirtothrips dorsalis* (Table 11.4) are listed below.

Table 11.1. Alternate hosts for *Frankliniella schulzei*

Botanical Name	Family	Reference
Canavalia gladiata	Leguminosae	Amin, (1983)
Vigna unguiculata	"	"
Glycine max	"	"
Vigna mungo	"	"
Vigna radiata	"	"
Crotalaria juncea	"	"
Vicia faba	"	"
Pasium sativum	"	"
Tephrosia purpurea	"	"
Desmodium triflorum	"	"
Ageratum conyzoides	Compositae	"
Chrysanthemum indicum	"	"
Zinnia elegans	"	"
Helianthus annuus	"	"
Acanthospermum hispidum	"	"
Carthamus tinctorius	"	"
Cosmos bipinstatus	"	"
Lagasca mollis	"	"
Tridax procumbens	"	"
Xanthium strumorium	"	"
Solanum melongena	Solanaceae	"
Solanum tuberosum	"	"
Lycopersicon esculantum	"	"
Nicotiana tabacum	"	"
Cassia tora	Caesalpiniceae	"
Cassia obtusifolia	"	"
Cassia occidentalis	"	"
Allium cepa	Liliaceae	"
Datura stromonium	Convolvulaceae	"
Leucas aspara	Labiateae	"
Papaver sp.	Papaveraceae	"
Tribulus triticus	Zygophyllaceae	"
Calotropis gigantean	Asclepiadaceae	"
Celosia argentea	Amaranthaceae	"

Table 11.2. Alternate hosts for *Thrips palmi*

Botanical Name	Family	Reference
Gossypium arboreum	Malvacae	"
Seasamum indicum	Pedaliaceae	"
Mangifera indica	Anacardiaceae	"
Carica papaya	Caricaceae	"
Psidium guajava	Myrtaceae	"
Punica granatum	"	"
Prunus serasus	Rosaceae	"
Citrus limonum	Rutaceae	"
Benincasa lispida	Cucurbitaceae	"
Momordicha charantica	"	"
Cucurbita moschata	"	"
Luffa acutangula	"	"
Citrullus vulgaris	"	"
Alysicarpus longifolius	Leguminosae	"
Vicia faba	"	"
Cyamopsis tetragonoloba	"	"
Vigna sinenesis	"	"
Glycine max	"	"
Vigna mungo	"	"
Vigna radiata	"	"
Crotalaria juncea	"	"
Alysicarpus monilifer	"	"
Alysicarpus rugsosus	"	"
Desmodium dicholomum	"	"
Cassia tora	"	"
Sesbanium Spp	"	"
Dolichos lablab	"	"
Moringa pterigosperma	"	"
Abelmoschus esculentus	Malvaceae	"
Nerium oleander	Apocynaceae	"
Calymma alliaceum	Begnoniaceae	"
Begnonia gracillis	"	"
Techomaria capensis	"	"
Canna indica	Cannaceae	"
Quisqualis indica	Combretaceae	"
Ageratum conyzoides	Compositae	"
Chrysanthemum indicum	"	"
Arnica Montana	"	"
Guizotia abyssinica	"	Vijalakshmi, (1994)
Helianthus annuus	"	"
Zinnia elegans	"	"
Ipomea carnea	Horsefalleaceae	"
Michelia champaca	Mangoliaceae	"
Punica granatum	Punicaceae	"
Phasalis spp.	Solanceae	"
Datura stromonium	"	"

Contd...

Table 11.2. Contd...

Petunia hybrida	"	"
Solanum melongena	"	"
Solanum wendlandii	"	"
Cleodendron imfortunatum	Verbenaceae	"
Verbena paciniata	"	"
Celosia argentea	Amaranthaceae	"
Digera arvensis	"	"
Almania nodiflora	"	"
Amaranthus viridis	"	"
Calotropis gigantea	Asclepiadaceae	"
Lagascea mollis	Asteraceae	"
Parthenium histerophorus	"	"
Acalypha hispida	Euphorbiaceae	"
Acalypha indica	"	"
Phyllanthus maderaspatensis	"	"
Commelina bengalensis	Commelinaceae	"
Ipomoea spp.	Convolulaceae	"
Cleome viscose	Capparaceae	"
Ocimum sanctum	Labiatae	"
Portulaca indica	Portilacaceae	"
Trianthema portulacastrum	"	"
Oldenlandea corymbosa	Rubiaceae	"
Tribulus terrestris	Zygophyllaceae	"
Trichoderma spp.	"	"

Table 11.3. Alternate hosts for *C. indicus*

Solanum melongena	India	Thanki and Patel (1991)
Vigna mungo	"	Subhash-Katare *et al.*, 1998
Vigna radiata	"	Selbha-Devesthali, (1994)
Cotton *(Gossypium spp.)*	Sudan	El-Amin and Ahmed (1991)
Achyranthes aspera Linn.	Amarantaceae	Daniel *et al.*, 1984

Table 11.4. Alternate hosts for *Scirtothrips dorsalis*

Delaware grapes	Japan,	Shibao, (1997)
Turnips (*Brassica rapa*)	Pakistan	Abro, (1997)
Carthamus tinctorius	"	Abro, (1997)
Cotton *(Gossypium spp.)*	"	Abro, (1997)
Okra	"	Abro, (1997)
mango *(Mangifera indica)* cv. Alphonso	India	Kumar *et al.*, (1994)
Rosa sp.	"	Onkarappa *et al.*, (1998)
Citrus aurantiifolia	"	Yadav and Rizvi, (1995)
Punica granatum cv. Ganesh	"	Bagle, (1993)
Citrus spp.	Japan	Masui and Ishigami, (1995)
Tea orchards	"	Tsuchiya *et al.*, (1995)
Viburnum awabuki	"	Tatara, (1994)
Thea sinensis [*Camellia sinensis*	"	Tatara, (1994)
Capsicum annuum	Sri Lanka	Dharmasena, (1998)
Pummelos	Thailand	Malaipan *et al.*, (1992)
Lotus (*Nelumbo nucifera)*	China	Wang-ChinLing *et al.*, (1999)

11.6.Biology

The characters which differentiate the four important species of thrips occurring on groundnut are given in Table 11.5.

11.6.1. *C. indicus*

A detailed study on the bionomics of *C. indicus* was carried out under laboratory condition at a constant temperature of 25 ± 1° C at Junagadh (Nandagopal personal communication).

Adult: The adult female is dark-black in colour, and male is completely paler and smaller than female. The fore and hind wings are were fringed. The forewings are marked with three narrow pale brown bands. The tip of abdomen of female is pointed, whereas it is bluntly rounded in male. Both male and female are very minute insects having average body length of 1.01 and 1.48 mm with the width of 1.05 and 1.52 mm respectively.

Oviposition: Immediately after eclosion from pupae, the adult female thrips climbs on the plant, makes movement to locate proper site for oviposition. The female thrusts eggs singly inside leaf tissues with the help of her pointed ovipositor. The eggs are kidney shaped and pale in colour. The egg embedded in leaf tissues may hatch in 6 to 9 days. However, the period is prolonged under field condition (6 to 15 days). A single female can produce on an average 65 progenies (as many as 150) on groundnut during 15 to 20 days of its adult longevity.

Larva: Prior to attaining the prepupal stage the larva passes through two successive larval instars. The first instar larvae is tiny, delicate, starchy coloured and looks like a segmented worm with red eyes. There is no indication of pegs and antennae, but the appendages can be clearly observed.

The second instar larva resembles the first instar in general appearance except body size and colour. It is whitish yellow and body divided into 10 segments. It becomes full grown in 5 to 9 days.

Pupa: The fully developed larva pupates inside the soil. Before pupation, the larva enters a pre-pupal stage and remains in this crescent stage for 1 to 2 days. The pupa, when completely formed shows slight movements. It becomes darker before emergence of adult. The pupal period varies from 1 to 2 days.

The thrips, passes through four life stages and the entire life-span (from egg to death of adult) varied found 29 to 41 days with an average of 38 days, when reared on groundnut (Acharya and Koshiya, 1987; Amin, 1984).

Table 11.5. Distinguishing characters of thrips species occurring on groundnut

Character	*F. schultzei*	*T. palmi*	*S. dorsalis*	*C. indicus*
Adult female	Pale in colour, 1 mm long	Straw yellow to pale brown, 0.9 mm long	Relatively small, yellow in colour, 0.7 mm long	Adult male smaller than female, female measures
Pronotum	With 2 pairs of setae on the anterior-lateral margin and 2 pairs on the posterio-lateral margin	2 pairs of setae on the posterio-lateral margin; no setae on the anterior-lateral margin	No setae. Dark patches on the dorsal side of abdominal tergites	lateral third of abdominal tergites with sculpture broadly or of ill formed reticulations with internal wrinkles
Forewings	With two complete rows of wing vein setae	With broken rows of wing vein setae	With few small setae on the vein, hide wings with 2 setae	Banded and dark brown in colour
Larvae	Pale yellow move slowly and bend their abdomen while changing the direction	Whitish	Whitish.	Whitish. Both larvae and adults active, moving in a darted fashion

11.6.2. *T. palmi*

Vijayalakshmi (1994) worked out the biology of this species in different temperature regimes (Table 11.6). There was a drastic reduction in the adult emergence at 35°C (5 adults) compared to 25 °C. The longevity of the adult increased with decrease in temperature from 25°C. The egg production increased from 22/female at 15 °C to 78 eggs at 25 °C and beyond that there was reduction in the egg production. When host plants such as cowpea, blackgram, greengram, soybean, groundnut and sunhemp were tried on the population development greengram favoured the population by producing 150 larvae/female , followed by cowpea, groundnut supported only 69 larvae (Table 11.7).

Table 11.6. Effect of temperature on the developmental biology of *T. palmi*

Temp(°C)	Mean developmental period (days)			Fecundity (total no.of eggs/ female)	Adult emergence (%)	Adult longivity (days)	Total no. of larvae produced/ female
	Egg	Larva	Pupa				
15	*	7.7	6.0	22.6	3	28.6	*
20	6.3	6.5	5.4	75.7	15	25.4	66.6
25	5.1	4.7	3.7	78.2	30	24.1	68.2
30	4.0	3.8	2.9	55.6	13	12.9	54.5
35	3.6	3.4	3.0	6.8	5	8.4	6.2
SE(m)	0.5	0.2	0.2	8.2	-	1.9	8.8
CV(%)	36	10	16	54	-	30	57
CD at 5%	1.6	0.5	0.7	23.2	-	5.4	23.5

*no egg hatching

Table 11.7. Effect of various legume hosts on the growth and development of *T. palmi*

Host	Larval period (days)	Pupal period (days)	No. of larvae produced / female	Adult longevity (days)	Adult emergence (%)
Cowpea	4.0	3.8	138.3	21.0	40.0
Black gram	4.7	4.5	117.3	22.0	50.0
Green gram	4.4	4.8	150.1	23.0	48.0
Soybean	4.3	3.9	63.5	15.7	47.0
Groundnut	4.5	4.0	68.9	19.9	20.0
Sun hemp	5.1	4.1	74.9	14.7	50.0
SE(m)	0.1	0.1	10.4	1.4	-
CV(%)	9.6	11.5	32.1	22.6	-
CD at 5%	0.2	0.2	29.4	3.9	-

11.7.Management

11.7.1.Cultural control

The effects of intercropping groundnuts with pearl millet, cowpeas or sorghum on populations of leafminers, leafhoppers and thrips, and their natural enemies were examined by Kennedy *et al.* (1990). Intercropping with pearl millet significantly reduced pest incidence and increased coccinellid numbers per plant compared with groundnut monocultures. The dominant predators were *Coccinella* sp. and *Menochilus sexmaculatus* (=*Cheilomenes sexmaculata*) and *Chelonus* sp. was the dominant parasitoid.

Fast growing cereals, such as pearl millet is reported to reduce the population of thrips (Sreenivasulu *et al.*, 1991). In an another experiment Singh *et al.* (1991) studied the effect of intercropping in groundnut (to control insect pests) on the abundance of natural enemies. In general the number of cicadellids were accompanied by a corresponding increase or decrease in the number of *Brumus* sp. Spiders [Araneae], *Laius malleifer*, and coccinellids, *Coccinella septempunctata, Orius* sp. and *Cheilomenes sexmaculata* effectively controlled thrips infesting groundnut. Shetgar *et al.* (1994) investigated the effect of sowing date on incidence of foliar pests and yield losses on groundnuts and reported that higher population were recorded on plants sown on 15 July, while the greatest numbers of the cicadellid and the thrips *F. schultzei, S. dorsalis* and *C. indicus* were recorded on plants sown on 15 June. Maximum pod yield (707 kg/ha) was recorded from plants sown on 15 June. Pod yield losses were 31.7, 23.9, 25.5 and 26.7 per cent for plants sown on 15 June, and 1, 15 and 30 July, respectively. In an interesting study, Senguttuvan (1999) studied the efficacy of neem extracts, indigenous plant extracts and 0.05 per cent monocrotophos against *Scirtothrips dorsalis* on groundnuts in Tamil Nadu, India, during kharif 1995, winter 1995-96 and *kharif* 1996. Monocrotophos was more effective than the plant products, resulting in a 45-50 per cent reduction of the pest population.

11.7.2. Chemical control

Spraying the plant growth regulators maleic hydrazide, chlormequat chloride and naphthalene acetic acid onto groundnut plants suppressed populations of the thrips, *Caliothrips indicus* considerably, but numbers were not reduced sufficiently to prevent damage to foliage. It is concluded that the chemicals were unsuitable for control of the pest population (Nandagopal and Vasantha, 1991).

Out of 9 insecticides tested for the efficacy against *Caliothrips indicus* in groundnut, lowest leaf damage was recorded with dimethoate (0.03%). (Acharya and Koshiya 1991). Earlier, systemic and contact insecticides like malathion 0.05 and dimethoate 0.03 per cent (Harding, 1961) DDT (0.24%) and phosphamiodon SC (0.1%) (Anon., 1967) have been used successfully to reduce thrips population on groundnuts (Ananthkrishnan, 1973). Seed treatment of groundnuts with 4% carbofuran and applying phorate at 3 kg a.i. or carbofuran at 1.5 kg a.i./ha effectively controlled *C. indicus* and increased yields significantly compared with untreated plots (Thimmaiah and Panchabhavi, 1973), dimethoate and monocrotophos (0.05%) (Ghewande *et al.*, 1987) and monocrotophos 0.05% (Patil *et al.*,1992).

Various newer insecticides were tested under field conditions at G.A.U., Junagadh for the economic and effective control of thrips in kharif and summer groundnut crops. Among different insecticides, dimethoate (0.03%), phosphamidon (0.03%), methyl-0demeton (0.03%), quinalphos (0.05%), and monocrotophos (0.04%) found effective in reducing the thrips damage with higher pod yield during both the seasons (Acharya and Koshiya, 1987). In field experiments in 1985-89 in Maharashtra, India, 14 insecticide treatments were evaluated for the control of thrips [Thripidae] on groundnut. It is recommended that 2 applications of monocrotophos (sprayed in solution with 0.05% a.i.) are used to control these pests (Shetgar *et al.*,1989).A study conducted at Bangalore and Hiriyur during kharif 1995-97 proved the effectiveness of Starthene 75 SP (acephate) against insect pests of groundnuts. At both locations, the crops also harboured low to moderate populations of thrips. Plants treated with Starthene at 750 and 1000 g/ha were vigorous and had significantly increased yields compared to controls. Observations over several seasons and locations confirmed the safety of Starthene to bees, other pollinators, parasitoids and predators. Starthene was not phytotoxic to the crops, and although marginally expensive (Rs 770/kg), its long persistence makes it economically viable. It is concluded that Starthene is suitable for use in integrated pest management of oilseed crops (Chakravarthy, 1999). Patil *et al.* (1992) conducted field studies in Maharashtra, India, during the kharif seasons, the greatest reduction in the number of thrips [Thysanoptera] infesting groundnuts was obtained by applying monocrotophos with a ULV sprayer.

Seed treatment of groundnuts with 4% carbofuran and applying phorate at 3 kg a.i. or carbofuran at 1.5 kg a.i./ha effectively controlled *C. indicus* and increased yields significantly compared with untreated plots (Thimmaiah and Panchabhavi, 1973).

11.7.3. Biocontrol

In a field studies during the summer and *kharif* seasons, Singh *et al.* (1991) quantified the abundance of their natural enemies such as spiders [Araneae], and coccinellids *Laius malleifer, Coccinella septempunctata, Orius* sp. and *Cheilomenes sexmaculata. Orius albidipennis* Router, *O. maxidentex* Ghauri, *Orius* (*Dimorphella*) spp., (Heteroptera:Antocoridae) *Thyene* sp., *Aelurrllus* sp., *Menomerus* sp., *Saitis* sp., (Araneae:Salticidae) have been recorded as predators of *C. indicus* (Nandagopal, 1992). *Paederus fuscipes* Curtis (Heteroptera: Anthocoridae) has also been reported as predator of *C. indicus* by Pophaly and Gupta (1996).

11.7.4. Host plant resistance

Bred by the bulk pedigree method from a cross between F334A-B14 and NC Ac2214, this Spanish type groundnut germplasm (PI561917) was released in 1991 as a source of resistance to *Thrips palmi, Empoasca kerri, Spodoptera litura, Aproaerema modicella* and bud necrosis virus. It is photoperiod insensitive and resistant to iron deficiency chlorosis, has erect growth habit with sequential branching, 4-8 primary and up to 2 secondary branches and matures in 105 to 110 days in the rainy season in peninsular India. It has 1- to 2-seeded, small-sized pods, rose tan seeds with a 100-seed weight of 39 g. Oil content averages 52 per cent and protein content 20 per cent. ICGV86031 is recommended for cultivation in areas where *Spodoptera* spp., *A. modicella* and bud necrosis disease are endemic (Dwivedi *et al.*, 1993). Genotype NCAc 2575 has some resistance to thrips and cultivar Robut 33-1 has shown some field resistance have been identified (Amin, 1985). Out of 28 genotypes, the lowest mean number of thrip eggs laid were noted in PBS 24005 (0.5) followed by PBS23003 (1.4), PBS 24006 (1.0), 24030 (1.0), PBS 24040 (2.0), Girnar 1 (2.1), PBS 21020 (2.6) and PBS 21031 (3.6), and these were considered resistant. With regard to aphid infestation, PBS 24005, PBS 24006, PBS24030, Girnar 1, PBS 21031 and PBS 21020 were categorised as moderately resistant. (Samdur *et al.*, 1999). Out of Twenty-eight genotypes of groundnut were evaluated for resistance to thrips (*Caliothrips indicus, Frankliniella schultzei, Scirtothrips dorsalis* and *Thrips tabaci*). Six genotypes (MH-11, MH-92, MH-46, MH-51, MH-53 and MH-55) showed resistance to thrips. The genotypes with small, green and hairy leaves were less infested and damaged (Rohilla *et al.*, 1999).

Out of seventeen wild species belonged to six sections (Arachis, Caulorhizae, Heteranthae, Rhizomatosae, Extranervosae and Erectoides) were screened, *A. cardenasii* and *A. diogoi* recorded resistance for the foliar diseases, *Scirtothrips dorsalis* and *Empoasca kerri* (Varman *et al.*, 2000).

11.7.5. Chemical control

The efficacy of 0.01% cypermethrin, 0.005% deltamethrin, 0.04% monocrotophos, 0.05% quinalphos and 0.15% carbaryl, alone and in combination, for the control of *Scirtothrips dorsalis, Frankliniella schultzei* and the gelechiid, *Aproaerema modicella* on groundnuts was evaluated by Murthy *et al.* (1988). Cypermethrin alone or in combination with monocrotophos or quinalphos, and deltamethrin with monocrotophos or carbaryl gave good control of *S. dorsalis* and *F. schultzei*. Cypermethrin or deltamethrin with carbaryl, cypermethrin with monocrotophos, and monocrotophos alone were effective against *A. modicella*. High yields were obtained following treatments with mixtures of cypermethrin and carbaryl (1275 kg/ha) and deltamethrin and carbaryl (1183 kg/ha). Maximum net returns were obtained with these treatments.

Table 11.8. List of natural enemies of thrips (*Scirtothrips dorsalis* Hood, *Thrips palmi* Karny, *Frankliniella schultzei* (Trybom), *Caliothrips indicus* (Bagnall) (Thysanoptera: Thripidae)

Natural enemy	Order : Family	Reference
Cheilomenes sexmaculata	Coleoptera: Coccinellidae	Singh *et al.* 1991
Franklinothrips megalops (Trybom)	Thysanoptera Aeolothripidae	Suresh kumar and Ananthakrishnan, 1987
Geocoris ochropterus	Hemiptera: Lygaeid	Suresh Kumar and Ananthakrishnan, 1985
Orius maxidentex Ghauri	Heteroptera Anthocoridae	Amin, 1987
Orius(=Dimorphella) albidipennis (Router)	Heteroptera Anthocoridae	Nandagopal, 1992

Monocrotophos was more effective than the plant products, resulting in a 45-50% reduction of the pest population. All the neem products significantly reduced the thrips population, with Achook, Neem Gold and 3% neem oil resulting in 29, 29 and 28% lower populations, respectively, than no treatment. Of the indigenous plant products, Vitex leaf extract and Eucalyptus leaf extract were the best treatments, resulting in reductions of 30 and 29%, respectively (Senguttuvan, 1999).

The bioefficacy of imidacloprid 200 SL (at 100, 150, 200 and 250 ml/ha), with dimethoate and methyl-o-demeton at 500 ml/ha as standards, was evaluated against thrips and aphids on groundnut cv. Co2 by Babu and Santharam (2001). Imidacloprid at 100 ml/ha effectively controlled the population of thrips and aphids on groundnut and was superior to the standard chemicals.

References

Abro, G. H. (1997). Comparison of knapsack, rotary atomiser and electrodynamic spraying for the control of pests in different crops. *International Pest Control*. 39(3): 77-79.

Acharya, M. F. and Koshiya, D. J. (1987). Bionomics and chemical control of groundnut thrips, *Caliothrips indicus* Bagnall (Thysanoptera: Thripidae). M.Sc., (Ag) thesis submitted to Gujarat Agricultural University, Sardar Krishinagar, Guajarat, pp 63.

Acharya, M. F. and Koshiya, D. J. (1991). Efficacy of different insecticides against thrips Caliothrips indicus Bagnal (Thysanoptera: Thripidae) on groundnut. *Gujarat Agricultural University Research Journal*. 16(2): 87-90.

Amin, P. W. (1987). Insect pests of groundnut in India and their management. In: *Plant Protection in Field Crops*, pp. 219-233. Veerabhadra Rao, M. and Sithanantham, S. (Eds). Plant Protection Association of India, Rajendranagar, Hyderabad, India.

Amin, P. W. (1984). Studies on arthropod vectors of groundnut viruses, their ecology and control. *Groundnut Improvement Programme*. ICRISAT, 1982. pp 166.

Amin, P. W. (1983). Major field and storage insect pests of groundnut in India and their control. Paper presented at the Summer Institute on *'Recent Advances in Groundnut Productivity'* held at Junagadh, 9-30, April 1983.

Amin, P. W. (1985). Apparent resistance of groundnut cultivar Robut 33-1 to bud necrosis diseases. *Plant Diseses*.69: 718-719.

Amin, P. W. and Palmer, J. (1985). Identification of groundnut Thysanoptera. *Tropical pest management*. 31: 286-91.

Ananthakrishnan, T. N. (1973). *Thrips: Biology and Control*, 120pp. Mc Millan Co. of India Ltd., New Delhi.

Anonymous. (1967). Pest control in Groundnuts. *Pans Manual* No.2. Centre for Overseas Pest Research, London, pp 50.

Ayyar, T. V. R. (1929). The economic status of Indian Thysanoptera. *Bull. Ent. Res.* 20: 77-80.

Babu, K. R. and Santharam, G. (2001). Bio-efficacy of imidacloprid against thrips and aphids on groundnut, *Arachis hypogaea* L. *Madras Agricultural Journal*. 87(10/12): 605-608.

Bagle, B. G. (1993). Seasonal incidence and control of thrips *Scirtothrips dorsalis* Hood in pomegranate. *Indian Journal of Entomology*. 55(2): 148-153.

Chakravarthy, A. K. (1999). Role of starthene 75 SP in integrated pest management. *Insect-Environment*. 4(4): 137-139.

Daniel, A. M., Bakthavatsalam, N. and Kumar, N. S. (1984). Weed-crop interaction with reference to *Caliothrips indicus* (Bagnall). (Thysanoptera: Thripidae) on *Arachis hypogaea* Willd. (Fabaceae) and an alternate weed host *Achyranthes aspera* Linn. (Amarantaceae). *Entomon.* 9(1): 47-51.

Dharmasena, C. M. D. (1998). Present status of managing chilli leaf curl complex in the North Central Province of Sri Lanka. *Tropical Agricultural Research and Extension.* 1(2): 154-158..

Dwivedi, S. L., Reddy, D. V. R., Nigam, S. N., Rao, G. V. R., Wightman, J. A., Amin, P. W., Nagabhushanam, G. V. S., Reddy, A. S., Scholberg, E. and Ramraj, V. M.(1993). Registration of ICGV 86031 peanut germplasm. *Crop Science.* 33(1): 220.

El-Amin, E. T. M. and Ahmed, M. A. (1991). Strategies for integrated cotton pest control in the Sudan. 1 - cultural and legislative measures. *Insect Science and its Application.* 12(5-6): 547-552.

Ghewande, M. P. (1987). Virus diseases of groundnut and their management. In symposium on Diseases of Oilseed and Pulse Crops and their control. (Eds. S. C. Vyas and V. N. Shroff), Jawaharlal Nehru Krishi Vishwa Vidhyalaya, Indore, (M.P.) pp.62-67.

Harding, J. A. (1961). Evaluation of insecticides for control of Tobacco Thrips on peanuts. *Journal of Economic Entomology.* 54(1): 200-201.

Kendre, M. S., Patange, N. R., Neharkar, P. S. and Telang, S. M. (2000). Occurrence of *Thrips palmi*, a vector of bud necrosis disease of groundnut in Marathwada region. Journal of Soils and Crops. 10(2): 226-230.

Kennedy, F. J. S., Rajamanickam, K. and Raveendran, T. S. (1990). Effect of intercropping on insect pests of groundnut and their natural enemies. Journal of Biological Control. 4 (1) : 63-64.

Kumar, S., Patel, C. B., Bhatt, R. I. and Rai, A. B. (1994). Population dynamics and insecticidal management of the mango thrips, *Scirtothrips dorsalis* Hood (Thysanoptera: Thripidae) in South Gujarat. *Pest Management and Economic Zoology.* 2(1): 59-62.

Malaipan, S., Kongpitak, P. and Sujitratanan, S. (1992). Comparative study on the effect of insecticides and neem extracts to the thrips and honey bee pollinators of pomelo cultivar Tongdee. *Kasetsart Journal Natural Sciences.* 26(2): 146-150.

Masui, S. and Ishigami, S. (1995). The local forecasting of pest occurrence - focus on citrus. *Agrochemicals*, Japan. 67: 8-10.

Murthy, G. V. S., Subbaiah, K. R., Rao, K. R. S. and Murthy, M. M. K. (1988). Efficacy of certain synthetic pyrethroids and conventional insecticides alone and in combination against thrips and leaf miner of groundnut. *Journal of Research-APAU.* 16: 6-9.

Nandagopal, V. (1992). Studies on Integrated pest management in Groundnut in Saurashtra, Ph. D. thesis, Saurashtra University, Rajkot, 246 pp.

Nandagopal, V. and Vasantha. S. (1991). Influence of plant growth regulators on the population of thrips *Caliothrips indicus* (Bagnall) and damage in groundnut, *Entomon.* 16: 87-89.

Onkarappa, S., Mallik, B., and Reddy, P. P. and Kumar, N. K. K. (1998). Distribution and management of *Scirtothrips dorsalis* Hood (Thysanoptera: Thripidae) on rose. *In* (ed.). Verghese- Advances in IPM for horticultural crops. *Proceedings of the First National Symposium on Pest Management in Horticultural Crops: environmental implications and thrusts*, Bangalore, India, 15-17 October 1997. 165-167.

Palmer, J. M., Reddy, D. V. R., Wightman, J. A. and Rao, G. V. R. (1990). New information on the thrips vectors of tomato spotted wilt virus in groundnut crops in India. *International Arachis Newsletter*. 7: 24-25.

Patil, N. M., Ghule, B. D., Dhumal, V. S. and Deokar, A. B. (1992). Critical time and frequency of insecticidal sprays for the control of groundnut sucking pests. *Journal of Maharashtra Agricultural Universities*.17(3): 477-478.

Poos, F. W. (1941). On the causes of peanut Pouts. *Journal of Economic Entomology*. 34: 727-728.

Pophaly, D. J. and Gupta, A. K. (1996). Preliminary studies on insect pests and associated parasitoids on *Lathyrus sativus*. *Current Research*, University of Agricultural Sciences Bangalore, 25(3): 51-55.

Ranga Rao, G.V. and Wightman, J.A. (1993). Groundnut thrips of economic importance their occurrence and seasonal distribution. *Journal of Entomological Science*. 4: 13.

Rao, R. D. V. J. P., Reddy, A. S., Reddy, S. V., Thirumala-Devi, K., Rao, S. C., Kumar, V. M., Subramaniam, K., Reddy, T. Y., Nigam, S. N. and Reddy, D. V. R. (2003). The host range of Tobacco streak virus in India and transmission by thrips. *Annals of Applied Biology*. 142(3): 365-368.

Rohilla, H. R., Harvir-Singh and Batra, G. R. (1999). Field evaluation of groundnut genotypes for resistance to thrips. *Annals of Biology Ludhiana*. 15(2): 219-221.

Saboo, K. C. and Puri, S. N. (1978). Effect of insecticides on incidence of sucking pests and yield of groundnut, *Arachis hypogea* Linn. *Indian Journal of Entomology*. 40: 311-315.

Samdur, M. Y., Nandagopal, V., Mathur, R. K. and Manivel, P. (1999). Evaluation of advanced breeding lines of groundnut for resistance to thrips and aphids. *International Arachis Newsletter*. 19: 19-21.

Selbha Devesthali, Joshi, M. and Devesthali, S. (1994). Infestation and varietal preference of insect pests in green gram. *Indian Agriculturist*. 38(4): 263-272.

Senapathi, B. and Patnaik, N. C. (1973). Occurrence of *Scirtothirps dorsalis* Hood on groundnut and preliminary evaluation of loss in pod yield. *Journal of Research of the Orissa University of Agriculture and Technology*. 3: 110–113.

Senguttuvan, T. (1999). Effect of plant products against thrips in groundnut. *International Arachis Newsletter*. 19: 36 – 39.

Shetgar, S. S., Bilapate, G. G. and Londhe, G. M. (1994). Effect of sowing dates on pests incidence and yield losses due to foliage pests on groundnut. *Indian Journal of Entomology*. 56(4): 441-443.

Shetgar, S. S., Bilapate, G. G., Puri, S. N., Patil, V. V. and Londhe, G. M. (1989). Bio-efficacy of different insecticides against foliage pests of groundnut. *Pestology*. 13 (11): 12–17.

Shibao, M. (1997). Effects of insecticide application on population density of the chillie thrips, *Scirtothrips dorsalis* Hood (Thysanoptera:Thripidae) on grape. *Applied Entomology and Zoology*. 32(3): 512-514.

Singh, T. V. K., Singh, K. M. and Singh, R. N. (1991). Influence of intercropping. III. Natural enemy complex in groundnut. *Indian Journal of Entomology*. 53(3): 333-368.

Sreenivasulu, P., Kuhn, C. W., Naidu, R. A., Demski, J. W., Reddy, D. V. R. and Nayudu, M. V. (1991). Viruses infecting peanuts: Taxonomy, identification, and disease management. *Research Bulletin 406*, The Georgia Agricultural Experiment Stations, The University of Georgia College of Agricultural and Environmental Sciences, p. 34.

Subhash Katare, Deshpande, R. R., Bhadauria, N. S., Bhadauria, N. K. S. and Katare, S. (1998). Response of black gram cultivars against their major insect pests. *Bhartiya Krishi Anusandhan Patrika*. 13(1-2): 89-91.

Suresh Kumar, N. and Ananthakrishnan, T. N. (1985). *Geocoris ochropterus* Fabr. as a predator of some thrips. *Proceedings, Indian National Science Academy, Biological Sciences.* 51 (2): 185-193.

Sureshkumar, N. and Ananthakrishnan, T. N. (1987). Biotic interactions in relation to prey-predator relationship with special reference to some thrips species (Thysanoptera: Insecta). *Journal of Entomological Research*. 11(2): 192-202.

Tatara, A. (1994). Effect of temperature and host plant on the development, fertility and longevity of *Scirtothrips dorsalis* Hood (Thysanoptera: Thripidae). *Applied Entomology and Zoology*. 29(1): 31-37.

Thanki, K. V. and Patel, J. R. (1991). Field evaluation of some brinjal varieties for resistance to insect pests and disease in north Gujarat conditions. *Gujarat Agricultural University Research Journal*. 16(2): 94-97.

Thimmaiah, G. and Panchabhavi, K. S. (1973). Insecticidal control of thrips (*Heliothrips indicus* Bagnall) on groundnut. *Mysore Journal of Agricultural Sciences*. 7: 334-336.

Tsuchiya, M., Furuhashi, K. and Masui, S. (1995). Behavior of yellow tea thrips *Scirtothrips dorsalis* (Hood) on a reflective sheet. *Japanese Journal of Applied Entomology and Zoology*. 39(4): 289-297.

Upadhyay, V. R. and Vyas, H. N. (1983). Yield loss in groundnut due to jassid *Empoasca kerri* (Pruthi) and thrips *Frankliniella schultzei* (Bagnall) at Junagadh (Gujarat State). *Indian Journal of Entomology*. (Special issues). 2: 322 - 26.

Varman, P.V., Ganesan, K. N. and Mothilal, A. (2000). Wild germplasm: potential source for resistance breeding in groundnut. *Journal of Ecobiology*. 12(3): 223-228.

Vijayalaxmi, K. (1994). Transmission and ecology of *Thrips palmi* Karny, the vector of peanut bud necrosis virus. Ph. D. thesis. Andra Pradesh Agric. University, Rajendra Nagar, Hyderabad, India. pp 99.

Wang ChinLing., Hsu MonYu., Young ChinHan., Jian YuLing., Wang, C. L., Hsu, M. Y., Young, C. H. and Jian, Y. L. (1999). Toxicity of insecticides for *Scirtothrips dorsalis*

Hood (Thysanoptera: Thripidae) on lotus. *Chinese Journal of Entomology*. 19 (4): 377-380.

Wightman, J. A. and Ranga Rao, G. V. (1994). Groundnut pests *in* The Groundnut Crop: (Smart, J., ed). London, 126 pp.

Yadav, L. B. and Rizvi, S. M. A. (1995). Studies on the insect-pests of lime, *Citrus aurantifolia* Swinh. in Eastern Uttar Pradesh. *Annals of Plant Protection Sciences*. 3(1): 34-37.

□□□

Minor Insects and Non-Insect Pests

M. JOHN SUDHEER AND V. NANDAGOPAL

12.1. Introduction

Depending upon the damaging capacity and population density, insect pest have been classified as key, major, minor, sporadic and potential pests. In order to arrive at definite and practicable concept, different categories of pests are defined in terms of the general equilibrium position (GEP), the economic injury level (EIL) and the damaging boundary (DB). GEP is the mean value of pest density around which the pest population tends to fluctuate as changes occur in the biotic and abiotic factors of the environment. The lowest level of injury where the damage can be measured is called the DB while the lowest number of insects that will cause economic damage is referred to as EIL.

In the case of a minor pest, GEP lies below the EIL and DB. When environmental factors are favourable the insect population may cross the EIL and DB for a short period which may not be significant. These minor pests are easily controlled by the available control measures or by a single pesticide application. There are several minor pests in groundnut crop, which sometimes may cause damage to the crop and may attain the status of major pests when the biotic and abiotic factors of the environment are favourable. A good example is, the tobacco caterpillar (*Spodoptera litura* F.) on groundnut. It was rated as minor pest two decades ago, but has become a major problem in groundnut growing areas.

12.2. Grass hoppers (Acridiidae, Orthoptera)

The acridiid grasshoppers recorded on groundnut are *Atractomorpha crenulata (F.), Cyrtacanthacris ranacea, Chrotogonus trachypterus, Acrida turrita* Linn., *Oxya chinensis* (Thunberg), *Orthacris* sp., *Oedaleus* sp. and *Trilophida sp.* (Amin,1988). Both adults and nymphs feed on the foliage. Males generally pass through six instars, whereas females apparently have seven instars. They prefer low vegetation and have no climbing habits.

12.3. Earwigs (Forficulidae, Dermaptera)

The earwig, *Euborellia stali* Dohrn (**Plate 1**) is a pest of groundnut in South India; feeding on the kernels causing up to 20 per cent yield loss. *E. stali* was first recorded as pest of groundnut at Coimbatore and South Arcot districts of Tamil Nadu (Cherian and Basheer, 1940). Its occurrence was also reported from Bombay and Manipur (Barwal, 1985). Recently, it was observed as the predominant pod borer of groundnut in Tamil Nadu (Senguttuvan and Dhanakodi, 1997).

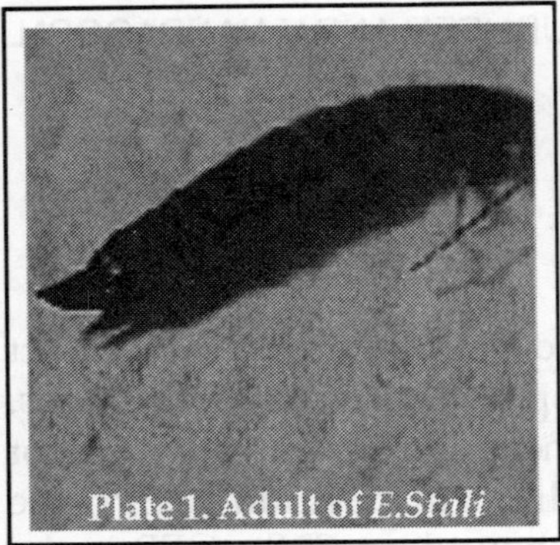
Plate 1. Adult of *E.Stali*

The insect mates 7-10 days after emergence and lays eggs after 10-23 days. Generally 20-100 eggs (average of 58) are laid in clusters in the soil and in damaged pods. Eggs hatch in a week. The larva has 5 instars and resembles adult in all instars. Complete life cycle takes 56-72 days. Adult life span lasts 106-252 days. The larvae and adults cause damage to maturing pods by making a hole. The damaged pods and kernels are prone to fungal pathogen attack.

The Forficulid *Euborellia stali* (Dohrn.) was found infesting pods and kernels in an experimental crop of an imported variety of groundnut at Tindivanam, Madras, India, harvested in December 1967. In five samples of about 446 pods each, the average rate of infestation was 46.6% [cf. RAE/A 29, p. 518]. The yield was only 238 kg/ha, instead of the normal 457 kg, and this represented a loss of oil of 114 kg/ha (Purushothaman *et al.*, 1970).

Lewin *et al.* (1971) studied the resistance of ten varieties of groundnut to earwig *Euborellia stali* (Dohrn) [cf. RAE/A 62, 619, etc.] [cf. 62, 3776, etc.] was greatest in two bunch varieties.However, these two varieties were least susceptible to attack by the earwig, damage to the pods (**Plate 2**) averaging 1.3-1.4%, as compared with 5.5-18.4% for the other varieties. Infestation by *Euborellia stali* (=*E. annulata*) under field conditions ranged from 0.81 (in TG19A) to 6.72% among 23 Virginia bunch varieties, and from 0.27 in C335) to 4.35% among 15 Virginia runner varieties. Varietal differences were significant in both groups (Giridharan *et al.*, 1985). Barwal (1985) reported *Euborellia annulipes* and the termitid *Odontotermes obesus* on groundnut crop in the north-eastern hills of India. Infestation by *Euborellia stali* (=*E. annulata*) under field conditions ranged from 0.81 (in TG19A) to 6.72% among 23

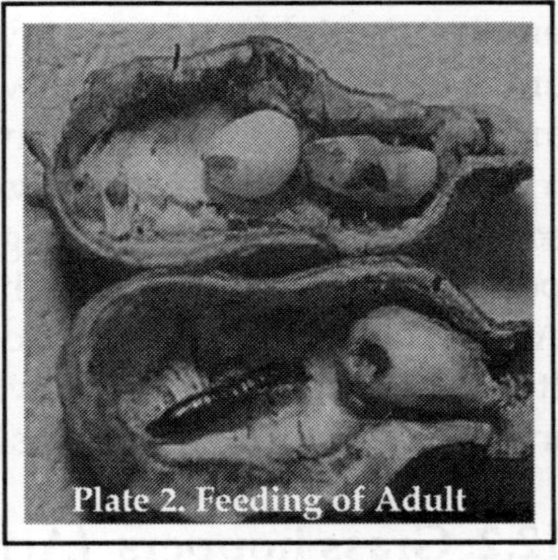
Plate 2. Feeding of Adult

Virginia bunch varieties, and from 0.27 (in C335) to 4.35% among 15 Virginia runner varieties. Varietal differences were significant in both groups (Giridharan *et al.*, 1985). Barwal and Gupta (1991) reported that when groundnut was sown fortnightly in April and early May, low incidence of *Euborellia annulipes* was observed. Basal application of heptachlor dust at 5% was highly effective in reducing infestation by *E. annulipes* and increasing yield, followed by dusts containing 5% carbaryl; each of the compounds was applied at 1.0 kg a.i./ha.

Cultural practices to control pod boring insects (including *Euborellia stali* (= *E. annulata*) and *Dorylus* sp.) were assessed for 4 seasons by Senguttuvan and Dhanakodi (1997) under red lateritic soil conditions at Vamban in Tamil Nadu, India, under irrigated conditions using the groundnut variety TMV7. Soil application of neem (*Azadirachta indica*) cake at 250 kg/ha at the time of sowing resulted in the lowest pod damage (4.95%), which was 23.75% less than in the control (8.85% damage). Maintaining weed free plots until 70 days after sowing (5.46%) and irrigation once in 5 days (6.08%) were also effective. These cultural measures resulted in a 19.0-22.0% yield increase over the control (906 kg/ha). Plots which were unweeded until maturity had a 4.65% increase in pod damage and a 41.27% yield decrease compared with the control.

Different formulations of insecticides were assessed for their toxicity against the pod borer complex, *Euborellia stali*, wireworm [Elateridae] and *Dorylus* sp., of groundnut by Senguttuvan and Dhanakodi (1997). Soil application of carbofuran 3G, phorate 10G at 10 kg/ha and quinalphos 1.5D at 25 kg/ha at the time of sowing and 45 days after sowing resulted in 60.94, 46.43 and 32.43% less pod damage and 33.58, 32.02 and 17.50% higher yields, respectively, than the untreated control.

Senguttuvan and Dhanakodi (1997) identified the pod borers of groundnut, their seasonal occurrence, and the nature and extent of damage caused to groundnuts on red lateritic soils. Wireworms [Elateridae] preferred mature groundnut pods and tunnelled the pulp of the husk before damaging the kernels. *Dorylus* sp. scraped the shell of mature pods, boring spherical holes and feeding on the kernels. *Euborellia stali* preferred young, developing pods and caused 80-90% of pod damage. *E. annulata* and Elateridae were present throughout the year, but total pod damage was lowest in crops sown from September to November. The greatest yield was achieved in groundnuts sown in late December (1440 kg/ha).

12.4. Pod sucking bug (Lygaeidae, Hemiptera)

Elasmolomus sordidus (Fab.) is regarded as a minor pest of groundnut in India but at times it assumes considerable importance. Kulkarni *et al.* (2001) studied the population fluctuation and feeding potential of predatory earwig

(*Labidura* sp.) and cricket (*Gryllus domesticus* [= *Acheta domesticus*]), natural enemies of pod bug, *E. sordidus*. The mean consumption by earwig on different stages of pod bug was 24.26, 20.93, 15.69, 7.90, 4.70, 3.09 and 2.63 for egg, 1st, 2nd, 3rd, 4th, 5th instar nymphs and adult, respectively. A similar trend was observed in case of cricket. Fifth-instar larvae and adults of *Elasmolomus sordidus* showed the highest consumption and weight gain on seeds of sesame, followed by those of groundnut, wheat and *Ficus religiosa*, a non-economic host. Fourth-instar larvae showed the highest consumption and weight gain on groundnut seeds (Mukhopadhyay and Saha, 1995).Chopra and Singal (1982) gave a taxonomic account of the 3 species of the lygaeid genus *Elasmolomus* that occur in India. The species include *E. sordidus* (F.), which occurs throughout the country and has been recorded attacking the grain and seeds of a variety of crops, either in storage or in the field; notes are given on its biology, nomenclature and known distribution, and the adult is described.

12.4.1. Distribution

This pest has been reported from Bihar, Maharashtra, Gujarat, Tamilnadu and Andhra Pradesh (Narayanan, 1962; Rai, 1976).

12.4.2. Damage

Damage to the groundnut is not easily visible. Both nymph and adult suck the oil from the kernel by feeding on the freshly harvested pod in field or in store. Such pod has shell spotted with excreta and kernels loose viability. The reduction in the oil quality has been reported in India (Kasargode and Deshpande, 1921). Gillier (1970) reported that in heavily attacked pods, the oil contents in the affected nuts fell by 4.6 per cent and oleic acid increased by 5.2 per cent. Above 3% acidity (free fatty acid), 1% less price for every extra percent of acidity is allotted up to 7% acidity, and for 7-10% free fatty acid the reduction in the price of extra 1% acidity is 2% (Gillier and Bockellee-Morvan, 1979). At the ICRISAT, the bugs were observed to suck the content of the cotyledons of newly emerged seedling but the effect on the seed growth has not been studied.

12.4.3. Biology

Eggs are laid in the soil. Freshly laid eggs are dull white but turn pink before hatching. They hatch in 9 days. In about 4 week, the nymph turns into adult (Pradhan, unpublished data). The insects assemble in large numbers under groundnut windrows or trash. Groundnut, sesame, bajra and tenai *Panicum italica* are reported as crop hosts (Rai, 1976). However, more information is required on the host that support large population of this bug before it moves on to groundnut. The details on bionomics are not available.

12.4.4. Control

Surface application of dust formulation commonly practiced for the control of this bug.

Trapping with rubbish heaps at various places and then such heaps may later be burnt.

12.5. Coreids (Coreidae, Heteroptera)

The Coreid bug nymphs and adults are similar, both being feeders on plant sap. The legs, especially the posterior pair, in many cases expanded in a leaf like fashion, a feature that gives rise to the name of 'leaf-footed bugs.'

12.5.1. Whitefly (Aleyrodidae, Homoptera)

The whitefly *Bemisia tabaci* Gennadius appears in large numbers on the under surface of the leaves (**Plate 3**). warranting insecticide spray. The occurrences were recorded from Orissa (Desh, 1989) and south Gujarat (Patel and Jhala, 1992). A heavy infestation of *Aleurodicus dispersus* was observed on TMV-2 groundnut plants in Coimbatore, Tamil Nadu, India, during May 1998 (Geetha *et al.*,1998).

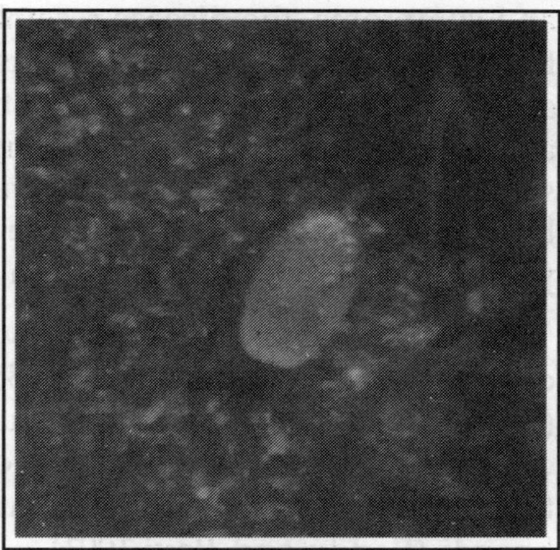

Plate 3. Whirafly

12.5.2. Host range

Host range includes sunflower, brinjal, sesame, Indian bean mustard (most preferred) and gram, cowpea, sweat potato and groundnut (least preferred) (Patel and Jhala, 1992); *Syzygium cumini* (Yadav and Rizvi, 1994); cotton (Borad and Puri, 1991); tomato (Ramappa *et al.*, 1998); green gram (Sahoo and Patanaik, 1994); *Sida* spp, *Euphorbia* spp., *Abutilon* spp., *Solanum nigrum,* and cluster bean (Dhawan *et al.*, 1998) have also been reported to be the host plants.

12.5.3. Natural Enemies

The natural enemies recorded against the whitefly are *Mallada boninensis* (Okamoto) (Joshi and Yadav, 1990); *Chrysoperla cornea* Steph (Chrysopidae: Neuroptera) (Kapdia and Puri, 1990) *Deraeocoris* sp. and *Campylomma nicolasi* (Miridae : Hemiptera) and *Geocoris ochropterus,* (Lygaeidae : Hemiptera) (Kapdia and Puri, 1991); *Serangium parcesetosum* (Coccinellidae: Coleoptera) (Kapdia and Puri, 1992); *Encarsia* sp. and *Eretmocerus mundus* (Ramappa *et al.*, 1998).

12.5.4. *As a vector*

In 1978, a yellow mosaic was observed on groundnut at Hyderabad in Andhra Pradesh, India, characterised by bright yellow patches with shortening, thickening and slight upward curling of the leaves. The disease proved to be transmissible by grafting and also by *Bemisia tabaci* (Gennadius), which acquired the virus from diseased plants in 2 days and (at the rate of 15 insects/plant) were able to transmit it in 2 days to *Vigna radiata* (*Phaseolus aureus*), *V. mungo* (*P. mungo*), *Lablab purpureus* (*Dolichos lablab*), *D. biflorus* and *Cajanus cajan*, all of which reacted with bright yellow mosaic, while a variety of groundnut different from that of the source plant reacted with necrosis, curling and thickening of the leaves. In transmission, host range and symptoms, the disease resembled mung bean yellow mosaic. This is the first record of a yellow mosaic transmitted by *B. tabaci* on groundnut (Sudhakar *et al.*, 1980). *Bemisia tabaci* transmits horsegram [*Dolichos*] yellow mosaic virus and cowpea mild mottle virus in India. Yellow mosaics occurring on *Voandzeia subterranea*, bean (*Phaseolus vulgaris*), groundnut, lima bean [*P. lunatus*], mung bean [*Vigna radiata*], pigeon pea and soyabean are all caused by *Dolichos* yellow mosaic virus (Muniyappa and Veeresh, 1984).

This species transmit many viral diseases such as tomato leaf curl virus disease (Borah and Bardoloi, 1998); groundnut mild mottle and severe mottle (Shivaprasad and Sreenivasulu,1996); yellow green mosaic disease of cucumber (Raj and Singh, 1996); cotton leaf curl virus (Nateshan *et al.*, 1996); tobacco leaf curl (Valand and Muniappya, 1992) and yellow vein disease of bhendi (Reddy and Ravi,1991).

12.5.5. *Population Dynamics*

The population build up was positively correlated with temperature and morning vapour pressure, whereas, it was negatively correlated with wind velocity (Borad and Puri, 1995). Jayanthi *et al.* (1993) observed positive correlation with temperature, rain fall, wind speed, evening relative humidity and negative correlation with morning relative humidity and sunshine. Prasad and Logiswaran (1997) also observed positive correlation of population with maximum temperature, relative humidity and wind velocity.

12.5.6. *Monitoring*

To monitor this insect, Diraviam and Uthamasamy (1993) and Nandhalli *et al.* (1993) used yellow sticky trap placed 30 cm above the ground.

12.5.7. *Control*

Dash (1989) reported that populations of the aleyrodid *Bemisia tabaci* were reduced by 74.68% in plots treated with dimethoate (0.2%) compared to a 19.05% increase in untreated plots. Higher pod yields were obtained in the treated plots. This insect can be managed by using the spray of dimethoate

(0.2%). Intercropping sorghum in groundnut reduce the population of *B. tabacci* (Singh *et al.*, 1991).

12.6. Blue striped Nettle-grub (Limacodidae, Lepidoptera)

Blue striped Nettle-grub, (*Latoia lepida* Cram.) larvae defoliate groundnut besides causing skin irritation to the workers. In India, it is a minor pest of different crops, but becomes sporadically serious on groundnut.

12.7. Pea blue moth (Lycaenidae, Lepidoptera)

Pea blue butterfly is a widespread pest of legumes, commonly encountered, but not often a serious pest. The larva is green in color and blends well with the general color of the foliage. Pupation takes place on the plant foliage. The adults show distinct sexual dimorphism in that the upper surface of the male's wings are bright blue but the female is predominantly brown with a blue infuscation near the body; both sexes have small 'tails' on the trailing edge of the hind wings.

12.8. Semilooper (Noctuidae, Lepidoptera)

The semilooper, *Plusia signata* F. is a polyphagous pest attacking many important cultivated crops. The adult moths have a golden patch in their forewings. The female lays 400-500 greenish-white eggs singly on the under surface of a leaf. The full-grown caterpillars measure 3-305 cm in length and are of leaf-green color. Pupation takes place in the folds of leaves webbed with silken strands.

12.9. Groundnut bud borer (Gelechiidae, Lepidoptera)

The groundnut bud borer *Anarsia ephippias* (Meyrick) occurs in northern India. Fully grown larvae are chocolate brown in colour and 10-15 mm long. Larvae prefer to bore into terminal buds and shoot. Brar *et al.*, reported that *Anarsia ephippias* population was markedly lower in the treated plots. Seed yield was higher in the treated than in the untreated plots (1.24 vs. 0.99 t/ ha). The pest density was similar in all cultivars; yield differences between the cultivars were not related to pest density. The high yielding cultivars were M13 (1.54 t/ha) and M 335 (1.42 t). *Anarsia ephippias* (Meyr.) feeds on the shoots and webs the leaves of groundnut in Punjab State, India. Up to 100% of the plants of some varieties of groundnut. When the gelechiid was reared in the laboratory, the egg, larval and pupal stages lasted 3-5, 11-21 and 10-21 days, respectively, and the adults lived for 2-6 days (Bakhetia, 1977). Bakhetia (1977) observed parasitism of the larvae *Anarsia ephippias* ranged from 7.7% to 24%, the parasites reared being *Bracon gelechiae* Ashm. and a species of Brachymeria. Of 8 compounds tested for the control of the larvae *Anarsia ephippias*, endrin and parathion retained their effectiveness the longest (at least 2 weeks), but on grounds of safety, fenitrothion at 0.05% is recommended

(Bakhetia, 1977). A foliar application of 1 l quinalphos or dimethoate, 1.5 l endosulfan, 2 kg carbaryl, 0.5 l fenthion or 1 l monocrotophos/ha to groundnuts at the flower initiation stage gave effective control of *Anarsia ephippias* and resulted in shelled nut yields of 735-983 kg/ha compared with 469 kg without pest control; quinalphos gave the highest yield (Singh and Rawat, 1981).

12.10. Tussock caterpillars (Lymantridae, Lepidoptera)

Tussock caterpillars, *Euproctis* spp. feed on groundnut leaves, bud and flower. *Euproctis* adults are medium sized moths with a 20 mm wingspan. The caterpillar are easily recognized by the presence of prominent, compact tuft of short hairs on the back and near the head.

12.11. Flower beetle (Cetoniinae, Scarabaeidae, Coleoptera)

The flower beetle, *Oxycetonia versicolor* L. feeds on flowers and tender shoots. The beetles are medium sized flat beetles and many of them have very beautiful colouring. The males have protuberances on head and thorax. They are diurnal. The larvae live in decaying matter and in roots. Adults of *Oxycetonia versicolor* (F.) were found feeding on the flowers of Citrus cotton and bhindi (*Hibiscus esculentus*), the earheads of sorghum and maize and the flowers and shoots of groundnut (Bhatnagar, 1970).

12.12. Blister beetle (Meloidae, Coleoptera)

Blister beetle, *Mylabris pustulata* T. is a flower feeder. Eggs are laid in batches in soil and hatch into very active triungulin larvae, which feed on egg pods of Orthoptera. The later larval stages are often sluggish with a large body and reduced legs. Pupation takes place in the soil. The adults are large beetles, 25-35 mm long, with bright conspicuous red (or yellow) and black patterned coloration. They are strong fliers.

12.13. Weevils (Curculionidae, Coleoptera)

The ash weevil, *Myllocerus undecimpustulatus maculosus* Derab.feeds on leaves, terminal buds and flower buds. This has also appeared as severe pest during *kharif* 1986 at NRCG. The surface weevil, *Atactogaster finitimus* Fst. also damages to the germinating groundnut seedlings in Anantapur district of Andhra Pradesh. Varieties found resistant to ash weevil are listed in (Table 12.1). A negative correlation was observed between K content and damage to leaves (r= -0.69**). Nitrogen and phenolics were not found to have any role on the resistance (Table 12.2). Singh *et al.* (1991) reported that intercropping with pigeon peas or *V. radiata* increased the incidence of *Myllocerus dentifer* apart from other insects like *Empoasca kerri* and *Caliothrips indicus* and reduced the incidence of and *Spilosoma obliqua* [*Spilarctia obliqua*]. Intercropping with sorghum reduced the incidence of *E. kerri, C. indicus, S. obliqua* and *Spodoptera litura* and increased the incidence of *M. dentifer*. Brar and Sandhu (1975) reported that varieties with a bunchy or semi-spreading growth habit were found to be more favourable to *Myllocerus* species, because of the good shelter afforded, than spreading varieties.

Table 12. 1. Groundnut varieties resistant to ash-weevil

	NRCG	ICG	VARIETY	Origin	HBT	OTHER TRAITS
1.	—	—	BG 2	INDIA	HYB	1,36
2.	—	—	CGC 5	INDIA	—	
3.	5014	5286	Makulu Red	ZMB	HYB	
4.	9465	3635	Spancross	USA	VUL	15,28
5.	8938	13380	TG 19 A	INDIA	HYB	28

Table 12.2. Biochemical basis of resistance to ash weevil

Particulars	Independent			Dependent
	Nitrogen % (X1)	Phenolics % (X2)	Potash mg/g (X3)	Damage % (Y)
Not Preferred				
TG 19A	1.6	1.3	1.0	0.8
Spancross	1.7	1.5	1.6	0.5
BG 2	1.9	1.5	1.0	0.1
Highly Resistant				
Spanimprove	1.8	1.5	0.9	7.8
Panjob 1	1.4	1.3	1.0	2.3
Latur 36	1.5	1.4	0.7	6.6
Moderately Resistant				
GAUG 10	2.0	1.4	0.8	13.2
TMV9	1.6	1.3	0.5	11.7
TMV10	1.8	1.6	0.9	10.9
Moderately Susceptible				
JL 24	1.8	1.9	0.7	27.7
TMV11	1.8	1.9	0.7	25.1
Chandra	1.8	1.7	0.6	27.1
Highly Susceptible				
GAUG 1	1.8	1.4	0.5	59.2
Starr	1.8	0.6	0.5	80.9
Kadiri 1	2	0.4	0.5	54.4
R	0.40 ns	0.42ns	-0.69**	

n=15, ** Highly significant at 1% level; ns = Non-significant

12.14. Groundnut root borer *Sphenoptera* perrotetii (Buprestidae, Coleoptera)

12.14.1. Distribution

Groundnut root borer has been reported to injure groundnut crop in India (Narayanan, 1962). The occurrence of *Sphenoptera perrotetii* on groundnut is reported for the first time by Yadgirwar *et al.* (1994) in Maharashtra, India (central Vidharba zone). Nearly all summer groundnut varieties were infested. It is important in some localized area in southern India (Jai rao *et al.,* 1976; Nair, 1986; Amin,1988)

12.14.2. Injury

The larvae bore into the basal portion of the stem just beneath the soil surface and tunnel through the main root resulting in death of the plant. Such plant can be easily pulled out from the soil (Jai rao *et al.,* 1976).

12.14.3. Host plant

Besides groundnut, cowpea *(Vigna unguiculata)*, agathi, *Agati grandiflora, Sesbania* sp., sickle senna *Cassia* sp. and wild sunhemp *Crotalaria* sp. have been reported to be the hosts (Jai rao *et al.,* 1976).

12.14.4. Biology

Eggs are laid on the main stem of the plant (Ayyar, 1940). The grubs on hatching bore into the stem. A full-grown grub measures about 3.5 cm in length. Grubs pupate in the larval burrow, the pupal period completed in 10 days. The infestation commences about 75 days after sowing in bunch type and 90 days in semi spreading type during *kharif* and 7 and 14 days earlier in summer. The maximum incidence was recorded in October in *kharif* and April in summer groundnut (Jai Rao *et al.,* 1976).

12.14.5. Natural Enemies

A braconid *Glyptomorpha smeenus* (Cameron) was reported parasitizing grubs (Rao and Thirumalachar, 1977).

12.14.6. Control

Soil insecticides such as chlordane and heptachlor can be useful in managing this pest.

12.15.Two spotted spider mites (Tetranychidae, Acarina)

In India, two species of the mites namely the two spotted white spider mite, *Tetranychus cinnabarinus* (Boisd)(Plate 6) and *Tetranychus hypogaea* Gupta (Amin, 1988; Nandagopal and Gedia, 1995,1999) are occurring on groundnut occasionally during hot periods (Gupta and Sandhu, 1969 and RangaRao *et al.*,1990)

12.15.1. Biology of T. cinnabarinus

12.15.2. Egg

The eggs are deposited on both the surface of the leaflet over which webbings are made probably to prevent exposure of eggs to their natural enemies such as parasites, like Trichogramma sps. The eggs are dull white in colour, round in shape measuring 0.12 mm diameter. The egg period lasts for 3.28 + 0.61 days for both male and female (Table 12.3)

12.15.3. Larva

The larval stage lasted for 1 and 1.04 days for male and female respectively to moult to protonymph. Newly emerged nymphs of both the sexes possessed three pairs of legs. Male nymph measured 0.1 mm x 0.07 mm and female with 0.2 mm x 0.15 mm of length and width respectively. No difference could be seen in shape and size between the sexes excepting the abdomen. The casting off the old exuviae has taken place when the nymph became sedentary and probes its stylet into the leaf tissue. When the old exuviae was completely detached from new, the mite exerts its body upward. First there was a crack at the dorsal side just below propodosomal and due to further pressure, the exuviae's got removed and moves further in search of site for feeding. The sedentary phase existed just for 5-6 h before moulting.

12.15.4.Protonymph

Female took about 3 h more in completing this stage compared to male that took 1.11 days. The size of the female was increased by 20 per cent from the protonymph and the increase in size of the male was 33 per cent.

12.15.5. Deutonymph

This instar was important for female as the female attains 10 per cent more increase in size of the protonymph. The abdomen was robust due to ovarion development. The male increased in size by 25 per cent and showed a clear distinction in terms of tapering abdomen towards anal end. It was 1.11 days and 1.08 days respectively, to complete this stage, for male and female.

12.15.6.Adult

The adult longevity was 3.17 days for males and 5.04 days for females which included pre-oviposition (1.12 days), oviposition (3 days) and followed by post-oviposition period of 1.22 days. The size of the female was 0.35 mm x 0.03 mm of length and width respectively. The female laid eggs immediately after becoming adult. Two types of reproduction were recorded. The female reproduce parthenogenetically as well as sexually. When the female become receptive to males, it stops moving and the male moves below the female in between legs of female and bends its anus upwards and transferred spermatophores with its spermatodactyle to the opening of the female's spermathecae. Males mate frequently with a few minutes interval, taking 17 to 24 seconds with an average of 20 seconds. The females, after egg laying, died instantly. The males predate on female of the same species and of *T. hypogaea.* The total life cycle was completed in 10.39 and 11.51 days for male and female respectively. Females measured 0.25 mm x 0.3 mm of length and width. Each female laid an on average of 35 + 5.46 eggs in its life span and maximum eggs were laid in the first 4 days during oviposition period of 5 days.

12.15.7. Nature of damage

The mite infested leaves show stippling followed by yellowing and finally turn to white. The mite infested plants show extensive webbing and tips of the plants may appear reddish, if red spider mite and white , if white spider mite assembled in large numbers. Ultimately the whole plant withers and dies. In the beginning the infestation of mite appears in patches, which in due time extends to other area. The dissemination of mites usually occurs through human and wind. The spray of insecticides induces mite population considerably. The application of dimethoate @ 300ml a.i./ha early in the season followed by quinalphos @ 500 ml a.i./ha in mid season with no fungicide caused out break of the spider mite in groundnut (Ranga Rao *et al.,*1990).

12.16.1. Biology of white mite (T. hypogaea)

Detailed biology of the white spider mite was studied by Nandagopal and Gedia (1999).

12.16.1.1. Egg

The eggs were laid singly on both the surfaces of a leaflet and covered partially with webbing. The eggs were dull white in colour, round in shape and measured 0.11 mm diameter. The egg period lasted for 3.04 ± 0.45 days (Table 12.4).

12.16.1.2. Larva

The larva is the stage between emergence of neonate nymph from egg and the first moulting. There was not much variation on the duration for both male and female. The larva of both male and female possessed three pairs of legs. The male measured 0.15 mm x 0.09 mm in length and width while female measured 0.25 mm x 0.20 mm length and width. There was no difference in size and shape of the instar of male and female at this stage.

12.16.1.3. Protonymph

The protonymph stage lasted 1.04 day and 1 days respectively for male and female. The size of the female nymph increased by 20 per cent while apparently there was no difference in size of the male nymph. The male nymph measured 0.18 mm in length and 0.12 mm in width while the female nymph measured 0.30 mm in length and 0.25 mm in width.

12.16.1.4. Deutonymph

The female nymph attained considerable increase in size (50%) as compared to its preceding stage. The abdomen of female was almost round probably due to ovarian development. The males, however, increased in size by 30 per cent. The male exhibited clear difference from female in size and abdomen tapering towards the anal end. The duration of this stage was 1.15 days and 1.61 days for female and male respectively.

12.16.1.5. Adult

The adult male lived for 3.07 days while the female lived on an average for 4.69 days. Male: female ratio was 0.03:1.00. The mating pattern was similar to that of *T. cinnabarinus* (Nandagopal and Gedia,1995). A single male mated several females and mating lasted from 25 to 50 seconds with an average of 36 seconds. The female laid eggs continuously for a maximum of 5 days with a mean of 3 days. Maximum number of eggs (85%) was laid during first three days. A mean of 43 ± 8.97 eggs ranging from 25 to 51 were recorded from a single female. Rarely the male pierces its stylet into the abdomen of the female and suck the content for quite some time (4 - 7 minutes) and eventually the female dies. Similarly the male of this species attack the female of *T. cinnabarinus* also. The total life cycle is completed in 11.57 days by females and 10.57 days by males. The adults of both male and female measured 0.3 mm x 0.2 mm and 0.5 mm x 0.35 mm in length and width, respectively (Table 12.4).

12.16.1.6. Management

Extensive spraying of acaricides may not be required as the spider mites occur in spots and sporadic in nature and also it would be a costly affair to organize a spray exclusively for mite control, unless, otherwise the mite occurs in endemic proportions. Phorate, disulfoton in split application and carbofuran at planting gave only a slight control of mite (Amin,1988). Corrective sprays with dicofol, pictran, carzol, omite,galectron and azodrin were very effective, carbaryl + Bravoer Carbaryl + fungispore tank mixes developed more sever mite infestation than Carbaryl + Du-Ter and Carbaryl +benlate mixes. A mixture of 2% crude neem oil + kelthane 15 ml/ 10 lit gave good control over both *T. cinnabarrinus* and *T. hypogaea*.. Nandagopal and Gedia (1996) reported that residual toxicity of synthetic insecticides (phosphamidon and endosulfan) persisted for a few hours and gave 45-50% mortality of *T. hypogaea*. The crude neem oil at 2% gave the highest LT50 values of 156 h for *T. hypogaea* and 182.4 h for *T. cinnabarinus*. Mixture of synthetic insecticides and kelthane with neem oil resulted in higher LT50 than alone.

12.17. Minor storage pests

The minor storage insects include the rust flour beetle, Khapra beetle, merchant grain beetle, moths such as rice moth, tropical warehouse moth, Indian meal moth and a pod sucking bug. Apart from these insects there are other species such as *Tenebroides mauritanicus* (L.)(Coleoptera:Trogossitidae), *Lasioderma sericorne* (F.) (Coleoptera: Anobiidae), *Latiticus aryzae* Waterhouse (Coleoptera:Tenebrionidae), *Cryptolestes* spp. (Coleoptera:Cucujidae), *Alphitobius* spp. (Coleoptera : Tenebrionidae), *Carpophilus* spp. (Coleoptera: Nitidulidae) are attacking stored groundnut. These species occasionally infest stored groundnut and are only considered to be of minor importance. They usually occur in association with one or more of the other pest described above by Dick (1987).

12.17.1. Rust-red flour beetle, *Tribolium castaneum* Herbst (Coleoptera: Tenebrionidae)

The rust-red flour beetle is found throughout the tropics and is a pest of shelled groundnuts. The adult measures 3-4 mm long, with chestnut brown in colour. It lives for several months. The adults are strong fliers and are often the first species to colonize in stored groundnuts. The female lays their eggs in the cracks on the testa or in crevices in the kernels. The neonate instar larvae, which could not bore the intact testae, are able to feed the cotyledons.

The larvae are cylindrical in shaped with two prominent 'horn' on the last abdominal segment. They pupate in the damaged kernel without forming the cocoon. The adult and the larvae are facultative predators of the eggs and pupae of the other storage pest and also cannibalistic. The mean development period from eggs to adult is approximately 32 days at 30°C and 90% RH but is twice as long at 70% RH.

12.17.2. Merchant grain beetle, *Oryzaephilus mercator* Fauvel (Coleoptera: Silvanidae)

This is a cosmopolitan species having similar ecology and behavior to *Tribolium castaneum.* The adults are 2.5-3.5 mm long with distinctive ridged prothorax bearing six large teeth on either side. The larvae are cream coloured and possess more prominent thoracic legs than those of *Tribolium castaneum.* The first instar larvae cannot penetrate the intact testae and must feed by the kernels damaged by the adults or by the other storage pests. Under optimal condition, 30-33°C and 70% RH, the life cycle is completed in 28-35 days. *O. mercator* is difficult to distinguish morphologically from its sibling species of *O. surinamensis,* the saw toothed grain beetle, which is usually associated within the cereal products.

12.17.3. Khapra beetle, *Trogoderma granarium Everts* (Coleoptera: Dermestidae)

Khapra beetle is able to tolerate hot and drier conditions compared to many other storage pest that are found in semi-arid areas of Africa, West Asia and northern India. This species has also been recorded from Southeast Asia, South America and Australia. The adults are oval in shape and 2-3 mm long and dark brown with black mottling. Their dorsal surface is covered with fine hairs. Adults live for 14 days and do not feed or fly. The larvae are straw coloured and from the 4th instar have dense tuft of hairs on abdominal and thoracic tergites.

The bionomics of this species on groundnut has not been studied. The mature prediapause larvae leave the stored commodity to enter cervices in storage structure where they remain without feeding for many months. Complete disinfestation is difficult and for this reason many countries reject consignments with khapra beetle infestation.

12.17.4. Pod bug, *Elasmolomus sordidus Fabricius* (Hemiptera:Lygaeidae)

This bug is widely reported from Africa and India. They occur on pod left for drying in the field and in the store. The adult is dark brown, approximately 10 mm long and 2 mm wide. In filed the females lay the eggs in the soil or on groundnut haulms but, in stores, eggs are laid loosely among the groundnuts or in sacking. The first instar nymph has bright red abdomen and later instars become progressively darker. All stages feed on kernels, perforating the pods with their rostrum. This causes the kernels to shrivel with increase in the free fatty acid content of the oil, producing a rancid flavour.

12.17.5. Rice moth, *Corcyra cephalonica* Stainton (Lepidoptera : Pyralidae)

This species can develop at low humidity. The adult is brown and 12-15 mm long with its wings folded. The labial palps point directly forward and are long and pointed in females, but short and inconspicuous in the males.

The adults do not feed. The females live for 1-2 weeks and scatter their eggs among the produce. The larvae are capable of damaging intact kernels and feed both of on the surface and within seeds. They spin a tough silken fibre, webbing together kernels and cast larval skin. This type of contamination is easily distinguishable from the fine dust that result from beetle infestation and may be of greater economic importance than the weight loss caused by larval feeding.

Pupation takes place either within the food source, in sacking or in crevices in storage structures. At 20°C and 70 % RH the life cycle from egg to adult takes 40-50 days.

12.17.6. Tropical warehouse moth or almond moth, *Ephestia cautella* (Walker) (Lepidoptera : Pyralidae)

This is a pest of many stored commodities throughout the tropics but is less common in arid areas. The adults are grayish brown with an indistinct pattern on the forewings. This moth is smaller than *C. cephalonica,* measuring 6-9 mm long at the resting position. The labial palps in both males and females point upwards. Its life cycle is similar to that of the other moths infesting stored products. The larvae are extremely mobile and move freely through the produce, contaminating it with fine silk webbing. The life cycle takes 40-50 days at 28°C and 70% RH.

12.17.7. Indian meal moth, *Plodia interpunctella* Hubner (Lepidoptera: Pyralidae)

This species occurs in cooler areas of tropics. Its habits and life history are similar to those of other moth species already described. The marking on there forewing are used to easily recognize the adult : the basal third is cream coloured while the rest is reddish brown. Adults are 8-12 mm long at rest, with labial palps that points directly forwards. Under extreme temperature or overcrowding, the life cycle may be prolonged by the larval diapause. During diapause, normal application of insecticides, especially fumigants, may prove ineffective.

12.18. Conclusions

When there is a pest that causes considerable amount of damage to the groundnut crop, the farmers take up control measures for that particular pest, which in turn suppress the minor pests also. In some cases, indiscriminate use of pesticides may lead to the resurgence and minor pests. With favorable atmosphere and absence of natural enemy complex of a minor insect or non-insect may assume a major proportion. Hence it is always advisable to have integrated approach to combat insect populations.

Table 12.3. Post embryonic observations on the two spotted white spider mite *T. cinnabarinus* on groundnut.

Instar	Number of individual observed	Range (Days)	Mean ± sd (days)
Egg	25	2 to 4	3.04 ± 0.54
Male:			
Larva	14	12h to 2.12	1.39 ± 0.56
Protonymph	14	12h to 2.12	1.04 ± 0.46
Deutonymph	14	12h to 3.00	1.61 ± 0.68
Adult	14	2 .00 to 5.00	3.07 ± 0.92
Total Life Span	14		7.11 ± 2.62
Female:			
Larva	26	12h to 2.12	1.27 ± 0.47
Protonymph	26	12h to 3.00	1.00 ± 0.61
Deutonymph	26	12h to 2.00	1.15 ± 0.44
Adult	26	3.00 to 8.00	4.69 ± 3.04

Table 12.4. Post embryonic observations on the two spotted white spider mite *T .hypogaea* on groundnut.

Instar	Number of individual observed	Duration (days) range	Mean ± sd (days)
Egg	25	2 to 4	3.28 ± 0.61
Male:			
Larva	12	--	within 24 h
Protonymph	12	1 to 2	1.09±0.12
Deutonymph	12	1 to 2	1.11±0.13
Adult	12	to 6	3.17±1.15
Total Life Span	12	--	7.11±2.15
Female:			
Larva	13	12h to 2 days	1.04±0.08
Protonymph	13	12h to 2 days	1.12±0.14
Deutonymph	13	1 to 2	1.08±0.10
Adult	13	3 to 7	5.04±2.14
Pre-oviposition	13	1 to 3	1.12±0.08
Oviposition	13	1 to 5	2.22±1.05
Post-oviposition	13	1 to 5	1.22±1.05

References

Amin, P.W. (1988). Insect and mite pests and their control. In *Groundnut* (Ed. P.S. Reddy). pp. 393-452. Indian Council of Agricultural Research, New Delhi.

Ayyar, T.V.R. (1940). *Handbook of Economic Entomology for south India* – Government Press, Madras. 516 Pages.

Bakhetia,D.R.C.(1977). *Anarsia ephippias* (Meyrick) (Lepidoptera: Gelechiidae) damaging the groundnut crop in the Punjab. Journal-of-Research,-Punjab-Agricultural-University. 14: 2, 232-233.

Barwal, R.N. (1985). Status of soil inhabiting insect pests of crops. *Journal of Soil Biology and Ecology.*, 5, 139-142.

Barwal,R.N. (1985). Status of soil inhabiting insect-pests of crops. Journal-of-Soil-Biology-and-Ecology. 1985, 5: 2, 139-142.

Barwal,R.N., and Gupta,D.K. (1991). Seasonal incidence and control of pod borer (*Euborellia annulipes*) and Cercospora leaf spots in groundnut. Indian-Journal-of-Plant-Protection. 1991, 19: 2, 156-160.

Bhatnagar,S.P.(1970). Records of new Cetonides pests in Rajasthan State. Labdev-Journal-of-Science-and-Technology., 8-B: 2, 119-120.

Borad, V. K. and Puri, S. N. (1991). Life tables studies of *Bemisia tabaci* (Gennadius) on cotton and brinjal under field conditions by artificial infestation. *Indian Journal of Ecology*. 18(2), 186-189.

Borad, V. K. and Puri, S. N. (1995). Dispersal flight activity of *Bemisia tabaci* (Gennadius) in relation to time of the day and weather parameters. *Gujarat Agricultural University Research Journal.*, 21(1), 64-68.

Borah, R. K. and Bordoloi, D. K. (1998). Influence of planting time on the incidence of leaf curl virus disease and white fly population on tomato. *Indian Journal of Virology.*, 14(1), 71-73.

Brar,K.S., and Sandhu,G.S. (1975). A note on the evaluation of resistance among different cultivars of groundnut against aphid and grey weevil. Science-and-Culture. 1975, 41:445-448.

Brar,K.S., Labana,K.S., Mohinder-Singh. (1995). Losses in yield of groundnut due to foliage feeding insects. Indian-Journal-of-Entomology. 1995; 57(4): 420-423.

Charati,S.N., Pokharkar,D.S., Ghorpade,S.A. (2003). Abundance of spiralling whitefly, a newly introduced pest in Maharashtra State. Journal-of-Maharashtra-Agricultural-Universities. 2003; 28(1): 83-84.

Cherian, M.C. and Basheer, M. (1940). *Euborellia stali* Doharn (Forficulidae) as a pest of Dash, P. C. (1989) Assessment of yield loss in groundnut due to whitefly in Orissa. *Madras Agricultural Journal.* 76(5), 280-81.

Chopra,N.P., Singal,S.K.(1982). Studies on the genus *Elasmolomus* (Hemiptera Lygaeidae: Rhyparochrominae) of India. Bulletin-of-Entomology. 1982, 23: 1-2, 16-23.

Dash,P.C.(1989). Assessment of yield loss in groundnut due to whitefly in Orissa. Madras-Agricultural-Journal. 76: 5, 280-281.

Dhawan, A. K., Simwat, G. S., Dhaliwal, G. S. (ed.)., Arora, R. (ed.)., Randhawa, N. S.(ed.) and Dhawan, A. K. (1998). Population dynamics of whitefly, Bemisia tabaci (Gennadius) on cotton: an ecobehavioural approach. Ecological agriculture and sustainable development: Volume 1. Proceedings of an International Conference on Ecological Agriculture: Towards Sustainable Development, Chandigarh, India, 15-17 November, 1997. 435-448.

Dick, K.M. (1987). Pest management in stored groundnut. *Information Bulletin* No.22. International Crop Research Institute for the Semi-Arid Tropics, Patancheru, India, 28pp.

Diraviam, J and Uthamasamy, S. (1993). Monitoring whitefly, *Bemisia tabaci* (Genn.) on sunflower with yellow sticky traps. Journal-of-Entomological-Research., 16: 2, 163-165.

Feakin, S.D. (1973). Pest control in groundnuts. *PANS mannual* No.2, 3rd edn., Centre for Oversees Pest Research, London, pp. 155-175.

Geetha,B., Loganathan,M., Swamiappan,M. (1998). Record of spiralling whitefly, *Aleurodicus dispersus* Russell on groundnut. Insect-Environment. 1998; 4(2): 55.

Gillier, P. (1970). Influence des attaques d' *Aphanus sordidus* sur la qualite des graines d'Arachide. *Oleagineux* 25,465-66.

Gillier, P. and Bockellee-Morvan, A. (1979). La protection des stocks d'arachidae contre les insects. *Oleagineux* 34, 137-37.

Giridharan,S., Palanisamy,G.A., Sivasubramanian,P., Subbulakshmi,B. (1985). Varietal reaction to earwig (*Euborellia stali* D.) in groundnut. Madras-Agricultural-Journal. 1985, 72: 11, 640.

Giridharan,S., Palanisamy,G.A., Sivasubramanian,P., Subbulakshmi,B. (1985). Varietal reaction to earwig (*Euborellia stali* D.) in groundnut. Madras-Agricultural-Journal. 1985, 72: 11, 640.

Gupta, J.C. and Sandhu,R.S. (1969). Out break and new records. F.A.O. *Plant Protection Bulletin,* 17, 17.

Jai Rao, K., Sindagi, S.S. and Krishna Rao, M. S. (1976). Observation on the extent of damage and seasonal abundance of the groundnut root borer *Sphenoptera (=Hoplistura) perroteti* Guer. (Coleoptera: Buprestidae), *Mysore Journal of the Agricultural Science*,10,454-457.

Jayanthi, M., Singh,K., M. Singh,R. N. (1993). Population build up of insect pests on MH 4 variety of groundnut influenced by abiotic factors. *Indian Journal of Entomology,* 55(2), 109-123.

Jayanthi,M., Singh, K.M. and Singh,R.N. (1993a). Pest complex of a high yielding groundnut variety Mh4 under delhi conditions. *Indian Journal of Entomology,* 55(1), 30-33.

Joshi, B. C. and Yadav, D. N. (1990). Biology and feeding of potential of *Mallada boninensis* (Okamoto), a chrysopid predator of white fly *Bemisia tabaci* Gennadius. Journal of Biological Control., 4(1), 18-21.

Kapadia, M. N. and Puri, S. N. (1990). Development, relative proportions and emergence of *Encarsia transvena* (Timberlake) and *Eretmocerus mundus* Mercet, important parasitoids of *Bemisia tabaci* (Gennadius). Entomon., 15(3-4), 235-239.

Kapadia, M. N. and Puri, S. N. (1990a). Feeding behaviour of *Chrysoperla carnea* (Stephens) on the parasitized pupae of Bemisia tabaci (Gennadius). Entomon., 15: 3-4, 283-284.

Kapadia, M. N. and Puri, S. N. (1991). Biology and comparative predation efficacy of three heteropteran species recorded as predators of *Bemisia tabaci* in Maharashtra. Entomophaga., 36: 4, 555-559.

Kapadia, M. N. and Puri, S. N. (1992). Biology of *Serangium parcesetosum* as a predator of cotton whitefly. Journal of Maharashtra Agricultural Universities., 17: 1, 162-163.

Kasargode, R. S. and Deshpande, V. G. (1921). The groundnut bug and its control. Bulletin of the Department of Agriculture, Bombay, No. 105, pp 20.

Kulkarni,V.S., Deshpande,V.P., Hugar,P.S. (2001). Population fluctuations and feeding potential of predators of groundnut pod bug *Elasmolomus sordidus* F. Karnataka-Journal-of-Agricultural-Sciences. 2001; 14(3): 813-815.

Lewin,H.D., David,A.L., Saroja, R., Natarajan,S., Janarthanam, R. (1971). A comparative study on the pests and diseases in different strains of groundnut. Madras-Agricultural-Journal. 1971, 58: 1, 15-20:2.

Mukhopadhyay, A.and Saha,B.(1995). Nutrient requirement of seed bug (*Elasmolomus sordidus*) (Heteroptera: Lygaeidae) on four host seeds. Indian-Journal-of-Agricultural-Sciences.; 65(2): 154-155.

Muniyappa,V. and Veeresh,G.K. (1984). Plant virus diseases transmitted by whiteflies in Karnataka. Proceedings-of-the-Indian-Academy-of-Sciences,-Animal-Sciences. 93 : 4, 397-406.

Nair, M.R.G.K. (1986). Insects and mites of crop in India. Indian Council of Agricultural Research, New Delhi, pp 408.

Nandagopal,V. and Gedia, M. V. (1999). Postembryonic observations on two spotted white spider mite *Tetranychus hypogaea* Gupta - a pest of groundnut, *Entomon*, 24(4), 359-362.

Nandagopal,V., Gedia,M.V. (1996). Residual effect of pesticides against two spider mites of groundnut. Karnataka-Journal-of-Agricultural-Sciences. 1996; 9(4): 601-605.

Nandagopal,V.and Gedia, M.V. (1995).Biology of the red spider mite *Tetranychus cinnabarinus* (Boised).- a pest of groundnut, *Entomon*, 20, 41-43.

Nandihalli, B.S., Patil, B.V. and Lingappa, S. (1993) . Population dynamics of cotton whitefly Bemisia tabaci Genn.). Karnataka-Journal-of-Agricultural-Sciences., 6: 1, 25-29.

Narayanan, E. S. (1962). Insect pest of groundnut and method of their control. Groundnut, a monogram. Indian central oilseed committee, India.

Narayanan, E. S. (1962a). Insect pest of groundnut and Methods of their control in groundnut- a Monograph, pp 222-8 Seshadri, C. R. (Ed.). Indian Central oilseed Committee, Hydrabad.

Nateshan, H. M., Muniyappa, V., Swanson, M. M. and Harrison, B. D. (1996). Host range, vector relations and serological relationships of cotton leaf curl virus from southern India. Annals of Applied Biology. 128(2), 233-244.

Patel, H.M. and Jhala R.C. (1992). Studies on host range, host preference and population dynamics of whitefly, *Bemisia tabaci* (Gennadius) in south Gujarat, India. Gujarat-Agricultural-University-Research-Journal. 17: 2, 76-81.

Prasad, G.S. and Logiswaran, G. (1997). Influence of weather factors on population fluctuation of insect pests on brinjal at Madurai, Tamil Nadu. Indian-Journal-of-Entomology., 59: 4, 385-388.

Purushothaman,G., Das,G.G., Venugopal,S., Govinda-Das,G. (1970). A note on *Euborellia stali* D., as a pest on Asiriya mwitund. Madras-Agricultural-Journal. 1970, 57: 1, 44-45.

Rai, K. (1976). Pests of oilseed crops in India and their control. Indian Council of Agricultural Research (ICAR)., New Delhi. 121 Pages.

Raj, S.K. and Singh, B.P. (1996). Association of geminivirus infection with yellow green mosaic disease of *Cucumis sativus*: diagnosis by nucleic acid probes. Indian Journal of Experimental Biology., 34: 6, 603-605.

Ramappa, H.K., Muniyappa, V. and Colvin, J. (1998). The contribution of tomato and alternative host plants to tomato leaf curl virus inoculum pressure in different areas of South India. Annals-of-Applied-Biology. 133: 2, 187-198.

Ranga Rao, G. V. Ranga Rao D. V., Prabhakara Reddy, M. and Wightman, J. A. (1990). Overhead irrigation and integrated management of spider mite in groundnut crop. *International Arachis Newsletter*, 9, 14.

Rao, K. J. and Thirumalachar, (1977). *Glyptomorpha smeenus* (Cameron) (Hymenoptera: Braconidae) a new natural enemy of groundnut root borer *Sphenoptera* (*Hoplistra*) perroteti Guer.(Coleoptera:Bupresitae). Current Research,6, 79-80.

Reddy, H.R. and Ravi, K.S. (1991). A leaf curl disease of *Cassia uniflora*. Current-Research -University-of-Agricultural-Sciences,-Bangalore., 20: 1, 16-17.

Sahoo, B. K. and Patnaik, N. C. (1994). Insect pests in greengram and blackgram in the South Coastal Region of Orissa with notes on their seasonal activity. Orissa Journal of Agricultural Research., Vol. 7: suppl, 74-76.

Senguttuvan, T. and Dhanakodi, C.V. (1997c). Seasonal occurrence of pod borers of groundnut in red lateritic soils. *Journal of Oilseed Research.*, 14(2), 312-314.

Senguttuvan, T., and Dhanakodi, C.V. (1997). Management of pod borer in groundnut through manipulation of cultural practices in alfisols. Journal-of-Oilseeds-Research. 14 (2): 269-273.

Senguttuvan, T., and Dhanakodi,C.,V. (1997). Seasonal occurrence of pod borers of groundnut in red lateritic soils. Journal-of-Oilseeds-Research. 1997; 14(2): 312-314.

Senguttuvan,T., and Dhanakodi,C.V. (1997). Control of pod borers of groundnut by soil application of insecticides in alfisols. Journal-of-Oilseeds-Research. 1997; 14(2): 274-277.

Singh, P. and Bhandari, R.S. (1991). Insect pests of leguminous forest tree seed and their control. *Indian Forester*, 14(2), 844-853.

Singh, T.V.K., Singh, K.M. and Singh, R.N. (1991). Host range of groundnut jassid, *Empoasca kerri* Pruthi. *Indian Journal of Entomology.*, 53(1), 1-17.

Singh,O.P., and Rawat,R.R. (1981). Field efficacy of some foliar insecticides against some major insect pests of groundnut in Madhya Pradesh. JNKVV-Research-Journal. 15: 1-2, 41-42.

Singh,T.V.K., Singh,K.M., Singh,R.N.(1991). Influence of intercropping. I. Incidence of major pests in groundnut (*Arachis hypogaea*) Linn. Indian-Journal-of-Entomology. 1991; 53(1): 18-44.

Sivaprasad, V. and Sreenivasulu, P. (1996). Characterization of two strains of cowpea mild mottle virus naturally infecting groundnut (*Arachis hypogaea* L.) in India. *Journal of Phytopathology.*, 144: 1, 19-23.

Sudhakar ,R.A., Rao,R.D.V.,Prasada, Reddy,P.S. (1980). A whitefly transmitted yellow mosaic disease on groundnut (*Arachis hypogaea* L.). Current-Science. 1980, 49: 4, 160; 5.

Valand, G. B. and Muniyappa, V. (1992). Epidemiology of tobacco leaf curl virus in India. Annals of Applied Biology. 120(2), 257-267.

Yadav, L. B., Rizvi, S. M. A. (1994). Studies on the insect-pests of lime, *Citrus aurantifolia* Swinh. in Eastern Uttar Pradesh. *Annals of Plant Protection Sciences.*, 3(1), 34-37.

Yadgirwar,P.V., Ghuguskar,H.,T., Naitam,N.,R., Bhoyte,S.H., Thakur,K.G. (1994). Occurrence of root borer *Sphenoptera perroteti* (Guer.) on groundnut in Central Vidarbha Zone, Maharashtra. Journal-of-Soils-and-Crops. 1994; 4(2): 175.

□□□

Pod Borers

T. SENGUTTUVAN

13.1. Introduction

Insect pests are one of the major constraints in groundnut production. The insect pests on foliage can be noticed but those underground go undetected often. These subterranean insect pests, however cause considerable yield loss. Among them, pod borers are serious pests as they cause direct economic loss (20-50per cent) by destroying the pods and kernels. Wightman and Amin (1988) reported that white grubs, termites, millipedes, wireworms, blind ants and earwigs bore holes in developing pods throughout the semi-arid tropics. The earwig, *Euborellia stalli* Dohrn (Cherian and Basheer, 1940; Senguttuvan and Dhanakodi, 1997a); termites, *Odontotermes obesus* Rambur (Kaushal and Deshpande, 1967b); false wireworm, *Penthicoides seriatoporus* Fair (Kapadia, 1994a) and blind ant, *Dorylus sp*. (Senguttuvan and Dhanakodi, 1997c) were reported as major pod borers in India, inflicting 20, 46, 25 and 30 per cent pod damage respectively.

13.1.1.Termites (Termitidae: Isoptera)

Of the 2000 termite species described worldwide, 300 species are known from the Indian region. Twenty species of termites infest groundnut (Verma and Kashyap, 1980), of which the important ones belong to the genera *Odontotermes* and *Microtermes*. *Odontotermes obesus* (Rambur) and *Microtermes obesi* Holmgren are the two species of termites reported to cause damage to groundnut in India (Kaushal and Deshpande, 1967b; Logan *et al.*, 1992).

13.1.2.Distribution

Termites are distributed all over the groundnut growing areas in India and are serious in West Bengal, Uttar Pradesh (Anon., 1962), Rajasthan (Srivastava *et al.*, 1962), Madhya Pradesh (Kaushal and Deshpande, 1967a), Gujarat, Haryana, Punjab (Amin and McDonald, 1981; Singh *et al.*, 1989) and Andhra Pradesh (Ranga Rao *et al.*, 1991; Logan *et al.*, 1992).

13.1.3. Economic importance

Termites caused 5-50 per cent plant mortality and 23 per cent pod damage (Kaushal and Deshpande, 1967b). Pod damage by termites was reported to the extent of 46 per cent in Madhya Pradesh (Kaushal and Deshpande, 1967a). In Orissa, termites have been found to cause pod damage to an extent of 6-11 per cent during rainy season and 2.5 to 28 per cent during post rainy (winter) seasons (Mohapatra *et al.*, 1995). Pod damage by termites increases considerably when there is a delay in harvesting (Amin, 1988) or irregular ripening of the pods. Termite attack on groundnut is more severe during dry seasons or when the rainfall is sparse during the growing season (Wightman *et al.*, 1990).

13.1.4. Nature of damage

Termites damage the groundnut plants from seedling stage to maturity. They gnaw and hollow out taproot causing wilting and premature death of plants especially in sandy and red soils (Rawat *et al.*, 1970). When the damage is on pod, they feed on pod shell by removing the corky material between the strands of vascular tissues, which is called 'scarification' (Ranga Rao *et al.*, 1991). The scarified pods are liable to shatter or crack during harvest. Breakage results in loss of kernels especially during mechanical harvesting. Cracking is more serious in the moist growing areas, where the kernels in such pods become more susceptible to invasion by afflotoxin fungi (*Aspergillus flavus*). This in turn exposes consumers to toxins and carcinogens (Gold *et al.*, 1989; Wightman and Ranga Rao, 1993). Termites also cut the outer shell of the pods, enter them and eat the kernels. Inside the empty shell, a mass of dry mud and cadavers of termites can be seen. Termites enter into the root system and burrow inside the root and stem.

13.1.5. Alternate hosts

The crops of agricultural importance which are commonly damaged by termites are, wheat, rice, sugarcane, maize, sunflower, coconut, groundnut, millets, pulses, cotton, vegetables, Apple, Cashew, sapota, citrus, pomegranate, grapevine, guava, mango, peach and others. Forest trees like casuarina, *eucalyptus* and silver oak are highly susceptible to termite attack particularly when the tree saplings are one to three years old (Rajagopal, 1988).

13.2. Earwig, *Euborellia (=Anisolabis) stali* Dohrn. (Forficulidae: Dermaptera)

13.2.1.Distribution

E. stali was first recorded as pest of pods and kernels of groundnut at Coimbatore and South Arcot districts of Tamil Nadu (Cherian and Basheer, 1940). It is an important pod borer in Southern India. However, its occurrence was also reported from Bombay and Manipur (Barwal, 1985). Recently, the earwig was observed as the predominant pod borer of groundnut in Tamil Nadu (Senguttuvan and Dhanakodi, 1997c).

13.2.2.Nature of damage

Both nymphs and adults bore into the tender developing pods and feed on kernels. Mature pods are also attacked. But young and developing pods are preferred more (Senguttuvan and Dhanakodi, 1997c). The bore holes due to earwig are mostly plugged inside with sand particles, excreta and decayed pulp (Purusothaman *et al.*, 1970; Amin, 1988). The infested pods were either empty or with partly fed kernels and rarely with sand and faecal materials. The insects are very active, fast moving and occurring at localised patches in and around the pods.

13.2.3.Damage potential

A maximum pod infestation of 13.5 per cent at Tindivanam and 20 per cent at Coimbatore of Tamil Nadu was recorded (Cherian and Basheer, 1940). Earwig damage on pods ranged from 26.4-65.0 per cent in an exotic Tanganyikan variety, Asiriya mwitunde at Tindivanam, Tamil Nadu, with an average of 46.6 per cent. About 44 per cent of matured pods and 52 per cent of immature pods were found bored (Purusothaman *et al.*, 1970). Both earwig and wireworms together caused 1.2 to 11.5 per cent pod damage at Pudukkottai, Tamil Nadu (Senguttuvan and Dhanakodi, 1997c). *Euborellia annulipes* was reported as a major pest of groundnut in Manipur, inflicting 63 per cent damage to pods (Barwal, 1985).

13.2.4. Population dynamics

Seasonal occurrence studies conducted at National Pulses Research Centre, Pudukkottai indicated that the earwig damage was noticed throughout the year. The damage was high on crops sown during second fortnight of January to first fortnight of March and first fortnight of June to first fortnight of September (Senguttuvan and Dhanakodi, 1997c). This indicated that crops sown during *kharif* (June-July), late *kharif* (August) and spring (February-March) suffer more from earwig attack. It calls for preventive measures against pod borers to avoid the yield loss during these seasons. In Manipur, the earwig incidence was high in crops sown during summer months (May-July) and low in crops sown during April (Barwal and Gupta, 1991).

13.2.5.Biology

Nymphs are white in colour in early stages and later turn brown. Adults are dark brown to black in colour with forceps like caudal cerci and white leg joints. The adults mate 7-10 days after emergence and female lays eggs 10-23 days after pairing. Eggs are laid in clusters in soil under stones or debris and in the pods. The eggs are globular and pearly white in colour. The fecundity ranged from 21-108 eggs with an average of 58 eggs. Generally female earwig guards the egg mass till hatching. Eggs hatch in a week and the nymphs which resemble the adults pass through 5 instars before becoming adults. The life cycle (from egg to adult) is completed in 56-72 days and the adults live upto 252 days (Cherian and Basheer, 1940).

13.2.6. Alternate hosts

Cabbage, cotton bolls, sorghum stems (Cherian and Basheer, 1940), groundnut ((Senguttuvan and Dhanakodi, 1997c), and onion bulbs are also infested by earwigs (Regupathy *et al.*, 1989).

13.3. Blind ant, *Dorylus orientalis* Westwood (Formicidae: Hymenoptera)

13.3.1.Distribution

The doryline or the blind ant (**Plate 1**), which is also known as root cutting or root eating ant, is a sporadic but serious pest on groundnut grown during winter. Fletcher (1914) reported first that it perforates the pods of groundnut and consumes contents and also the roots in Sri Lanka. Later in 1920, its attack on underground parts of vegetables (potatoes, cauliflower) and early sown groundnuts was noticed in Bihar and Uttar Pradesh in India (Fletcher, 1920). Now, it is distributed throughout India especially in Uttar Pradesh, Bihar, Orissa, West Bengal, Karnataka and Maharashtra except the western and north western parts (Roonwal, 1975). Severe occurrence of blind ants on groundnut pods was reported from ICRISAT, Hyderabad (Wightman and Ranga Rao, 1993) and in red loamy, red lateritic and red sandy soils of Pudukkottai district in Tamil Nadu (Senguttuvan and Dhanakodi, 1997a,b,c).

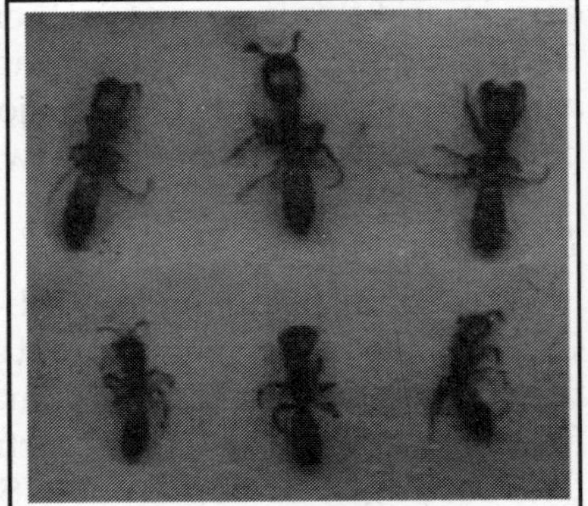

Plate 1: Blind ants

13.3.2.Nature of damage

The blind ants scrap the shell of mature pods **(Plate 2)** and bore either oblong or round holes of 3-5 mm in size more often at the beak or abdominal area of pods and feed on kernels. The infested pods were either empty or filled with sand particles (Senguttuvan and Dhanakodi, 1997c). Wightman and Ranga Rao (1993) described that blind ants make several neat, round holes of 0.5-2mm in dia. and remove the seed completely without leaving any soil inside the empty shell.

13.3.3. Economic importance

Blind ants attack was severe on crops sown during December – April and its damage on pods ranged between 1.82 per cent in crops sown in first fortnight of May and 30.21per cent in crops sown during first fortnight of February, while pods were completely free from ant damage when sown during other periods (Senguttuvan and Dhanakodi 1997c). An isolated groundnut crop at ICRISAT Centre, Hyderabad was completely destroyed during the 1989 post rainy season (Wightman and Ranga Rao, 1993). In northern India, blind ants swarm at the end of cold weather in late February (Lefroy, 1909), but swarming in April also occurs at Dehra Dun (Roonwal, 1975). Sridhar and Mahto (2000) reported that groundnut cv. ICGS-1 was attacked by as many as 37 insect and mite pests, belonging to different orders, associated with six natural enemies. Two soil inhabiting pests found to cause significant damage to the crop as well as maturing pods were identified as Oriental army ant, *Dorylus orientalis* (Westwood).

Plate 2: Ants damage

13.3.4. Host plant resistance

Singh and Singh (1992) reported that the damage by *D. orientalis* was also greater in sorghum intercropped with groundnut. The holes made by *D. orientalis* are reported for the first time from this part of the world. The varietal susceptibility of 63 groundnut varieties to *D. orientalis* was studied in Delhi, India, in 1989. Damage was highest (52per cent) in Ah-7903, and nil in VR-3317, U/4/4/38, NS-78 and NCAC-17840 (Mahto, 1991)

13.3.5. Biology

Male doryline ants are winged, brownish yellow with dark reddish brown head. They are carnivorous and are 17-25mm long. Details of females are unknown. Workers do not have wings and eyes, but cause severe damage to plant parts. They are dark brownish in colour with elongated dorsoventrally flattened abdomen. They lead a subterranean life like termites making tunnels and galleries through soil to reach the underground parts of plants. There are two forms namely major (5-11mm) and minor (2.5-3mm) in workers (Roonwal, 1975). Exact biology of the pest is not known.

13.3.6.Other host plants

Blind ants also attack coconut seedlings, citrus and sugarcane setts (Roonwal, 1975) and potato (Kishore *et al.*, 1990; Borah, 1994).

13.4. Wireworms

13.4.1. True wireworm (Elateridae: Coleoptera)

The grubs of true wireworm are long, cylindrical, shiny and slightly flattened and hence called 'wireworms'. The fully grown grub are 2.5-3.5cm long and 2mm thick with sclerotised dorsal plate margined with teeth and hooks on the ninth abdominal segment. An undetermined species of wireworm is reported to cause damage to groundnut pods at ICRISAT Center, Andhra Pradesh (Wightman and Amin, 1988) and National Pulses Research Centre, Vamban, Pudukkottai district of Tamil Nadu (Senguttuvan and Dhanakodi, 1997a,b). The grubs pierce the pods and damage kernels (Nayar *et al.,* 1976; Wightman and Amin, 1988). The pods infested with wireworms have bore holes of different sizes either plugged or unplugged and pods filled with faecal and decayed materials. On piercing, wireworms first tunnel the pulp of the shell and then damage the kernels (Senguttuvan and Dhanakodi, 1997c). Pods damaged by earwig and wireworms can not be distinguished from outside unless it is opened and seen. The wireworms and earwigs together caused 1.2 to 11.5 per cent pod damage and noticed throughout the year. About 80-90 per cent of the total damage was due to earwigs and the remaining was due to wireworms.

13.4.2 False wireworm, Penthicoides seriatoporus, Gonocephalum sp. *and Opatroides frater* (Tenebrionidae: Coleoptera)

The grubs of false wireworm are light yellowish or dusky brown elongate, cylindrical and indistinguishable from grubs of true wireworms. But they have two dorsal hooks on hind end and a ventral retractile process. Grubs occasionally cause damage to groundnut pods. False wireworms make smaller holes of 1-2 mm dia in the pods and burrow into the kernels (Logan *et al.,* 1992). About 5 per cent of mature pods were found damaged by the species in black soils of Andhra Pradesh (Reddy *et al.,* 1992). The false wireworm, *Penthicoides seriatoporus* Fair, *Gonocephalum* sp. and *Opatroides frater* were reported as a major pod borer of groundnut in Gujarat, causing 25 per cent pod damage (Kapadia, 1994a). Crop losses due to *Penthicoides seriatoporus* and *Anisolabis* [*Euborellia*] *annulipes* in groundnut was studied (Kapadia, 1996). Percentage seed damage in infested pods was 63.5 and percentage weight loss of damaged seeds was 59.2. in groundnuts.

13.5. Millipede, *Spirostreptus* sp. (Spirostreptidae: Spirostreptida: Myriapoda)

It is a non-insect pest of groundnut pod, occasionally damage the roots of vegetable plants and damaging the pods of groundnut (Ayyar, 1963). The millipede attacking groundnut was identified as *Spirostreptus sp.* in Tamil Nadu. Both adults and nymphs dig down into the soil, make holes and feeds on the kernels of developing groundnut pod by inserting its head inside. It completely empties the inner content of the pods (Mohanasundaram, 1986; Regupathy *et al.,* 1989).

The millipedes can easily be identified by their elongated cylindrical body with annular ring like segmentation with two pairs of legs in each segment (**Plate 10**). Adult is completely brown in colour and measures 6-7.5 cm in length, while young ones are pale red in colour. The young ones are gregarious and assemble into masses during night and disperse after sunrise (Mohanasundaram, 1986). Biology of millipede on groundnut is not known. The damage potential, distribution and seasonal occurrence are not yet well-studied. Studies on pod borers of groundnut during 1994 to 1997 revealed that millipedes were not found to attack groundnut pods at Pudukkottai, Tamil Nadu.

13.5.1. Spodoptera litura (Noctuidae: Lepidoptera)

The larvae are usually foliage feeders, when the population is high, in such cases, larvae migrate in large groups from one field to another in search of food. In areas where, groundnut is grown on light soils, the caterpillars have been seen boring into the pods and causing damage.

13.6 Management

As the pod borers inhabit in soil and the damage on pod can not be visualised, it is not possible to take up control measures after assessing the damage as in the case of foliage feeders. Thus, it calls for preventive measures against pod borers wherever they are problem in groundnut. The different control measures against the pod borers are described below.

13.6.1.Cultural control

Delay in harvesting predisposes the crop to more termite attack. Harvesting of pods as soon as they mature will reduce the termite damage to pods. Repeated inter-cultivation operations reduce termite population and attack on plants. Frequent irrigation also helps in reducing the termite attack. It was suggested to have 2 cm thickness layer of neem cake to safely dry the groundnut pods from termites, which reduced 80-90 per cent scarification (Gold *et al.,* 1991). When the same dose was tested as soil mulch in the field, there was no much reduction in termite population and 14 per cent reduction in the number of scarified pods but was less economic (Ranga Rao *et al.,* 1991).

Soil application of neem cake @ 250 kg/ha at the time of sowing was found to be effective against the pod borer complex such as *E. stali, D. orientalis* and wireworms. Maintaining weed free till 70 days and irrigation once in 5 days were also effective cultural measures in controlling the pod borers. These measures recorded 19.0-22.0 per cent high yield over control (Senguttuvan and Dhanakodi, 1997b).

13.6.2. Host plant resistance

Growing resistant varieties of groundnut would be a long term and economic control measure against pod borers. ICRISAT has identified few resistant entries against pod scarifying termites, which include the North Carolina accessions 2243, 2232, 2242, 343 and 10033 (Anon., 1982,1983; Amin *et al.*, 1985). JL 24, J11, TMV 2 and CO 1 were relatively less susceptible to termite attack, while TMV 7, Dh 22, BGP 512, VG 55 and Kisan were highly susceptible (Mohapatra *et al.*, 1995).

The spreading and semi spreading varieties were found more susceptible to earwig than the bunch varieties (Lewin *et al.*, 1971). Infestation by *E. stali* under field conditions ranged from 0.81 (in TG 19A) to 6.72 per cent among 23 virginia bunch varieties and from 0.27 (C 335) to 4.35 per cent among 15 virginia runner varieties (Gridharan *et al.*, 1985).

13.6.3. Chemical control

Control of pod borers depends much on the use of insecticides, as the pests are subterranean in nature. The insecticides which were found to provide good protection against soil pests were persistent organochlorines (Melamed-Madjar and Shalomo, 1970; Padmanabhan *et al.*, 1973; Rao *et al.*, 1978; Barwal and Gupta, 1991). Heptachlor 6D applied along the furrows before sowing @ 25 kg/ha was found to be effective in reducing the plant mortality due to termites (Rawat *et al.*, 1970). This was the first study on chemical control of termites on groundnut in India. However, there was no mention about the effect on pod damage. Soil application of controlled release formulation of chlorpyriphos @ 4 kg a.i./ha along the furrows at the time of sowing recorded less pod damage due to termites and earwig (Logan *et al.*, 1991). However, the expensive formulation was unlikely to be cost-effective with unacceptable residue level in kernels.

Against earwig, chlordane 6D was found to be more effective (Melamed-Madjar and Shalomo, 1970). Subsequent findings showed that DDT 5D was the most effective insecticide when applied in soil twice at sowing and 40 DAS @ 50 kg/ha and heptachlor 6D, and chlordane 5D did not give economic returns (Purusothaman *et al.*, 1970). But application of heptachlor 6D @ 25 kg/ha into soil was suggested for the control of earwig (Nayar *et al.*, 1976). Basal application of heptachlor 5D was found significantly superior in reducing pod borer infestation and enhanced the yield (Barwal and Gupta 1991).

The combined dangers of handling chlorinated hydrocarbons and probability of persistence as residues accumulating in the nuts forced the necessity for a safer and more effective insecticide or control measure. The organochlorines, which were found to be effective against pod borers, are now either completely banned from use or banned from use in oil seeds. Thus, a study was conducted to evaluate different formulations of insecticides which are under current use against pod borer complex such as earwig, wireworms and blind ants at National Pulses Research Centre, Vamban, Pudukkottai, Tamil Nadu. Soil application of either carbofuran 3 G or phorate 10 G @ 10 kg/ha or quinalphos 1.5 D @ 25 kg/ha twice at the time of sowing and 45 DAS were effective in controlling the pod borer complex and recorded 61, 46 and 32 per cent less pod damage and 34, 32 and 18 per cent higher yield over untreated control (Senguttuvan and Dhanakodi, 1997a). Soil application of malathion 5D or endosulfan 4 D @ 25 kg/ha twice at prior to sowing and 40th day along with gypsum application and earthing up is suggested for the control of earwig. One time application of methyl parathion 2D @ 25 kg/ha during gypsum application and earthling up is recommended against millipede (Mohanasundaram, 1986; Anon., 1999). Applying carbofuran 3G in furrows @ 1kg a.i./ha is suggested as an effective prophylactic measure against soil pests of groundnut (Wightman and Ranga Rao, 1993).

13.7. Conclusion

Termites, earwig, blind ant, wireworms and false wireworms are some times considered economically important pests of groundnut, as they cause 20-50 per cent yield loss. The losses caused by the pod borers are higher in southern India than in central or northern India. Termites are distributed throughout India and cause damage to both plants and pods in sandy and red lateritic soils. They cause 2-46 per cent pod damage both by scarifying the pods and making bore holes. The activity of termites is high during dry periods. The earwig is a pest in southern India feeding on young and maturing pods of groundnut. The damage potential of the pest ranged from 1.0-65.0 per cent and its occurrence is noticed throughout the year in Tamil Nadu. The dorylines or blind ants are sporadic pod borers of groundnut reported from southern India. Its activity is noticed during post rainy season especially on crops sown during December to April with the damage potential of 1.82-30.21 per cent.

True wireworms (Elateridae) and false wireworms (Tenebrionidae) occasionally cause damage to pods. The grubs of these wireworms can not be distinguished from each other easily. The damage due to wireworms is often wrongly interpreted with earwig damage, unless it is opened and seen (wireworm attacked pods are filled with decayed and faecal materials). Wireworms cause about 5-25 per cent pod damage. Millipede, a non insect pest, was reported as a pod borer from Tamil Nadu, but of very minor in pest status.

Being subterranean in nature, the pod borers require preventive measures for minimising the yield loss in areas where they are a problem to groundnut. However, maintaining the field weed free, frequent irrigations and harvesting of pods as soon as they mature are some of the effective cultural measures against pod borers. Soil application of neem cake @ 250 kg/ha at the time of sowing is found to be effective in reducing the pod damage due to pod borer complex. Although few studies are made on resistance in groundnut to pod borers, they are only academic. Regarding chemical control, most of the organochlorine compounds reported as effective against pod borers are now banned from use. Hence, it is suggested that soil application of dusts such as quinalphos 1.5D, Malathion 5D, endosulfan 4D, methyl parathion 2D @ 25 kg/ha or granules such as carbofuran 3G or phorate 10G @ 10 kg/ha twice at the time of sowing and during gypsum application and earthing up {40-45 days after sowing (DAS)} may be followed for the effective management of pod borer complex in groundnut, wherever they pose problem. In other areas, one time application of any one of the above insecticide may be done at 40-45 DAS during gypsum application. Intensified works should be carried out on cultural measures, identification of resistant sources and use of plant products for the control of pod borers in groundnut. An integrated approach is very imperative against pod borers.

References

Amin, P.W. (1988). Insect and mite pests and their control. In *Groundnut* (Ed. P.S. Reddy). pp. 393-452. Indian Council of Agricultural Research, New Delhi.

Amin, P.W. and McDonald, D. (1981). Termites as pests of groundnuts. pp. 273-278. In. Progress in Soil Biol. Ecol. in India. *UAS Technical Series* No. 37.

Anonymous, (1962). Information collected by the Indian Central Oilseeds Committee from Indian States regarding pests of oilseeds, their status, time of appearance and control measures, etc. ICOC Report, 1962.

Anonymous, (1982). Annual Report, All India Coordinated project on Pulses 1981-82 University of Agriculture science, Bangalore pp. 118.

Anonymous, (1983). Entomological Research in Tamilnadu. Annual Report 1982-83. TNAU Coimbatore 235p

Anonymous, (1999). Crop Production Guide. Agricultural information unit press, Commissioner of Agriculture, Chennai, 300 p.

Ayyar, T.V.R. (1963). *Hand book of Economic Entomology for South India.* Govt. Press, Madras, 526 p. (2nd ed.).

Barwal, R.N. (1985). Status of soil inhabiting insect pests of crops. *Journal of Soil Biology. and Ecology.* 5:139-142.

Barwal, R.N. and Gupta, D.K. (1991). Seasonal incidence and control of pod borer (*Euborellia annulipes*) and *Cercospora* leaf spots in groundnut. *Indian Journal of Plant Protection.* 19(2):156-160.

Borah, R.K. (1994). Efficacy of some insecticides and plant materials against red ant, *Dorylus orientalis* (Westwood) on potato. *Journal of the Agricultural Scienc, Society of North EastIndia.* 7(1):41-44.

Cherian, M.C. and Basheer, M. (1940). *Euborellia stali* Doharn (Forficulidae) as a pest of groundnut in South India. *Indian Journal of Entomology.* 2:155-158.

Fletcher, T.B. (1914). Some South Indian Insect and other Animals (of Importance Considered Especially from an Economic Point of View). Govt. Press, Madras. 565p.

Fletcher, T.B. (1920). Annotated list of Indian crop-pests. *Proceedings of the 3rd Entomological meet* (Pusa, Feb. 1919). Calcutta, 1, 35.

Gold, C.S., Wightman, J.A. and Pimbert, M. (1989). Mulching effects on termite scarification of drying groundnut pods. *International Arachis Newsletter.* 6:22-23.

Gold, C.S., Wightman, J.A. and Pimbert, M. (1991). Effects of mulches on foraging behaviour of *Microtermes obesi* and *Odontotermus* spp. in India. *Insect Science Application.* 12(1-3):297-303.

Gridharan, S., Palaniswamy, G., Sivasubramanian, P. and Subbulakshmi, B. (1985). Varietal reaction to earwig (*Euborellia stali* D.) in groundnut. *Madras Agricutulral Journal.* 72(11):640.

Kapadia, M.N. (1994a). Record of three false wireworms infesting groundnut in Gujarat. *GAU. Research Journal.* 19(2):128-129.

Kapadia,M.N. (1996). Estimation of losses due to pod borers in oilseed crops. Journal-of-Oilseeds-Research. 1996; 13(1): 139-140.

Kaushal, P.K. and Deshpande, R.R. (1967a). Losses to groundnut by termites. *Jawaharlal Nehru Krishi Viswa Vidyapeeth Research Journal.* 1:92-93.

Kaushal, P.K. and Deshpande, R.R. (1967b). Termites need no longer threaten groundnut in Nimar. *Indian Farming.* 17(7):17.

Kishore, R., Ram, G. and Misra, S.S. (1990). Red ant *Dorylus orientalis* Westwood - an insect pest of potatoes in Bihar. *Journal of Entomological Research.* 14(1):87-88.

Lefroy, H.M. (1909). Indian Insect Life. Thacker Spink and Co., Calcutta, 396 p. *In* Legumes: Ecology and Control (eds.R.Singh, H.F. van Emden, and T.Ajibola Taylor), London, U.K.: Aademic Press.

Lewin, H.D., David, A.L., Saroja, R., Natarajan, S. and Janarthanan, R. (1971). A comparative study of pests and diseases of different strains of groundnut. *Madras Agricultual Journal.* 58:15-20.

Logan, J.W.M., Rajagopal, D., Wightman, J.A. and Pearce, M.J. (1992). Control of termites and other soil pests of groundnuts with special reference to controlled release formulations of non persistent insecticides in India and Sudan. *Bulletin of Entomological Research.* 82(1):57-66.

Mahto, Y. (1991).Varietal susceptibility of *Dorylus orientalis* Westwood (Hymenoptera: Formicidae) in groundnut varieties. *Journal of Entomological Research.* 15 (2):144-148.

Melamed - Madjar, V. and Shalomo, T. (1970). Tests with alternative insecticides in the control of earwigs in peanuts. *Int. Pest Control.* 12(5):29-3.

Mohansundram, M. (1986). Insect pest management in groundnut. In Pest and Disease Management- Oilseeds, Pulses, Millets and Cotton (ed. S.Jayaraj). Tamil Nadu Agril. Univ. Coimbatore. pp 252.

Mohapatra, H.K., Padhi, J., Samalo, A.P. and Patra, G.J. (1995). Screening promising groundnut varieties against termite damage at Bhubaneshwar, Orissa, India. *Intenational Arachis Newsletter.* 15:59-60.

Nayar, K.K., Ananthakrishnan, T.N. and David, B.V. (1976). General and Applied Entomology. Tata McGraw-Hill Publishing Company Limited, New Delhi. p 589.

Padmanabhan, M.D., Lewin, H.D. and Saroja, R. (1973). Soil treatment to control groundnut pod borer (*Euborellia stali* Dohn.). *Pesticides.* 7(8):34-35.

Purushothaman, G., Das, G.G. and Venugopal, S. (1970). A note on *Euborellia stali* D. as a pest on *Asiriya mwitunde. Madras Agricultural Journal.* 57(1):44-45.

Rajagopal, D. (1988). Termites. In: Applied Soil Biology and Ecology (Veeresh, G.K. and Rajagopal, D., Eds.). Oxford and IBH Publishing Co. Pvt. Ltd., New Delhi. pp375.

Ranga Rao, D.V., Singh, K.N., Wightman, J.A and Ranga Rao, G.V. (1991). Economic status of neem cake mulch for termite control in groundnut. *International Arachis Newsletter.* 9:12-13.

Rao, B.H.K., Narayana, K.L. and Rao, B.N. (1978). White grubs problems in Andhra Pradesh and their control. pp. 206-209. In: Soil Biology and Ecology in India (Edwards, C.A. and Veeresh, G.K., Eds.), University of Agricultural Sciences, Bangalore.

Rawat, S.R., Jakhmola, S.S and Sohuch, R.H. (1970). Assessment of losses of hybrid sorghum (S.H.1) to earhead caterpillar and comparison of insecticidal control. *PANS.* 16:367-379.

Reddy, M.V., Kumar, V.P.K. and Reddy, B.K. (1992). Occurrence of *Gonocephalum* sp. (Tenebrionidae:Coleoptera) as a pest of groundnut in India. *Tropical Pest Management.* 38(1):108.

Regupathy, A., Palanisamy, S., Chandramohan, N. and Gunathilagaraj, K. (1989). A Guide on Crop Pests. Rajalakshmi publications, Nagercoil. Pp 242 .

Roonwal, M.L. (1975). Plant pest status of root eating ant, *Dorylus orientalis* with notes on taxonomy; distribution and habits (Insecta: Hymenoptera). *Journal of Bombay Natural History Society.* 72(2):305-313.

Senguttuvan, T. and Dhanakodi, C.V. (1997a.) Control of pod borers of groundnut by soil application of insecticides in alfisols. *Journal of Oilseed Research.*, 14(2).: 274-277.

Senguttuvan, T. and Dhanakodi, C.V. (1997b). Management of pod borer in groundnut through manipulation of cultural practices in alfisols. *Journal of Oilseed Research.* 14(2):269-273.

Senguttuvan, T. and Dhanakodi, C.V. (1997c). Seasonal occurrence of pod borers of groundnut in red lateritic soils. *Journal of Oilseed Research.* 14(2):312-314.

Senguttuvan, T. and Dhanakodi. C.V. (1997). Control of pod borers of groundnut by soil application of insecticides in alfisols. *Journal of Oilseed Research.* 14:274-277.

Singh, H., Rohilla, H.R., Kalra, V.K. and Singh, H. (1989). Control of termites (*Odontotermes obesus* Ramb.) in groundnut through seed treatment with aldrin. *Indian Journal of Plant Protection.* 17:263-264.

Singh, T.V.K. and Singh, K.M. (1992). Effect of different intercrops on termites and oriental army ant, *Dorylus orientalis* Westwood damage to groundnut. *Indian Journal of Plant Protection.* 20 (2) :129-132.

Sridhar,V., and Mahto,Y. (2000). Pest and natural enemy complex of groundnut (*Arachis hypogaea* L.) cv.ICGS-1 in the agro-ecosystem of Delhi. Indian-Journal-of-Entomology. 2000; 62 (4) : 335-340.

Srivastava, A.S., Vyas, H.K., Rathore, Y.K. and Khan, R.M. (1962). Survey of injurious insects of oilseed crops in Rajasthan and suggestions for control of the major pests. Utter Bharati journal of Research, University of Utter Pradesh 8:111-117.

Verma, A.N. and Kashyap, R.K. (1980). Termites - their damage and control in field crops. *Memoirs of the Entomological Society of India.* 8:51.

Wightman, J. and Ranga Rao, G.V. (1993). Pheromone trapping in south Asia:Review and Requirments. *IOBC Wprs Bulletin.*16:149-156.

Wightman, J.A. and Amin, P.W. (1988). Groundnut pests and their control in the Semi-Arid Tropics. *Tropical Pest Management.* 34:218-226.

Wightman, J.A., Dick, K.M., Ranga Rao, G.v., Shanower, T.G. and Gold, C.S. (1990). Pests of groundnut in the semi-arid tropics. In: Insect Pests of Food Legumes (Ed. by S.R.singh, John Wiley & Sons, New York. pp.243-322.

❑❑❑

Storage Pests of Groundnut and Their Management

V. NANDAGOPAL, T. V. PRASAD AND M.V. GEDIA

14.1. Introduction

In India, groundnut is stored as pods as well as kernels. Post harvest groundnuts, in all forms are subject to attack by insects and mites with subsequent damage, contamination and deterioration from the time they are harvested until they are utilized or consumed. The amount of deterioration caused by insects and mites depends largely upon the moisture content of groundnut and the form in which they are stored. The treatment of groundnut during harvesting, handling and drying, affects the loss in quality from insects. If stored as in shell, the percentage of cracked and damaged pods, loose shelled kernels, degree of seed maturity at time of harvest and amount of foreign material are some of the major factors that will have a direct bearing on the groundnut quality after storage. Sound, unbroken pods, having a moisture content of below 9 % will store for long periods (up to 1 year) without insect damage. Groundnut produced in India is mostly utilized within the country; about 81.6 per cent is used for oil extraction, 12 per cent as seeds and 5.3 per cent for edible purposes. At present only 1.1 per cent of total production is exported to other countries mostly in the form of hand picked selected (HPS) kernels for edible purpose. Insects destroy approximately 6- 10 per cent of the groundnut kernels stored in bags.

The losses in groundnut varied according to the storage period and may vary from 19 to 60 per cent, when the groundnut was stored for five months (Dick, 1987; Matokot *et al.*, 1987; Wightman *et al.*, 1987). A survey conducted at Gujarat revealed that about 84 per cent damage in groundnut was recorded in the variety GG-2. Similar survey conducted in Maharashtra indicated a damage level ranging from 25 to 90 per cent (Anon., 1991). If the storage of groundnut pod is longer i.e., about 90 days, the damage in seven different groundnut varieties varied from 64 to 93 per cent (Ghorpade *et al.*, 1998). The weight loss may go up to 65 per cent in cv.GAUG 10 and affected the germination by 48 per cent (Kapadia, 1994).

The various insect pests effecting stored groundnut are given in Table 14.1.

Table 14.1. List of storage pests of groundnut

Species	Common name	Family:Order
Aglossa caprealis (Hubner)	Murky meal moth caterpillar	Pyralinae: Lepidoptera
Ahasverus advena (Waltl)	Foreign grain beetle	Silvanidae : Coleoptera
Alphitobius diaperinus (Panzer)	lesser meal worm	Tenebrionidae: Coleoptera
Alphitobius laevigatus F.	Flour beetle	Tenebrionidae: Coleoptera
Araecerus fasciculatus (De Geer)	Coffee bean beetle	Anthribidae: Coleoptera
Attagenus megatoma (L.)	Black carpet beetle	Dermestidae: Coleoptera
Carpophilus dimidiatus (F.)	Corn sap beetle	Nitidulidae:Coleoptera
Caryedon serratus (Olivier)	Ground seed beetle	Bruchidae: Coleoptera
Cathartus quadricollis (Guerin-Meneville)	Squarenecked grain beetle	Silvanidae : Coleoptera
Caulophilus oryzae (Gyllenhal)	Broadnosed grain beetle	Bostrychidae: Coleoptera
Corcyra cephalonica (Stainton)	Rice moth	Galleridae: Lepidoptera
Cryptolestes pusillus (Schoenherr)	Flat grain beetle	Cucujidae: Coleoptera
Cynaeus angustus (LeConte)	Largerblack flour beetle	Tenebrion idae: Coleoptera
Dermestes ladarius	Larger beetle	Dermestidae: Coleoptera
Ephestia cautella (Walker)	Almond moth	Phycitidae: Lepidoptera
Gnathocerus cornutus (F.)	Broad horned flour beetle	Tenebrionidae: Coleoptera
Gnathocerus maxillosus (F.)	Slenderhorned flour beetle	Tenebrionidae: Coleoptera
Lacmiphelus pussilus Oliv.	Grain beetle	Cucujidae: Coleoptera
Lasioderma serricorne (F.)	Cigarette beetle	Anobiidae: Coleoptera
Latheticus oryzae Waterhouse	Longheaded flour beetle	Tenebrionidae: Coleoptera
Liposcelis sp.	Book louse	Liposcelidae: Psocoptera
Lophocateres pusillus (Klug)	Siamese grain beetle	Lophocateridae: Coleoptera
Necrobia rufipes De Geer.	Copra beetle	Cleridae: Coleoptera
Oryzaephilus mercator (Fauvel)	Merchant grain beetle	Silvanidae : Coleoptera
Oryzaephilus surinamensis (L.)	Sawtoothed grain beetle	Cucujidae: Coleoptera
Palorus ratzeburgi (Wissmann)	Small eyed flour beetle	Tenebrionidae: Coleoptera
Plodia interpunctella (Hb.)	Indian meal moth	Phycitidae:Lepidoptera
Prostephanus truncatus (Horne)	Larger grain borer	Bostrychidae: Coleoptera
Pyralis farinalis (L.)	Meal moth	Pyralinae: Lepidoptera
Rhizopertha dominica (F.)	Lesser grain borer	Bostrychidae: Coleoptera
Sitophilus oryzae (L.)	Rice weevil	Curculionidae: Coleoptera
Sitophilus zeamaiz Motschulsky	Maize weevil	Curculionidae: Coleoptera
Sitotroga cerealella (Olivier)	Angoumois grain moth	Gelechiidae: Lepidoptera
Stegobium paniceum (L.)	Drug store beetle	Anobiidae: Coleoptera
Tenebrio molitor L.	Yellow meal worm	Tenebrion idae: Coleoptera
Tenebroides mauritanicus Linn.	Cadella	Trogosstidae: Coleoptera
Tribolium audax Halstead	Black flour beetle	Tenebrionidae: Coleoptera
Tribolium castaneum (Herbst)	Rust red flour beetle	Tenebrionidae: Coleoptera
Tribolium confusum Jacquelin du Val	Confused flour beetle	Tenebrionidae: Coleoptera
Trogoderma bicolour	Khapra beetle	Dermestidae: Coleoptera
Trogoderma granarium Everts.	Khapra beetle	Dermestidae: Coleoptera
Trogoderma inclusum Le Conte	Larger cabinet beetle	Dermestidae: Coleoptera
Typhaea stercorea (L.)	Hairy fungus beetle	Mycetophagidae: Coleoptera

Source: Peanut Science and Technology edited by Ed. Harold, E. Patte and Clayde T Young, 521-560 pp.

The major insect pests of stored groundnut in India are rust red flour beetle *Tribolium castaneum* Herbst (Kalra and Singal 1991; Srivastva,1970)the saw toothed beetle *Oryzaephilus surinamensis* Linn.; The almond moth *Ephestia cautella* (Walker) and the rice moth *Corcyra cephalonica* (Stainton). All are secondary pests and are not capable of infesting sound pods. The only primary pest of groundnut is the bruchid *Caryedon serratus* Olivier (Coleoptera : Bruchidae).

The minor storage insects include the Khapra beetle, merchant grain beetle, moths such as rice moth, tropical warehouse moth, Indian meal moth and a pod sucking bug. Apart from these insects there are other species such as *Tenebroides mauritanicus* (L.) (Coleoptera: Trogossitidae), *Lasioderma sericorne* (F.) (Coleoptera: Anobiidae), *Latheticus oryzae* Waterhouse (Coleoptera: Tenebrionidae), *Cryptolestes* spp. (Coleoptera: Cucujidae), *Alphitobius* spp. (Coleoptera: Tenebrionidae), *Carpophilus* spp. (Coleoptera: Nitidulidae) are recorded on stored groundnut elsewhere. These species occasionally infest stored groundnut and are only considered to be of minor importance.

14.1.1. Rust-red flour beetle, Tribolium castaneum Herbst *(Coleoptera: Tenebrionidae)*

The rust-red flour beetle is found throughout the tropics and is a pest of shelled groundnuts. The adult (Plate 1) measures 3-4 mm long, with chestnut brown in colour. It lives for several months. The adults are strong fliers and are often the first species to colonize in stored groundnuts. The female lays their eggs in the cracks on the testa or in crevices in the kernels. The neonate instar larvae, which could not bore the intact testae, are able to feed the cotyledons.

The larvae are cylindrical in shaped with two prominent 'horn' on the last abdominal segment. They pupate in the damaged kernel without forming the cocoon. The adult and the larvae are facultative predators of the eggs and pupae of the other storage pest and also cannibalistic. The mean development period from eggs to adult is approximately 32 days at 30ºC and 90% RH but is twice as long at 70% RH.

14.1.2. Merchant grain beetle, Oryzaephilus mercator Fauvel *(Coleoptera: Silvanidae)*

This is a cosmopolitan species having similar ecology and behavior to *Tribolium castaneum.* The adults (Plate 2) are 2.5-3.5 mm long with distinctive ridged prothorax bearing six large teeth on either side (Singh, 1991). The larvae are cream coloured and possess more prominent thoracic legs than those of *Tribolium castaneum.* The first instar larvae cannot penetrate the intact testae and must feed by the kernels damaged by the adults or by the other storage pests. Under optimal condition, 30-33ºC and 70% RH, the life cycle is completed in 28-35 days. *O. mercator* is difficult to distinguish morphologically from its sibling species of *O. surinamensis,* the saw toothed grain beetle, which is usually associated within the cereal products.

Plate 1. Rust-red flour beetle

Plate 2. Merchant grain beetle

14.1.3. Khapra beetle, Trogoderma granarium Everts *(Coleoptera: Dermestidae)*

Khapra beetle is able to tolerate hot and drier conditions compared to many other storage pest that are found in semi-arid areas of Africa, West Asia and northern India. This species has also been recorded from Southeast Asia, South America and Australia. The adults (Plate 3) are oval in shape and 2-3 mm long and dark brown with black mottling. Their dorsal surface is covered with fine hairs. Adults live for 14 days and do not feed or fly. The larvae are straw coloured and from the 4th instar have dense tuft of hairs on abdominal and thoracic tergites. The bionomics of this species on groundnut has not been studied. The mature prediapause larvae leave the stored commodity to enter cervices in storage structure where they remain without feeding for many months. Complete disinfestation is difficult and for this reason many countries reject consignments with khapra beetle infestation.

14.1.4. Pod bug, Elasmolomus sordidus Fabricius *(Hemiptera:Lygaeidae)*

This bug is widely reported from Africa and India. They occur on pod left for drying in the field and in the store. The adult is dark brown, approximately 10 mm long and 2 mm wide. In filed the females lay the eggs in the soil or on groundnut haulms but, in stores, eggs are laid loosely among the groundnuts or in sacking. The first instar nymph has bright red abdomen and later instars become progressively darker. All stages feed on kernels, perforating the pods with their rostrum. This causes the kernels to shrivel with increase in the free fatty acid content of the oil, producing a rancid flavour.

14.1. 5. Rice moth, Corcyra cephalonica Stainton *(Lepidoptera : Pyralidae)*

This species can develop at low humidity. The adult (Plate 4) is brown and 12-15 mm long with its wings folded. The labial palps point directly forward and are long and pointed in females, but short and inconspicuous in the males. The adults do not feed. The females live for 1-2 weeks and scatter their eggs among the produce. The larvae are capable of damaging intact kernels and feed both of on the surface and within seeds. They spin a tough silken fibre, webbing together kernels and cast larval skin. This type of contamination is easily distinguishable from the fine dust that result from beetle infestation and may be of greater economic importance than the weight loss caused by larval feeding (Sosamma-Jacob *et al.,* 2002). Pupation takes place either within the food source, in sacking or in crevices in storage structures. At 20°C and 70 % RH the life cycle from egg to adult takes 40-50 days.

Plate 3. Khapra beetle

Plate 4. Rice Moth

14.1.6.Tropical warehouse moth or almond moth, Ephestia cautella (Walker) *(Lepidoptera : Pyralidae)*

This is a pest of many stored commodities throughout the tropics but is less common in arid areas (Pandey *et al.,* 1977). The adults are grayish brown with an indistinct pattern on the forewings. This moth is smaller than *C. cephalonica,* measuring 6-9 mm long at the resting position. The labial palps in both males and females point upwards. Its life cycle is similar to that of the other moths infesting stored products. The larvae are extremely mobile and move freely through the produce, contaminating it with fine silk webbing. The life cycle takes 40-50 days at 28°C and 70% RH.

14.1.7. Indian meal moth, Plodia interpunctella Hubner *(Lepidoptera: Pyralidae)*

This species occurs in cooler areas of tropics. Its habits and life history are similar to those of other moth species already described. The marking on there

Plate 5. Indian meal moth

forewing are used to easily recognize the adult : the basal third ıs cream coloured while the rest is reddish brown. Adults (Plate 5) are 8-12 mm long at rest, with labial palps that points directly forwards. Under extreme temperature or overcrowding, the life cycle may be prolonged by the larval diapause. During diapause, normal application of insecticides, especially fumigants, may prove ineffective.

14.1.8. Groundnut bruchid, Caryedon serratus

The adult beetle is reddish brown to reddish yellow in colour with small black markings on the elytra (Plate 5 a-dorsal & b-ventral). Antenna is serrate type, 11 segmented and longer in male than female. Metathoracic femur is well developed with two rows of spines. The pygidium is exposed in case of female, while concealed in male. It has large prominent eyes and readily distinguished from the other storage pests by its very broad hind femur, which bears a conspicuous comb of spines. Elytra do not cover the abdomen completely. The egg of *C. serratus* is translucent, white or creamy in colour, oval in shape with somewhat pointed ends (Plate 2). The eggs were laid singly on the pods as well as on seeds of its hosts or it may stuck to the depressions on the shell of the nut or on the smooth testa of the Kernels. Larvae are white, but turn to pink just before pupation (Plate 3) or it may be deeply pink with strongly chitinised oval head, possessing mandibulate mouth parts which were retractile in to prothorax as in the early instars. The abdomen was ten segmented with well defined uromers. There are four moults recorded in this bruchid. The cocoons are made of coarse grainy white threads which are papery and tough white (Plate 4). The colour of the oval shaped cocoon, primarily depends on the host material it is pure white when reared on *Cassia fistula* but reddish brown when reared on *Tamarindus indica* (Mital, 1968).

14.2. Synonyms of *C. serratus*

Fabricius (1798) first described this bruchid as *Bruchus gongara*. Subsequently, after 62 years, Eliditt (1860) changed the name as *Caryoborus*

(=Bruchus) gongara Fab. Decaux (1894) referred it as *Caryoborus tamarindii* to give weight age to tamarind as its main host. Subsequently, many other names were applied to this insect. Within a span of 45 years since 1913 its name was changed 18 times (Pic, 1913). It was Mc Farlane (1962) who coined the name *Caryedon serratus* Olivier for this insect and the synonyms are listed in Table 14.2.

Table 14.2. Synonyms of *C. serratus*

Synonym	Author
Bruchus gongara Fab.	Fabricius, 1798
Caryoborus (=Bruchus) gongara F.	Eliditt, 1860
Caryoborus tamarindii	Decaux, 1894
Caryedon notativentris	Pic, 1913
Caryoborus gongara Fab.	Fletcher, 1914
Pachymerus gongara F	Hubenthal, 1915
Pachymerus (=Caryoborus.) gongara F.	Hall,1918; Beeson,1918
Caryoborus (= Pachymerus) gongara F.	Beeson,1919
Pachymerus acaciae Gylh	Feytaud, 1924
Pachymerus (=Caryedon) notativentris Pic	Zacher,1932
Pachymerus longus Pic	Corby, 1941
Pachymerus languidus Gylh	Mukerji and Chatterjee, 1951
Pachymerus (=Caryedon) fuscus Goeze.	Howe, 1952
Caryedon cassiae Gylh	Decelle, 1951
Caryedon fuscus Goeze.	Cotterell, 1952
Caryedon gongara F	Cancela da Fonseca, 1956
Caryedon (=Caryoborus=Pachymerus) fuscus Goeze.	Southgate and Pope, 1958
Caryedon (= Pachymerus) acaciae Gylh	Southgate and Pope, 1958
Caryedon (= Pachymerus) longus Pic	Southgate and Pope, 1958
Caryedon serratus Olivier	Mc Farlane, 1962

14.3. Distribution

In India, it has been reported from South India Fletcher, (1914) Rajasthan, Gujarat, Madhya Pradesh, Maharashtra and Uttar Pradesh (Mital and Khanna, 1974) Punjab, Haryana, Jammu and Kashmir, Himachal Pradesh, Tamil Nadu and Kerala and some of the islands in the Indian ocean (Arora and Singhal, 1978). At present, it has become a serious pest of stored groundnut wherever groundnut is grown in India.

14.4. Nature and extent of damage

Both intact pods and kernels of groundnut are damaged by *C. serratus*, which is the only primary pest of stored groundnut (**Plate 6**). *Caryedon serratus*

Plate 6. Damaged pods and kernels

has been recorded as a serious pest on pods and seeds of *Acacia nilotica* in the Thar desert. The insect remained in the field during the hottest months of May and June and has also been reported to infest pods and seeds of *Prosopis cineraria* and *P. juliflora*. These 3 tree species thus act as the secondary host for a population buildup of *C. serratus*, and from them it spreads to its primary host, the groundnut (Satya-Vir and Jindal, 1996).

In Andhra Pradesh, it was observed that after 5 months, the total dry weight loss of the kernels was approximately 20 per cent caused by *Caryedon serratus* and it was responsible for nearly all the damage (Dick, 1987). The extent of losses to the groundnut varieties Improved Spanish, RSB-87, TMV-3, Exotic-5, Samarella, RS-1 and TM-2 caused by the bruchid *Caryedon serratus* was determined under storage conditions and described by Mital (1991). The percentage of pods damaged by *C. serratus* varied from 64.40 (ICGS-11) to 92.82 (RVB-1). The loss of pod weight in cv. ICGS-11 and SB-11 was 3.17 and 3.60%, respectively, compared with 22.93% in RVB-1 (Ghorpade *et al.*, 1998).

A percentage pod damage of 6.8 in *Acacia nilotica* by the bruchid *Caryedon serratus* in the field in Haryana. It also acts as secondary host for the population build-up of *C. serratus*, which spreads to its primary host, groundnut (Singal and Toky, 1990). Venugopal Reddy (1990) reported that bruchid's attacking the stored groundnut kernels resulted in heavy loss up to 87 per cent. Devi and Rao (2000) observed 50- 70 % dmage in pods and 80- 85 % damage in kernels of groundnut (cvs. Jl-24 and ICGS-29) and 7-9% loss in pod weight.

14.4.1. Qualitative loss

The free fatty acid value of the oil increased with the increase of infestation of *C. serratus* in groundnut (Davey *et al.*, 1959). The percentage loss in protein content varied from 0.20 to 8.76% in different varieties of groundnut. The percentage loss in oil content varied from 1.98 to 6.54 per cent with increase in feeding intensity (Radadia, 1995). Similarly, the percentage of free fatty acid was estimated to vary between 1.4 and 2.1 in damaged groundnut kernels. Anitha Kumari *et al.* (2002) reported that extent of damage on cv. TMV2 by bruchid was 77.1% in pods and 67.8% in kernels. Weight loss in pods and kernels was 55.1 and 52.3%, respectively. The reduction in oil content of infested pods and kernels was 2.9 and 3.5%, respectively. The acid value, which gave free fatty acid content, was higher in infested pods and kernels than in uninfested pods and kernels by 0.201 and 3.088%, respectively.

Radadia (2003) conducted a survey in the Junagadh area (Gujarat) to assess the extent of infestation of *C. serratus* on stored groundnut. The maximum pod infestation was observed in old stock from Visavadar taluka, Junagadh, (73% had larval holes and 100% contained eggs). Infestation levels in new stocks increased from January onwards, and the maximum infestation was obtained in April. The lowest percentage of larval holes (1.33%) was obtained in August, and the highest (43.71%) was observed in April. The lowest percentage of pods containing *C. serratus* eggs was obtained in March (10.22%), while the highest was obtained in April, May, June and December (100%). Among the stocks of 22 different cultivars, those of GG-2 and GAUG-10 had the highest infestation.

14.4.2. Host Range

Fifth-instar larvae and adults of *Elasmolomus sordidus* showed the highest consumption and weight gain on seeds of sesame, followed by those of groundnut, wheat and Ficus religiosa, a non-economic host. Fourth-instar larvae showed the highest consumption and weight gain on groundnut seeds (Mukhopadhyay and Saha, 1995).

It is evident from the published information that *C. serratus* is has a wide host range (Table 14.3). *Caryedon serratus* was found in seeds of tamarind (*Tamarindus indica*), *Acacia farnesiana, A. farnesiana* and *A. arabica* (Mital and Khanna, 1968), rice (Arora and Singal, 1978), *Acacia nilotica, A. tortilis* and *Prosopis cineraria* (Singal and Toky, 1988), and *Pongamia pinnata* (Singal and Toky, 1989).

Table 14.3. Host range of *C .serratus* in India

Name of host (s)	Author(s)
A. nilotica	Pruthi & Singh, 1950; Singal and Toky, 1990
A. pinnata	Pruthi & Singh, 1950; Singal and Toky, 1990
A. ontana	Pruthi & Singh, 1950; Singal and Toky, 1990
A. tortilis	Pruthi & Singh, 1950; Singal and Toky, 1990

Contd...

Table 14.3 Contd...

C. fistula	Pruthi & Singh,1950; Singal, S. K. and Toky, 1990
Acacia nilotica (L.) Willd ex Del.	Satya-Vir and Jindal, 1996
Acacia arabica Willd.	Mital & Khanna, 1974
Acacia chundra Willd	Pruthi & Singh, 1950
Acacia sp.	Fletcher, 1914
Acacia donaldi	Chatterjee, 1916
Albizzia lebbek Benth	Arora & Singal, 1978; Waterston, 1922
Bauhinia monandra Kurz.	Beeson, 1918
B. tomentosa	Beeson, 1918
B. malaborica	Beeson 1918
B. racemosa	Beeson 1918
Bauhinia sp.	Bagheri-Zenous, 1995
Caesalpinia pulcherrima (Linn.) SW	Beeson, 1918
Cassia fistula Linn.	Beeson 1918
C. grandis	Beeson, 1918
C. nodosa	Beeson, 1918
Cassia sp.	Fletcher, 1914; Beeson 1919
Cassia renigera Wall	Radadia, 1995
Cassia auriculata Linn.	Singh & Bhandari, 1991
Cassia ontana Hayne ex Roxb.	Pruthi & Singh, 1950; Singal and Toky, 1990
Cassia moschata	Romero and Johnson, 2002
Caesalpinea bocalla (Linn.) Flim	Raina, 1972
Casuarina equisetifolia Linn	Beeson, 1918
Hardwickia binata Roxb.	Pruthi & Singh, 1950; Singal and Toky, 1990
Helianthus annus Linn	Rangarajan *et al*, 1978
Pongamia pinnata (Linn.) ierre	Singal &Tokky, 1991
Prosopis cineraria (L.) Druce	Satya-Vir and Jindal, 1996
Prosopis juliflora (SW.)	Beeson 1918; Satya-Vir and Jindal, 1996
Prosopis spicigera	Bagheri-Zenous, 1995
Tamarindus indica Linn.	Mital and Khanna, 1968

14.4.3. Biology

Adults of *Caryedon serratus* (Ol.) avoided light; mating took place at dusk and oviposition usually in darkness (Pajni and Gill, 1974). Female lays eggs on the pods (**Plate 7**). On hatching, larvae of the bruchid invariably cut through the shell of groundnut pod and later burrow into the kernel. The small hole in the shell of the nut could usually be seen with the aid of a lens, through transparent egg shell (Davey, 1958). The hole is covered by a yellow granular material. When the egg shell shrivels or is knocked off, it is difficult to locate these holes. There are no other external signs of the presence of a developing larva in the kernel of the groundnut. The larva does not leave an unshelled nut before it is fully grown. It then makes a hole of 2 to 3 mm in diameter in the shell.

The incubation period ranges from 6 to 15 days. The variations in the incubation period may be due to the variations in the temperature inside the storage structure during the study period. The days to complete hatching varied ranging from 2.63 days to 58.59 days. Balaganur (1988) reported to have 5 to 11 days for hatching of eggs of *C. serratus*. According to the temperature prevailing, the period varies from 5.4 to 14.4 with an average of 9.8 days (Sontakke *et al.*, 1992). Similarly Kapadia (1995) reported 2 and 6 days for hatching at 30°C ± 2.52°C and 75.2 ±17.2 % Rh respectively.

The total period of the grub (**Plate 8**) varied from 11 to 49 days which depends on the temperature, quality of the food. Recently, Nandagopal *et al.*, (2001) reported on the moulting behaviour of this grub in-vivo. There were four grub instars taking a total of about 51.8 days wherein the last instar took the 17 days followed by the first and third instars with 13 days.

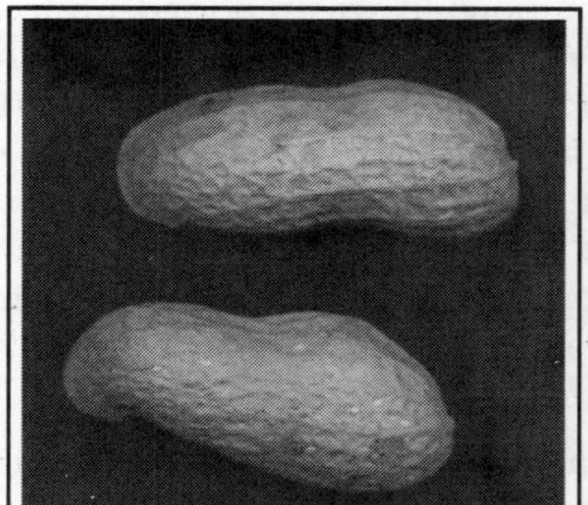

Plate 7: Eggs on pods

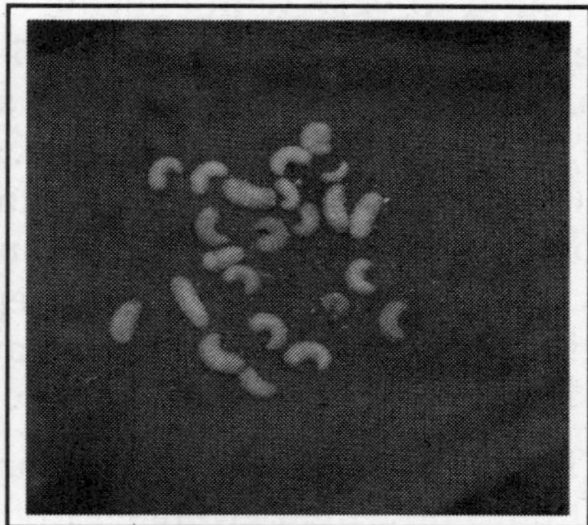

Plate 8: Eggs on pods

The pupa is enclosed in cocoon (**Plate 9**). The pupal period varied from 6 to 90 days, however, in majority cases it was of about 13 to 23 days. The total developmental period varied from 32 days to 142 days irrespective of the sex. It was also found to be minimum (66.22 days) in cv. SB-11 In a recent study, the duration of development was 10 to 49 days for male and 19 to 80 days for female (Nandagopal *et al.*, 2001). The sex ratio ranged from 1: 0.66 to 1.78. The lowest ratio was in GG-2 and a higher ratio was recorded when the insect was reared on GAUG-10 (table 13.7). These reports clearly indicated that the varietal differences greatly influenced the sex ratio.

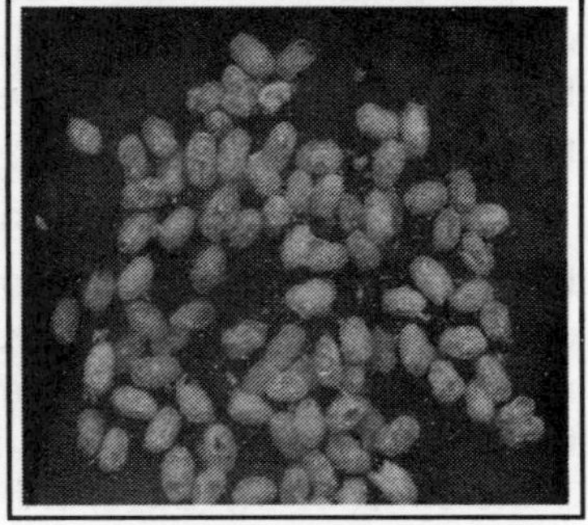

Plate 9: Pupae

The pre-oviposition period was generally short and varied from 24 hrs. to 5 days. In a recent study reported that the pre-oviposition period was 1 to 4 days (Nandagopal *et al.,* 2001). Oviposition was continued for 6 to 15 days Corby (1941). The oviposition period lasted for 16 to 19 days (Davey, 1958; Calderon *et al.,* 1967; Mital,1968; Balaganur, 1988) with maximum egg laying on the 5th and 6th day. Some time, the oviposition period may go up to 23 mean days with a range of 8 to 40 days on groundnut variety GAUG-10 (Kapadia,1995). The results reported so far revealed a wide variation in the oviposition period at different localities, however, it was around 5 to 10 days in majority of studies whereas Nandagopal *et al.* (2001) reported that the oviposition period was 2 to 20 days. Large variation was found in oviposition capacity of female bruchid. The variation was wide from 2 to 153 eggs/female.

The post oviposition period varied from 2 to 37 (Av.17.10) days, when reared on groundnut variety GAUG-10 (Kapadia, 1995) and 8.55 days at constant temperature of 30°C (Radadia, 1995) whereas post oviposition period of 2 to 20 days was reported by Nandagopal *et al.* (2001). Longevity of adults was found to vary with climatic conditions. It was reported up to 50 days. Male and female survived 12 to 45 (Av.27.70) days and 11 to 69 (Av.42.20) days, respectively (Kapadia, 1995) and 18.75 and 19.55 days at 30°C (Radadia, 1995). From the literature it is clear that adult bruchid survived generally for 10 to 40 days, however, maximum (6 months) duration was also reported. Among the various parameters, only temperature influenced the survival of insects. Nandagopal *et al.* (2001) reported that the longevity of the male was ranging from 4 to 27 days with a mean of 20 days and it was from 7 days to 25 days with a mean of 16 days for female.

The average duration of the larval stage at 30 deg C varied from 60.25 to 33.50 days and the percentages of larvae that became full-fed ranged from 22.98 to 60.77. The average amount of food consumed varied from 0.0242 to 0.1804 g/larva. The amount tended to be less in varieties deemed highly susceptible, but the larger populations in these varieties led to greater total losses in weight. TMV-3 was more effective than the others in limiting population increase and prolonging larval development, and one was definitely inferior. (Mital, 1969). D-mannose, fed adult males and females lived for 63 and 70 days, respectively, as compared with 19 and 24 days with no food (Mital, 1971).

Patel and Koshiya (1994) studied the life tables and age-specific fecundity for *Caryedon serratus* on groundnuts cv. GG 2. They also provided information on the mean generation length, innate capacity for increase and the finite rate of increase. It has four instars. The first instar was a non-feeding stage that survived for 1-2 days, second instar grub consumed 0.0026 g of kernel in a life span of 5.5 days, third instar grub consumed 0.0079 g in six days and fourth instar grub consumed 0.1310 g of kernel in 7-8 days. The first instar grub bored on the shell with the help of its prothoracic hooks and entered the kernel to settle underneath the seed coat [testa]. The total quantity of food consumed by a single grub with four instars was 0.1415 g of kernel (Vijay-Singh *et al.*, 2002).

The comparative biology of this pest was studied on various hosts by Halle *et al.* (2002). It is suggested that the testa of the seed is an important factor which provides the stimulus for egg laying (Mukhtar-Ahmad and Sushil-Kumar, 1990).

14.5. Population dynamics

There was not much information available on population dynamics. However, recently, Motka (1997) reported that the number of *C. serratus* increased from 162 to 758 in 250 g of kernels in a storage period of 180 days. It showed 16,55 and 76 folds of multiplication after 60 days, 120 days and 180 days respectively. Subsequently, the trend of population buildup was slowed down due to scarcity of food.

Host plant Resistance

On the basis of the numbers of eggs laid, larval preference, duration of development, growth index, adult emergence and the damage caused by the *Ephestia cautella* (Wlk.) (*Cadra cautella*) to the seeds, variety T28 was the most and M13 the least susceptible (Pandey *et al.*, 1977). Genotypes with resistance to stored products pests *Cadra cautella* [*Ephestia cautella*] and *Plodia interpunctella* included NC 6, GP-NC 343 X NC 5, PI467307, NC 7 X NC 10247, GP-NC 343 X NC 17367 and breeding line 10-P10-B1-B1-B1-B1-B2. Resistance to preharvest and postharvest insect complexes suggests some common chemicals may be present in the green crop and the harvested seed that elicit ovipositional response (Kashyap and Campbell, 1990). Variety Chitra, followed by ICGS-5, proved to be highly suitable for growth and development of the anobiid, *Lasioderma serricorne* (Fabricius) and ICGS-11 was the least suitable. T-28, M-13, Kaushal, ICGS-1 and Chandra were moderately suitable for growth and development (Sachan and Chandel, 1991). Kalra and Singal (1991) screened twenty varieties or cultivars of groundnut against the stored products pest *Tribolium castaneum*. The reduction in weight of kernels was highest (22.8%) in M 145 and C 336, and lowest (3.9-4.9%) in MII 2 and ICGS 52. The amount of frass left in each vial was proportional to the number of beetles that emerged. MII 2 and ICGS 52 were comparatively resistant to the tenebrionid. Singh (1991) studied the effect of natural infestation by the stored products pests *Oryzaephilus surinamensis* and *Tribolium castaneum* on the quality of 3 groundnut varieties. Increase in kernel moisture content, kernel damage, kernel weight loss, oil content loss, iodine value loss and fatty acid value in varieties JL 24 and TMV 2 were more than in Kadin 3. Variety Kadin 3 was found to be the least susceptible to both pests and retained its good quality. Out of nine varieties of groundnut (M 13, M37, M 197, SG 84, PG 1, JL 24, ARK II 9001 (red & white-red pericarps) and ARK II 9005) evaluated for their resistance under storage conditions to *Tribolium castaneum* varieties ARK II 9001, ARK II 9005 and M 37 while JL 24 and M 13 were the most preferred (Chander *et al.*, 1993). Forty-two groundnut test entries were screened to study the storage losses caused by different storage pests and to identify the entries showing tolerance to attack by storage pests. Rice moth (*Corcyra cephalonica*), red flour beetle (*Tribolium castaneum*) and flat grain beetle (*Cryptolestes pusillus*) were found damaging stored groundnuts. The storage pests caused 1-30% pod damage for all entries. Fourteen entries were free from pod damage caused by *Corcyra cephalonica*, 16 were not affected by *T. castaneum*, and 23 were free from *Cryptolestes pusillus* damage. *Corcyra cephalonica* was found to be the most serious pest of stored groundnut. Among the entries, ASK-2008 and IVK-1-2006 showed no weight loss indicating their tolerance towards storage pests. Fifteen of the 42 entries suffered 1-5% weight loss, while 11 entries showed 5-10% weight loss and the rest showed 10-50% (Sosamma-Jacob *et al.*, 2002).

Of seven varieties tested against *Caryedon serratus*, TMV3 showed maximum resistance. None was immune (Mital, 1969). Delayed larval development and adverse effect on larval/pupal and adult emergence has been attributed to nutritional and antibiotic factors (Mital, 1968). Maximum developmental period of 60 days was observed when the grub was allowed to feed on the cv.TMV-2. The growth index was highest on cv. GG-2 (1.98) followed by GAUG 10 with 1.82 (Pancholi, 1993). The total developmental period of the insect from first instar grub till adult emergence took 123 to 142 days (Mackie, 1943) indicating the adverse effect of the antibiotics present in the groundnut seeds.

Resistance in groundnut against the bruchid can be attributed to the observations made on the damage caused, the larval and pupal period, total development period, mortality and growth index of the insect. Based on this criteria seeds of ICGS-10 were found resistant against *C. serratus* (Balaganur, 1988). Maximum larval period of 37 days was observed on SG-84 variety (Pancholi, 1993). The total developmental period was 57 days when the insect was allowed to feed on the variety SG 84 indicating the antibiosis effect on the insect. Similarly, a maximum larval period of 31 days was observed when the grub was feeding on the variety J- 11 (Kapadia, 1995). On the basis of oviposition, pod damage, loss of pod weight and developmental period the cv. ICGS-11 and SB-11 were classified as least susceptible (Ghorpade *et al.*, 1998). In laboratory studies, *Caryedon serratus* completed its life cycle on groundnuts within 82 days. The preferred varieties for development were JL-24, GG-2 and GAUG-10, with J-11 being the least preferred (Kapadia, 1995).

The pods of cv. ICGS-11 and ICGS-76 exhibiting low fecundity and high mean developmental period of the beetle with low index of susceptibility, low weight loss and damage were grouped as least susceptible and remaining cultivars were either moderately susceptible or highly susceptible (Haritha *et al.*, 1999). The percentage oviposition was lowest in cv. ICGS-11 and SB-11 and developmental period was also found to be minimum (66.22 days) in cv. SB-11. The percentage of pods damaged varied from 64.40 (ICGS-11) to 92.82 (RVB-1). The loss of pod weight in cv. ICGS-11 and SB-11 was 3.17 and 3.60%, respectively, compared with 22.93% in RVB-1. On the basis of oviposition, pod damage, loss of pod weight and developmental period, cv. ICGS-11 and SB-11 was classified as least susceptible (Ghorpade *et al.*, 1999). Cultivars TGCS-61, TMV-2 and TPT-3, which had moderate reticulation, were less preferred by *C. serratus* for egg laying. The number of grubs that pupated into percent adult emergence, the number of damaged pods and kernels and the percent weight loss were low in the cultivars TCGS-61 and TCGS-88. (Devi and Rao, 2000).

14.5.1. Viability

Seed viability was significantly higher in both the plastic silo and the polyethylene-lined gunny bag with CaCl2 than in the ordinary gunny bag (Patra *et al.*, 1996). Groundnut cv. JL-24 pods stored in polyethylene-lined bags containing anhydrous calcium chloride (CaCl2) or silica gel gave seed viability above 50% after 12 months of storage. Seeds stored in airtight containers without any chemicals (closed tins or plastic silos) maintained above 50% viability in storage for up to 10 months, while those stored in gunny bags or earthen pots failed to germinate after this period. Retention of high seed viability in storage with CaCl2 or silica gel could be attributed to low moisture percentage of seeds during storage. The seeds stored in polyethylene-lined gunny bag containing CaCl2 had the highest field emergence and produced the highest pod yield (Tripathy *et al.*, 1996). Polyethylene bag was much better than the cloth bag for maintaining seed viability. Shelled seeds of Big Japan and M-13 stored in polyethylene bags maintained high viability for 14 and 10 months, respectively. The corresponding seed germination percentages were 82 and 89% (Sinha *et al.*, 1997).

14 .6. Storage options and Protection

The work on efficacy of different types of receptacles/storage structures for safe storage of groundnut is relatively scanty, however, the available reports clearly showed that metal bin and high density polyethylene bags were the most effective storage structures for minimizing storage losses due to groundnut bruchid.

14.6.1. Botanicals

The longevity of *Caryedon serratus* and the duration of the pupal periods were drastically reduced following exposure to 75-100 ppm neem derivative, Fortune Aza 0.15%, compared to the untreated control (Murugesan *et al.*, 1998). Citronella oil and palmarosa oil gave total protection to groundnut pods by inhibiting oviposition by the bruchid for 6 months with an efficacy equal to that of malathion dust (Malathion 5D) and *Lantana camara* had a good oviposition deterrent activity but lost effectiveness gradually after one month (Kumari *et al.*, 1998). Leaf extract of *Azadirachta indica* was found to be the most effective repellent against *T. castaneum* on groundnut seed, followed by those of *Vitex negundo*, onion and *Calotropis gigantean* (Sahayaraj and Paulraj, 2000). Volatile constituents (essential oils) isolated from leaves of 32 higher plants were tested for their repellent activity against *T. granarium* by Narendra-Kumar and Tripathi (2004). The oil of *P. roxburghii* showed 100% repellency at 0.02 ml while others exhibited lower level of repellency. The repellency of *P. roxburghii* oil was not destroyed by autoclaving, storage (120 days) and temperature treatment (100 degrees C). During in vivo investigation, the oil protected the seeds completely from *T. granarium* development up to 3 and 6 months separately and was found more efficacious than the synthetic fumigants used. The tin containers proved superior to gunny bags for the storage of groundnut seeds. The oil did not show any adverse effect on seed germination, seedling growth, nodulation, general health and morphology of the host plant. Neem oil (0.5 and 0.75%), neem leaves (2.5 and 5% w/w), and custard apple leaves (2.5 and 5% w/w) were effective against *C. serratus* at the concentrations used (Manjula, 2003).

14.6.2. Biocontrol

Verma (1993) observed *Pachyneuron* sp. a parasite on *Caryedon serratus.* Recently, Anon (1998) reported a cosmopolitian parasite *Anisopteromalus calandrae* (Howard) parasitising pupae of *C. serratus* to maximum 100%. Delobel (1989) reported trichogrammatid *Uscana caryedoni* (Hym.: Trichogrammatidae) a parasitoid of the eggs of both *Caryedon congensis* (=*C. crampeli*) and *C. serratus*. Dick (1987b) suggested dissemination of insect pathogens like *Bacillus thuringiensis* Berliner or NPV or Granulosis virus either by direct application onto the stored commodity or by attracting insects to traps containing a source of the diseases. There was no trials conducted and a favorable results obtained on the use of these biocontrol agents to recommend for the use either in small scale or in large scale storage of groundnut.

14.6.3. Chemical control

Gupta *et al.* (1976) evaluated various insecticides against *Caryedon serratus* and reported that fenthion (Lebaycid), endosulfan, gamma -BHC (lindane), malathion, aldrin, dichlorvos (Nuvan), trichlorphon (Diptrex) and carbaryl were 82.41, 53.72, 48.24, 39.67, 11.605, 11.47, 7.92 and 0.025 times as toxic, respectively, as BHC. The LC50s (in mg/litre) were 14.36 for ethylene dibromide, 56.62 for acrylonitrile and 83.37 for carbon tetrachloride. Fumigation with phosphine from aluminium phosphide tablets weighing 3 g each, applied at 0.5 tablet/ton, killed adults of *C. serratus* (Mital and Gupta, 1978). Two alkylating agents (metepa and tepa) and 2 miscellaneous compounds (thiohempa and hempa) were tested against *Caryedon serratus* by Misra *et al.* (1980) and reported that the safety margins metepa, tepa, thiohempa and hempa applied by dipping were 11.90, 12.74, 1.41 and 0.66, respectively, and by topical application 4.19, 1.70, 0.88 and 0.57. Thus they concluded that dipping method had the higher safety margin, and in addition metepa and tepa were safer for practical purposes than thiohempa and hempa. An attapulgite-based clay dust mixed with groundnut seeds at 10 g/2 kg seed controlled *Corcyra cephalonica* and *Tribolium castaneum* for up to 120 days. The treatment of pods infested with *Caryedon serratus* with 0.5% (w/w) dust afforded some control (Mital and Wightman, 1989). Effective control of eggs, larvae and adults of *Caryedon serratus* in stored groundnuts was achieved with vacuum fumigation with methyl bromide (16 g/msuperscript 3 for 4). Under normal atmospheric pressure (NAP), phosphine fumigation at 0.5, 1.0 and 2.0 g a.i./msuperscript 3 for 24 h, or at 0.25 g a.i. for 72 h gave effective control of eggs and larvae. Phosphine (1.0 g a.i./40 kg burlap bag for 120 h) gave 100 and 93% control of larvae and adults, resp. Neither fumigant impaired the viability of peanut seeds (Rao *et al.*, 1993). Fenvalerate at 4 g/kg seed showed quick action and killed all the *Corcyra* eggs, and *Caryedon* larvae and adults, within 48 h. chlorpyrifos and carbaryl were less effective, although the former showed high persistence. Fenvalerate and carbaryl were significantly superior to a range of rates of chlorpyrifos up to 180 days after treatment, and were on a par with aldrin. For seed exports that require storage of 60 days or more, chlorpyrifos could be used as an effective replacement for aldrin. The chemicals tested had no deleterious effects on seed viability of pearl millet (*Pennisetum glaucum*) and groundnut up to 180 days at ambient temperature during storage (Ghanekar *et al.*, 1996). The residual toxicity of BPMC [fenobucarb], carbosulfan (each at 20, 40, 60, 80 ppm) and fipronil (10, 20, 30 and 40 ppm) against seed beetle *Caryedon serratus* infesting stored groundnut indicated that the residual toxicity in the order of fipronil>carbosulfan>BPMC. Fipronil successfully checked the growth and multiplication of the beetle after 1 year of treatment. Third dose of each insecticide (60 ppm for BPMC and carbosulfan and 30 ppm for fipronil) was found to be superior than lower two doses and as effective as the higher dose (Tripathy *et al.*, 2000). Methyl parathion [parathion-methyl] dust was the most effective in suppressing *C. serratus* (Manjula, 2003).

14.6.4. Receptacles

Groundnuts are stored in oil mills, by farmers and by merchants and short time storage by exporters. About 81% of the produce is stored by oil millers followed by farmers (12%), 7% by merchants and 1 % by the exporters. Different methods are employed to store the groundnut in different levels. The efficacy to protect groundnut from bruchid attack varies according to the materials they used for the storage structures and the method of storing (Table 14 .4) and protectants (Table 14.4a).

Table 14.4. The efficacy of different storage structures

Storage structure	Material	Observation	Reference
Receptacles	A course aeroplane cloth of heavy cotton counting 60 threads by 104 in an inch	Minimum insects (74 insects)	Bissell and Pree, 1946
	Polyvinyl chloride covered pyramid	Less infestation	Halliday, 1967
	D.W. single warp bag D.W.plain weave bag	161.33 eggs/20 females. 320.66 eggs/20 female	Mital, 1968
	metal drums	minimum losses	Ramasivan *et al.*, 1968
	muslin cotton cloth	minimum losses	Khaire and Shukla, 1979
	gunny bags lined with polythene	best for storage	Anon., 1989
	metal bin	minimum of 3.9 % loss	Awakanvar *et al.*, 1989
	plastic bag container tin container	5.0 % 5.8 %	Anon., 1991
	Cloth bag	4%, 6%,7.5% and 11%	Anon., 1991
	polyethylene lined bag	did not show infestation even after 180 days of storage	Radadia *et al.*, 1992
	S.W Tarpaulin Canvas bags	minimum loss	Radadia, 1995

Table 14.4a. Protectants against the buchid

Oils	groundnut oil coconut oil palmoil (0.1,0.3,0.5 and 1.0 ml per 50 g of seed)	all the oils were more effective against *Trogoderma granarium*	Odeyemi, 1991
	neem oil (@ 10 and 25 ml/kg of seeds and pods)	neem oil was more effective	Senguttuvan *et al.*, 1995
Inert materials			
	attapulgile based clay dust	controlled *C. cephalonica*, *T. castaneum* up to 120 days	Mital and Wightman, 1990
Plant products			
	Tephrosia vogelii Hook	gave 90 and 99 % mortality of adults within 13 days	Delobel & Malongo, 1987
	nochi leaf powder neem @ 10 g/kgleaf powder @ 25 g/kg pods	proved more effective in stored groundnut	Senguttuvan *et al.*, 1995

14 .7. Control of established infestations

If the sacks are already infested when stored, the only effective method would be to go for disinfections by fumigation (table 14 .5).

Table 14 .5. Fumigation for the established infestation

How to use	Insecticides	Rate of application
Fumigation of bagged or bulk stocks in gas-light stores, or under gas-tight sheeting	Methyl bromide + chloropicrin (gas)	10-15 g m^{-2} for 24h
	Phosphine (solid aluminium phosphide)	3-5 tablets or 15-25 pellets 57% a.i.t^{-1} for 7-10 days
	Phosphine (solid)	2-3 pellets 100 kg^{-1} for 7-10days
Fumigation of small quantities of pods or kernels in polythene-lined sacks or containers such as drums	Ethylene bromide (liquid)	3ml 100 kg^{-1}
	Ethylene dibromide + carbon tetrachloride, 1:8 mixture (liquid)	12ml 100 kg^{-1}

14.8. Small scale storage

For small scale storage of groundnut either in the form of kernels or as pods, the best way to use metal bins or polythene lined gunny bags. Kernels in sacks can be protected as in the case of non-shelled stocks by applying one of the recommended insecticides as a spray or dust to the out side surface of the sacks. If there is a need store for longer period in the sacks, repeat the spray or dusting after every 30 days. Often phosphine is used to treat small quantities of infested groundnuts, provided if stored in gas –tight containers. For Indian conditions, one pellet (0.6g) placed on middle layer of groundnut in polythene–lined gunny bags resulted good control of the insects.

14 .9. Large scale storage

Large scale storage where huge quantity is to be stored requires well planned strategy. If it were to be a go-down, it should be cleaned thoroughly and fumigated or sprayed with recommended insecticides. The infested groundnut must be covered with a gas –proof sheet; either it may be PVC of 0.25 mm thick or polythene with minimum of 0.13 mm thickness. The recommended fumigant is Methyl bromide, when large quantity of groundnut has to be disinfected. The dosage recommended is 10-15g m^{-3} for 24 h. Since methyl bromide is retained in the oil seeds, before taking for crushing it is advisable to ventilate thoroughly so that the residues will be minimized. Similarly, Phosphine is usually recommended to even small to large quantity of storage of groundnut both for prevention and as well for disinfestion. The rate of quantity recommended is 3-5 tablets or 25 pellets t^{-1} .If large number of sacks are to be treated, it is better the tablets enclosed in a piece of paper, folded, and placed between the sacks at the sides and the entire bulk has to be covered air tight. Within 24 h of treatment the tablet melt to become powder and after 2/3 days the powder has to be removed and the produce has to be ventilated thoroughly so that the residues in the oilseeds is minimized. To prevent the infestation, the insecticides to be sprayed as given in Table 14 .6.

Table 14 .6. Insecticides to be sprayed to prevent the infestation

How to use	insecticides	Rate of application
Prevention of infestation		
spray to interior surfaces of infested stores before refilling	Malathion (w.p.)	250g 25% a.i. in 5L water 100 m^{-2}
	Fenitrothion (e.c.)	200 ml 50% a.i. in 5L water 100 m^{-2}
	Chloropyriphos-methyl (EC)	"
	Pirimophos-methyl (EC)	"
	Iodofenphos (WP)	200 g 50% a.i. in 5L water 100 m^{-2}
	Deltamethrin (WP)	50 g 2.5 % a.i. in 3.3 L water 100 m^{-2}
Space treatment of empty stores before refilling	Dichlorovos (resin strips)	1 stripe 30 m-2
Direct application of spray to pods or to sacks containing pods or kernels	Malathion (WP) Fenitrothion (EC) Chloropyriphos-methyl (EC) Pirimophos-methyl (EC) Iodofenphos (WP)	Apply at half the rate recommended above
	Deltamethrin + piperonyl butoxide(e.c.)	10 ml in 990 ml water t-1
Surface application to, or admixture of insecticidal dust with pods	Malathion (dust)	500g 2% a.i.t^{-1}
	Deltamethrin (dust)	500g 0.2% a.i.t^{-1}
	Fenitrothion +carbaryl (dust)	500g 1.2% + 0.8 % a.i.t^{-1}
	Bromohos (dust)	200g 2% a.i. of surface
Control established infestations		
Fumigation oi bags or bulk stocks in gas tight stores or under gas tight sheeting	Methyl bromide +Chloropicrin (gas)	10 to 15g m^{-3} for 24h (doses increase for control of *T. granarium*)
	Phosphin (solid aluminium phosphid)	3 to 5 tablets or 15-25 pellets a.i./t^{-1} for 7-10 days
Fumigation of small quantities of pods and kernels in polythene lined sacks or containers such as oil drums	Phosphine (solid aluminium phosphide)	2-3 pellets/100 kgs
	Ethylene dibromide (liquid)	3 ml/100 kgs
	Ethylene dibromide + Carbon tetrachloride, 1:8 ratio mixture (liquid)	12 ml/100 kgs

After Dick (1987b) and Ranga Rao *et al.* (1993)

14.10. Integrated pest management

The need for development of IPM in the stored groundnut has been increasingly felt important. The information on the resistance development for phosphine and methyl bromide which are used extensively lead to the pressure of minimal use of conventional pesticides against the storage insects. Dick (1987b) and Mital (1994) through communication after a meeting at Directorate of Plant Protection and Quarantine & Storage, Faridabad have given the steps to be followed in the storage pest management as below: Clean up the warehouse/go-downs inside and outside before storing groundnuts. Burn or remove all refuse. Remove storing seed or pod of different host of *C. serratus* as tamarinds with groundnut. Spray the insecticides such as dichlorvos 0.5% or malathion 1.25% or fenitrothion 2% or chlorpyriphos-methyl 2%, primiphos-methyl 2% or deltamethrin 0.04%@ 5 lit. spray solution for 100 sq.m. area on the walls, floor and roof of the warehouse or go-down, before any groundnut stored. The maximum moisture content for the safe storage of unshelled nuts should be below 9% and for kernels 7%. The groundnut should be dry in the sunlight for reducing moisture content at safe limits.

As eggs are laid in the field condition during the drying of the pods, they should be fumigated with aluminum phosphate@ 3-5 tablet/tones of groundnut before storing it in warehouse or go-down. Used gunny sacks should be checked for the presence of the insect before they are filled again. If necessary sacks can be disinfested by rolling them up together and placing them in sealed oil drum with single aluminum phosphide tablet. Avoid the use of D. W. plain weave and D. W. single wap (hessian) sacks because females of the *C. serratus* are able to lay eggs on the pod/kernels through interweaving space in these type of bags. Bags with spiral type of stretching shows emergence of the adults mature grubs. A-Twill, A-Twill with bitumen paint with polythene lining inside or B-Twill types of jute bags with two straight sewing on the side prevents the infestation by *C. Serratus*. Impregnated jute bags with deltamethrin 0.025 % or malathion 0.25 % or fenitrothion 0.25 % before using should be dried in shade. After providing proper dunnage materials the stack should not have a height more than 2 and 3 meters in case groundnut. In bulk storage, adult of *C. serratus* are unable to copulate in between the inter space of the nuts. This behaviour of the pest is very well utilized in reducing the population of the pest by providing 3 to 4 cm sand layer on the top of the groundnut in the container. As adult are found to penetrate up to 50 cm in the bulk storage, thus the surface of the bulk storage can be sprayed or dusted with any of the insecticide recommended for residual application on the store. Bromophos dust (2 % a.i.) applied at a rate of 200gm/ sq. m. to the surface and base of the large heaps of pod in the open air, effectively control *C. serratus*.

To reduce the adult population, wall and stacks be sprayed with Dichlorovos 0.5 % at evening hours as adults come out from the stacks for copulation during this time. If infested groundnut is taken into store the only effective method of disinfections is by means of fumigation under gas proof sheet or in a gas tight store.

14.11. Conclusions

Information on the biology of *C. serratus* has been generated with respect to groundnut. Comprehensive information is required on the population dynamics, factors that leads to population buildup. Life table studies will be very useful to understand the roll of resistant and susceptible varieties in relation to the key mortality factors that operate on this insect. The information given on the use of storage structures and the insecticides would be useful for small scale and large scale storage of groundnut for longer periods.

References

Ahmad, M. and Sushil Kumar. (1990). Ovipositional response of *Caryedon serratus* Olivier (Coleoptera: Bruchidae). *India Journal of Forestry*. 13(3): 192-198.

Anitha Kumari, D., Vijay-Singh., Reddy, V. S., Tejkumar, S. (2002). Quantitative and qualitative losses caused by pod bruchid, *Caryedon serratus* Olivier (Bruchidae : Coleoptera) in stored groundnut. *Indian Journal of Plant Protection*. 30(2): 213-214.

Anonymous, (1989). Annual Report, National Research Center for Groundnut (Indian Council of Agricultural Research), Junagadh, Gujarat. 75 pp.

Anonymous, (1998). Annual Report, National Research Center for Groundnut (Indian Council of Agricultural Research), Junagadh, Gujarat. 85 pp.

Anonymous, (1989). Groundnut. Publications and Informations Directorate, CSIR, New Delhi. 55 pp.

Anonymous, (1989). Twenty-seventh meeting of the Research Council held on Dec 22, 1989 at TNAU, Coimbatore. Agenda notes. Directorate of Research, TNAU, Coimbatore. 137p.

Anonymous, (1991). Annual progress report Groundnut 1990. 38th annual Kharif Oilseed research worker's group meeting. All India Co-ordinated Research Project on Oilseeds. pp 16-18.

Anonymous, (1991). Annual Progress Report (1990-1991). of AICRP on Rodents Control. (Junagadh: Gujarat Agricultural).

Appert, J. (1956). La bruche des arachides. Bull. Agron. Sect. tech. Agric. trap., No.13, 181-190. (Fide:R.A.E.,45, 492-493).

Arora, G. L. and Singal, S. K. (1978). *Oryza sativa* Linn. (paddy) as a new host-plant record of *Caryedon serratus* (Olivier) (Coleoptera: Bruchidae) from India. Indian Journal of Entomology. 40: 86.

Arora, G. N. and Singal, S.K. (1978). *Oryza sativa* Linn. (paddy). as a new host plant recorded of *Caryedon serratus* (Oilvier) (Coleoptera: Bruchidae) from India. India J. Ent., 40(1), 86.

Awakanvar, J. S., Krishnamurthy, K. C., Babu, C. K. and Ramkumar, M. V. (1989). Different storage structures for safe storage of pulses. *Mysore Journal of Agricultural Science*. 23: 526-532.

Bagheri-Zenous, E. (1995).The groundnut seed beetle (*Caryedon palaestinicus* Stout. Coleoptera: Bruchidae). A new pest in Iran. *Iranian Journal of Agricultural Sciences*. 26: 65-69.

Balaganur, S. H.(1988). Bio-ecology and control of groundnut seed beetle, *Caryedon serratus* (Olivier) (Coleoptera: Bruchidae).Karnataka Journal of Agricultual Science, 4(3-4), 193.

Beeson, C. F. C. (1918). Forest Entomology. Reprint from Annual Report of Scientific Advicer for India, 1916-17, Calcutta, Economic Zoology, pp 1-3 (Fide:R.A.E., 6:519).

Beeson, C.F.C.(1919). The food plants of Indian forest insects. Part I and II. Indian Forestry, 14 (2-3), 49-56. and 139-153.

Bisell, T. L and M. Du Pree. (1946). Insects in shelled peanuts in relation to storage and bagging. *Journal of Economic Entomology*, 39, 550-552.

Bridwell, J. C. (1918). Notes on Bruchidae and their parasites in the Hawaiian Islands. Proceedings of Hawaiian Entomological Society, 1917. Hanolulu, III no-5, April, 1918..465-505 pp.

Calderon, M., Donahaye, E. and Navarro, S. (1967). The life cycle of the groundnut seed beetle, *Caryedon serratus* (Ol.) in Israel. *Israel Journal of Agricultural Research* 17(3): 145-148.

Cancela da Fonseca, J. P. (1956). Phytosanitary aspects of groundnuts stored in Portuguese Guinea. I. mounds and stores. *An Jta. Miss. Geor.*, 2(2): 53-82.

Chander, H., Bakhetia, D. R. C., Sehgal, R. C., and Berry, S. K. (1993). Evaluation of some varieties of groundnut for resistance against rust red flour beetle, *Tribolium castaneum* (Herbst) in storage. *Journal of Insect Science*. 6(2): 219-222.

Chatterjee, N. C. (1916). Forest Entomology. Reprint from Annual Report. Scientific Adviser for India 1915-16, Calcutta, Economic Zoology, 1-4 pp.

Chitra, N. and Soundararajan, R. P. (2001). Incidence of *Caryedon serratus* (Olivier) (Bruchidae: Coleoptera) on tamarind seeds in Tamil Nadu. *Insect Environment*. 7 (1): 8-9.

Conway, J. A. C. (1974). Investigations into the origin, development and control of *Caryedon serratus* (Oi.). (Coleoptera: Bruchiidae). attacking stored groundnuts in the Gambia. Crop Protection Unit, Yundun Experiment Station, The Gambia, pp.554- 566.

Corby, J. D. R. (1941). Report of a study of a pest (*Pachymerus longus* Pic.) causing damage to groundnuts in the Wurkum Dep. Samary. Zaria. No. 9302/167:115 (Quoted by Davey, P.M. (1958)., *Bulletin of Entomological Research* 49 (2): 385-404.

Cotterell, G. S. (1952). The insects associated with export produce in Southern Nigeria. *Bulletin of Entomological Research* 43: 145-152.

Davey, P. M., Hall, D. W., Coveney, P. L. K., Raymond, W.D. and Sauires, J.A. (1959). The effect of insect infestation on the quality of decorticated groundnuts with special reference to high and low humidity. *Trop. Sci.*, 1(4): 296-307.

Davey, P. M. (1958). The groundnut bruchid, *Caryedon gonagra* (F). *Bulletin of Entomological Research.* 49(2): 385-404.

Devi, D. R. and Rao, N. V. (200). A note on reaction of groundnut varieties to the bruchid, *Caryedon serratus* (Olivier). *Journal of Research, ANGRAU.* 28: 41-43.

De Jonghe d'Ardoye,E. (1935). Note sur la bruche de 1'arachide *Pachymerus acaciae* Gill. Bull. Ann. Soc. ent. Belg.,75(11-12), 421-422.

Decaux, C. (1894). Metamorphosis du Caryoborus tamarindi. Le Naturallste, 16, 129-131.

Decelle, J. E. (1951). Contribution a 1'etute des Bruchidae du Congo belge (Col: Phytophaga). Rev.Zool. Bot afr, 45:172-192. (Quoted by Davey, P. M. (1958) *Bulletin of Entomological Research*, 49(2): 385-404.

Delfosse, P., Legreve, A., Devi, P.S., Reddy, A. S. and Maraite, H. (1999). Indian peanut clump virus is highly infectious on graminaceous plants. Zeitschrift-fur-Pflanzenkrankheiten-und-Pflanzenschutz. 106(4): 418-424.

Delfosse, P., Reddy, A. S., Legreve, A., Devi, P.S., Devi, P.S., Maraite, H. and Reddy-D. V. R. (1999). Indian peanut clump virus (IPCV) infection on wheat and barley: symptoms, yield loss and transmission through seed. *Plant Pathology*. 48 (2) : 273-282.

Delfosse, P., Reddy, A. S., Devi, P.S., Devi, P.S., and Maraite, H. (1999). Pearl millet as a bait crop for the management of the Indian peanut clump virus. *Indian Journal of Plant Protection.* 25: 70-72.

Delfosse, P., P. S. Devi, A.Legreve, S. V. Wesley, A. S. Reddy, H. Maraitr and D. V. Reddy. (1996). Strategy for management of Indian Peanut Clump Virus in Groundnut virus diseases in the Asia -Pacific Regions: summary and Recommendations of the fourth Meeting of the International Working Group, 12-12 March, Khon Kaen University, Thailand. Patancheru 502 324, Andhra Pradesh, India: International Crops Research Institute for the Semi-Arid Tropics. 44 pp.

Delobel, A. (1997). Two new species of Caryedon, seed-beetles of *Terminalia macroptera* (Combretaceae) in the Senegal country (Coleoptera: Bruchidae). *Bulletin de la Societe Entomologique de France.* 102: 4, 391-396.

Delobel, A. and Malonga. P. (1987). Insecticidal properties of six plant materials against *Caryedon serratus* (Ol). (Coleoptera: Bruchidae). *Journal of Stored Products Research.* 23(3): 173-176.

Delobel-A, (1995), The shift of *Caryedon serratus* Ol. from wild Caesalpiniaceae to groundnuts took place in West Africa (Coleoptera: Bruchidae). *Journal of Stored Products Research.* 31: 101-102.

Devi, D. R. and Rao, N. V. (2000). A note on reaction of groundnut varieties to the bruchid *Caryedon serratus* (Olivier). *Journal of Research ANGRAU.* 28(3): 41-43.

Dick, K. M. (1987b). Pest management in stored groundnut. Information Bulletin No.22. International Crop Research Institute for the Semi-Arid Tropics, Patancheru, India, 28pp.

Dick, K. M.(1987a). Losses caused by insects stored in a warehouse in India. Tropical Science, 27(2), 65-75.

Dick, K. M. (1987). Losses caused by insects to groundnuts stored in a warehouse in India. *Tropical Science.* 27: 65-75.

Donahaye, E., Navarro, S. and Calderon, M. (1966). Observations on the life cycle of *Caryedon serratus* (Ol.) on its natural hosts Israel, *Acacia spirocarpa* and *A. tortilis. Tropical Science.* 8(2): 85-89.

El-Atta, H.A.(1993). The effect of *Caryedon serratus* Olivier (Coleoptera: bruchidae). on viability and germination of seeds of *Acacia nilotica* (L. Willed. ex.Del.) in the Sudan. *Forest Ecology and Management.* 57(1-4): 169- 177.

Eliditt, H. L.(1860). Die metamorphose des Caryedon (Bruchus). gonagra F. Konigsberg, Rathke,pp.12. *Bulletin of Entomological Research.* 49(2): 385-464.

Fabricius, J. C.(1798). *Bruchus gonagra* F. Ent.Syst. Suppl., 159.

Feakin, S. D. (1973). Pest control in groundnuts. PANS mannual No.2, 3rd edn., Centre for Oversees Pest Research, London, pp. 155-175.

Feakin, S. D.1973. (Ed). Pest control in groundnuts. PANS Manual No.2. Overseas Development Natural Resourses Institute London U.K 197 p.

Feytaud, J. (1924). Les Insects del' Arachide. Rev. Zool. agric. and appl., 23 (4): 85 - 92.

Fletcher, T.B. (1914). Some South Indian Insect and other Animals (of Importance Considered Especially from an Economic Point of View). Govt. Press, Madras. 565p.

Friendship, R. (1973). The effect on *Caryedon serratus* (Ol.) of exposure for various periods to two insecticidal dusts. Tropical Product Institute, L.32, 1-7.

Gagnepain, C. and Rasplus, J. Y. (1991). *Caryedon serratus* and its parasitoids in the Savanna around Lamto, Ivory Coast. *Entomophaga* 34(4): 559-567.

Gahukar, R.T. (1992). Groundnut Entomology. Retrospect and Prospect. *Agricultural Zoological Reviews* 5: 139-199.

Ghanekar, A. M., Rao, G. V. R., Murthy, K. S. Surender, A. and Saheb, S. B. (1996). Seed protectants for healthy exports. *Indian Journal of Plant Protection.* 24(1/2): 37-43.

Ghorpade, S. A., Ghule, S. L. and Thakur, S. G. (1999). Relative susceptibility of some groundnut cultivars to pod borer, *Caryedon serratus* Olivier in storage. *Seed Research.* 26(2): 174-177.

Ghorpade, S. A., Ghule, S. L. and Thakur, S. D. (1998). Relative susceptibility of some groundnut cultivars to pod borer *Caryedon serratus* Olivier in storage. *Seed Research.* 26 (2), 174-177.

Ghosh, C. C.(1926).. Report of the entomologists, Mandalay and Sericultural Work for the year ended 30th June, 1925, Rangoon. pp 8.

Green, A. A. (1960). The control of insects infesting groundnut after harvest in the Gambia. II. Field trails on the control of groundnut borer, *Caryedon gonagra* (F.). *Tropical Science* 2(1-2): 44-54.

Gupta, H. C. L., Misra, R. P. and Mital, V. P. (1976). Evaluation of insecticides against *Caryedon serratus* F. (Bruchidae: Coleoptera). *Bulletin of Grain Technology*. 14: 68-69.

Hall, C. J. J.(1918). Ziekten en plagen der cultuurgewassen in Nederlandsch Indie in 1917. Medal Laboratorium Plantenziekten, Buitenzorg. pp.44.

Hall, D. W.(1954). The quality of groundnuts from the Gambia with special reference to insect infestation. Colon.Pl. Anim. Prod., 4(2): 227- 235.

Halle, D. N., Awaknavar, J. S. and Somashekhar. (2002). Biology of tamarind beetle *Caryedon serratus* (Olivier) on groundnut and other hosts. *Insect Environment*. 8(2): 67-69.

Halliday, D.C. (1967). A preliminary trail to determine the suitability of plastic envelopes for coverage of groundnut pyramids. Stored Product Research Institute, pp. 61-65.

Harada, T. (1940). On *Pachymerus gonagra* Fab. Najikairyoshiryo. Ministry of Agriculture, Japan. 152: 108-110.

Haritha, V., Vijayalakshmi, K., Murthy, M. M. K. and Rao, P. A. (1999). Relative preference of *Caryedon serratus* (Olivier) (Coleoptera: Bruchidae) for selected groundnut genotypes in storage. *Journal of Entomological Research*. 23(1): 71-74.

Howe, R. W. and Freeman, J. A. (1955). Insect infestation of West africa produce imported into Britain. *Bulletin of Entomological Research*, 46(3): 643-668.

Howe, R.W. (1952). Entomological problems of food storage in Northern Nigeria. *Bulletin of Entomological Research*, 43(1): 111-144.

Hubenthal, W. (1915). Ueber einige in Deuts, chaland eigischleppte exotischae kafer (Col.).. *Entom. Mith.*, 4: 128-130.

Jack, R.W. (1937). Division of Entomology, Annual Report for years 1936, Agricultural. *Rhod. Agric. J.*, 34(7): 570 - 576.

Kalra, V. K. and Singal, S. K. (1991). Screening of groundnut, *Arachis hypogaea* Linn. cultivars/varieties against red flour beetle, *Tribolium castaneum* (Herbst). *Journal of Insect Science*. 4(1): 80-81.

Kapadia, (1995) Biology and varietal preference of groundnut seed beetle, *Caryedon serratus* (Olivier). *Gujarat Agricultural University Research Journal*. 20(2): 170-172.

Kapadia, M.N. (1994a). Record of three false wireworms infesting groundnut in Gujarat. *GAU. Research. Journal*. 19(2): 128-129.

Kapadia, M. N. (1994b). Losses in weight and germination in groundnut due to seed beetle, *Caryedon serratus* Olivier. *Journal of Applied Zoological Research*. 5(2): 134.

Kapadia, M.N.(1995). Biology and varietal preference of groundnut seed beetle, *Caryedon serratus* Olivier. *GAU. Research. Journal*. 20(2): 170-172.

Kashyap, R. K. and Campbell, W. V. (1990). Ovipositional preference and larval establishment of the Indian meal moth and almond moth on selected peanut genotypes. *Peanut Science*. 17: 12-15.

Khaire, B. P. and Shukla, R. P.(1979). Penetration of different packing materials by some stored product insects. *Indian Journal of Agricultural Science*. 49(2): 117.

Kingsolver, J. M. (1970). Insects not known to occur in the continental United States. Groundnut bruchid, *Caryedon serratus* (Olivier). U.S. Deptt. Agri. Coop. Econ. Ins. Rpt., 20 (18): 803- 804.

Kingsolver, J. M. (1992). *Caryedon serratus* (Olivier) new to continental United States (Coleoptera: Bruchidae). *Insect mundi*. 6(1): 22.

Kulkarni, V. S., Deshpande, V. P. and Hugar, P. S. (2001). Population fluctuations and feeding potential of predators of groundnut pod bug *Elasmolomus sordidus* F. Karnataka *Journal of Agricultural Sciences*. 14(3): 813-815.

Kumari, D. A., Kumar, S. T. and Reddy, V. S. (1998). Management of the groundnut bruchid, *Caryedon serratus* (Olivier) with some plant products. Pest Management and Economic Zoology. 6(2): 127-131.

Lepesme, P. (1944). Les coleoptera des denrees alimentaires etdes produits industries entreposes. Encycl. Ent. (A)., 22:335. (Quoted by Davey, P.M.(1958). *Bull. Ent. Res.*, 49 (2): 385- 404.

Lever, R. J. A. W.(1942). Entomological notes. Agricultural Journal, Fiji, 13(1): 24-26.

Lienard, V., Seek, D., Haubruge, E. and Gaspar, C. (1992). Biology of *Caryedon pallidus* (Oliv.1790). (Col: Bruchidae)., pest of *Cassia occidentalis* (L.). seeds. *Bulletin et annules de la societe royale belgy Ol Entomolgie.*, 128(10-12): 339-342.

Mackie, J. R.(1943). Annual report on the Agricultural department (Nigeria). for the year 1943. Lagos. pp.34.

Mackie, F. R. (1946). Notes on the biology of the groundnut bruchid, *Caryedon fuscus* Goeze. B.Sc. thesis, Univ. Glasgow. (Quoted by Davey, P.M.(1958). *Bull. ent. Res.*, 49(2): 385-404.

Manjula, K. (2003). Management of groundnut bruchid, *Caryedon serratus* (Olivier) with botanicals in stored groundnut. *Pest Management and Economic Zoology*. 11 (2): 115-118.

Matokot, L., Mapangoa-Divassa, S. and Delobel, A. (1987). Evaluation of *Caryedon serratus* (Coleoptera: Bruchidae) population in stored groundnut in Congo. L'Agronomie Tropicale, 42(1): 69-74.

Mc Farlane, J.A.(1962). A note on the field infestation of pods of the tamarind tree, *Tamarindus indica* L. in Jamaica. *Entomologist mon. mag.*, 97: 198-199.

Mishra, R. P., Mital, V. P. and Gupta, H. C. L. (1976). Use of gelatin capsule for isolation of sexes of bruchids. *Bulletin of Grain Technology*. 14: 66-67.

Misra, R. P., Yadav, P. R. and Mital, V. P. (1980). Safety margins of some chemosterilants against groundnut bruchid, *Caryedon gonagra* Fabr. *Indian Journal of Entomology.* 42: 170-172.

Mital, V. P. (1969). Studies on relative resistance and susceptibility of some important groundnut (*Arachis hypogaea* L.) varieties to groundnut Bruchid, *Caryedon gonagra,* Fabricius. (Bruchidae, Coleoptera). *Bulletin of Grain Technology.* 7: 75-79.

Mital, V. P. (1971). Observations on the influence of a few food media on the life span of *Caryedon gonagra* Fabricius (Bruchidae: Coleoptera). *Madras Agricultural Journal.* 58: 904.

Mital, V. P. and Gupta, H. C. L. (1978). Efficacy of some fumigants against adults of tamarind bruchid, *Caryedon serratus* F. *Bulletin of Grain Technology.* 16: 147-148.

Mital, V. P. and Khanna, S. S. (1968). A note on tamarind Bruchid (*Caryedon gonagra* Fabricius) (Bruchidae, Coleoptera), a serious pest of stored tamarind (*Tamarindus indica* L.) and other leguminous seeds of economic importance. *Agra University Journal of Research.* 16: 99-101.

Mital, V.P. and Khann, S.S. (1974). A note on tamarind bruchid *Caryedon gonagra* Fabricius). (Bruchidae: Coleoptera). a serious pest of stored tamarind (*Tamarindus indica* L.). and other leguminous seeds of economic importance. *Agra University Journal of Research,* 16(2): 99-101.

Mital, V. P. (1991). Extent of losses to groundnut varieties by groundnut bruchid, *Caryedon serratus* Ol. (Bruchidae: Coleoptera). in storage. *GAU. Research Journal,* 17(1): 43-48.

Mital, S. and Wightman, J. A. (1989). An inert dust protects stored groundnut from insect pests. *International Arachis Newsletter.* (6): 21-22.

Mital, V. P. (1991). Extent of losses to groundnut varieties by groundnut bruchid, *Caryedon serratus* Fab. (Bruchidae: Coleoptera) in storage. *Gujarat Agricultural University Research Journal.* 17(1): 43-48.

Mital, V. P.(1994). A note on Management techniques for minimizing the damage in groundnut caused by Bruchid (*Caryedon serratus* Ol). A hand note submitted to Department of Plant Protection, Quarantine & Storage, Faridabad, pp. 4.

Mital, V. P.(1968). Studies on genus *Caryedon* (Bruchidae: Coleoptera). on some important legumes and spices in storage in U.P., Ph.D. thesis submitted to Agra University, Agra.

Motka, M. N.(1997). Biology, Population Dynamics, Varietal screening and Management of Groundnut beetle *Caryedon serratus* (Olivier). Ph.D. thesis submitted to the Gujarat Agricultural University, Junagadh pp. 260.

Mucunguzi-P. (1995). Bruchids and survival of Acacia seeds. African Journal of Ecology. 33: 175-183.

Mukerji, S. and Chatterjee, S. N. (1951). Morphology of the genital structure of some of the bruchidae (Lariidae) of India and Ceylon and their taxonomic importance. *Indian Journal of Entomology* 13(1): 1-28.

Mukhopadhyay, A. and Saha, B. (1995). Nutrient requirement of seed bug (*Elasmolomus sordidus*) (Heteroptera: Lygaeidae) on four host seeds. *Indian Journal of Agricultural Sciences*. 65(2): 154-155.

Mukhtar-Ahmad and Sushil-Kumar. (1990). Ovipositional response of *Caryedon serratus* Oliver (Coleoptera: Bruchidae). *Indian Journal of Forestry*. 13(3): 192-198.

Murugesan, S., Balu, A., Durairaj, S., Pankajam, S. and Sunitha, B. (1998). Insecticidal performance of a neem product in control of two major seed pests of forestry tree species. *Entomon*. 23(3): 197-201.

Narendra Kumar and Tripathi, N. N. (2004). Repellent property of volatile oil isolated from *Putranjiva roxburghii* against *Trogoderma granarium* associated with stored groundnut seeds. Proceedings of the National Academy of Sciences India Section B, Biological Sciences. 74(2): 179-187.

Nilsson, J. A. and Johnson, C. D. (1992). New host *Bauhinia variegata* L. and new locality records for *Caryedon serratus* (Olivier). in the new world (Coleoptera: Bruchidae: Pachymerinae). *Pan-pacific Entomologist*, 68(1): 62-63.

Odeyemi, O. O. (1991). Control of khapra beetle, *Trogoderma granarium* Everts. in decorticated groundnut with vegetable oils. *Applied Entomology and Phytopathology*. 58(1-2): 31-38.

Pajni, H. R. and Gill, K. M. (1974). Effect of light on the pests of stored products. *Bulletin of Grain Technology*. 12: 151-153.

Pancholi, K. R. (1993). Relative resistance, susceptibility and extent of losses of groundnut varieties to *Caryedon serratus* (Olivier) in storage. M.Sc., (Agri) thesis submitted to the Gujarat Agricultural University, Sardar Krishinagar, Gujarat.

Pandey, N. D., Misra, R. L. and Tripathi, R. A. (1977). Relative susceptibility of some groundnut varieties to the almond moth, *Cadra cautella* Walker (Phycitidae: Lepidoptera). *Bulletin of Grain Technology*. 15: 105-110.

Patel, C. C. and Koshiya, D. J. (1994). Life tables and intrinsic rate of increase in number of *Caryedon serratus* (Olivier) (Coleoptera : Bruchidae) on groundnut. *Gujarat Agricultural University Research Journal*. 20(1): 174-177.

Patra, A. K., Tripathy, S. K., Samui, R. C. and Nanda, M. K. (1996). Effect of potassium nutrition and type of storage container on seed quality of stored groundnut pods. *International Arachis Newsletter*. 16: 56-58.

Pic, M. (1913). Bruchidae: Coleoptera. Cat. pars.55, pp.74. (Quated by Davey, p.m.(1958). *Bulletin of Entomological Research*, 49(2): 385-404.

Pierre, D. and Huignard, J. (1991). The biological cycle of *Caryedon serratus* (coleoptera: Bruchidae) on one of its hosts *Bauhinia rufescens* Lam. (Caesalpinioidae) in a Sahelian zone. *Acta Oecologia*, 11(1): 93-10.

Prevett, P. F. (1953). Studies on the fecundity and longevity of the groundnut bruchid, *Caryedon fuscus* Goeze and the external morphology of immature stages. M.Sc. thesis, Univ. London.

Prevett, P. F. (1954). Some aspects of the biology and control of the groundnut bruchid (*Caryedon fuscus* Goeze). D.I.C. thesis, Univ. London.

Prevett, P. F. (1967). The larva of *Caryedon serratus* (O1): The groundnut seed beetle (Coleoptera: Bruchidae). *Journal of Stored Products Research*, 3: 117-123.

Prevett, P. F. (1966). Observations on biology in the genus *Caryedon schonherr* (Coleoptera)., Bruchidae). in Northern Nigeria, with a list of associated parasitic Hymenoptera. Proceeding .of Entomological Research Society. London. (A). 41 (1-3): 9-16.

Pruthi, H.S. and Mohan singh. (1950). Pests of stored grains and their control. Manager of publication, Govt. of India, New Delhi. pp. 88.

Radadia, L.B., Patel, N.C. and Chauhan, P.M. (1992). Studies on storage characteristics of groundnut. *Bulletin of Grain Technology* 30(3): 231-235.

Radadia, G. G. (2003). Assessment of infestation caused by *Caryedon serratus* (Olivier) in stored groundnut. *Insect Environment*. 9(3): 121-122.

Radadia, G.G. (1995). Bio-ecology and management of groundnut bruchid, *Caryedon serratus* (Olivier). in stored groundnut. Ph.D. thesis submitted to Gujarat Agricultureal University, Sardarkrushinagar (Unpublished).

Raina, A.K. (1972).. Observations on bruchids as field pests of pulses. *Indian Journal of Entomology* 33(2): 194-197.

Ramasivan, T., Krishnamurthy, K. and Pingale, S. V. (1968). Studies on preservation of food grains in rural storages. Part-II. Storage of grains in villages near Hapur. *Bulletin of Grain. Technology.*, 6(2): 69-75.

Ranga Rao, G. V., Ravi, U., Surender, A., Murthy, K. S. and Joshi, N. C. (1993). Effect of phosphine and methyl bromide fumigation of different life stages of peanut bruchid, *Caryedon serratus* Oliver. *Indian Journal of plant protection.*, 21(1): 72-74.

Rangarajan, A.V., Mahadevan, N.R. and Iyemperumal, S. (1978). Pest complex of sunflower (*Helianthus annus* Linn.). in Tamil Nadu. *Indian Journal of Entomology*, 37(2): 188-191.

Rao, G. V. R., Upendra-Ravi., Surender, A., Murthy, K. S., Joshi, N. C. (1993). Effect of phosphine and methyl bromide fumigation of different life stages of peanut bruchid, *Caryedon serratus* Olivier. *Indian Journal of Plant Protection*. 21(1): 72-74.

Ritchie, A.H. (1916). Report of Entomologist for year 1915-16. Jamaica Deparment of Agriculture. Kingston. pp. 31-34.

Romero, J. and Johnson,C.D.(2002). *Cassia moschata* H.B.K., new host for *Caryedon serratus* (Olivier) in the New World (Coleoptera: Bruchidae: Pachymerinae). Coleopterists-Bulletin. 2002; 56(1): 95-96.

Rouboud, E. (1916). Insects and the deterioration of groundnuts in Senegal. Separate, undated from L'Ann. et. Mem du com d'Etudes hist et scien del. Afrique occ., pp 77.

Sachan, G. C and Chandel, S. K. (1991). Susceptibility of some groundnut varieties to *Lasioderma serricorne* (Fabricius). *Indian Journal of Entomology*. 53(4): 559-563

Sagot, R. and Bouffil, F. (1936). Etudes sur la bruche de 1' Arachide (*Pachymerus acaciae*). *Bull Com. A.O.F.,* 18(1): 79-91.

Sahayaraj, K and Paulraj, M. G. (2000). Impact of some plant products on the behavior of *Tribolium castaneum* in groundnut seed. *International Arachis Newsletter*. (20): 75-76.

Sardesai, J. B. (1961). Effects of the density of the adult population on the oviposition of *Caryedon gonagra F. Garcia de orta,* 9(2): 227-233.

Satya-Vir and Jindal, S. K. (1996). Field infestation of *Caryedon serratus* Olivier (Coleoptera: Bruchidae) on the pods and seeds of *Acacia nilotica* in the Thar desert of *India. Journal of Tropical Forest Science*. 9(2): 189-193.

Sembene,M. and Delobel, A.(1996) Morphometric identification of sudano-sahelian populations of the groundnut seed-beetle, *Caryedon serratus* (Olivier) (Coleoptera Bruchidae). *Journal of African Zoology*. 110(5): 357-366.

Singal, S.K. and Tokky, O.P. (1991). *Pongamia pinnata* (L). pierre; a new record of *Caryedon serratus* (Oliver). (Bruchidae: Coleoptera). from India. Research and development report, HAU, Hissar, 6(2): 91-92.

Singal, S.K. and Tokky, O.P. (1990). Carry over of bruchid, *Caryedon serratus* (Oliver). (Coleoptera). from field to stores through seeds of Acacia nilotica (L). willd. in India. *Tropical Pest Management*, 36(1): 66-67.

Singal, S. K. and Toky, O. P. (1988). New host plants of *Caryedon serratus* (Olivier) (Bruchidae: Coleoptera) from India. *Research and Development Reporter*. 5: 91-92.

Singal, S. K., Toky, O. P. (1989). *Pongamia pinnata* (L.) Pierre: a new record of *Caryedon serratus* (Olivier) (Bruchidae: Coleoptera) from India. *Research and Development Reporter*. 6(2): 91-92.

Singal,S.K., and Toky,O.P.(1990). Carry over of bruchid, *Caryedon serratus* (Olivier) (Coleoptera) from field to stores through seeds of *Acacia nilotica* (L.) Willd. in India. Tropical Pest Management. 36: 66-67.

Singh, P. and Bhandari, R.S. (1987). Insect pests of *Acacia tortilis* in India. *Indian Forester,* 113(11): 734-743.

Singh, P. and Bhandari, R.S. (1991). Insect pests of leguminous forest tree seed and their control. *Indian Forester*. 14(2): 844-853.

Singh, V. (1991) Effect of natural infestation of *Oryzaephilus surinamensis* Linn. and *Tribolium castaneum* Herbst on the kernel quality of groundnut varieties. *Journal of Insect Science*. 4(1): 82-84.

Sinha, R. P., Vijay-Kumar., Singh, H. N. and Jha, B. N. (1997). Storability of groundnut with and without shell. *Journal of Applied Biology*. 7(1/2): 26-27.

Sohi, G. S. (1940). *Pachumerus gonagra* F. as a store pest. *Indian Journal of Entomology* 2: 241-246.

Sontakke, B. K., Mishra, P. R. and Rout, G. D. (1992). Observation on biology and habits of *Caryedon serratus* Olivier (Coleoptera: Bruchidae) infesting stored groundnut pod. *Bulletin of Grain Technology*. 30(3): 228-230.

Sosamma Jacob, Vilasini, T. N., Suja, G and John, S. (2002). Loss assessment in groundnut test entries due to storage pest. *Insect Environment*. 8(2): 93-94.

Southgate, B. J. and Pope, R.D. (1958). The groundnut seed beetle, a study of its identity and taxonomic position. *Ann. Mag. Nat. Hist.*, 10 (117) : 669-672.

Southgate, B.J. (1976). A new subspecies of *Caryedon* (Coleoptera: Bruchidae). from the Middle East. *Israel Journal of Zoology.*, 25(4): 194- 198.

Srivastva,A.S.(1970). Important insect pest of stored oilseeds in India. International Pest control 12(3):18-20.

Strong, L.A. (1922). Bureau of plant quarantine, a synopsis of work for months of March, April, May, June and July, 1922. Monthly Bulletin of California, Department of Agriculture. 11(10): 775-780.

Sundria, M. M. and Akhil-Kumar. (2004). Biology of groundnut bruchid, *Caryedon serratus* (OL), on different test hosts. Annals of Plant Protection Sciences. 12(1): 9-12.

Tripathy, M. K., Mohanty, S. K. and Das, B. C. (2000). Residual toxicity of different pesticides against seed beetles, *Caryedon serratus* (Bruchidae, Coleoptera) attacking stored groundnut at Bhubaneswar. *Environment and Ecology*. 18(3): 761-764.

Tripathy, S. K., Patra, A. K., Samui, R. C. and Nanda, M. K. (1996). Effect of storage containers on storability of groundnut seeds and their performance in the field. Indian Journal of Plant Physiology. 1(3): 180-184.

Vayssiere, P.(1939). Au sujet de I'orgaisation de laiutte centre les insects nuisibles au Senegal et au Sudan. Actes Association of Colonaration Sciences, 15, 65-70.(Fide: R.A.E., 27:139).

Venugopal Reddy, V. (1990). Estimation of losses and management of *Caryedon serratus* (Olivier) on groundnut. M.Sc. Thesis, Andhra Pradesh Agricultural University, Hyderabad (AP).

Verma, B. R. (1989). Observations on three parasites of bruchids on leguminous host plants. *Bulletin of Entomology*. 30(2): 246-247.

Verma, B. R. (1993). Observations on three parasites of bruchids on leguminous host plants. *Indian Journal of Entomology*. 55(3): 328-329.

Vijay-Singh, Kumari, D. A., Reddy, V. S. and Tejkumar, S. (2002). Feeding potential of pod bruchid, *Caryedon serratus* Olivier (Bruchidae: Coleoptera) on stored groundnut. *Indian Journal of Plant Protection*. 30(2): 177-179.

Wali-ur-Rehman (1993). Studies on seed insect pests of forest trees. Pakistan Journal of Forestry., 43(1), 27-31. (Fide: R.A.E., 83:1023).

Wightman, J.A.; Amin, P.W.; Rao, G.V.R. and Dick, K. M. (1987). Second groundnut workshop for Southern Africa, ICRISAT. pp. 103-114.

Williams, J.O. (1980). Note on Bruchidae associated with stored products in Nigeria. *Tropical Grain Legume Bulletin.,* 21: 5-10.

Zacher, F. (1932). Die tierischen Samenschad linge in Freil land and large spimmentiers, Kafer and Haulfligter. Wiss. U. Tech. *Gartenb Heft.,* 5: 78.

Zacher, F. (1936). Beitrog zur Nahrpflan-zenkenntnis der Samen kafer (Col:Bruch: Lariidae).(Contribution to a knowledge of the food plants of the beetles attacking seeds). 7(1): 10-13.

Zutzhi, M. K. (1966). Storage of wheat by farmers in Delhi. *Bull. Grain Tech.,* 4(1): 143-145.

❑❑❑

Nematodes

D. J. PATEL AND S. K. PATEL

15.1. Introduction

More than 55 pathogens including nematodes have been reported to attack groundnut crop in different regions of the world (Sharma 1985). Of these phyto-nematodes, root-knot nematodes (*Meloidogyne* spp.) attack groundnut crop severely in India.

15.2.Root-knot nematodes (*Meloidogyne* spp.)

15.2.1. Meloidogyne arenaria

Root-knot nematodes are the most destructive soil dwellers limiting groundnut yield. It was originally described on groundnut in Florida, USA and is widely distributed throughout the groundnut growing areas of the world. In India, it was reported for the first time from Bihar by Sen (1960) and later from Delhi (Prasad e*t al.*, 1964) Punjab (Sharma e*t al.*, 1978; Mahajan and Chhabra, 1979) and Gujarat (Vaishnav and Dhruj, 1981). Because of the conspicuous root 'galls' and 'warts' caused by these neamtodes, the seriousness of the damage was recognized early. Yield losses due to root-knot nematodes ranged from 14-20 %. Groundnut plants infected with nematodes commonly develop enlarged roots and pegs which develop into galls of various sizes (**Plate 1**). The galls result from an internal swelling of the root tissue. Thus, they can be distinguished from *Rhizobium* nodules which usually are appended to the side of the root. Pods also become infected and develop knots, protuberances or small warts. Galls on roots, pegs and pods sometimes begin to deteriorate by the time of peanut maturity. Development of the root system is generally much reduced and plants show variable degree of stunting and chlorosis. The root system is much restricted and vascular elements of infected tissues are disrupted resulting in poor flow of water and nutrients from roots to the shoots. Infected plants show a tendency to wilt during dry periods or when subjected to water stress (Porter e*t al.*, 1984).

Plate 1: Galls

Vaishnav *et al.* (1985) reported that groundnut plant infected by root-knot nematode (*M. arenaria*) remained stunted because of reduction in various plant growth parameters due to increased nematode population. Total chlorophyll content of the leaves and leaf lamina was also reduced by 76.72 and 33.83 % respectively over normal plants (Table 15.1). As a result, infected plants look stunted with smaller chlorotic leaves. In field, the infected rows appear thinner compared to healthy rows and give the patchy appearance. The root system of such infected plants shows severe galling with numerous egg-masses. The nematode completes its life cycle within 28 days at a temperature range of 27-33°C.

The Pathogenicity study of root-knot nematodes on groundnut conducted by Patel (1983) indicated that initial inoculum of 100 J2/kg soil was detrimental to the growth and development of groundnut cv. GAUG 1 while it was 1,000 J2/kg soil for GAUG 10 and JL 24. Regarding interaction studies between *M. arenaria* and *Fusarium solani*, Patel *et al.* (1985) observed concomitant inoculation of both the organisms induced wilt symptoms earlier on groundnut and suppressed plant growth than fungus alone, whereas interaction effect of *M. arenaria* with *Aspergillus niger* had neutral type for plant growth and nematode multiplication. However, synergistic interaction effect between *M. arenaria* and *A. flavus* in reducing shoot growth of groundnut cv. GAUG 10 was observed with the concomitant inoculation of four gram mycelium of *A. flavus* and either 1,000 or 2,000 *M. arenaria* J2/plant. Adverse effect of concomitant inoculation of *M. arenaria* with either *A. niger* or *A. flavus* was observed on number of galls/plant as well as final nematode population. The presence of *M. arenaria* significantly reduced numbers of *Rhizobium* nodules/plant on groundnut (Table 15.2) (Patel, 1983). Some of the resistance varieties which have been reported against the nematode are given in table 15.2a.

Garcia and Mitchell (1975) reported that *Pythium myriotylum* interacted synergistically with *Fusarium solani* and *M. arenaria* in causing pre-emergence damping-off of peanut seedlings. Pod rot was also more in presence of these organisms.

15.2.2.*Meloidogyne javanica*

Root-knot nematodes (*Meloidogyne javanica*) was first time recorded on groundnut cv. JL 24 and J 11 in Kapadwanj areas of Kheda district in middle Gujarat (Patel *et al.,* 1988). It has been also reported on groundnut in Uttar Pradesh, Punjab, Orissa, Maharashtra and other states of India. Infected plants exhibit similar symptoms as induced by *M. arenaria*. However, root galling pattern of *M. javanica* is quite conspicuous. Root galling by nematode is observed on all the roots of plant including tertiary roots like root galling pattern on vegetable crops. Whereas in case of *M. arenaria*, profuse galling (comparatively smaller in size) is seen mainly on tertiary roots and very few galls are produced on main root portion.

Host differential study conducted by Patel *et al.* (1993) indicated that generally *M. incognita* and *M. javanica* do not infect groundnut crop. But *M. javanica* reported on groundnut cvs. JL 24 and J 11 produced heavy root galling and nematode was reported as *M. javanica* pathotype 2 to cause discernable damage. This *M. javanica* pt.2 infects Cotton *(Gossypium* spp.*)*, mungbean, urdbean, fennel, pigeonpea crops severely besides groundnut in the area. Nematode infection drastically reduces plant growth and thereby crops production is poor (Patel *et al.*, 1996).

Pathogenicity study conducted by Joshi (1989) revealed that an inoculum level of 1000 *M. javanica* J2/kg soil/plant drastically reduced plant growth of groundnut cv. GG 2 & JL 24 . Nematode infection on groundnut cv. GG 2 significantly reduced plant height, fresh as well as dry shoot and root weights, leaf area, no. of internodes & branches and produced more galls with higher nematode population (Table 15.3). Groundnut plants infected by *M. javanica* had increased leaf temperatures and transpiration rate and decreased diffusion resistance than healthy non infected plants. Similarly nematode infected plants recorded less peroxidase and polyphenol oxidase activities and higher total phenol contents over healthy plants.

Patel *et al.* (1998) reported the results of experiment on management of root-knot nematodes indicated that soil application of carbofuran @ 3 kg a.i./ ha increased dry pod yield by 42.8 % and dry fodder yield by 28.8 % and reduced root-knot disease by 45.2 % over control. There was 36.3 % increase in pod yield and 36.5 % reduction in RKI due to phorate @ 3 kg a.i./ha, lower dose of 2 kg a.i./ha of the nematicides was less effective than higher dose of 3 kg a.i./ha (Table 15. 4).

Significant differences in enzymatic activities viz. peroxidase, polyphonel oxidase and total phenol contents were observed in groundnut roots due to *M. javanica* at varying inoculum levels (Fig.1). Higher peroxidase activity was recorded in 10,000 J_2/plant followed by 1,000 J_2/plant, both differing significantly from each other. Inoculum levels of 10 and 100/J_2 plant, though had increased peroxidase activity, but did not differ significantly from each other. Uninoculated control plants had significantly less peroxidase activity. In case of polyphenol oxidase activity and total phenol contents similar trends were noticed(Patel *et al.*, 1998 b).

The results obtained were in agreement with the findings of Joshi (1989) who also reported increase in enzymatic activities in root-knot infected groundnut. Significant reduction was recorded in chlorophyll a,b and total chlorophyll contents of leaves due to nematode inoculum levels of 100, 1000 and 10,000 J_2/plant. Findings in present study are in confirmation with Joshi

(1989) who reported increased enzymatic activity in groundnut due to *M. javanica* infection. Patel (1983) also reported reduction in chlorophyll content of groundnut leaves due to infection by *M. arenaria*.

Experiment on Pathogenicity of *Meloidogyne javanica* pt.2 on groundnut cv. GG 20 conducted by Patel & Patel (1998) revealed that an inoculum level of 100 J_2 and above/plant significantly reduced plant growth parameters viz. plant height, fresh and dry shoot weight. Fresh and dry root weights progressively increased with an increase in nematode inoculum level from 10 to 10,000 J_2/plant due to heavy root galling and profuse egg sac formation on roots. Nematode inoculum level of 10,000 J2/plant had significantly minimum plant height, fresh and dry shoot weights. Total nematode population increased with an increase in nematode inoculum levels while nematode reproduction rate decreased with corresponding increase in inoculum levels and it was maximum of 4782.2 times at the level of 10 J_2/plant and minimum of 10.5 times at the highest level of 10,000 J_2/plant.

15.3. Other nematodes

Kalahasti Malady disease caused by *Tylenchorhynchus brevilineatus* is characterized by small brownish yellow lesion appearance on the pegs, pods, stalks as well as on young developing pods. The margin of the lesions are slightly elevated because of the proliferation of host cells around the lesions. Pod stalks are much reduced in length and pod surface becomes discoloured on advanced stages of the disease. Root discoluoration is also observed. Affected plants remain stunted and greener than normal foliage. The root lesion nematodes (*Pratylenchulus* spp.) are migratory endo-parasites that attack peanut roots, pegs as well as pods and feed within the parenchymatous tissues. The nematode can enter or live in the roots, pods or pegs during all of its developmental stages. Both mechanical and chemical damage result from the nematode feeding within the tissues. Root lesions develop and on large nematode population, these lesions coalesce, causing extensive discoluoration and damage that result in reduction in growth and pod production. Both degree of damage and pod less at harvest are positively correlated with nematode numbers which can increase to over 3,000 /g tissue in the roots, pegs and pods. In addition, nematode feeding within the pegs weakens them, resulting in pod loss at harvest. Lesion nematode numbers are positively correlated with percentage of damaged seeds and negatively correlated with yields of pods, percentage of sound mature seeds, percentage of total seeds and value. Other microorganisms may colonize the damaged tissue and penetrate through the pod, causing seed damage. The combined effect of the nematodes and associated micro-organisms can result in significant reduction in yield, quality and value of the crop. *Pratylenchus* spp. also causes disfiguration of the roots and pods of groundnut in USA (Endo, 1959; Graham,

1949; Steiner, 1919), Graham (1955) reported *Criconemoides* spp. causing stunting and loss in green weight of tops and roots in groundnut crop. While studying host-parasite relationship and Pathogenicity of *P. brachyurus* on groundnut, Good *et al.* (1958) observed that nematodes cause dark coloured necrotic lesions on the shell and cause peg rot condition in the crop. Heyhs (1962) reported *Xiphinema vanderlindei* in the rhizosphere of groundnut in South Africa, whereas in Senegal, Germani (1981) while studying Pathogenecity of *Scutellonema cavenassi* on peanut showed that the parasite affected all parts of the plant and symbiotic nitrogen fixation apparatus.

15.4. Management

Soil application of aldicarb, carbofuran and phenamiphos effectively reduced root-knot infestation in groundnut (Anon., 1985; Sakhuja and Sethi 1986 a). The losses in groundnut yield due to *M. arenaria* was estimated to the tune of 59.46 %. Similarly *M. arenaria* induced 22.1 % loss in pod yield of groundnut cv. GAUG 10 in Saurashtra area of Gujarat State. Seeds treated with either aldicarb sulfone 75 WP or carbofuran flowable @ 6 % had minimum larval penetration to the roots which ultimately resulted in significantly more fresh and dry shoot weights of groundnut plants, finally giving significantly less root galls and final nema population (Anon., 1985). Marshal 25 STD and Oncol 3 G used as seed dressers were found effective for root-knot management in ground nut at lower concentrations (1 and 2 %) as compared to higher concentrations (3 and 4 %) at Delhi (Anon., 1986).

Seed soaking in monocrotophos, phenmaniphos, oxamyl and phosalone significantly reduced penetration of root-knot nematodes in groundnut roots and thereby improved plant growth with reduced root-knot disease. Monocrotophos and phenamiphos were more effective. Higher concentrations (500 ppm) of each of these chemicals effectively reduced nematode infection as compared to lower ones. Organic amendments viz., castor, mustard and neem cakes also significantly reduced *M. arenaria* population and thereby increased groundnut yield at Junagadh (Anon., 1987). Varietal screening (227 genotypes) against *M. arenaria* did not yield any genotype resistant to nematode. Simultaneous inoculation of either *Rhizoctonia bataticola* of *Fusarium solani* with *M. javanica* on groundnut indicated that both the fungi reduced multiplication of *M. javanica* subsequently giving less galling on groundnut roots. *R. bataticola* proved more antagonistic to *M. javanica* than *F. solani* (Sakhuja and Sethi, 1986 b).

Infection of *M. javanica* on groundnut cv. Samrala reduced 36.6 % pod yield in Kapadwanj area of Kheda district in Gujarat. At Kanpur (UP), castor, neem and mustard cakes effectively managed *M. javanica* on groundnut and thereby increased crop production (Table 6). Higher doses (3 kg/ha) of carbofuran, phenamiphos and phorate proved effective in reducing *M. javanica* infection and helped to improve plant growth (Joshi, 1989).

Soil application of phenamiphos @ 2 kg/ha significantly increased plant growth and developments over control in field. Dry pod and fodder yield loss was 30 and 20.4 % respectively over control. Total pod and fodder yield loss was estimated to the extent of 50.4 % which resulted in monitory loss of Rs. 6400/ha (Patel, 1998).

15.5.Future thrust

In our opinion, following aspects need to be attended on priority basis.

- Intensive and rigorous screening of germplasm should be done in order to locate resistance source(s) in peanut.
- Resistant genotypes should be used for transferring resistant factors into susceptible high yielding varieties
- Screening of large germplasm of groundnut crop yielded very few entries/lines to be resistant to root-knot nematodes and hence work on mutation breeding through mutagenic agents like Sodium azide, EMS (Ethyl Methane Sulfonate), irradiation etc. should be initiated
- Use of soil solarization (slow soil pasteutration) should be explored for management of nematode diseases, wherever possible.
- More emphasis should be given on use of organic amendments and crop rotations.
- Integrated Management Practices adopting cultural practices, biological, agents, resistant varieties, low rate of nematicides, organic amendments etc. should be planned for economic management of nematode disease in groundnut crop.
- Intensive survey work should be carried out to get awarness about the damage by different plant parasitic nematodes beside root-knot.
- Role of nematodes in aggravating fungal disease like stem/collar rot, root-rot, pod-rot, etc. should be studied and management practices be evolved.

Table 15.1. Effect of root-knot nematode infection (*Meloidogyne arenaria*) on groundnut cv. GAUG 10.

Character	Inoculated	Control	%increase decrease over control
1. Shoot length, cm	24.67	36.80	-32.96
2. Shoot fresh weight, g	11.03	17.80	-38.03
3. Shoot dry weight, g	2.47	4.20	-41.19
4. Root length, cm	19.93	26.93	-25.99
5. Root fresh weight, g	4.75	3.09	-53.72

Contd...

6. Root dry weight, g	1.12	0.70	+60.00
7. Leaf lamina size (10th leaf), cm^2	11.56	17.47	-33.83
8. Petiole length (10th leaf), cm^2	5.29	6.01	-11.98
9. No.of internodes on main shoot	12.00	14.00	-14.29
10. Fifth internode length, cm	1.92	2.81	-31.67
11. Chlorophyll content, mg/g			
Chlorophyll a	2.11	0.68	-75.69
Chlorophyll b	0.70	3.39	-79.35
Total chlorophyll	2.81	12.07	-76.72
12. Number of galls/plant	2305	—	---
13. Final populations	136432	—	—

Table 15.2. Effect of *Meloidogyne arenaria* on formation of Rhizobium nodules on groundnut cv. GAUG 10.

Treatment	Plant ht., cm	Fresh wt; g		Dry wt., g		No. of nodules/plant
		Shoot	Root	Shoot	Root	
R	41.00	19.33	3.42	4.85	0.90	145.33
N_1	26.00 (36.59)	12.00 (32.92)	4.55 (33.04)	3.00 (37.89)	1.20 (33.33)	0
N_2	24.50 (40.24)	11.25 (41.80)	4.75 (38.89)	2.82 (41.61)	1.25 (38.89)	0
RN_1	33.42 (18.49)	15.45 (20.07)	4.65 (35.96)	3.87 (19.88)	1.23 (36.67)	96.67 (33.48)
RN_2	31.00 (24.39)	14.50 (24.99)	4.83 (41.23)	3.63 (24.84)	1.28 (42.22)	87.17 (40.02)

R = 25 ml Rhizobium bacteria (13×10^9 cells/ml)
N_1 = 1000 *M. arenaria* J_2; N_2 2000 *M. arenaria* J_2
Figures in parentheses indicate per cent disease over R.

Table 15.2a. Resistant varieties to root-knot nematode

	NRCG	ICG	VARIETY	SYNONYM	ORG	HBT	OTHER TRAITS
1	--	--	13-30		UNK	--	
2	203	1463	17-5		IND	VUL	
3	206	1471	55-437	EC 109271	SEN	VUL	
4	5729	673	2799		USA	VUL	
5	101	1155	Ah-32 A		IND	VUL	
6	3509	1268	Ah 7253		SDN	VUL	
7	371	6591	AI 373		MWI	HYB	
8	360	6577	AM 1		MWI	HYB	
9	3149	3701	Apaxue		SUN	VUL	
10	--	--	BG 1		IND	HYB	
11	--	--	BG 2		IND	HYB	1,35
12	5867	831	C 41		IND	HYR	*
13	42	881	C 145-12-P-7		IND	HYR	30
14	--	--	EC 85994		UNK	--	

contd...

Table 15.2a. contd....

15	365	6584	Georgia Bunch		USA	HYB	*
16	3754	3376	JH 223		ARG	FST	
17	3225	--	Jolpin-220-15		UNK	HYR	*
18	401	6750	Kanyoma		MWI	HYB	24
19	4414	3863	KG/61-22	NFG 638	IND	HYB	
20	3237	--	Local Ambadi		IND	HYR	
21	3242	--	Local Wadmedani (R)		SDN	HYR	
22	3236	--	Local-256		IND	HYR	*
23	346	6554	Mabwa Runner		TZA	HYB	*
24	--	--	NC 707		UNK	VUL	13
25	5198	6429	NC-6		USA	HYR	*
26	1067	262	NCAc 50		ARG	FST	
27	1069	264	NCAc 56		NGA	FST	*
28	1009	5927	NCAc 460		USA	HYB	
29	1077	284	NCAC 527		USA	VUL	*
30	1081	--	NCAC 548		UNK	FST	*
31	785	305	NCAC 592		ARG	VUL	13
32	1099	336	NCAC 763		URY	VUL	*
33	870	351	NCAc 832		IND	HYB	
34	5001	6022	NCAC 927		USA	HYB	18,21, 26,28,32
35	816	--	NCAC 1313		UNK	VUL	
36	1060	6136	NCAC 2196		USA	HYR	*
37	1061	6146	NCAc 2298		USA	HYB	
38	1027	6767	NCAC 2565		UNK	-	
39	901	--	NCAC 2736		USA	VUL	
40	1031	6684	NCAc 9937		ZWE	HYR	
41	1064	--	NCAC 16578		SEN	HYR	
42	1537	--	PI 270787		ZWE	VUL	
43	5836	794	Punjab 1-10-1		IND	HYR	
44	336	6542	S.Am.Coll.106		SAM	HYB	
45	18	104	Small Japan		JPN	FST	4
46	4166	808	Talod-34-7		IND	HYR	
47	184	1389	U 2-1-19	EC 21040	SDN	FST	
48	481	7161	VRR 307		IND	VUL	

Table 15.3. Effect of root-knot nematodes (*Meloidogyne javanica*) infection on plant growth and nematode multiplication on groundnut cv. GG 2.

Character	Inoculated	Control	% increase/ decerease over control
1. Shoot length, cm	13.6	17.10	-20.46
2. Shoot fresh weight, g	8.50	12.00	-29.16
3. Shoot dry weight, g	2.60	4.10	-36.58
4. Root length, cm	7.10	9.60	-26.04
5. Root fresh weight, g	5.10	3.60	+41.66
6. Root dry weight, g	1.08	0.76	-42.10
7. Leaf area, cm^2	3.00	5.25	-42.85
8. Number of internode in man shoot	9.00	11.0	-18.18

Table 15.4. Management of *M. javanica* in groundnut field through organic amendments

Organic amendment	Yield, q/ha	RKI (1-5)*
Castor cake @ 1 t/ha	67 (+28.85)	2.00 (-38.46)
Neem cake @ 1 t/ha	73 (+40.38)	2.00 (-38.46)
Mustard cake @ 1 t/ha	62 (+19.23)	2.75 (-15.38)
Carbofuran @ 2 kg/ha	75 (+44.23)	1.75 (-46.15)
Control	52	3.25

* 1= Free, 5= Maximum disease intensity.

Table 15.5. List of nematodes

Species	Reference
Aphelenchoides sp.	Sharma and McDonald, 1990
Aphelenchus sp.	"
Criconemoides sp.	"
Ditylenchus sp.	"
Helicotylenchus arachisi Mulk and Jairajpuri	"
H. dihystera (Cobb) Sher	"
H. egyptiensis Tarjan	"
H. retusus Siddique and Brown	"
Helicotylenctus sp.	Sharma and McDonald, 1990
Heterodera sp.	"
Hischmaniella sp.	"
Hoplolaimus arachidis Maharaju and Das	"
H. indicus Sher	"
H. seinhorsti Luc	
H. sesadrii Mulk and Jairajpuri	"
Hoplolaimus sp.	"
Macroposthonia curvata (Raski) de Grisse and Loof	"
Meloidogyne arenaria (Neal) Chitwood	Sharma and McDonald, 1990
M. hapla Chitwood	"
M. incognita (Kofoid and White)	"
M. javanica (Treub) Chitwood	"
Pratylenchus brachyurus (Godfrey) Filipjev and Sch. Stekh	"
P. coffeae (Zimmermann) Filipjev and Sch. Stekh.	"
P. delattrei Luc	"
P. indicus Das	"
P. nizamabadensis Maharaju and Das	"

Contd...

Table 15.5 Contd...

P. zeae Graham	"
Radopholus similis (Cobb) Thorne	"
R. reniformis Linford and Oliveria	"
Siddiqia citri (Siddiqi) Khan, Chawla and Saha	"
Tylenchorhynchus brevilineatus Williams	"
Tylenchorhynchus mashhoodi Siddiqi and Basir	"
T. phaseoli Sethi and Swarup	Sharma and McDonald, 1990
T. vulgaris Upadhyay, Swarup andSethi	"
T. zeae Sethi and Swarup	"
Tylenchorhynchus sp	"
Xiphinema insigne Loos	"

References

Anonymous, (1985). Studies on pathogenicity of root –knot nematodes on groundnut with respect in crop losses. Biennial Report AICRP (Nematodes)., Anand Centre, Gujarat. P. 81 and 83.

Anonymous, (1986). Control of root –knot nematode by seed treatment in groundnut field. Consolidated reports, AICRP (Nematodes), Indian Agricultural Research Institute, New Delhi.

Anonymous, (1987). Plant Protection in groundnut, Technical Bulletin, National Research Centre for Groundnut, Junagadh, Gujarat, p.59.

Endo, B.Y. (1959). Response of root knot nematodes, *Pratylenchus brachyurus* and *P. zeae* to various host plants and soil types. *Phytopathology*. 49:866-68.

Garcia, R. and Mitchell, D.J. (1975). Interaction of *Phythium myriotyhum* with *Fusarium solani, Rhizoctonia solani* and *Meloidogyne arenaria* in pre- emergence damping off of peanut. *Plant diseases Reporter*. 40:866-868.

Germani, G. (1981). Pathogenecity of the nematode *Scutellonema cavenessi* on peanut and soyabean. *Revue de Nematologie*. 4:203-8.

Good, J.M., L.W. Boyle and Hammons, R.O. (1958). Studies of *Pratylenchus brachyurus* on peanuts. *Phytopathology*. 48:530-37.

Graham, T.W. (1949). Nematode root tot of tobacco and other crops. *Phytopathology*. 39:8.

Graham, T.W. (1955). Pathogenecity of *Rotylenchus brachyurus* on tobacco and of Criconemoides sp. on tobacco and peanuts. *Phytopathology*. 45:347.

Heyhs, J. (1962). A reoprt on South African nematodes of the families Longidoridae, Belondiridae and Alaimidae (Nemata: Dorylaimoidea) with descriptios of three new species. *Nematologica*. 8:15-20.

Joshi, P.R. (1989). Root rot nematode *Meloidogyne javanica* on groundnut. M.Sc. (Agri.). Thesis, Submitted to Gujarat Agricultural University, Sardar Krushinagar (GS).

Mahajan, R. and Chhabra, M.K. (1979). Occurrence of plant parasitic nematodes in Punjab. India. *PANS*. 25:46-49.

Patel, B.A., Patel, D.J., Sharma, S.B., Patel, H.V. and Patel, S.K. (1996). Nematode problems of groundnut and their management in Gujarat, India. *International Arachis News letter*. 16:38-39.

Patel, D.J, B.A. Patel, J.C. Chavda and H.V. Patel. (1988). Record of *Meloidogyne javanica* on groundnut in Gujarat, India. *International Arachis Newsletter*. 3:16-17.

Patel, D.J., B. A. Patel and H.V. Patel (1993). Pathotypes of *Meloidogyne javanica* in India. *Nematologia Mediterranea*. 21:207-8.

Patel, H.R. (1983). Studies on root-knot nematodes of groundnut Ph. D. Thesis, Gujarat Agricultural University, Sardar Krushingar, Gujarat, pp125.

Patel, H.R., M.U. Vaishnav and I.U. Dhruj (1985). Interaction of *Meloidogyne arenaria* and *Fusarium solani* on groundnut, *Indian Journal of Nematology*. 15:98-99.

Patel, H.V., Patel, D.J. and Patel, S.K. (1998). Management of root knot nematodes with seed treatment and soil application of granular nematicides in groundnut fields. Nematoalogical society of India, National Symposium held at BACA, GAU, Anand. pp.56-57.

Patel, S.K. (1998). Ph.D. Thesis, Gujarat Agricultural University, Sardar-Krushinagar, Gujarat.

Patel, S.K. and Patel, D.J. (1998). Determination of threshold level of *Meloidogyne javanica* pt2 attacking groundnut cv. GG 20. Nematological Society of India, National Symposium, held at BACA, GAU, Anand P. 82-85.

Patel, S.K., Patel, D.J. and Talati, J.G. (1998). Nematolaogical Society of India. National Symposium held at BACA, GAU, Anand P. 76-78.

Porter, D.M., Smith, D.H. and R. Rodriguez Kabana (1984). Compendium of Peanut Diseases. *American Phytopathological Society*. P. 38-41.

Prasad, S.K., Dasgupta, D.R. and Mukhopadhyaya, N.C. (1964). Nematodes associated with commercial crops in Northern India and host range of *Meloidogyne javanica* (Treub,1885) Chitwood 1949. *Indian Journal of Entomology*. 26:438-446.

Sakhuja, P.K. and C.L. Sethi. (1986 a). Multiplication of *Meloidogyne javanica. Indian Journal of Nematology*. 16:23-26.

Sakhuja, P.K. and Sethi, C.L. (1986 b). *Indian Journal of Nematology*. 16:1-3.

Sen, A.C. (1960). *Indian Journal of Entomology*. 22:261-64.

Sharma, N. K., Sharma, S.K. and Midha, (1978). *Indian Farming*. 28:19-21.

Sharma, S. B. (1985). A world list of nematodes Pathogens Associated With Chickpea, groundnut, pearl Millet, pigeonpea and sorghum. ICARISAT, Hyderabad. Pp 20.

Sharma, S. B. and D. McDonald. (1990). A world list of plant-parasitic nematodes associated with groundnut. *International Arachis News Letter*. 7: 13-18.

Steiner, G. (1919). Nematodes, the grower should know. Soil Science Society & Plant Proceedings, 4/B:72-117.

Vaishnav, M.U. and I.U. Dhruj. (1985). *Indian Journal of Nematology*. 15:235-36.

Vaishnav, M.U. and I.U. Dhruj. (1981). Nematological Society of India, Symposium held at TNAU, Coimbatore on Feb. 13-15.P.33.

□□□

16

Rodents

H. J. VYAS AND T.V.PRASAD

16.1. Introduction

Good amount of research information is available on the insect pests of groundnut, however, limited information is available with regard to the extent of losses caused by the rodents in field and storage, their species composition, population dynamics and control measures in-spite of the fact that the damage to groundnut, rodents is a matter of concern at times. In India groundnut is attacked by a number of rodent species, such as *Bandicota bengalensis,* **(Plate 1)** *Millardia meltada, Tatera indica, Rattus meltada, Mus booduga, Mus platythrix, Mus musculus* and *Fenambulus pennanti* in Gujarat, Andhra Pradesh, Tamil Nadu, Karnataka, Maharashtra, Madhya Pradesh and Punjab, which covers about 85 per cent of groundnut growing area of the country.

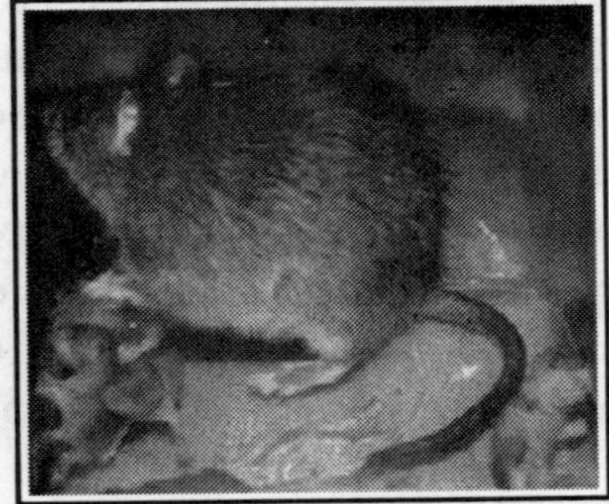

Plate 1. Rats

16.2. Nature of damage

Rodent damage in groundnut growing areas occur from the sowing to harvesting of the crop. They inhabit bunds, barren/waste lands and in stored fodder for cattle for use in off season. After the sowing of groundnut, the rodents come to the sown area, dig out the seeds and consume them. This loss is comparatively more, when there is less rain and the fields are in a dry condition after the sowing operation. With receipt of rains, the seed/seedling damage by the rodents decreases due to water logging of the field. Later on, the field rodents damage the groundnut crop by cutting the branches and remain in the burrows on the bunds **(Plate 2)**. Farmers avoid the interculturing operations after peg formation in groundnut. This undisturbed condition of groundnut fields enhance the rodent activity and at the same time crop canopy also plays an important role as it provides shelter to them. Thus, the activity of field rodents increases after pod formation in groundnut and maximum damage is found at pod maturity. The population of field rodents and damage due to them also increase when un-threshed groundnut crop is kept in the field for drying after its harvest. At this time rodents feed on the groundnut pods showing half eaten pods with exposed kernels and/or stored them in the burrows for their future need.

Plate 2. Damage by rats

16.3. Species composition

The species composition of rodents varies from one habitat to another because of their habitat specificity. Very often the same species may be found in two different habitats, but their relative abundance may not be the same in these two locations. Neelanarayanan *et al.* (1995) studied the species composition of rodents in groundnut fields of Nagapattinam area, Tamil Nadu, and the extent of their depredation during vegetative, budding and flowering, pod formation and pod maturation stages by diagonal and quadrat methods. Of the three species (including *Mus booduga*), *Bandicota bengalensis* was the predominant pest followed by *Millardia meltada.* Damage occurred during the later stages of crop development, pod formation (11.8% damage observed using both methods) and pod maturation (21.6% and 19.7% for diagonal and quadrat methods, respectively). The species composition of rodents in different states is given in the table 16.1. List of rodents in residential areas and Houses and go downs and warehouses are given in Table 16.2 and 16.3 respectively.

Table 16.1. The species composition of rodents recorded on groundnut in different states

State	Predominant species	Reference
Gujarat	*Bandicota bengalensis* Kok (Gray)	Anon, 1989
	Millardia meltada meltada (Gray)	"
	Tatera indica indica (hardwicke)	"
Karnataka	*Bandicota bengalensis* Kok (Gray)	Anon, 1988
	Rattus meltada (Gray)	
	Tatera indica indica (Hardwicke)	
	Mus platythrix Bennett	
Madhya Pradesh	*Millardia meltada* (Gray)	Anon,1984c
	Bandicota bengalensis Kok (Gray)	"
	Funambulus pennanti Wroughton	"
Punjab	*Mus musculus* Linnaeus	Sood and Ubi,1975
	Mus booduga (Gray)	"
	Tatera indica (Hardwicke)	"
	Bandicota bengalensis Kok (Gray)	"
	Rattus meltada (Gray)	"
Tamil Nadu	*Mus booduga* (Gray)	Neelanarayanan *et al.*,1995
	Bandicota bengalensis Kok (Gray	"
	Millardia meltada (Gray)	"

Table 16.2 List of Rodents in residential areas and Houses

S. No.	Species	Reference
1.	*Funambulus pennanti*	Rina Chakrabaorty, 1993; Chopra et al., 1996
2.	*Rattus rattus*	Rina Chakrabaorty, 1993; Chopra *et al.*,1996
3.	*Bandicota indica*	Rina Chakrabaorty, 1993; Chopra *et al.*, 1996
4.	*Mus musculus*	Rina Chakrabaorty, 1993; Chopra *et al.*, 1996
5	*Bandicota bengalensis*	Rina Chakrabaorty, 1993;
6	*Tatera indica*	Rina Chakrabaorty, 1993;
7	*Mus cervicolor*	Rina Chakrabaorty, 1993;

Table 16.3 List of Rodents in godowns and warehouses

S. No.	Species	Reference
1.	*Rattus rattus*	Rina Chakrabaorty, 1993; Chopra *et al.*, 1996
2.	*Bandicota bengalensis*	Rina Chakrabaorty, 1993; Chopra *et al.*, 1996
3.	*Bandicota indica*	Rina Chakrabaorty, 1993; Chopra *et al.*, 1996
4.	*Mus musculus*	Rina Chakrabaorty, 1993; Chopra *et al.*, 1996
5	*Tatera indica,*	Rina Chakrabaorty, 1993;
6	*Funambulus pennati*	Rina Chakrabaorty, 1993;
7	*Mus cervicolor*	Rina Chakrabaorty, 1993;

16.4. Relative abundance

The relative abundance of rodent species may vary with soil types, seasons, irrigation facilities and other ecological conditions of different locations. Our studies indicated the predominance of *Bandicota bengalensis* followed by *Millardia meltada* and *Tatera indica* in almost all types of soils in groundnut growing areas of Gujarat during *kharif* and summer seasons except in coastal alluvial, medium black and shallow black soil in the *kharif* season and in mixed red and shallow black soil in the summer season. Due to the diverse ecological conditions, the rodents species may vary with soil type, seasons, irrigations, and ecological conditions (Prakash, 1975 & 1981). In Madhya Pradesh, soft furred rat *M. meltada* was predominant (Anon, 1984b). Rodent pest species in groundnut fields around Ludhiana were *Tatera indica* 4.2 and 15.0; *Bandicota bengalensis* 6.5 and 0; *Rattus meltada* 4.2 and 7.5; and *Mus* spp., 13.9 and 17.5, resp., in irrigated and non-irrigated fields (Parshad *et al.*, 1987). In Punjab, 43% population of rodents belong to *M. meltada* while in Gujarat, *B. bengalensis* was found to be the predominant species (upto 60%) followed by *M. meltada* or *T. indica* (Anon, 1990b). Type of irrigation appears to influence the relative abundance of rodents. *M. meltada* was found comparatively higher in number than *T. indica* in groundnut growing areas having canal irrigation facility in Gujarat, whereas the population of *T. indica* was higher than that of *M. indica* where well water irrigation was carried out during summer season (Mittal and Vyas, 1992).

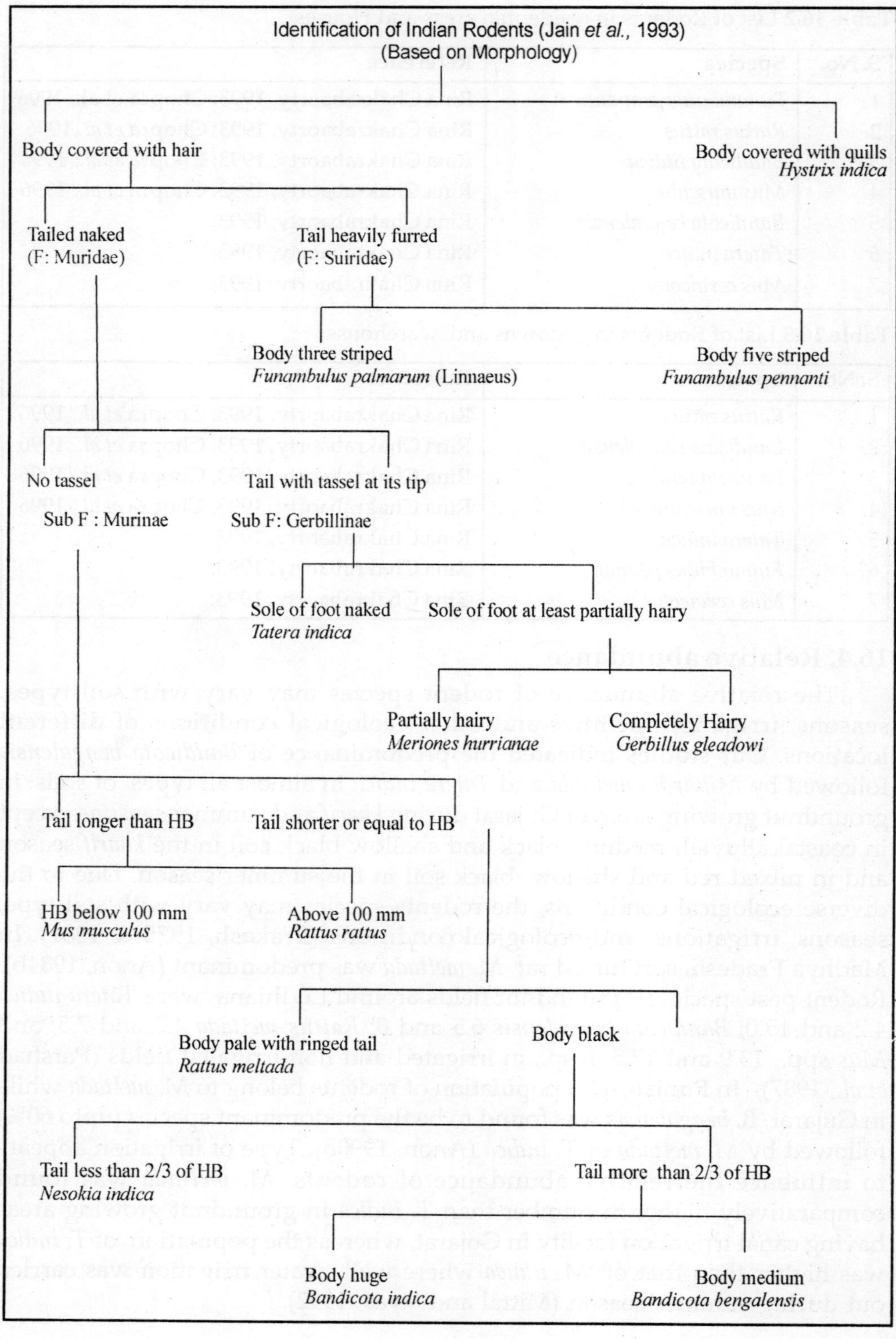
Identification of Indian Rodents (Jain et al., 1993)
(Based on Morphology)
Body covered with hair
Body covered with quills
Hystrix indica
Tailed naked
(F: Muridae)
Tail heavily furred
(F: Suiridae)
Body three striped
Funambulus palmarum (Linnaeus)
Body five striped
Funambulus pennanti
No tassel
Sub F : Murinae
Tail with tassel at its tip
Sub F: Gerbillinae
Sole of foot naked
Tatera indica
Sole of foot at least partially hairy
Partially hairy
Meriones hurrianae
Completely Hairy
Gerbillus gleadowi
Tail longer than HB
Tail shorter or equal to HB
HB below 100 mm
Mus musculus
Above 100 mm
Rattus rattus
Body pale with ringed tail
Rattus meltada
Body black
Tail less than 2/3 of HB
Nesokia indica
Tail more than 2/3 of HB
Body huge
Bandicota indica
Body medium
Bandicota bengalensis

Population dynamics

Population dynamics of rodents species carried out in Gujarat during summer and *kharif* seasons of 1992 indicated that *Bandicota benaglensis* Kok was trapped throughout the summer season, whereas, *Millardia meltada meltada* was observed from second fortnight of February and *Tatera indica indica* was observed from second fortnight of March. Thus, *B. bengalensis* was found predominant species. The pest was more active from pod formation to harvest stage of the crop and highest number of rodents were trapped during the harvest stage of crop. Almost similar trends were also observed in burrow census of these three species. In all the three species, the male was found predominant. The overall male female ratio was 58.70:41.30. During *kharif B. bengalensis* Kok and *M. meltada meltada* were observed throughout, amongst which *B. bengalensis* Kok was found predominant one. The third species, *T. indica indica* was observed from first fortnight of September to harvest stage of the crop. The rodent activity was increased from pod formation in groundnut and was highest at harvest stage of the crop. Similarly the results of burrow census also showed that the highest number of live burrows were also found at the harvest stage of the crop. The male was also found predominant in all three species as the overall male female ratio was 61.34:38.66 (Anon., 1992-93).

16.5. Biological characteristics of rodents

The biological characteristics of rodent are important so that based on their biology and behavioral aspects the management strategy can be planned. The rodents have well developed sense of smell, hearing and touch. They are good swimmers. They are colour blind but distinguish shades. Usually field rodents are mostly burrowing types, however, squirrels and rodents in plantation crops are arboreal. For movements they always use runways. These rodents transmit several diseases. They are mostly nocturnal in habit and cannot vomit the poisons taken in. They exhibit both omnivorous/cannibalism. They are also Neophobic and neophilic and highly adaptive to the situation. The life span is between 1-2 years. The incisors grow very fast @ 0.33 mm/day or 12 cm/year. The female attain puberty between 6-16 weeks and has oestrous cycle for 3-7 days. The heat period last for 9-14 hours and has the promiscuous mating habit. The gestation period last for 18-30 days. The breeding season is for all through the year. Each female gives about 1-22 litter size and varies with sps. The post-partum heat is for 4-96 hours. Under the conducive atmosphere, one pair of rodent may become 800-1200 rodents in a year.

16.6. Economic importance and extent of loss

Rodents constitute the largest order of existing mammals. In its multiplicity of taxa and in the enormous swarm of individuals, this group stands out among all the mammals. Bindra and Sagar (1968) reported that loss due to rodents in four villages varied from 30 to 76 kg/ha. Parshad *et al.* (1987) reported a partial loss due to damage to plants up to 9% and most damage occurred 80-120 days after sowing. Prakash and Mathur (1987) indicated that the per cent loss in groundnut due to rodents varied from 4.1 to 25.8.

Sudden multiplication of rodents, and their population assuming menacing numbers, happened in Gujarat during the year 1975-76 and 1989-90. Rodent damage in field crops increased from *kharif* 1988 to *rabi* 1989-90 and the extent of yield loss in one isolated field surrounded by barren/fallow land was as much as 85.42 per cent during summer, 1989. Because of heavy damage due to rodents in *rabi* crops of 1989-90, farmers avoided sowing the summer groundnut during 1990 in most of the groundnut growing areas of the Saurashtra region of Gujarat. The loss due to rodent damage in groundnut crop in Gujarat has been estimated to the tune of Rs.243, Rs.815, and Rs.669 per hectare during the year 1988, 1989 and 1990, respectively (Mittal and Vyas,1992).

Pradhan (1980) reported that total damage caused by *Rattus rattus, Bandicota bengalensis* varied from 5 –25 %. Ranjan Advani and Mathur (1981) reported that as a result of three year of rodent control, the field rodent population was reduced by 95% and losses to various vegetables, wheat and bajra crops declined by 88%. The cost benefit ratio was worked out to be 1 : 900. However, in the residential premise, the rodent population was reduced by 85% with a corresponding decline in losses to stored grain, vegetables etc and the cost benefit ratio being 1 : 220. Jain *et al.* (1993) reported that *R. rattus* and *M.musculus* inflinct cause serious losses to different food grains at farmers level or godowns and warehouses. The post harvest losses due to rodents may amount to 25-30 per cent. Other sources quote 1-12.5 m tones. In a 100 x 100 m^2 space in godown at Howrah, the loss of grains due to rodents was reported to be 4200 kg. The loss due to rodents in groundnut is given in the table 16.4.

16.7. Management techniques

There are several methods of rodent control which are needed to be integrated in a manner to manage the rodent population to 5-10 per cent of its original level, because a single method may not achieve this level of success. It is more or less impossible to eradicate a species which is a prolific breeder. The common methods used for the management in India are summarized below.

16.7.1.Trapping

It is one of the oldest practice of rodent control reported from earliest civilization. There are many types of traps in use today. Some commonly

Table 16.4. Losses due to rodents in groundnut in different states

State	Stage	% damage	Reference
Andhra Pradesh	Seedling	6 to 9	Jotwani and Beri, 1968
	Plants	18.2 to 25.8, 14.3	Rajasekharan and Dharmaraju, 1975
	Pod	4.1 to 7.6, 47.2	"
Gujarat	Seedlings	4 to 51.9, 4.4 to 9.5	Anon, 1990b
	Plants	3 to 10.2, 2.9 to 26.8 3.3 to 7.4	Anon, 1989 Anon, 1990b
	Pod	4.9 to 10, 8.2 to 14.6 5 to 8.5	Anon, 1989 Anon, 1990b
Karnataka	-	2 to 5	Hopf *et al.*, 1976
Madhya Pradesh	Plant	8.8 to 17.7	Anon, 1984
	Pod	4.6 to 28.6, 12.7	Anon, 1984
Punjab	-	25.8	Hopf *et al.*, 1976
	Pod	7.1	Anon, 1984b
		3.9 to 19	Parshad *et al.*, 1987
Rajasthan	-	18	Anon, 1984a
Tamil Nadu	-	1 to 1.5	Hopf *et al.*, 1976
	Pod	11.8 to 21.6	Neelanarayanan *et al.*, 1995

used traps are sherman traps, wonder traps, snap traps and other locally made traditional traps. Among these traps, wonder traps are found effective. Mostly adults can be trapped in this method. It can be an effective mean for control of small population. It can also be used as follow up action after chemical control operation. Peanut butter or any other locally prepared bait may be used for these traps. These traps should be placed on rodent runways.

Muktha Bai and Majumder (1989) evaluated the glue trap supplied by Ecosafe Traps (India) Ltd., Bombay for their efficacy in food stores, houses and other complex environment and stated that roof rat, *Rattus rattus* and hose mouse, *Mus musculus* come to the traps. However, mice were trapped more easily than rats. In one trap a maximum of 14 mice were caught within 14 hours. Comparatively, rats took little longer time to come to traps and the maximum catch was observed when traps were left overnight and undisturbed. Neelnarayanan *et al.* (1994) reported that were 30 rodents were trapped in pit fall traps with in three days of which 16 were *Bandicota bengalensis*, 10 *Millardia meltada* and *Mus booduga*. Mathur *et al.* (1994) reported that *Rattus ratus* and *Mus musculus* were easily trapped in sicky or glue trap supplied by Eco safe trap (India) Ltd.. However, mice were trapped more frequently than rat due to their fast movement and less realization for trap laid. Ranga Reddy (1994) designed and developed a device for the subterranean rodent in paddy field. The unit consists of hollow cylinder of 10″ diameter and 8″height made of 15 guage M S sheet. There is an inlet on the top made of 1.5″ G.I. pipe of 4″ length . The outlet is at the bottom made of 3.8″ G.I. pipe of 5″ length . Handle is provided at the side to carry the unit from place to place. The cylindrical portion of the unit is stuffed with paddy straw through the outlet and ignited. As the traps catches fire, air is blown slowly by rotating the hand blower inserted in to the top inlet. The smoke enters in to the burrow and within a short time all the tunnels of the burrow are filled with smoke which kill all the rats with their litters inside. Neelnarayanan *et al.* (1995) reported that three species of rodents viz; *Badicota bengalensis, Millardia meltada and Mus booduga* were successfully trapped in Tanjore bow trap and it is simple, safe and effective way of trapping rodents.

Godowns

Pradhan (1980) reported that total damage caused by *Rattus rattus, Bandicota bengalensis* varied from 5 –25 % of the total cost. One nicely constructed godown of the capacity of 65000 bags neatly arranged where proper care was taken to prevent the entry of rodents and which simultaneously gave contract to a private company showed average damage 5% of total cost. Razvi *et al.* (1981) reported that mechanical method was more effective followed by fumigation with Aluminum phosphide and poison baiting with zinc phosphide . It was suggested that integration of methods can effectively manage the rodent pests during the period between *kharif* harvest and *rabi* sowing.

Residential areas

Yashoda L. Urs (1996) used an ultrasonic unit which emitted a variable (24-28 KHz) intermittent noise with an intensity of about 120 db were successfully promising for the control of *Rattus rattus*. Similarly, results on the behavioral studies through remote sensing system indicated that ultrasound unit installed 2-3 m distance repelled rats for 2-3 days. Rats, when scared and diverted by ultrasound, in the process of running away were trapped near rat run and also while hiding under cover and getting inside the card board station. Rana (1998) recommended the use of local bamboo traps be encouraged both in crops and homestead.

Botanicals

Neelu Singh *et al.* (1995) reported that rats died within 3-8 days after consumption of 3-5 jetropha, *Jetropha carsus* seeds per rat.

16.7.2.Biological control

Biological control is suppression of a post population either by increasing predation or by introducing diseases.

(a) **Predation:** Cats, owls, snakes, fox, Mongoose and Kites are the major predators of rodent. These predators are useful in maintaining the natural ecological balance but do not prove to be an effective tool for rodent management because these predators are opportunist animals and prey on whatever is easily available.

(b) **Diseases:** Salmonella disease caused by *Salmonella typhimurium* and *S. enteritidis* were successfully used against mice, voles and rats in foreign countries. In India, 16-18 per cent kill in house rat and bandicoots were reported with this bacterium. The use of pathogens in rodent control has very little practical value on one side and has great potential for public health hazards on other side. That is why WHO and FAO have stated its least usefulness in rodent management.

Rina Chakrabaorty and Roy (1993) reported that *Rattus rattus,* were regularly sacrificed and first helminth infecion was recorded from 45 days onwards. However, only 10% sub-adults were found infected as against 92 % adult rats. The helminthes, *Cysticerus fasciolaris* and *hymeolepis diminuta* were identified from the liver and intenstine of rats , respectively.

Residential areas

Madhu Bhadauria *et al.* (1995) reported that majority of people used biological means i.e. cat. For the control of *Rattus rattus, Mus musculus* and *Bandicota bengalensis* can be more effective and successful if control operations adopted on a whole village or mohalla basis.

16.7.3 Habitat manipulation or Environmental Management

In any habitat food, shelter and water are major governing factors in rodent environment. By removing any of these factors, one can reduce the carrying capacity of the habitat holding the rodents. The reduced food and shelter will increase intra and inter specific competition and aggressiveness among rodents which may result into decrease in population either through mortality or migration.

These methods are very easy, effective and require no extra expenditure hence can be easily adopted.

(a) **Deep ploughing :** Deep ploughing of the fields at the time of land preparation helps in destruction of rodent burrows which exposes the newly born to the predators and adults migrate to other areas.

(b) **Reduction in bund size :** Rodents inhabit high bunds present around the crop fields, which remain undisturbed. Such bunds should be kept at minimum possible level to reduce rodent infestation.

(c) **Planting of non-preferred crops :** It is suggested that if *Opuntia* plantings are done on the bunds, rodents population decreases considerably. Similarly a band of low preferred crop like cluster bean which is difficult for rodents to de-husk may be grown in 5-10 m strip around main crop, can also reduce the entry of rodents.

(d) **Weed management :** Weeds serve as food of rodents during lean period. They also act as hiding places. Regular weed management practices have been reported to reduce the rodent infestation, significantly.

16.7.4. Chemical Control

Among various methods of rodent control, chemical control is the most common expedient and humane. Following types of chemicals have been effectively used for the control of rodent management.

(a) **Chemosterilants:** Researchers have evaluated a number of chemosterilants to produce sterility among rodents. The use of such agents against rodent pests have yet to be studied in greater detail before being introduced for public use. Rodents being mammals, such sterility agents may pose great threat for human beings and other non-target species. Moreover, the vertebrate pest control through chemosterilant seems less practicable because, the sterile individuals will present and exert pressure on the crop and lands, because of their longer life and hence it is not feasible to wait for their life span before the noxious individuals are removed.

(b) **Repellents** : A number of compounds are found repellent to rodents. Malathion and cyclo hexamides were effectively used against Norway rats in U.S.A. More studies are needed and at this stage it is not possible to make any firm recommendations.

(c) **Rodenticides** : An ideal rodenticide should have three basic characteristics viz., toxicity, acceptability and safety in management operations.

Rodenticides are mainly classified as : 1) Acute poison (single dose and quick acting) eg. Zinc phosphide and aluminium phosphide and (2) Chronic poison (single or multiple dose and slow acting poisons) eg. warfarin, fumarin (first generation multiple dose anticoagulants) and bromadiolone, brodifacoum, flocoumafen and diphethialone (second generation single dose anticoagulants). The major drawback with the acute poison, being the induction of bait shyness behaviour in the rodents and prebaiting requirement makes the problem of rodent management.

"Bait-shyness" can then be mitigated or eliminated by using for baits: (i) cereals in an alternative textural form; (ii) cereal baits blended with strong-tasting substances like chocolate; (iii) cereal mixed with groundnut oil; (iv) or an equivalent wt/ wt mixture of cereals containing groundnut oil (5%) (Naheed and Khan, 1989).

Residential areas

Shamsuddin and Koya (1983) reported that with 50g of 0.005% bromadiolone bait altogether 18 *Mus musculus* and 6 *Rattus rattus* were found dead from the 3rd day to 6th day of poison baiting. The maximum rodents died on 5th day. The results indicated that all the houses were relieved of rodent nuisance from the fourth day of poisoning. Further, it was also found that both the species of rats preferred bromadiolone bait than the available food articles in the house hold. Dwivedi *et al.* (1990) reported that Ratobar caused 90% mortality of *Rattus rattus,* with an average consumption of 185 mg/kg. Contradictorily, the mortality was quite low under choice test, showing poor acceptability of the cake. The average consumption of poison during first 20 days was 23.3 mg/kg resulting in 50% mortality. AAT the end of 60 days, it was evident that plain food (wheat floor) was highly preferred to Ratobar cake.

Poor acceptability of Ratobar cake was further manifested when these were placed in 40 houses of a village having an estimated rodent population of 3.97 rat/house. This treatment resulted in only 25.8% control success leaving about 3 rats/house uncontrolled. Thus, it may be concluded that Ratobar cake is not an effective formulation for containing the rodent menace under field situation. Sharma (1994) placed three plates at hunt sites of the mouse *Mus musculus, Rattus rattus,* Powdered Makhan, Bad Sweet and lightly sprinkled over that. One mouse was located dead in three hours at a pot, two other mice and large female house rat (*Rattus rattus*) were observed dead in the early morning the next day. Rana *et al.* (1994) reported that rodents were killed twice a year in the month of March and September. The poison treatment were carried out with 0.26% warferin for 15 days.

The potential of barium carbonate as a safe rodenticide for use in animal and human dwellings was evaluated against *Rattus rattus* in feeding and pen trials. At a concentration of 10% in a bait prepared in a mixture of wheat flour, sugar and groundnut oil (96:2:2), barium carbonate caused 100% mortality of *R. rattus* in 48-h no-choice and choice feeding trials. In pen trials, baiting with 10% barium carbonate for 2 days caused 55% mortality of *R. rattus* compared with 100% with 2% zinc phosphide bait for 1 day and 65% with 0.005% bromadiolone bait for 3 days. Most of the rats poisoned with barium carbonate died within 3-5 days in the open field of the pen and were easily retrieved. Rats poisoned with zinc phosphide died within 1-3 days and were located in both the open and concealed areas of the pen. The period of mortality of the rats after poisoning with bromadiolone ranged from 4 to 15 days and most of them died in concealed locations (Malhi *et al.*, 1993).

Rodenticide baiting and trapping are the common methods for the control of the house rat, *Rattus rattus,* which is a predominant commensal and agricultural pest throughout the tropical and subtropical regions of the world. A synthetic form of carbon disulphide (CS_2), attracted both male and female *R. rattus* in bio-choice feeding tests with individual rats and pen experiments with groups of 10 rats and also in field experiments in poultry house premises. Addition of 1% CS_2 in 2% zinc phosphide, 0.005% bromadiolone and 0.0025% difethialone baits significantly improved bait acceptance by *R. rattus.*

(Corresponding values of per cent acceptance of scented bait over the unscented bait were 77.3, 74.2 and 65.9.) Similarly a greater number of rats were trapped with cereal bait containing 1% CS_2 than without it in pen and field experiments. In poultry houses CS_2-scented bait was effective in improving the trapability of rats during the first 3 days, after which it did not have a significant effect on the residual population (Parshad, 2002). The variation in food consumption of rodents are dependent on the type of food material and species of grain available. It is suggested that wheat flour can be used as baiting material for the control of rat populations, especially in agricultural areas (Bajpai, 2002).

Rodent Management In Groundut

Rodent damage of even less than 1 per cent justifies rodent control operations in groundnut. Baiting with Zinc phosphide in Punjab and fumigating the burrows by Aluminium phosphide in Punjab and Karnataka meet with fair success (about 50%) in rodent control in groundnut fields. Bromadiolone 0.005% bait gave 92.01 and 95.49 per cent control success at 3 and 5 days after its application, respectively, in summer groundnut in Gujarat State. In irrigated fields of Punjab, 58 and 67 per cent rodent control in groundnut was achieved with Zinc phosphide and bromadiolone baits, respectively. A single treatment in non-irrigated fields 80-90 days after sowing with 2.4% Zinc phosphide, 0.005% brodifacoum or 0.005% bromadiolone in cereal baits at 1 kg/ha resulted in 58, 42 and 41% rodent control, resp. In irrigated fields, 58 and 67% rodent control was achieved with Zinc phosphide and brodifacoum baits respectively. Significantly higher rodent control was obtained with 2 treatments of either brodifacoum or bromadiolone than with a single treatment of any rodenticide. Wax blocks containing 0.005% brodifacoum were less effective than cereal baits containing the same rodenticide. Two applications at 10-day intervals of either 0.005% brodifacoum or bromadiolone between 80-100 days after sowing was suggested for rodent control in groundnut fields (Parshad *et al.*, 1987). Vyas *et al.* (1985) revealed that the rodent population in groundnut crop can be effectively managed by two applications of bromadiolone 0.005% wax cake, first at the time of flowering and second at pod maturity or first application of zinc phosphide 2% poison bait at the time of flowering and second application of bromadiolone 0.005% wax cake at pod maturity each @ 10 g poison bait/live burrow. Prashad and Mahli (1995) stated that 0.75% tracking powder and 0.037% bait of Racumin (coumatetralyl) were suitable for the control of *B. bengalensis,* although efficacies varied under different field condition.Given a choice, the Indian mole rat, *Bandicota bengalensis* preferred rice over other grains such as wheat, barley and maize in the laboratory. However, its preference and consumption of millet was statistically at par with that of rice. Addition of 2% groundnut oil and 5% cane sugar increased the palatability and consumption of the grains (rice and millet). Moist and boiled rice, and boiled millet were preferred over their dry baits, and were ingested in significantly larger quantities. In the fields, boiled rice (plain or sweetened with 5% cane sugar) was accepted the most, followed by boiled millet; the consumption of the two differing significantly. Amongst the dry cereal bait, rice smeared with 2% groundnut oil and having 5% sugar was the most accepted bait material (Chander-Shekher, 1998).

Vyas *et al.* (2000) evaluated four rodenticides for their efficacy against field rodents in groundnut during four years from 1989-90 to 1992-93 and concluded that two application of bromadiolone 0.005% wax cake, first at the time of flowering and second at pod maturity stage (Cost benefit ratio 1:49.5) or first application of zinc phosphide 2% poison bait at the time of flowering and second application of bromadiolone 0.005% wax cake at pod maturity stage (cost benefit ratio 1 : 24.8) each @ 109 poison bait/live burrow are recommended for effective rodent management in groundnut crop. In another study Vyas *et al.* (2000) compared three rodenticidal combinations of two existing rodenticides for their efficacy against field rodents in groundnut during *kharif* season from 1993-94 to 1996-97 and concluded that two applications of bromadiolone 0.005% wax cake, first at the time of germination and second at pod formation stage (ICBR 1 : 52.08) or first application of zinc phosphide 2% poison bait at the time of germination and second application of bromadiolone 0.005% wax cake at pod formation stage (ICBR 1: 22.71) each @ 10 g poison bait/live burrow are recommended for effective rodent management in groundnut crop.

Farmer's motivation through training and education for rodent management and mass control campaign under experts supervision was found to have great impact on reduction in rodent population as well as crop damage in groundnut crop. A Pilot Project was undertaken and operated in 12 districts of Gujarat covering two village in each talukas. In all there were 232 villages covered under this pilot project. The mass awareness were also made by using different mass media like news paper, radio, television and also by imparting training to extension staff and farmers. In mass control campaign organized in these villages under this pilot project, overall 69 per cent control was observed with the use of zinc phosphide in the first phase control and for residual control overall 84 and 80 percent control was observed with the use of aluminium phosphide and bromadiolone, respectively. The total overall 95 per cent reduction in rodent population was noticed in this campaign.

16.8 Conclusion

Significant progress has been achieved on the biological and ecological studies on rodent species which damage groundnut crop and on efficacy evaluation of different rodenticides. Limited literature is available on the threshold levels which is essential feature of pest management as to keep the damage control operations under economically viable condition. For effective rodent control over large area, effective population reduction methods and a decision making process are essential. Population reduction is practical only through environmental manipulation and rodenticide application. These should be considered as complimentary rather than alternative approaches. However, some thrust is required to be given on monitoring of rodent pests, interaction among farmers of adopting the technology and on dissemination of the available technology to the policy makers at different levels in order to make rodent control one of the integral part of modern agriculture.

References

Anonymous, (1992-93). Annual Report, AICRP on Rodent Control, Central Arid Zone Research Institute, Jodhpur. pp. 81

Anonymous, (1984). Annual Progress Report (1983-1984). of AICRP on Rodents Control. (Ludhiana: Punjab Agricultural University).

Anonymous, (1984a). Annual Progress Report (1983-1984) of AICRP on Rodents Control. (Ludhiana: Punjab Agricultural University).

Anonymous, (1988). Annual Progress Report (1987-1988) of AICRP on Rodents Control. (Bangalore: University of Agricultural Sciences).

Anonymous, (1989). Annual Progress Report (1987-1988 and 1988-1989). of AICRP on Rodents Control. Junagadh: Gujarat Agricultural University.

Anonymous, (1989-1997). Annual Progress Reports (1988-89 to 1996-97) of AICRP on Rodent Control. Gujarat Agricultural University, Junagadh.

Anonymous, (1990). Annual Progress Report (1989-1990) of AICRP on Rodents Control. (Junagadh: Gujarat Agricultural).

Anonymous, (1990). Rodent control measures in Gujarat-Report of Expert Team, (Ahmedabad: Directorate of Agriculture, Gujarat State.

Anonymous, (1986). Annual Progress Report (1983-1984) of AICRP on Rodents Control. (Jabalpur: Jawalal Nehru Krishi Vishwa Vidyalaya).

Bajpai, R. (2002). Feeding preference of house rat (*Rattus rattus*) in Plus Maze. *JNKVV-Research Journal.* 35(1/2): 86.

Bindra, O.S. and Sagar, P. (1968). Study on the losses to wheat, groundnut and sugarcane crops by the field rats in Punjab. *Proceedings of International Symposium on Bionomics and Control of rodents*, Kanpur 28-31.

Chander Sheikher. (1998). Selection of bait material for rodenticide application. *Pest Management and Economic Zoology* 6(1): 7-14.

Chopra, G., Kaur, P. and Guraya, S. S. (1996). Rodents: Ecology, biology and control. R. Chand and Co., New Delhi.

Dwivedi, P. K., Agarwal, R. K. and Bansode, P. C. (1990). Evaluation of Ratobar (Fumarin) against house rat. *Rodent Newsletter* 14: 10-11.

Hopf, H. S., Morley, G. E. J. and Humphries, J. R. O. (1976). Rodent damage to growing crops and to farm and village storage in tropical and sub-tropical regions (London: Centre for Overseas Pest Research and Tropical Products Institute).

Jain, A. P., Tripathi,R. S. and Rana, B. D. (1993).Rodent management–The state of art. Project Coordinator, AICRP on Rodent control (ICAR), Jodhpur. pp.38

Jotwani, M. G. and Beri, Y. P. (1968). Economic losses due to rats. *Proceedings of International Symposium on Bionomics and Control of rodents*, Kanpur: 9-16.

Madhu Bhadauria, Neelam Kanwar and Bhadauria, A.S. (1995). Survey of rodent menace and their control in the residential area. *Rodent News letter* 19 (1):10.

Malhi, C. S., Parshard, V. R. and Ahmad, N. (1993). Determination of potential of barium carbonate for the control of house rat *Rattus rattus. Zeitschrift fur Angewandte Zoologie.* 80(1): 42-49.

Mann, G. S. and Bindra, O. S. (1978). Food consumption and food choice in *Mus* spp. in the laboratory. *Bulletin of Grain Technology*. 16(1): 10-14.

Mathur, Y. K., Bhadauria, A. S. and Singh, V. N. (1994). Employment of glue trap in the commensal rodent control. *Rodent News letter*, 18 (3): 12.

Mittal,V. P. and H. J. Vyas.(1992). Groundnut *in* Rodents in Indian Agriculture Vol.1. (eds.Iswar Prakash and P.K.Ghosh), 249-264.

Muktha bai, K. and Majumder, S. K. (1989). Eco-safe traps for controlling rodent pests. *Rodent News letter*,13: 7-8.

Naheed, G. and Khan, J. A. (1989) "Poison-shyness" and "bait-shyness" developed by wild rats (Rattus rattus L.). I. Methods for eliminating "shyness" caused by barium carbonate poisoning. *Applied Animal Behaviour Science* 24: 89-99.

Neelanarayanan, P., Nagarajan, R. and Kanakasabai, R. (1995). Intensity of groundnut pod loss due to rodents. *Journal of Ecotoxicology and Environmental Monitoring.* 5(1): 35-38.

Neelnarayanan, P., Nagarajan,R. and Kanakasabai, R. (1994). Pit fall trap for the control of field rodents in cotton fields. *Rodent News letter*, 18 (3): 7.

Neelnarayanan,P., Nagarajan,R. and Kanakasabai, R. (1995). Tanjore bow trap: An eco-friendly tool for rodent management. *Rodent News letter*, 19(2): 8-9.

Neelu Singh, Patel , R.K. and Alka Bhargav. (1995). Toxicity of jetropha against *Rattus rattus. Rodent News letter*, 19 (1) 9-10.

Parshad, V. R. (2002). Carbon disulphide for improving the efficacy of rodenticide baiting and trapping of the house rat, *Rattus rattus* L. *International Biodeterioration and Biodegradation*. 49(2/3): 151-155.

Parshad, V. R., Malhi, C. S., Ahmad, N. and Gupta, B. (1987). Rodent damage and control in peanut fields in India. *Peanut Science*. 14 (1): 4-6

Parshad, V. R., Mahli, C. S., Ahmad, Nand Gupta, B. (1987). Rodent damage and control in peanut fields in India. *Peanut Science.* 14:4-6.

Pradhan, M. S. (1980). Rodent damage survey in twenty five food grains godown at Pune. *Rodent Newsletter,* 4(1),2-3.

Prakash, I. and Mathur, R. P. (1987). Management of Rodent Pests. (New Delhi: Indian Council of Agricultural Research), pp 133.

Prakash, I. (1975). The population ecology of the rodents of the Rajasthan desert, India: *in* Rodents in Desert Environment, Eds I. Prakash and P. K. Gjosh, (The Hague : Dr.W.Junk Publ.b.v. Vrelag Publ.), 75-116.

Prakash, I. (1981). Ecology of Indian desert gerbril, *Meriones hurrianae. Central Arid Zone Research Institute Monograph* No. 10, 1-87.

Prashad, V. R. and Malhi, C. S. (1995). Rodent control with Racumin poison baiting and tracking powder. *Rodent News letter,* 19(2) 13-14.

Purushotham, K. R. and Mohana Rao, A. M. K. (1979). The distribution of field rodents and their burrowing habits around Tirupati (Andhra Pradesh). *Andhra Agricultural Journal.* 26 (3-4): 85-87

Rajasekharan, M. R. and Dharmaraju, E. (1975). Studies on field rodents in Andhra Pradesh; *Proceedings of All India Rodent Seminar,* Ahmedabad, p 54-56.

Rana, B. D. (1998) Recommendation for rodent pest management in NEH region. *Rodent Newsletter,* 22(1-2), 6.

Rana, B. D., Jain, A. P. and Tripathi, R. S. (1994). Fifteen years of coordinated research on rodent control. Technical Bulletin No.3, AICRP on Rodent Control, CAZRI, Jodhpur, pp 66.

Ranga Reddy, A. (1994). Burrow fumigator - A new device for rodent control. *Rodent News letter,*18 (4): 5-6.

Ranjan Advani and Mathur, R. P. (1981).Economics of rodent pest management in rural environment in Rajasthan. *Rodent Newsletter,* 5(1): 7.

Razvi, S. M. A., Gupta, S. P. and Singh, R. A. (1981). Integrated rodent pest management. *Rodent News letter,* 5(1): 6-7.

Rina Chakrabaorty and Roy, A. (1993). Infection of flatworm (*Cysticerus fasciolaris* and *hymeolepisdiminuta)* in *Rattus rattus. Rodent News letter,* 17 (1-2), 9.

Rina Chakrabaorty, (1993) Species composition of rodents in different ecological region of Gujarat. *Rodent News letter,* 17 (1-2): 1.

Shamsuddin, V.M. and Koya,R.I.A. (1983). Bromadiolone in house hold rat control in Agatti Island and Lakshdeep. *Rodent Newsletter* 4: 22.

Sharma, I.K. (1994). Excavating of house plant pot for breeding holes by house mouse (*Mus musculus*) in house. *Rodent Newsletter,* 18(2): 6.

Singh, Y.P., Kumar, D. and Gangwar, S.K. (1994). New record of some rodent species from the states of North Eastern Hill region. . *Rodent Newsletter,* 18 (3): 12.

Sood, M. L. and B. S. Ubi. (1975). Population fluctuations of sympatric murids occurring in the field of PAU Ludhiana: I. Effect of biotic factors. Proceedings All India Rodent Seminar, Ahmedabad. 76-81.

Sood, M. L. and Dilber, D. S. (1978). A note on food consumption and preference of northern palm squirrel, *Funambulus pennanti* Wroughton. *Journal of Research*, Punjab Agricultural University. 15(1): 132-133.

Vyas, H. J.; Vyas, H. N. and Singh P. M. (1985). Evaluation of Bromadiolone - a single dose anticoagulant - against field and domestic rats. *Indian Journal of Entomology*. 47: 45-46.

Vyas, H. J., Kotadia, V. S. and Butani, P. G.(2000) Glimpses of research on rodent management in Gujarat (1988-89 to 1998-99) . AICRP on Rodent Control, Department of Entomology, College of Agriculture, Gujarat Agricultural University, Junagadh. pp.74.

Yashoda L. Urs.(1996). Control of *Rattus rattus* in an aquarium through eco-safe traps and ultasonic repeller – a new approach. *Rodent Newsletter*, 20 (2-4), 9-10.

□□□

Vector-Virus Relationship and Their Management

V. NANDAGOPAL AND M. P. GHEWANDE

17.1. Introduction

The diseases caused by viruses significantly reduce the plants/unit area and the economic yields. All-to-gather 23 viruses of the 62 taxonomically characterized viruses known to infect peanuts naturally or under experimental conditions (Sreenivasulu *et al.*,1991). Among these Peanut bud necrosis virus, Indian peanut clump virus, Peanut mottle virus and Peanut Stripe Virus are economically important virus diseases in India The latter two viruses and cucumber mosaic virus is reported to be seed-borne in nature. During 2000, a new virus disease called peanut streak necrosis disease (PSND) caused by tobacco streak virus was reported on groundnut from Ananthapur area of Andhra Pradesh in India (Padma Raju, 2001). The losses caused by important virus diseases in groundnut ranges from 1.2 % to 100%.

Identification of groundnut virus diseases is a first necessary step in planning management programme. The occurrence of groundnut virus diseases so for documented have been based on symptoms, host range and biological properties, however, characterization of virus should be based on serology, electron microscopy, transmission and physio-chemical properties. Typical symptoms of economically important virus diseases like peanut bud necrosis, peanut mottle virus, Peanut Stripe Virus and peanut clump virus can be used to identify these diseases under field conditions. Vectors play a dominant role in spreading the virus diseases carrying viruses from one plant to another (Reddy *et al.*,1983). The insects/vectors involved in the spread of virus and virus like organisms associated with groundnut mostly belong to the Orders Homoptera, Heteroptera, Coleoptera, and Thysanoptera. It is also reported that a fungus transmits Indian peanut clump disease (Table 1). Special attention should be paid to virus diseases that occur consistently and cause heavy losses, as farmers will directly benefit economically from their control. The information available on different aspects like occurrence, distribution, relative importance, losses caused, symptoms and management of vectors in general and economically important virus diseases in particular are discussed in this chapter.

Table 17.1. Vectors of virus diseases of groundnut (After Sreenivasalu et.al.,1991)

Virus group	Virus	Distribution	Vector	Symptom	References
Bromovirus	Cowpea chlorotic virus	U.S.A	*Creatoma ruficornis, C. trifurcota, Diabrotica balteota, D. undecimpunctata* (Coleoptera)	SI	Kuhn and Demski,1987
Carlavirus	Cowpea mild mottle	India, Ghana, Indonesia Ivory Coast, Nigeria	*Bemisia tabaci* (Homoptera)	LNL,LCB, LP CHL, VN, ST	Iizuka *et al.,* 1984
	Groundnut crinkle	Ivory Coast	*Bemisia tabaci* (Homoptera)	CR	Dubern and Dollet,1981
Caulimovirus	Peanut Chlorotic Streak	India	*A.craccivora, Bemisia tabaci M.persicae* (Homoptera)	CST	Reddy *et al.,* 1993
Cucumovirus	Cucumber mosaic	China	*Macrosiphum euphorbiae* (Homoptera)	CS, LR, M, MT, VCH,ST	Xu and Barnett,1984
	Peanut stunt	U.S.A	*A.craccivora, A.persicae M.persicae* (Homoptera)	CHL, MT, GVB,LR, DF, ST	Miller and Troutman,1966; Fisher and Lockhart,1978
Furovirus	African peanut clump	Burkino Faso, Ivory Coast, Senegal	*Polymyxa graminis* Ledingham (Fungus)	MT, CRS, ST	Thouvenel *et.al.,1976* Ratna *et al.,* 1991
	Indian peanut clump	India	*Polymyxa graminis* Ledingham (Fungus)	M, MT, CRS, ST	Reddy *et.al.,1983* Nolt *et.al.,* 1988
Geminivirus	Groundnut yellow mosaic	India	*B.tabaci* (Homoptera)	YM, DF	Sudhakar Rao *et.al.,*1980
Ilavirus	Tobacco streak	Brazil	*Frankliniella* sp., *Thrips* tabaci (Thysanoptera)		Costa and Carvalho,1961 Sdoodee and Teakle,1987
Luteovirus	Groundnut rosette assistor	Africa	*A.craccivora, M.persicae* (Homoptera)	SI	Hull and Adams,1968; Reddy *et.al.,*1985a; Rajeswari & Murant, 1988
	Groundnut rosette	Africa, India, Indonesia and Phillipines	*A.craccivora,* (Homoptera)	PYGV, CHL,DL	Anon,1997
	Sunflower yellow blotch	Malawi	*A.gossypii* (Homoptera)	SK, SN	Theuri *et.al.,1987*
Potexvirus	Groundnut chlorotic spotting	Ivory Coast	*A.craccivora A. spiraecola* (Homoptera)	CS	Fauquet *et.al.,*1985; Dollet *et.al.,* 1987
Potyvirus	Bean yellow mosaic	U.S.A.	*A.craccivora* (Homoptera)	CRS	Bays and Demski,1986
	Groundnut eyespot	Ivory Coast	*A.citricola, A.craccivora* (Homoptera)	CS, CRS, LP	Dubern and Dollet,1980

	Passion fruits woodiness	Australia	*M persicae, A.gossypii* (Homoptera)	Not described	Boswell and Gibbs,1983
	Peanut green mosaic	India	*M persicae, A.gossypii* (Homoptera)	MT, M, VCL, RS, ST	Sreenivasulu *et.al.*,1991
	Peanut mottle	World wide	*A.craccivora M.persicae A.gossypii Rhopalosiphum padi* (Homoptera)	MT, M, LR, IVD, SN, LP, ST	Kuhn,1965; Behncken,1980;Bock, 1973 Paguio and Kuhn,1973; Rajeswari *et.al.*, 1983; Reddy *et.al.*, 1978
Tospovirus	Peanut yellow spot	India, Thailand	*Scirtothrips dorsalis* (Thysanoptera)	LCRS, YS	Wongkaew,1986
	Peanut Bud Necrosis Virus	World wide	*Thrips palmi* Karny (Thysanoptera)	CRS, SN, DF, LP, PR, RO, MT,M, ST	Vijayalakshmi,1994 and Reddy *et.al.*,1995
Ilarvirus	Peanut steam necrosis virus	India	*Frankliniella schulzei (Thysanoptera)*	NLTL,CS N,NSP,TN EP	Padmaraju, 2001

B= blotch, BL= black lesion, CHL= chlorosis, CL= chlorotic lesion, CR= crinkling, (L)CRS= (local) chlorotic ring spot, CS= chlorotic spotting, CST= chlorotic streaks, DF= systemic leaf deformation, GVB=green vein banding, IVD=interveinal depression, LI= latent infection, LP= line pattern, LR= leafrolling, M= mosaic, MT= mottle, (L)NL= (local)necrotic lesion, NRS= necrotic ring spot, PR= proliferation, RO= rosette, RS= ring spot, S= systemic, SI= symptomless infection, SK= streaking, SN= systemic necrosis/severe necrosis, SNS=systemic necrotic spotting, ST=stunting, STP= stripes, VB=vein banding, VCH=vein chlorosis, VCL= vein clearing, VN= veinal necrosis, YM= yellow mosaic, YMT= yellow mottle, YS= yellow spot, NLTL= necrotic lesion on terminal leaflets, CSN= complete stem necrosis, TNEP= total necrosis of entire plant and NSP= necrotic spot on pod, PYGV= Pale yellow with green veins, DL= distorted leaflets

17.2.1. Peanut Bud Necrosis Virus Disease

17.2.1.1. Occurrence

Plate 1: Peanut Bud Necrosis

The peanut bud necrosis disease was considered to be caused by tomato spotted wilt virus (TSWV) until 1990 (Reddy *et al.*, 1991). However, it was cleared by 1991 that the casual virus of PBND (**Plate 1**) in India was shown to be a serologically distinct tospovirus, which now referred to as Peanut bud necrosis virus (PBNV) mostly transmitted by *Thrips palmi* (Vijayalakshmi, 1994; Reddy *et al.* 1995). Earlier, the occurrence of this disease under different names was reported by many workers in India (Annual Report of the Indian Agricultural Research Institute in 1949; Capoor, 1964; Reddy *et al.*,1968; Chohan,1972; Chohan, 1974, Ghanekar *et al.*, 1979; Reddy 1988). Based on the surveys the disease was found to be economically important in all major groundnut growing states of India (Reddy *et al.*,1995).

17.2.1.2. Distribution

The distribution of PBND includes all major groundnut growing areas world wide. PBND was a minor groundnut disease in India until the mid 1960s (Reddy,1988). Since then it has occurred in epidemic proportions (Reddy et al., 1991; Vijayalakshmi, 1994). It has been recognized as one of the most important devastating virus diseases of groundnut in Andhra Pradesh, Uttar Pradesh, Madhya Pradesh, Tamil Nadu, Karnataka, Gujarat and Maharashtra on kharif groundnut crop. It has also been occurring on groundnut crop during post rainy season in Saurashtra region of Gujarat, Nizamabad, Nalgonda and Mahaboob nagar districts of Andhra Pradesh and Northern and Vidarbha regions of Maharashtra and North Eastern parts of Karnataka. In India, PBND is one of the most damaging virus disease in groundnut causing 30-90% yield losses depending upon the time of infection on plant growth stage (Ghanekar, 1979; Basu, 1995).

Surveys conducted in Saurashtra region of Gujarat State during summer, 1985 revealed that in some pockets of Jamnagar district, the incidence of the PBND was as high as 92% in Spanish bunch varieties of groundnut. It has been reported that the incidence of PBND ranged from 1.2 to 100% in different parts of the country (Chohan, 1974; Ghanekar *et al.*, 1979; Mayee 1987; Singh and Gupta, 1989; Anonymous, 1985-1998; Singh and Srivastava, 1995, Dharamraj *et al.*, 1995). The hot spot locations for PBND are: Jagtiyal and Hyderabad in Andhra Pradesh; Latur in Maharashtra, Tikamgarh in Madhya Pradesh; Raichur in Karnataka, Mainpuri in Uttar Pradesh and Saurashtra in Gujarat.

17.2.1.3. Symptoms

- PBND produces a variety of symptoms in groundnut (Reddy *et al*.,1995).
- The initial symptoms appear on young quadrifoliates
- Mild chlorotic mottle or spots which develop into necrotic and chlorotic
- Necrosis of the terminal bud, a characteristic symptom occurs on crops grown in the rainy and post-rainy seasons.
- The secondary symptoms are stunting,
- Auxiliary shoot proliferation and malformation of leaflets.
- if plants are infected early, they are stunted and bushy

- if plants older than one month are infected, the symptoms may be restricted to a few branches or to the apical parts of the plants.

Seeds obtained from early-infected plants appear small, shriveled, mottled and discolored, while, late infected plants may produce seed of normal size, but the testae on such seeds are often mottled and cracked.

17.2.1.4. Transmission

The disease was reported to be transmitted by grafting (Reddy *et al.*,1968; Chohan,1972). This disease is not transmitted by either seed or by aphids. Reddy *et al.* (1991) reported that TSWV is transmitted in a persistent manner by the vector thrips which, under optimum conditions, retains the virus throughout its life. Amin *et al.* (1981) reported that the virus causing PBND in India is transmitted by two species of thrips, *Frankliella schutzei* and *Scirtothrips dorsalis*. However, subsequent investigations showed that *Thrips palmi* transmit PBNV and not *F. schultizei* or *S. dorsalis* (Vijaya Lakshmi, 1994; Reddy *et al.*, 1995). *T. palmi* adults transmit the virus to a maximum of 100% when there were 10 adults per plant (Vijayalaxmi 1994). Cowpea was found to be the best host for rearing and multiplying *T. palmi* under laboratory conditions (Vijaya Lakshmi 1994).

17.2.1.5. Disease cycle

Both PBNV and the vector have wide host ranges in crop plants, ornamentals and weeds (Reddy *et al.*, 1983, 1991;Vijayalakshmi, 1994). The virus may, survive in these hosts and can provide an inoculum's source for vector thrips. In India, tomato, Brinjal (egg plant), green gram, black gram, beans and pea are highly susceptible to both virus and vector. Many ornamentals such as Zinnia, cosmos and sunflower harbor large numbers of thrips and Zinnia and cosmos are also hosts of the virus. Weeds such as *Ageratum conyzoides, Cassia tora, Acanthospermum hispidum, Desmodium triflorum* and *Lagasca mollis* are susceptible to both virus and vector.

Thrips are being passive fliers and are mainly carried by the winds. Temperatures around 30°C and a wind speed of 10 km/h have been shown to favour migration of thrips. The population of thrips increases rapidly and reaches its peak in late August and September in the rainy season. If the rain is withdrawn for a week to 10 days during the early vegetative stage of the crop, the thrips population buildup rapidly. The population of thrips builds up again during January and February and as such, crops sown in December-January suffer the most (Reddy *et al.*, 1983).

17.2.1.6. Management

Present day pest management depends on the integration of four areas of activity-insecticide application, crop management (cultural practices), biological control and host-plant resistance. Reddy *et al.* (1991) have also reported that since the disease (PBND) is not seed borne, legislation in the form of quarantine regulations will not help in preventing its spread to PBND-free areas.

17.2.1.6.1.Cultural practices

(a) Adjustment of sowing dates : The lower incidence of PBND was reported in early sown (June) crop of groundnut (Ghewande, 1983). Reddy *et al.* (1983) reported that at ICRISAT centre, Andhra Pradesh, India, groundnut crops sown at the onset of the rains in mid-to-late-June have a much lower rate and intensity of attack than crops sown one or two months later. They have also reported that the main vector flights usually occur in July and August in Andhra Pradesh, India. Early sown plants are sufficiently developed to withstand the infestation by migratory thrips.

Singh and Srivastava (1995) reported that under early-sown conditions, the disease pressure was much higher than when the crop was sown on later dates during the 1985 and 1988 rainy seasons at groundnut research station, Mainpuri in Uttar Pradesh of India. This situation is in contrast with the one in Karnataka State, where maximum PBND was reported in late sown crops (Patil, 1993). Dharmaraj *et al.* (1995) based on eight years experimentation at Regional Research Station, Raichur, Karnataka, reported that in the rainy season, the early sown crop, (sown in the first half of June) shows a lower incidence of PBND (< 10%) than the crop sown in late June (up to 30%). They have also reported that as the sowing date advanced further to the first half of July, the PBND incidence ranged from 20 to 45%. The highest incidence of 90% was recorded in the crop sown around the first of August and dry weather during the crop growth period promoted the disease to a greater extent.

(b) Plant Spacing : The lower incidence of PBND was reported in closer spacing of 22.2 x 7.5/10 cm) (Ghewande 1983). It has been reported by Reddy et al., (1983) that plant density can be increased to decrease the proportion of infected plants of Peanut Bud Necrosis disease. Reddy *et al.* (1991) reported that the incidence of PBND in the crops that were sown at low densities was proportionately higher than in those sown at high densities. Close spacing of 20x10 cm consistently gave lower incidence of PBND compared with wider spacing (Dharmaraj *et al.,* 1995).

(C) Intercropping : Groundnut intercropped with pearl millet/sorghum and early sowing showed less incidence of PBND at ICRISAT (Ghenekar, 1980). Similarly Reddy *et al.* (1991) reported that intercropping with one row of a fast-growing cereal, such as sorghum or pearl millet with three rows of groundnut can reduce the incidence of PBND. Reddy *et al.* (1995) opined that intercropping with fast-growing cereal crops such as maize and pearl millet can reduce the incidence of PBND.

(d) Rogueing : Rouging of infected plants, especially during early stages of plant growth, should be avoided because this practice creates gaps in the field and can increase PBND incidence (Reddy, *et al.*, 1991, 1995).

17.2.1.6.2.Host-plant resistance

Very good progress has been made in the identification of sources of field resistance to PBND under All India Coordinated Research Project on Groundnut AICRP(G), at ICRISAT and elsewhere as well. The important resistant/tolerant lines to PBND reported are: Spanish 5512, Spanish 67-5, ICGS 18, ICGV 86699, J 14, R 8821, R 7015, R 9021, ICG 1703, ICG 2711, EC 2215, ICG 5042, ICGV 98304, RSG 1 (Basu, 1995), CSMG-12. ICG 869, ICG 6317, CSMG-15, (Singh and Srivastava, 1995). R 8806, R 8970, R 8976, R 9621, R 9251, R 9214, R 9227, R 9204, ICGV numbers 86029, 86030, 86031, 89304, ICG 2271 (Dharamraj, *et al.*, 1995), ICG 239, (Ghewande and Desai, 1995). Reddy *et al.*, (1995) reported that although many high yielding PBNV resistant varieties have been developed, they are medium-maturing types. Some of the field resistant genotypes such as ICGV 86388, show resistance to PBNV and less colonization by vector thrips compared with susceptible genotypes (Buier et al., 1995, Dwivedi *et al.*, 995), Cultivars such as ICGS 11, Kadiri 3, ICGS 44, ICGS 1, ICG(FDRS) 10, ICGS 76, Dwivedi *et al.*, 1995, Dharmaraj *et al.*, 1995), Ah 35, 6677, 7403, 7284, Ex 1-1, NCAc 2575 (Narayanaswamy and Ramaiah, 1980), *Arachis chacoense, A.glabrata, A.skp.* (PI 262848), *A. pusilla* (Kolte 1984, Reddy, 1988), ICGV 86430, 2192-8(50) and 2169-5(9), (Buiel et al., 1995). CSMG 12, EC 21070, ICG 98, GBFDS 92, GBPRS 4, GBPRS 15, U 4-7-7, 28/207, Ah 7215, Ah 7286, Ah 7913, BPG 511, Chandra, CSMG 5, CSMG 5-1, CSMG 9, CSMG 15, CSMG 17, EC Nos., 1246, 20957 and 21161 (Jyotisingh *et al.*, 1998).

17.2.1.6.3. Biological control

It has been reported by Reddy *et al.* (1991) that the slow-moving soft bodied thrips larvae are easy prey for many predators, including anthocorid bugs (e.g., *Orius* spp), mirids, wasps, ladybird beetles, syrphids, spiders and predatory thrips. At ICRISAT one vector species, *Scirtothrips dorsalis* was found parasitized. Murlidharan and Ananthakrishnan (1978) have reported a bug *Carayonocoris indicus* feeding on *Frankllinella schultzei*. Three species belonging to Heteroptera and four belonging Salticidae (Spiders) were recorded predating on *Caliothrips indicus* occurring on groundnut (Nandagopal, 1992).

17.2.1.6.4 Use of insecticides and natural pesticides

Amin (1988) provided evidence that high and frequent doses of dimethoate have to be applied to groundnut before there is any reduction in the incidence of PBND. Singh and Srivastava (1995) reported that two sprays of quinalphos 0.02% or water extract of oak leaf 1.00% within 40 days after germination are found effective in minimizing the incidence of PBND and in increasing pod yield of groundnut in Uttar Pradesh. The use of natural pesticides such as sorghum or Coconut leaf extract proved to be more effective in reducing PBND incidence, thereby increasing groundnut yields (Dharmaraj *et al.*, 1995; Kulkarni *et al.*, 1997).

17.2.2.1. Occurrence

Peanut mottle virus was reported from India by Reddy et al. 1991. Subsequently this disease has been reported to occur on rabi/summer groundnut mainly in the States of Andhra Pradesh. Maharashtra and Gujarat. Surveys conducted during rabi/summer of 1983 revealed that the incidence of the disease ranged from 3.7 to 40 per cent in cultivated bunch varieties (Ghewande,1983). The disease can cause upto 30 per cent loss in yield (kuhn and Demski,1975).

17.2.2.2. Symptoms

Newly formed leaves show mild mottling and vein clearing, whereas older leaves show upward curling and interveinal depression with dark green islands. Infected plants are not severely stunted and older plants seldom show typical disease symptoms. Some pods from plants infected with PMV may be smaller than normal and have irregular, gray to brown patches. Seeds from such pods are discoloured.

17.2.2.3. Transmission

The virus is seed transmitted in a range from 0.1 to 3.5 per cent depending on the type of cultivar. PMV is transmitted in a stylet borne (non-persistent) manner by *Aphis craccivora, Myzus persicae, A. gossypii, Hyperomyzes lactucae, Raphalosiphum padi* and *R.maidis.*

17.2.2.4. Host range

PMV occurs in several important legume crops, including peanut and soybean, and weeds.

17.2.2.5. Disease cycle

The PMV occurs in nature on several important legume crops. The virus has been isolated from a few weed hosts (*Cassia obtusifolia, C. leptocorpa, C. occidentalis* and *Desmodium canum*) in nature. Transmission through seed appears to be the most important source of PMV for groundnut. Most commercial peanut seed lots have a low frequency (less than 15 % of seed infection. Aphids are efficient vectors of PMV and will transmit the virus rapidly to nearby plants (Kuhn and Demski,1984).

17.2.2.6. Management

No control schedules are developed for this disease. However, PMV-free seeds should be used as it is reported to be seed-borne. Several lines of *Arachis* species like *A. glabrata* are reported to be resistant to this disease. Germplasm lines viz., NCAc-2240 and NcAc 2243 have been reported to be tolerant to peanut mottle virus at ICRISAT.

17.2.3. Peanut Clump (Peanut Clump Virus- PCV)

Peanut clump has been reported from Punjab, Gujarat, Andhra Pradesh, Uttar Pradesh and Rajasthan on crops grown in sandy soils. Yield losses up to 60% have been recorded in late infected plants. The virus has a wide host range and several weeds commonly occurring in groundnut fields are also infected by the virus. Indian PCV isolates are transmitted through seed (Nolt and Reddy,1984). The virus has also been reported to be soil-borne and possibly transmitted by nematodes and fungus (Ghanekar,1980).

17.2.3.1. Symptoms

The disease occurs in patches in the field and reappears in the same position in progressively enlarged patches, in succeeding years. Typical symptoms are apparent first on newly emerged quadrifoliates of seedlings two to three weeks old. Plants are stunted, and the new quadrifoliates exhibit mosaic mottling and chlorotic rings. Subsequently, infected leaves turn dark green with faint mottling. Infected plants ultimately appear bushy and have small, dark green leaves and usually produce several flowers on erect petioles. The number and size of pods are reduced, resulting in seeds of reduced size. Root systems of infected plants show marked reduction in size. Roots become dark and the epidermal layer peals off easily (Nolt and Reddy,1984).

17.2.3.2. Host range

This disease has wide host range, which includes many monocot and dicot plants. Wheat and barley were reported to be alternate hosts for this disease (Delfosse *et al.*, 1999). A large number of weeds were collected from different groundnut fields infested with Indian peanut clump virus (IPCV) regularly throughout the year in the Jaipur district of Rajasthan, India. The weeds were tested for IPCV infection using DAS-ELISA. The results indicated that IPCV was detected in 8 out of 22 weed species tested. These included *Avena fatua, Cynodon dactylon, Digitaria ciliaris, Chenopodium murale, Cyperus rotundus, Sorghum halepense, Sorghum bicolor, Sorghum sudanensis, Dactyloctenium aegyptium, Cenchrus ciliaris, Eleusine coracana, Eragrotis ciliaris, E. tremula, E. uniloides, Pennisetum glaucum, Setaria italica, Triticum aestivum, Chenopodium album, Celosia aegentea and Oldenlandia corymbosa* (Bhargava and Sobti, 2000). Doucet *et al.*(1999) recorded barley, *Eleusine coracana*, groundnut, maize, rice, sorghum and wheat as host plants and finger millet.

17.2.3.3. Control

Seed should be obtained from areas free from PCV. Since the disease is transmitted by a fungus *Polymyxa* spp. it is possible to reduce the disease incidence by growing a crop non-preferred by *Polymyxa* spp. before raising groundnut in the rainy season. Such a crop would induce the germination of the resting spores, leading to a reduction in the inoculum of the fungus in the soil. Weed free, clean cultivation is very important in reducing the PCV. Solarization can effectively minimize the disease incidence but the effect is not permanent (Delfosse *et al.*, 1997). Soil application of Nemagon and Temik, one week before planting, is effective in reducing the PCV incidence. Germplasm accessions viz., NCAc 17099, NCAc 17133 (RF) and NcAc 17536 have been reported to be resistant to PCV at ICRISAT. In field trials in India, the incidence of peanut clump furovirus was reduced in groundnut plots where pearl millet had previously been grown (Delfosse *et al.* 1999).

17.2.4. Groundnut Rosette

The groundnut rosette virus was first reported from Tanzania in 1907. Several epidemics were recorded in Africa, one among them erupted in 1975 in Nigeria causing $250 million loss in crop plants. This has been reported in Asian countries such as Indonesia, India and in Philippines.

17.2.4.1. Symptoms

Chlorotic, mosaic and pale yellow with green veins. Early infected plant shows shunting with thickened stems also the pod produced frown such plant show reduction in numbers and size.

17.2.4.2. Host range

Groundnut is the natural host for this disease and so far no other hosts are reported.

17.2.4.3. Transmission

The viruses associated with all the three types of rosette disease are transmitted by the cowpea aphid *A. craccivora* in a persistent or circulative manner.

17.2.4.4. Management

The disease can be managed effectively by cultural practices such as destruction of volunteer and un-harvested infected plant, planting early in the season, use of high seed rate, maintaining plant stand and applying need based insecticides. Excellent source of resistance to rosette disease are reported. Merwe *et al.* (1997) reported ICG12988 and ICG12991 produced significantly greater yields (1577-1688 kg/ha) under all disease pressures than the other breeding lines, and recorded among the lowest disease incidence (1%) at high disease pressure.

Ozigi and Olorunju (1997) evaluated SAMNUT 10, 14 and 16, ICG5725 and ICG5240 for resistance to *A. craccivora* in the greenhouse during 1996 and found resistant. Olorunju *et al.* (1991) conducted field experimental and established to determine the effect of groundnut green rosette (caused by groundnut rosette virus (GRV and its satellite RNA) and groundnut rosette assistor luteovirus (GRAV)) on susceptible and resistant groundnut genotypes. Test genotypes (8) were planted in alternate rows with susceptible genotype F 452.4 in which rosette-diseased plants, infested with viruliferous aphids (*Aphis craccivora*), were transplanted every 1.5 m. When disease incidence was high (90-100%) but severity was moderate (measured on rating scale of 1-5) in susceptible genotypes in 1988, only 9-10% of the resistant plants of RMP 12 and RG 1 showed very mild mottle. In 1989, when incidence in susceptible plants was 100% and disease was very severe, c. 87% of the resistant plants showed symptoms but only 2-5% had severe symptoms. In the rosette environment, seed yield was 33 times higher in resistant genotypes than in susceptible ones, and yields in general were comparable in the 2 groups under rosette disease-free conditions. It is concluded that RMP 12 and RG 1 are highly resistant to GRV and its satellite but not to GRAV. A number of disease resistant varieties were developed, including M25.68 and M554.76 (resistant to groundnut rosette virus (GRV) Olorunju *et al.* (1992).

Merwe *et al.* (1999) screened short-duration groundnut cultivars under high, medium and low rosette disease pressures and evaluated for rosette disease incidence, presence of groundnut rosette assistor luteovirus and other agronomic characteristics. Heritable of these characteristics and interaction plus variance components across the three disease pressures were calculated. Significant differences among the entries were observed for both agronomical parameters and disease resistance. Broad-based heritable for pod and kernel yields under high disease pressure were higher than the values for rosette resistance. It is expected from the results that the genetic improvement in rosette disease resistance will be higher using pod yields as a selection criterion. The interaction components between cultivars and disease pressures were significant, suggesting different responses of the cultivars across the three disease pressures. JL24 and ICG (E) S56 and RG1 were introduced from ICRISAT, Malawi, and evaluated with IAR released early-maturing lines, SAMNUT 18 and SAMNUT14, and late-maturing rosette resistant SAMNUT 10 and found that RGI was resistant to GRV (Olorunju *et al.*, 1994).

17.2.5. Cowpea mild mottle (Cowpea Mild Mottle Virus- CMMV)

Although cowpea mild mottle virus (CMMV) occurs on Cowpea *(Vigna unguiculata)* and soybeans in East and West Africa and Thailand, its natural occurrence on groundnut plants has been reported only from India (Punjab,

Andhra Pradesh, and Uttar Pradesh) where the disease is of minor importance (Reddy and Rajeswari, 1984). The whitefly. *Bemisia tabaci* Genn, transmits the virus in a non-persistent manner.

17.2.5.1. Symptoms

Field infected plants are severely stunted with older leaves showing necrosis and younger leaves showing vein banding symptoms. In India, the disease usually occurs on the groundnut crop grown in the post-rainy season.

17.2.5.2. Host range

The virus produces local lesion and systemic symptoms on many hosts.

17.2.5.3. Transmission

It is readily sap transmissible. The whitefly *Bemisia tabaci* transmits the virus in a non- persistent manner It is not seed transmissible.

17.2.5.4. Control

Avoid planting groundnut near cowpea and soybean fields. Hosts highly susceptible to the virus and frequently colonized by *Bemisia tabaci* should not be intercropped with peanut.

17.2.6. Peanut green mosaic (Peanut Green Mosaic Virus –PGMV)

Groundnut green mosaic, caused by peanut green mosaic virus (PGMV), occurs in Chittor district of Andhra Pradesh in India. Infected plants show chorotic spots and vain clearing on young quadrifoliates, followed by severe mosaic (Reddy,1984).

17.2.6.1. Symptoms

Witch Broom: Witches' broom is caused by a mycoplasma like organism. This disease is characterized by excessive proliferation of auxiliary shoots, which results in bushy plants. Leaflets on affected plants are pale yellow and much smaller than usual. Pegs on plants tend to grow upward, and yields are greatly reduced. Witches' broom occurs in India, Thailand, Indonesia, Japan People's Republic of China and Taiwan (Reddy,1984).

Chlorotic spot: This disease was reported from Tirupati in Andhra Pradesh(Hargopal and Nayadu, 1971). The disease is characterized by development of chlorotic leaf spots in terminal leaves of auxiliary shoots followed by appearance of mottle.

Mosaic: Mosaic disease of groundnut was described in India (Niriani and Dingra, 1963). The leaves of the diseased plants show circular chlorotic rings in early stages of infection followed by the appearance of dark green blister-like chlorotic patches. More often the leaves are malformed, puckered with narrow and pointed tips. Reduction in leaf lamina to form filiform leaves is a common feature. In later stages, the cluster of such malformed leaves give the plant a bunchy top appearance. Chenulu *et al.* (1966) reported yield losses from groundnut mosaic ranging from 29 –100 per cent in kernel weight and 22-97 per cent in pod weight depending on disease intensity.

17.2.7. Peanut Stripe Virus Disease (PStV)

In India, the first report of PStV came from Prasada Rao *et al.* (1988 a,b) who surveyed 12 locations during the 1987 rainy season, and observed stripes, green blotches, and oak leaf mosaic symptoms typical of those of PStV at six locations in the Initial Evaluation Trial (IET) of All India Coordinated Research Project on Oilseeds (AICORPO). Moses and Ranga Reddy (1989) reported PStV from Southern Telangana zone of Andhra Pradesh, India. The information given on this disease is mostly from a review by Singh *et al.* (1993).

After the initial report of Peanut Stripe Virus from USA in 1984 (Demski *et al.*, 1984 b) it has been reported from Indonesia (Demski and Lovell, 1985; Saleh and Tantera, 1988), China (Xu, 1988), Thailand (Wongkaew *et al.*, 1988), Taiwan (Green *et al.*, 1986), Philippines (Adalla and Natural, 1988), Java (Green *et al.*, 1988), India (Prasada Rao *et al.*,1988 a), Japan (Ohki *et al.*, 1989; Senboku, 1989), Myanmar (Prasada Rao *et al.*, 1989 b; Prasada Rao and Chakrabarty, 1990) and Vietnam (Subrahmanyam *et al.*, 1990; ICRISAT, 1990). Reddy *et al.* (1985) surveyed many locations in Thailand and Philippines in 1984, and detected PStV incidence up to 100%.

17.2.7.1. Symptoms

PStV belongs to the potyvirus group which has been subdivided into three based on differences in the morphology of their cylindrical inclusions (Edwardson, 1974). Rechcigl *et al.* (1989) compared two symptom variants of PStV, viz, stripe (PStV- S) and blotch (PStV- B) and proposed that PStV should be placed in subdivision-III, characterized by cylindrical inclusions containing pinwheels and both scrolls and laminated aggregates.

The typical symptoms of PStV are banding and dark green stripes along the lateral leaflet veins that in many cases resemble "sergeant stripes" (Demski and Lovell, 1985). Saleh *et al.* (1989) reported that the disease was characterised by green blotches either randomly distributed over the entire leaflets or concentrated along the veins and sometimes surrounded by light green rings. The other variations of PStV symptoms include blotches (irregular green islands within the chlorotic tissue), chlorotic rings surrounding the blotching of young leaflets (Fukomoto *et al.*, 1986), oakleaf pattern (Mangaban *et al.*, 1989 a) and vein clearing, a mosaic pattern of dark green islands (Adalla and Natural, 1988), initial symptoms - chlorotic spots and ring spots followed by wild mottle with green stripes (Xu, 1988), severe mosaic and systemic ecrotic symptoms by PStV-Ts (PStV strain reported from Taiwan) (Chang *et al.*, 1990), systemic necrosis, stunting to severely stunting, mild mottle (Wongkaew and Dollet, 1990).

Symptoms of PStV tend to disappear at leaf maturity, but blotch symptoms persist on mature leaves (Middleton and Saleh, 1988), and mosaic pattern of dark green islands (Adalla and Natural, 1988). Symptoms of PStV vary due to virus strain (Chang *et al.*, 1990; Wongkaew and Dollet, 1989, 1990), crop variety (Adalla and Natural, 1988), and presence of other viruses (Wongkaew *et al.*, 1988). However, Middleton and Saleh (1988) observed no variation due to cultivar. Xu *et al.*, (1991) reported that peanut cultivars and lines of the spanish typė usually had pronounced symptoms of PStV.

Symptoms of PStV resemble those of Peanut Mottle Virus (PMV) and some other viruses. However, these viruses can be distinguished serologically.

17.2.7.2. Host range

PStV infects many plants systematically or locally as reported by Demski *et al.* (1984 a,1984 c), Demski and Lovell (1985), Demski and Warwick (1986), Xu (1988), Kameya-Iwaki(1988), Mangaban and Natural (1988), Natural *et al.*, (1989), Wongkaew (1986,1989) and Reddy, 1991. In the year 1993 a survaey was conducted to identify some of the possible alternate hosts for the disease. Among the crops belonging to cultivated legumes, *Cajanus cajan, Lablab purpureus, Vigna unguiculata, Cicer arietinum* have recorded positive apart from groundnut. Among the non-cultivated legumes *Cassia siamea, Samanea saman, Prosopis juliflora, Parkinsonia aculata,* and *Clitoria ternatea* resulted positive (NRCG, 1993).

17.2.7.3. Transmission

Natural transmission of PStV occurs through seed and by vectors. The virus is also transmitted mechanically and by grafting. Brunt et al.(1990) reported that transmission of potyviruses is rarely through pollen to seed. Although, no such report for PStV is available, yet its possibility can not be ruled out. As groundnut is a highly self-pollinated crop, transmission of virus from infected to healthy plants through pollen will not be of great concern even if it is demonstrated.

17.2.7.4. Seed transmission

Seed transmission of PStV is the primary source of disease spread, and helps in perpetuating disease incidence every year. Much variation in seed transmission has been reported e.g. 0.2-12.5% (Xu, 1988), 0-7% (Wongkaew *et al.,* 1988), 1-32% (Demski and Reddy, 1988 a, 0.4-5% (Culver and Sherwood, 1988), 0-33% (Natural *et al.,* 1989), 0.5-4% (Saleh *et al.,* 1989 b), 5-20% (Xu *et al.*, 1989 a), 0.1-5% (Reddy, 1991). Lynch *et al.*, (1988) observed less than 2% seed transmission under field conditions. Demski *et al.*, (1984 a, 1984 c) and Demski and Lovell (1985) reported seed transmission up to 37%. Xu *et al.*, (1991) observed 1-50 % seed transmission of PStV in the 935 peanut germplasm lines tested, but majority of the genotypes had 5-20 % seed transmission rate. The average seed transmission reported by Chang *et al.*, (1990) was 12.5%. Prasada Rao *et al.* (1989 b) reported variation in seed transmission freqency ranging from 29% for Robut 33-1 to 12% for JL 24.

Seed transmission is influenced by the age of plant at infection (Pinnow *et al.*,1990), genotype (Saleh *et al.*, 1989b, Natural *et al.*,1989; Mangaban *et al.*, 1990; Reddy, 1991; Xu *et al.*, 1991), virus isolate, environmental factors (Reddy, 1991) and duration of infection (Xu *et al.*,1991). Pinnow *et al.* (1990) observed less than 2% seed transmission when plants were inoculated at emergence, but no seed transmission observed when plants were inoculated later. Seed transmission rate may differ in controlled and field conditions. Xu (1988) and Xu *et al.* (1989 b and 1991) reported a higher incidence in shrivelled seeds compared with large ones. Moreover, Spanish type cultivars and lines had greater seed transmission rates than others (Xu *et al.*,1991). PStV is not seed transmitted in two of its host plants:- soybean (Demski and Warwick, 1986; Warwick and Demski, 1988; Gillaspie *et al.*, 1989; Gillaspie and Hopkins, 1991) and lima bean (*Phaseolus limensis*) (Gillaspie *et al.*, 1989). However, Green and Lee (1989) observed 2% seed transmission in cv. 'KS 8' of G. max when inoculated at an early growth stage. Surprisingly Peanut Stripe Virus was also detected in one of the soybean accessions imported from China (Chakrabarty *et al.*, 2004). Work at NRCG on the latency of the PStV in groundnut showed a wide variability among the varieties tried (our unpublished data) (table 21.2). The latency period varied between 3 to 60 days in different groundnuts. There is also variability within the variety, i.e., there was from 8 days to 32 days in the cv. GG 11.

17.2.7.5. Vector transmission

So far seven aphids have been reported to transmit PStV and also cause direct yield losses to peanuts. The main vector is *Aphis craccivora* (Demski *et al.*, 1984 a, 1984c; Adalla and Natural, 1988; Wongkaew *et al.*,1988; Sreenivasulu and Demski, 1988; Mangaban *et al.*, 1989a; Natural *et al.*, 1989; Saleh *et al.*,1989b). Other aphid vectors are :- *A. gossypii* (Adalla and Natural, 1988; Choopanya and Kittipakorn, 1989; Natural *et al.*,1989), *A. glycines* and *Hysteroneura setariae* (Saleh *et al.*, 1989 b), *A. citricola, Rophalosiphum maidis* (Natural *et al.*,1989), *Myzus persicae* (Sreenivasulu and Demski,1988; Warwick and Demski, 1988; Natural *et al.*, 1989; Sreenivasulu *et al.*,1991).

Vector transmission efficiency may depend on the virus variants tested, test conditions, genotype of cultivars, species of vectors etc. Wongkaew *et al.* (1988) observed 3-100% transmission by *A. craccivora* depending on the

Table 17.2. Latent infection of PStv in groundnut under poly house condition at 23.88 to 42.56°C during Rabi/summer, 1992-93

S. no.	Genotype	Latency Period (days)
2	JL 24	6
4	GG 2	60
5	Kadiri 3	14 -21
7	GG 11	8 - 32
8	M 13	8 -11
9	FDRS 10	24
11	CGC 4018	3
15	Code 1	3
16	PI 341879	3
19	ICG 239	3

virus variants and test conditions. Choopanya and Kittipakorn (1989) observed differences in transmission rates due to vector species and groundnut variety. Mangaban (1989) and Mangaban *et al.* (1989 a) observed *A. craccivora* as the most efficient vector among five aphid species. Sreenivasulu and Demski (1988) reported that *M. persicae* was more efficient than *A. craccivora* for transmitting PStV and PMV (Peanut Mottle Virus).Transmission rates with *M. persicae* were 29% for PStV and 9% for PMV, and with *A. craccivora* were 17% for PStV and 3% for PMV. It was also observed that PStV is more efficiently transmitted than PMV by both *M. persicae* and *A.craccivora*. Warwick and Demski (1988) reported that *M. persicae* trasmitted PStV from groundnut to soybean more efficiently (i.e. 16%) than *A.craccivora* (i.e. 1%). PStV is transmitted in a non-persistent manner (i.e. virus retained by vectors for a short period, usually a few hours or less).

PStV transmission was enhanced to 35% for *A. craccivora* and 45% for *M. persicae*, when the aphid vectors fed on PMV- infected plants before feeding on PStV source plants (Sreenivasulu and Demski, 1988).

17.2.7.6. Yield loss

Reports on economic losses from other countries indicated variable yield losses. As with other diseases, yield losses due to PStV are likely to depend on many factors, viz., environmental factors, genotype of crop, virus strain, vector species, age of plants at infection, length and severity of infection. Demski and Reddy (1988 b) gave protocols for estimating PStV yield losses. Variation in yield losses of (0.5 - 24%) depending upon the cultivar was observed by Adalla and Natural (1988) using infected and virus-free potted plants. Varietal effects on yield loss were also reported by Demski *et al.*, (1984 c) with 23% yield losses in 'Argentine' and 21% in 'Florunner'. Saleh *et al.*, (1989 b) reported that yield losses due to PStV varied depending on age of plant when infected: 2 weeks, 50%; 3-8 weeks, over 25%; and 9-10 weeks, 14%. In a second experiment, yield loss varied from 10 to 40% depending on the age at which the plants were infected. Wakman *et al.* (1989) observed 15-

50% yield losses due to PStV depending on length of infection and age of plants. Mangaban (1989) and Mangaban *et al.* (1989b) reported high yield reduction in plants inoculated at seedling stage but at flowering stage did not drastically affect the yield. Demski and Warwick (1986) observed no statistically significant yield reduction due to PStV in two green-house and two screen house tests.

Reddy *et al.*(1988 b) surveyed East and Southeast Asia and observed PStV to be economically important. Demski and Reddy (1988 a) expected frequent epidemics of PStV in USA due to the long duration of commonly grown cultivars, the omnipresence of aphid vectors, the susceptibility of all commercially grown cultivars, relatively high seed transmission rate, and the wide range of alternative hosts to PStV. Yield losses may reflect in quantity and quality of kernels. Demski and Lovell (1985) observed that fewer and smaller peanuts were grown from PStV - infected plants than from uninfected plants of the cv. 'Florunner' in green-house studies.

Contrasting report of yield loss came from Lynch *et al.,* (1988) based on field studies in 1985 and 1986 observing that PStV inoculation at emergence and 20, 40, 60 days after emergence had no influence on measured components of growth, yield and grade of Florunner peanut. Likewise other reports of no significant yield reduction by PStV are available (Demski and Warwick 1986; Wynne, 1986; Choopanya *et al.,* 1989). Wynne (1986) viewed that even though PStV has not spread in USA since its discovery and there are no report of significantly yield reduction, but the virus had a drastic impact on peanut breeding, because, a) exchange of germplasm was severely reduced, b) cooperative testing arrangements which were important in formulating cultivar release decisions, were reduced, c) field plots were destroyed, d) experimental plot and breeder seeds were discarded, e) winter nurseries were eliminated, and f) the limited resources available were diverted to PStV research. Reports on yield losses caused by PStV in soybean (*Glycine max*), are conflicting. While Gillaspie *et al.* (1989) and Gillaspie and Hopkins (1991) reported no losses in yield of soybeans infected by PStV; Green and Lee (1989) reported up to 90% yield reduction in susceptible cultivars in Taiwan. Significant yield reduction due to PStV was also reported in Henderson bush lima bean (*Phaseolus limensis*) (Gillaspie *et al.,* 1989).

The PStV is present in many countries much before 1982, the year in which it was first observed in USA. It seems that environmental factors, variation in genotypes of cultivars and vectors, stage of plants at infection play an important role in causing yield losses, and thus more intensive research is needed on various factors to draw any meaningful conclusions.

17.2.7.7. Management

Avoid primary infection in seed by :

- exchange of only PStV - free germplasm material,
- eliminating contaminated seeds, and growing only virus - free seeds,
- sowing only groundnut genotypes which have no or negligible seed transmission of PStV,
- strictly observing quarantine regulations.

Avoid secondary infection by aphids by control of vectors through chemical, biological or cultural methods. Wongkaew (1989) reported that insecticide sprays reduced PStV incidence considerably. For the biological control of aphids (*Aphis craccivora*), natural enemies, viz., Coccinellidae and Syrphidae, may be used. Misari *et al.* (1987) reported that natural enemy populations of *A. craccivora* attained their peak one week after the aphid population had reached their maximum limit, and disappeared one week after the aphids disappeared. This indicated that the relationship between natural enemies and aphids is density dependent. For cultural (agronomical) methods, mixed cropping and intercropping may be used to decrease vector populations, and to check their mobility.

Prevention of transmission by vectors through the following measures has been successfully done in the case of Peanut Mottle Virus. These may also be useful in controlling PStV.

- Sowing aphid -resistant genotypes. Growing aphid -resistant cultivars is an indirect form of virus resistance. The aphid -resistant genotypes available in groundnut are, ICG 5240 and 'Mwitunde'.
- Sowing trap crop around groundnut, that attracts aphids,
- Growing crop barriers to minimise vectors mobility for long distances.
- Minimising aphid transmission by manipulating sowing dates of groundnuts, wherever possible.

17.2.7.8. Controlling weed hosts

Weed hosts of PStV which remain in or surrounding groundnut fields should be controlled. Demski and Reddy (1988 a) reported *Desmodium* sp. (beggarweed)and *Indigofera* sp. as the likely inoculum sources. Hence, these weeds may also be managed.

17.2.7.9.Crop rotations

Crop rotations may be useful methods to reduce the PStV inoculum as well as vector populations. However, Demski and Lovell (1985) observed that when 1000 virus-free seeds were sown in 1984 at Georgia Experiment

Station, where PStV-infected peanuts were present in the two previous seasons none of the plants became PStV infected indicating that either weed hosts were not infected or they did not serve as virus sources.

17.2.7.10. Breeding for PStV resistance

Breeding Objectives include to vectors, resistance to PStV, non-transmission of PStV through seed, tolerance to PStV i.e. negligible yield losses despite showing PStV symptoms. Resistance breeding programme should combine virus-and vector resistance. The main emphasis should be on avoiding seed-borne inoculum of PStV, since this is convenient method of control. Breeding methods Pedigree methods coupled with backcrossing may be used for transferring desirable character(s) namely, vector resistance/PStV resistance/non-seed transmission of PStV, from donor parents into agronomically suitable varieties. For screening of advanced material against PStV, inoculation of PStV through mechanical means and or vectors should be done at proper stage of plants under screen houses, and care should be taken that screen house conditions do not differ greatly from field conditions for vector mobility etc.

17.2.7.11. Genetics of resistance

The genetics of PStV resistance in groundnut has not yet been worked out. However, in soybean line AGS 129 Choi *et al.* (1989a) postulated incompletely dominance to a soybean isolate of PStV (i.e. PSTV-strain PN). Further, Choi *et al.* (1989 b) studied linkage and reported that in soybean, reaction to soybean mosaic virus (SMV-strain G 1) and PStV were conditioned by genes with 9 + 2.4 percent recombination as coupling phase. The groundnut genotypes, reported to be resistant/tolerant to vectors and PStV, and showing less seed transmission are given in the table 17.3.

Table 17.3. Groundnut genotypes, reported to be resistant/tolerant to vectors and PStV, and showing less seed transmission

Genotypes	Character/remarks	Reference
cv. Mwitunde	Antibiosis to aphids	Jones (1987)
ICG 5240	Aphid - resistant	Adalla and Natural (1988); ICRISAT(1990)
Arachis cardenasii *A. otavi, A. khulaminii, A. kemph -mercadoi, A. stenosperma , A. batizocoi, A. duranensis, A. paraguariensis, A.correntina, A.chacoense, and A.stenophylla*	Aphid - resistant highly aphid resistant	ICRISAT (1990)
Xu Zhou 68-4, Hua 37, *A.glabrata*, PI 262801 and PI 262794	Tolerant/resistant to PStV	Xu (1988)

Contd...

Table 17.3. Contd...

ICG Nos. 4983, 8215, 8958 8973, 11558, 11560, 11562 & 12168	Resistant to PStV immune	Prasad Rao *et al.*, (1989b) Nigam *et al.*, (1991
PI 468141 (*A.diogoi* Hoehne) PI 468142 (*A. diogoi*, (PI 468144 *A. halodes* Mart, ex Hoehne)), PI 468345 (*Arachis* sp.)	Resistant to PStV	Culver *et al.*, 1987
PI 468174 (*Arachis* sp.) PI 468363 PI 468366		
RCM 387	Tolerant to PStV	Wongkaew and Kantrong (1989 a)
ICG 11710 and ICGV 88249	Mild symptom of PStV	ICRISAT (1990)
Inter- specific derivatives viz., 476-10, 478-3, 83/147-3-61and 83/372-5-10	Showed mild symptoms when inoculated with PStV	"
Interspecific derivatives, 85/ 165-19, 85/166-4,85/172-1 & 1093-2	Slightly affected by PStV	"
ICG 4806, ICGV 8786/7 & ICGV 8825/9	Showed less than 30% infection when adjacent test lines were all infected	"
China PI 1420334/433348, PI 433349, PI 420335, VA 81 B	Appeared to be non-seed transmitted when over 300 seeds were tested	Tolin and Rechcigl (1985)
F 87-158 and F 87-157	Negligible seed transmission of PStV	Xu *et al.*, (1989 b)
Tainan 9	Seed transmission	Wongkaew *et al.*, 1988
F 87-358	A non PMV transmission	Xu *et al.*, (1989 a)
(148-7-4-3-12-8 x NCAc 17090)	PStV transmission through seeds	"

17.2.8. Peanut Stem Necrosis Disease (PSND) (After Padma Raju,2001)

17.2.8.1. Introduction

A virus disease caused nearly 300 crores rupees worth of crop losses to groundnut in Andhra Pradesh during kharif in the year 2000. It was presumed to be caused by peanut bud necrosis virus (PBNV), a virus known to be widely distributed in India. Research done has shown conclusively that the disease found to be caused by a totally new virus. As a result the disease is named " Peanut Stem Necrosis" (PSND) and is caused by an isolate of tobacco streak virus (TSV). The information presented on this disease are as given by Padma Raju (2001). In some respect peanut bud necrosis disease (PBND) and PSND resemble. However, there are many distinguishing features.

17.2.8.2. Symptoms

Initial symptom appear as large necrotic lesions on young quadrifoliates. These coalesce and cover the entire leaf let, leading to complete necrosis of young quadrifoliates. This symptom will be followed by necrosis of the entire stem located below the necrosed quadrifoliates. If young plants are affected (less than one month old), the entire plant is often necrosed. In the case of older plants one or more branches will show necrosis. These plants are stunted and do not show any auxiliary shoot proliferation.

Necrotic spots are observed on majority of pods. Size of the pods is severely reduced and kernel are not marketable. Limited tests conducted on cv. TMV-2 indicate that the virus is not seed transmitted. However, seed transmission of virus is a complex phenomenon. It depends on when the infestation occurred, genotype, as well as environmental factors. Therefore additional tests are necessary to conclude that the virus is not seed transmitted in groundnut.

17.2.8.3. Host range of virus tobacco streak virus (TSV)

It has been conclusively shown that PSND is caused by tobacco streak virus (TSV). The virus infects many plants, including the most common weeds in groundnut fields. On the majority of the hosts it produces either necrotic or chlorotic lesions, followed by systemic necrosis.

- *Cowpea* : Necrotic lesions and venial necrosis followed by total necrosis of leaves and stem
- *Black gram* : Necrotic lesions followed by total necrosis of leaves and stem.
- *Sunflower* : Chlorotic lesions on leaves followed by necrosis . Stem shows necrotic lesions and streaks. Plants are severely stunted and head droops, thus making the infected plants conspicuous in the field.

17.2.8.4. Distinguishing features of PSND

Symptoms on Groundnut: Necrotic lesions on terminal leaflets, complete stem necrosis and often total necrosis of entire plant. Necrotic spots on pods. Testa are not discoloured or mottled. No evidence of seed transmission.

17.2.8.5. Strategies for the management

(a) Discourage cultivation of JL 24 and TMV 2 groundnut varieties and promote k 134 variety Among the two popular varieties cultivated in Anantpur district, the incidence of bud necrosis under high to severe category was slightly more in JL 24 when compared to TMV 2. Both these varieties were susceptible to bud necrosis disease and suggested that in areas where disease ie severe, highly susceptible cultivars such as TMV 2 should not be grown. The per cent disease incidence was observed to be less on K 134 variety. As of now, no resistant variety to this disease is identified. Till such time, K 134 may be recommended for cultivation in the areas prone to PSND. Supply of sufficient quantity of K 134 seed may be ensured.

(b) Although the groundnut crop is grown on red soils, limited area of black soils is under crop in Uravakonda and other black soil belts including Karnool district. The disease incidence was high to severe in black soils when compared to red soils during kharif 2000. Therefore, groundnut cultivation in black soil may be discouraged and the black soil area (previously under groundnut crop) may be sown with crops like sorghum, bajra, maize, ragi (*Eleusine coracana*), redgram etc.

(c) *Promote intercropping in groundnut (groundnut + redgram + bajra or groundnut + redgram + castor)*: During kharif 2000, sole crop of groundnut was severely affected as compared to groundnut with intercrops like sorghum, Bajra, Castor, redgram etc. Among the various intercrops, groundnut + redgram + bajra and groundnut + redgram + Castor appeared relatively better to withstand the disease. Bajra , sorghum may be grown as border crop (3 or 4 rows) on all the four sides of the groundnut fields.

(d) *Maintain optimum plant population of the crop:* The farmer need to be advised to adopt the recommended seed rate of 100 to 125 kg/ha and spacing of 30cm x 10 cm and to maintain optimum plant population because low population was found to be more infested with the disease.

(e) *Promote clean cultivation without weed either in field or on bunds:* Both virus and thrips (vector) have wide host range of crop plants, ornamentals and weeds. Thrips were found harboured on more than fifteen plant species near groundnut fields. Particularly weed plants like *Parthenium, Commalina bengalensis, Achanthospermum hispidum, sida sp., Acharanthus aspera* were found to be as collateral hosts for the disease. Frequent weeding and maintaining fields clean will help in reducing the disease incidence.

(f) *Discourage monocropping of groundnut:* As mono cropping results in repeated occurrence of pests and diseases, besides several other nutritional deficiency problems, it is advised to go for alternate crops like sorghum, Bajra, maize, ragi, pulses, castor, sesamum etc. at least once in 3 years.

(g) *Advocate judicious use of fertilizers and pesticides and encourage organic farming:* In the light of the field experience of the scientists during kharif 2000, judicious and need based application of fertilizers and pesticides are recommended to reduce the disease incidence.

(h) *Promote adoption of farming system:* Diversification of agriculture will help the farmers to absorb such sudden shocks and outbreak of disease. Farming systems like Agriculture + Horticulture, Agriculture + Dairy, Agriculture + poultry, Agriculture + Sylviculture, Agriculture + Sericulture etc are to be advocated.

(i) *Monitoring of the crop for the occurrence and spread of the diseases from June to November 2001:* The groundnut growing areas in Anantpur, Cuddapah, Kurnool and Chittoor districts are to be closely monitored for the incidence of PSND and pattern of its spread during the coming cropping season from June to November 2001 and the farmers should be alerted timely for the management of the disease.

(j) *Create awareness among farmers:* Wide publicity has to be given to the farmers on item a to i above through leaflets, pamphlets, broaches, press, doordarshan and private T.V.channels. The officials of the department of agriculture and the scientists of the SAUs should make the farmers adopt the above listed practices in their fields with utmost care.

(k) *Ensure the availability of spraying equipments:* The availability of the required number of spraying equipments to the farming community in all the villages of the district may be ensured. The equipment may be supplied at subsidized rates and/ or arrange through custom- hiring.

(l) *Training to Agricultural Officers and Agricultural Extension Officers and Farmers:* The Agricultural Officers and Agricultural Extension Officers should be given periodical training on the above aspects to create awareness and expertise on management of Peanut Stem Necrosis Disease by the scientists of the SAUs, ICRISAT and NRC for Groundnut. Also, farmers need to be trained on the disease symptom, control measures etc.

17.2.9 Future strategies

- Developing laboratories where virus antisera could be produced, purified, maintained and supplied to the research centres
- Strengthen programmes for training extension workers, farmers and researchers
- Establish survey and surveillance systems for groundnut virus diseases
- Establish seed producing units where virus-free seeds could be produced and supplied to farmers
- Strengthen quarantine regulations
- Assess yield losses due to virus diseases in the field. - Durable and dependable sources of resistance to virus diseases, vis-a-vis vector are

limited. Therefore, these have to be identified in relation to host x pathogen x environment interaction and possible occurrence of biotypes of pathogen

- Develop groundnut varieties with no or low seed transmission and resistance to both virus and vector
- Factors influencing virus diseases incidence and build-up with the aim of formulating effective and economic control measures
- Information on cultural and biological management of virus diseases are to be strengthened to formulate integrated disease management programmes for groundnut.

17.3. Conclusions

Virus diseases are considered to be economically important in the groundnut based cropping system. In recent years they have become increasingly problematic for the groundnut farmers, scientist and policy makers as well. Through multi-pronged approaches such as efficient vector management, developing resistant varieties, utilizing precise and rapid methods like ELISA which have many advantages are available for detecting virus diseases which could be utilized by research workers operating with limited facilities.

References

Adalla, C.B. and Natural, M.P. (1988). Peanut stripe virus disease in the Philippines. p.9.In First Coordinators' Meeting on Peanut Stripe Virus, 9- 12 June, 1987, held at Malang, Indonesia.ICRISAT, Patancheru, India.

Amin P.W. (1982). "Recent Advances in Groundnut Productivity Research", Proc. Summ. Inst, NRCG (ICAR), Junagadh, India, 145 pp.

Amin P.W., Reddy, D.V.R. Ghanekar, A.M., Reddy, M.S. (1981). Plant disease 65: 663.

Amin, P.W. (1988). Insect and mite pests and their control. In *Groundnut* (Ed. P.S. Reddy). pp. 393-452. Indian Council of Agricultural Research, New Delhi.

Anonymous, (1949). Annual Report of the Indian Agricultural Research Institute.

Anonymous, (1997). Compendium of Peanut Diseases, The American Phytopathological Society, 3340 Pilot Knob Road, St. Paul, Minnesota 55121- 2097,USA

Anonymous, (1985 to 1998). Annual progress report, All India Coordinated Research Project on Groundnut (ICAR), National Research Center for Groundnut, Junagadh-362 001, Gujarat, India.

Basu, M. S. (1995). Peanut Bud Necrotic Disease: Activities in the Indian national program. Pp 61-63 In Recent studies on peanut bud necrotic disease: Proceedings of a meeting, 20 March, 1995, ICRISAT Asia Center, India (Buiel, A. A. M., Parleviliet, J. E. and Lenne, J. M. eds.) Patancheru 502 0324, Andhra Predesh, India: International Crop

Research Institute for Semi- Arid Tropics; and P. O. Box 386, 6700 A. J. Wasgeningen, The Netherlands: Department of Plant Breeding, Agriculture University of Wasgeningen.

Bays, D. C. and Demski, J. W. (1986). Bean yellow mosaic virus isolate that infects peanut (*Arachis hypogaea*). *Plant Diseases*. 70, 667-669.

Behncken, G. M. (1980). The occurrence of peanut mottle virus in Queensland. *Australian Journal of Agriculture Research*., 21, 465-472.

Bhargava, A.K. and Sobti, A.K. (2000). Detection of Indian peanut clump virus in the weeds present in the infested fields of Arachis hypogaea. *Journal-of-Mycology-and-Plant-Pathology*. 30(1), 114-115.

Bock, K. R. (1973). Peanut mottle virus in East Africa. *Annals of Applied Biology*., 74, 171-179.

Boswell, K. F. and Gibbs, A. J. (1983). Virus identification data exchange: Virus of Legumes. Australian National University, Canberra. pp 139.

Brunt, A., Crabtree, K. and Gibbs, A., 1990. Potyvirus Group, p.668. In Viruses of Tropical Plants, CAB International.

Buiel, A. A. M. and Parlevliet, J. F. (1995). Epidemiology of peanut bud necrosis disease in groundnut in India. Pages 41-46. In Recent studies on peanut bud necrotic disease: Proceedings of a meeting, 20 March, 1995, ICRISAT Asia Center, India (Buiel, A. A. M., Parleviliet, J. E. and Lenne, J. M. eds.) Patancheru 502 0324, Andhra Predesh, India: International Crop Research Institute for Semi- Arid Tropics; and P. O. Box 386, 6700 A. J. Wasgeningen, The Netherlands: Department of Plant Breeding, Agriculture University of Wasgeningen.

Capoor, S. P. (1964). Shoot necrosis of groundnut. *Journal of Agricultural Research*, 41, 10 .

Chakrabarty,S.K., Anitha,K., Prasada Rao,R.D.J.V., Mohammed Ismail, Sarat Babu,B., Babu Abraham, Varaprasad,K.S.and R.K.Khetrapal. 2004. Pest Intercepted in oilseeds Germplasm imported during 1986-2003. Indian Journal of Plant Protection Vol,32: 2,120-124.

Chang, C.A., Purcifull, D.E. and Zettler, F.W. (1990). Comparison of two strains of Peanut Stripe Virus in Taiwan. *Plant Diseases*, 74(8), 593-596.

Chenulu, V.V., Munjal, R.L., Hora, T.S., Singh, A. 1966. Estimation of loses due to groundnut mosaic. Indian Phytopath. 19: 194-197.

Chohan J.S. (1972). Final Progress Report, ICAR Scheme for research on important disease of groundnut in the Punjab for the period 1957-1667. Department of Plant Pathology, Punjab Agricultural University, Ludhiana, India 117pp.

Chohan J.S. (1974). Recent advances in diseases of groundnut in India. Pages 171- 184 in current Trends in Plant Pathology (Eds, S.P.Raychanduri and J.P.Verma), Lucknow University, India.

Choi, S.H., Green, S.K. and Lee, D.R., (1989b). Linkage relationship between two genes conferring resistance to peanut stripe virus and soybean mosaic. Euphytica, 44 (1-2), 163-166.

Choi, S.H., Green, S.K., Lee, D.R, and Yoon, J.Y., (1989 a). Incompletely dominant single resistance gene for peanut stripe virus in soybean line AGS 129. Euphytica, 40 (3), 193-196.

Choopanya, D. and Kittipakorn, S. (1989). Transmission of peanut stripe virus by some aphids. pp. 203-205. In *Proceedings of the Seventh Thailand National Groundnut Meeting for 1987,* Pattaya, Chonburi, Thailand, 16-18 March 1988. Khon Kaen, Thailand : Khon Kaen University, Faculty of Agriculture.

Choopanya, D., Buapan, S. and Watayanont, A. (1989). Distribution of groundnut virus diseases in North-East Thailand and their effects on yield. pp. 200- 202. In *Proceedings of the Seventh Thailand National Groundnut Meeting for 1987,* Pattaya, Chonburi, Thailand, 16-18 March 1988. Khon Kaen, Thailand, Khon Kaen University, Faculty of Agriculture.

Costa, A. S. and Carvalho, A. M. B. (1961). Studies on Brazilian tobacco streak. *Phytopathology. Z.,* 42, 113-138. ICRISAT (1990). ICRISAT Annual Report 1989, ICRISAT, Patancheru, India ICRISAT 2002a On the wild side, SATrends, 16. ICRISAT, Patancheru, India

Culver, J.N. and Sherwood, J.L., 1988. Detection of peanut stripe virus in peanut seed by an indirect enzyme - linked Immunosorbent assay using a monoclonal antibody. Plant Disease, 72 (8), 676-679.

Culver, J.N., Sherwood, J.L. and Melouk, H. A., 1987. Resistance to peanut stripe virus in Arachis germplasm. Plant Disease, 71, 1080-1082.

Delfosse P Reddy A S Devi P S Devi P S and Maraite H 1997 Pearl millet as a bait crop for the management of the Indian peanut clump virus. *Indian Journal of Plant Protection.* 1997, 25, 1, 70-72.

Delfosse, P., Legreve, A., Devi, P.S., Reddy, A. S. and Maraite, H. (1999). Indian peanut clump virus is highly infectious on graminaceous plants. *Zeitschrift-fur-Pflanzenkrankheiten-und-Pflanzenschutz.* 106(4), 418-424.

Delfosse, P., Reddy, A. S., Legreve, A., Devi, P.S., Devi, P.S., Maraite, H. and Reddy-D. V. R. (1999). Indian peanut clump virus (IPCV) infection on wheat and barley: symptoms, yield loss and transmission through seed. *Plant Pathology.,* 48(2), 273-282.

Delfosse, P., Reddy, A. S., Devi, P.S., Devi, P.S., and Maraite, H. (1999). Pearl millet as a bait crop for the management of the Indian peanut clump virus. *Indian Journal of Plant Protection.* 1997, 25: 1, 70-72.

Demski, J.W. and LOVELL, G.R., (1985). Peanut Stripe Virus and the distribution of peanut seed. Plant Disease, 69 (9), 734-738.

Demski, J.W. and Reddy, D.V.R., (1988 b). Peanut stripe virus yield loss studies. pp. 18-19. In First Coordinators' Meeting on Peanut Stripe Virus, 9-12 June, 1987 held at Malang, Indonesia. ICRISAT, Patancheru.

Demski, J. W., Reddy, D.V.R., Sowell, G. Jr. (1984 b). Peanut Stripe, a new virus diease of peanut (Abstr.) Phytopathology 74, 627.

Demski, J. W., Reddy, D.V.R., Sowell, G. Jr. and Bays, D. (1984 c). Peanut Stripe Virus - a new seed-borne potyvirus from China infecting groundnut (Arachis hypogaea). Annals of applied Biology, 105, 495-501.

Demski, J.W. and Reddy, D.V.R. (1988 a). Peanut Stripe Virus in the USA. pp. 10-11, In First Coordinators' Meeting on Peanut Stripe Virus, 9-12 June 1987, held at Malang, Indonesia. ICRISAT, Patancheru, India.

Demski, J.W. and Warwick, D. (1986). Peanut Stripe Virus in Arachis hypogaea L.Proceedings of American Peanut Research and Education Society, 18,77.

Demski, J.W., Reddy, D.V.R. and Bays, D.C. (1984 a). Peanut Stripe. pp. 51-52. In Compendium of Peanut Diseases, (D. Morris Porter, . Smith and R. Rodriguez-Kabana, Ed.). Minnesota: The American Phytopathological Society.

Demski, J.W., Reddy, D.V.R., WONGKAEW, S., KAMEYA- IWAKI, M., SALEH, N. and Xu, Z. (1988). Naming of peanut stripe virus. Phytopathology, 78 (6), 631-632.

Dharmaraj, P. S., Naragund, V. B. and Somasekhar. (1995). Peanut Bud Necrotic Disease in Karnataka. Pp 69-72, In Recent studies on peanut bud necrotic disease: Proceedings of a meeting, 20 March, 1995, ICRISAT Asia Center, India (Buiel, A. A. M., Parleviliet, J. E. and Lenne, J. M. eds.) Patancheru 502 324, Andhra Predesh, India: International Crop Research Institute for Semi- Arid Tropics; and P. O. Box 386, 6700 A. J. Wasgeningen, The Netherlands: Department of Plant Breeding, Agriculture University of Wasgeningen.

Dollet, M., Reddy, D. V. R., Wongkaew, S., Kameya-Iwaki, M., Sahel, n. and Xu, Z. (1987). Les viruses de l'arachide en Afrque de l'quest. *Oleagineux.*, 42, 291-297

Doucet, D., Legreve, A., Delfosse, P., Devi, P.S. and Maraite, H. (1999). Distribution of Polymyxa graminis Ledingham in different host plants.Proceedings, *51st international symposium on crop protection, Gent,* Belgium, 4 May 1999. Part II. *Mededelingen-Faculteit-Landbouwkundige-en-Toegepaste-Biologische-tenschappen,-Universiteit-Gent.*, 64(3b), 627-635.

Dubern, J. and Dollet, M. (1980). Groundnut eyespot virus, anew member of the potyvirus group. *Annals of Applied Biology.*, 96, 193-200.

Dubern, J. and Dollet, M. (1981). Groundnut crinkle virus, a new member of carlavirus group. *Phytopathology. Z.*, 101, 337-347.

Dwivedi, S. L., Nigam, S. N., Reddy, D. V. R., and Ranga Rao, G. V. (1995). Progress in breeding groundnut varieties resistance to Peanut Bud Necrotic Virus and Vector. Pp 35-40. In Recent studies on peanut bud necrotic disease: Proceedings of a meeting, 20 March, 1995, ICRISAT Asia Center, India (Buiel, A. A. M., Parleviliet, J. E. and Lenne, J. M. eds.) Patancheru 502 324, Andhra Predesh, India: International Crop Research Institute for Semi- Arid Tropics; and P. O. Box 386, 6700 A. J. Wasgeningen, The Netherlands: Department of Plant Breeding, Agriculture University of Wasgeningen.

Edwardson, J. R. (1974). Some properties of the potato virus-Y group. Fla. Agriculture Experiment Station Monograph, 4, 398 pp.

Fauquet, C. Thouvenel, J. C. and Fargette, D. (1985). Une nouvelle maladie virale de l'arachide en cote d'Ivoire, la maladie des taches chlorotiques de l'arachide. C. R. Acad. Sci. Paris (cited by Dollet *et al* 1987).

Fischer, H. U. and Lockhart, B. E. L. (1978). Host range and properties of peanut stunt virus from Morocco. *Phytopathology.*, 68, 289-293.

Fukumoto, F. Thongmeearkom, P., Iwaki, M., Choopanya, D., Sarindu, N., Deema, N. and Tsuchizaki, T. (1986). Peanut chlorotic ring mottle virus occurring on peanut in Thailand. Technical Bulletin, Tropical Agriculture Research Centre, Japan, 21, 150-157.

Ghanekar, A.M., Reddy D.V.R., Uzuka, N., Amin, P.W., Gibbons, R.W. (1979). Bud Necrosis of Groundnut (*Arachis hypogaea* L.) in India Caused by tomato spotted wilt virus. *Annals of Applied Biology* 93, 173-179.

Ghanekar, A.M., Reddy D.V.R., Uzuka, N., Amin, P.W., Gibbons, R.W. (1979). Ann.,Appl. Biol. 93 : 173-179.

Ghanekar, A.M. (1980). Groundnut Virus Research at ICRISAT (International Crops Research Institute for the Semi-Arid Tropics), 1980. Proceedings of the International Workshop on groundnuts, 13-17 October, 1980, Patancheru, A.P., India.

Ghewande, M. P. and Desai, S. (1995). Annual Report 1995-96, National Research Centre for Groundnut, Junagadh- 362 001, Gujarat, India.

Ghewande, M.P. (1983). Effect of cultural practices on the disease (bud necrosis, collar rot, and stem rot) incidence and yield of groundnut. *Indian Botanical Reporter.*, 3(1), 98.

Ghewande, M.P. (1983a). Effect of cultural practices on the disease (budnecrosis, collar rot, stem rot) incidence and yield of groundnut. *Indian Botanical Reporter*, 2, 176-177.

Gillaspie, A.G. and Hopkins, M.S. (1991). Spread of peanut stripe virus from peanut to soybean and yield effects on soybean. *Plant Diseases*, 75(11), 1157-1159.

Gillaspie, A.G. Jr., Wright, J.M.and Hopkins, M.S. (1989). Epidemiology of Peanut Stripe Virus (PStV) on soybean and lima-bean (Abstr.) *Phytopathology*, 79, 372-373.

Green, S.K. and Lee, D. R., (1989). Occurrence of Peanut Stripe Virus (PStV) on soybean in Taiwan- Effect on yield and screening for resistance. Tropical Pest Management, 35, 123-126.

Green, S.K., Lee, D.R., Lesemann, D.E. and Vetten, H.J. (1986). The occurrence of Peanut Stripe Virus on peanut and soybean in Taiwan. Plant Protection Bulletin, Taiwan (ROC) 28, 445-446.

Green, S.K., Lee, D.R. and Horne, N.M. (1988). Survey for viruses of soybean, mung bean and peanut in Java, Indonesia. Plant Disease, 72 (11), 994.

Hargopal, T., Nayadu, M.V. (1971). A new sap transmissable groundnut chlorotic spot virus. Phytopathology Z., 71: 33-41.

Hull, R. and Adams, A. M. (1968). Groundnut rosette and its assistor virus. *Annals of Applied Biology*, 62, 139-145.

IARI. (1949). Annual Report, Indian Agricultural Research Institute. New Delhi.

ICRISAT (1990). Groundnut, pp. 139-192. ICRISAT Annual Report 1989, ICRISAT, Patancheru, India.

ICRISAT, (1988). Coordination of research on peanut stripe virus: summary proceedings of the First Meeting to Coordinate Research on Peanut Stripe Virus Disease of Groundnut, 9-12 June 1987, Malang, Indonesia. Patancheru, A.P. 502 324, India.

Iizuka, N., Rajeshwari, R., Reddy, D. V. R., Goto, T., Muniyappa, V., Bharathan, N. and Ghanekar, A. M. (1984). Natural occurrence of the strain of cowpea mild mottle virus on groundnut (*Arachis hypogea*) in India. *Phytopathology. Z.*, 109, 254-253.

Jones, A.T. (1987). Control of virus infection in crop plants through vector resistance : A review of achievements, prospects and problems. Annals of Applied Biology, 111, 745-772.

Joshi, P.R. (1989). M.Sc. (Agri.). Thesis, Gujarat Agricultural University, Sardar Krushinagar

Jyotisingh, Srivastava, S. K. and Singh, A. B. (1998). Screening of groundnut germplasm for resistance to peanut bud necrosis. *Journal of Oilseed Research.*, 15(1), 160-162.

Kameya-Iwaki, M. (1988). Virus program in Southeast Asia sponsored by the Tropical Agriculture Research Centre, Japan, p. 11. In First Coordinators' Meeting on Peanut Stripe Virus 9-12 June 1987, held at Malang, Indonesia. ICRISAT, Patancheru, India.

Kolte, S. J. (1984). Diseases of annual edible oilseed crops. *Peanut Diseases*, Vol.I, Florida, CRC Press Inc. Usa, pp. 29.

Kuhn, C. W. and Demski, J. W. (1987). Latent virus in peanut in Georgia identified as cowpea chlorotic mottle virus. *Plant Diseases*. 71, 101.

Kuhn, C.W., Demski, J.W. (1975). The relationship of peanut mottle virus to peanut production. Ga. Agric. Exp. Stn. Res Rep 213. 19pp.

Kuhn, C.W., Demski, J.W. (1984). In Compendium of Peanut Disease (Eds. Porter, D.M.Smith, D.H.Rodrignez. Kabana, R.) 73pp.

Kuhn, C.W., Demski, J.W., Reddy, D.V.R., Benner, C.P. and Bijaisoradat, M. (1984). Identification and incidence of peanut viruses in Georgia. Peanut Science, 11, 67-69.

Kuhn, C.W. (1965). Symptomatology,host range and effect on yield of a seed trasmitted peanut virus. Phytopathology, 55,880-884.

Kulkarni, M. S., Anahosur, M. S. and Kulkarni, S. (1997). Management of groundnut bud necrosis virus disease using coconut and sorghum leaf extracts. In *International Conference on Integrated Plant Diseasesease Management for Sustainable Agriculture* (Abstract) pp 353, 10-15 Nov. 1997, Indian Phytopathological Society, ICAR & IARI, New Delhi, India.

LI, R.H., Zettler, F.W., Elliott, M.S., Petersen, M.A., Still, P.E., Baker, C.A. and Mink, G.I. (1991). A strain of peanut mottle virus seed borne in Bambarra groundnut. Plant Disease, 75 (2), 130-133.

Lynch, R.E., Demski, J.W., Branch, W.D., Holbrook, C.C. and Morgan, L.W., (1988). Influence of Peanut Stripe Virus on growth, yield and quality of Florunner Peanut. Peanut Science, 15 (2), 47-52.

Mangaban, F.L. (1989). Characterization of peanut stripe virus and its effect on growth and yield of peanut (*Arachis hypogaea* L.). M.S. in Plant Pathology Thesis. Philippines University, College, Laguna, Philippines. 59 pp.

Mangaban, F.L., Malabanan, M.J.C. and Natural, M.P. (1990). Screening of peanut lines and accessions for low peanut stripe virus (PStV) seed transmissibility. Bacolod City, Philippines 1 p. Summary only presented at Pest Control Council of the Philippines Annual Convention, 21st, Bacolod City, Philippines, 7-10 May 1990.

Mangaban, F.L. and Natural, M.P. (1988). Occurrence, symptomatology and host range of peanut stripe virus in Philippines, 1 pp. Summary only. Presented at, 19th Pest Control Council of Philippines, Cebu City, Philippines, 3-7 May, 1988.

Mangaban, F.L., Natural, M.P. and Valdez, R.B. (1989 a). Host range, method of transmission and properties of peanut stripe virus in the Philippines. 1 pp. Summary only; presented at 20th Pest Control Council of the Philippines, Baguio City, Philippines, 9-12 May 1989.

Mangaban, F.L., Valdez, R.B. and Natural, M.P. (1989 b). Effects of peanut stripe virus on the growth and yield of peanut (Arachis hypogaea L.). Philippines Journal of Crop Science 14 (Supplement No. 1): p 15. Summary only, Presented at 5th Scientific Meeting of the Federation of Crop Science Societies of the Philippines, Iloilo city, Philippines, 26-29 April 1989.

Mayee, C. D. (1987). Disease of groundnut and their management. In " *Plant Protection in field crops*" (eds. Rao, M.A.B. and Sithanathan, S.) Plant Protection assistance of India, Hyderabad, pp. 235-243

McBride, K.E., Svab, Z., Schaef, D.J., Hogan,P.S., Stalker, D.M. and Maliga, P. (1995). *Botechnolgy*,13, 362-365.

Merwe, P.J. A-van-der., Reddy, L.J., Subrahmanyam, P., Naidu, R.A., Van-der-Merwe, P.J.A. (1999). Criteria for selecting groundnut varieties in breeding for resistance to rosette disease. *South African Journal of Plant and Soil.*, 16(1), 56-58.

Merwe, P.J.A, Subrahmanyam, P. and Van-der-Merwe, P. J. A. (1997). Screening of rosette-resistant short-duration groundnut breeding lines for yield and other characteristics. *International Arachis Newsletter.*, 17, 23-24

Middleton, K.J. and Saleh, N. (1988). Peanut Stripe Virus disease in Indonesia and the ACIAR Project, pp. 4-6. In First Coordinators' Meeting on Peanut Stripe Virus 9-12 June 1987, held at Malang, Indonesia. ICRISAT, Patancheru, India.

Miller, L. I. and Troutman, J. L. (1966). Stunt disease of peanut in Virginia. *Plant Diseases Report.*, 50, 139-143

Misari, S.M., Ansa, O.A., Demski, J.W., Kuhn, C.W., Casper, R. and Breyel, E. (1987). Reaction of peanut genotypes to the rosette virus and its vector *Aphis craccivora* Koch. *Proceedings of American Peanut Research and Education Society*, 19, 29.

Misari, S.M., O.A. Ansa, J.W.Demski, C.W.kuhn, R.Casper and E.Breyel.(1987). Reaction of peanut henotypes to the rosette virus and its vetor Aphis craccivora Koch. Proceedings of American Peanut Research and Education Society, 19,29.

Moses, G.J. and Ranga Reddy, R. (1989). New viral and fungal diseases on groundnut in the Southern Telangana zone of Andhra Pradesh, India. *International Arachis Newsletter*, 5, 10-11.

Muralidharan,N and Ananthakrishnan,T.N.(1978).Bio-ecology of four species of Anthocoreidae (Hemiptera:Insecta) predacious on thrips, with key to genera of Anthocorieds from India. *Record of the Zoological Survey of India* 11, 1-32.

Nandagopal, V. (1992). Studies on Integrated pest management in Groundnut in Saurashtra, Ph. D. thesis, Saurashtra University, Rajkot, pp 246.

Narayanaswamy, P. and Ramaiah, M. (1980). In *Proceedings, National Seminar on disease resistance in crop plants*, Tamil Nadu Agric. University, Coimbatore, India, pp 42- 45

Natural, M.P., Mangaban, F.L. and Valencia, L.D. (1989). Groundnut virus research in the Philippines,. In *Second Coordinators' Meeting on Peanut Stripe Virus*, 1-4 Aug. 1989, held at ICRISAT Centre, Patancheru, India. p. 12.

Nigam, S.N., Reddy, D.V.R., Moss, J.P., Dwivedi, S.L. and Reddy, L.J. (1991). Breeding for resistance to groundnut virus diseases at ICRISAT Centre, In *Groundnut Virus Diseases in Africa*, ICRISAT, Patancheru, India. pp. 17-18.

Niriani, T,K. and Dhingra, K.L. (1963). A mosaic disease of groundnut (*Arachis hypogea*) *Indian Journal of Agricultural Sciences.*, 33: 25-27.

Nolt, B.L., Reddy, D.V.R. (1984). In Compendium of Peanut Disease (Eds. Porter, D.M.Smith, D.H.Rodrignez Kabana, R.) 73pp.

Nolt, B. L., R. Rajeswari, D. V. R. Reddy N. Bharathan and S. K. Manohar., (1988). Indian peanut clump virus isolates: host range,symptomatology, serological relationships,and some physical properties. Phphytopathology, 78,310-313.

NRCG. (1993). Work done report at the National Research Centre for Groundnut on the penut Stripe Virus Disease from 1987 to 1993.Hand out , pp 9.

Ohki, S.T., Nakasuji, T. and Inouye, T. (1989). Peanut Stripe Virus, a seed borne potyvirus isolated from peanut plant in Japan. Annals of the Phytopathological Society of Japan, 55 (1), 72-75.

Olorunju, P. E., Kuhn, C.W., Demski, J. W., Misari, S.M. and Ansa, O.A. (1991). Disease reactions and yield performance of peanut genotypes grown under groundnut rosette and rosette-free field environments. *Plant Diseases.*, 75(12), 1269-1273.

Olorunju, P. E., Misari, S.M. and Rao, R. C. N. (ed.); Subrahmanyam, P. (1992). Groundnut breeding in Nigeria: past and present achievements. *Proceedings of the fifth regional groundnut workshop for southern Africa*, 9-12 March 1992, Lilongwe, Malawi. pp 39-41.

Olorunju, P. E., Misari, S.M. Demski, J. W. (1994). Performance of introduced early- and late-maturing groundnut varieties at Samaru, Nigeria. *International Arachis Newsletter.*, 14, 16.

Ozigi, I.A., Olorunju, P.E. (1997). Evaluation of aphid resistance in some groundnut genotypes. *International Arachis Newsletter*. 17, 46-47.

Padma Raju. (2001). Strategy paper for the manamegement of Peanut Stem Necrosis Disease (PSND). Achrya N. G. Ranga Agric. University, Rajendra nagar, Hyderabad-500 030.

Paguio, O. R. and Kuhn, C. W. (1973). Strains of peanut mottle virus. *Phytopathology.*, 63, 976-980.

Patil, S. A. (1993). Collaborative research in India on breeding groundnut for resistance to bud necrosis disease. *Proceedings of a meeting* , 28 Sep. 1992, ICRISAT Center, India, Patancheru 502 324, Andhra Predesh, pp 28.

Pinnow, D.L., Chalkley, J.H. and Demski, J.W. (1990). A practical method for the detection of peanut stripe virus in peanut seed. Research Report 584. May 1990. The Georgia Agricultural Experiment Stations, College of Agriculture, The University of Georgia. pp 12.

Prasada Rao, R.D.V.J., Reddy, A.S., Chakrabarty, S.K., Sastry, K.S., Reddy, D.V.R., Ramnath and Moss, J.P. (1989 b). Peanut stripe virus research in India. pp. 8-9, In Second Coordinators' Meeting on Peanut Stripe Virus, 1-4 August 1989, ICRISAT, Patancheru, India.

Prasada Rao, R. D. V. J. and Chakrabarty, S.K. (1990). Outbreaks and new records. India. Interception of peanut stripe virus in groundnut seeds imported from Myanmar. FAO Plant Protection Bulletin 38 (1), 48-49.

Prasada Rao, Reddy, R.D.V. and Chakrabarty, A.S. (1988b). Survey for peanut stripe virus in India. *Indian Journal of Plant Protection,* 16 (1), 99-102.

Prasada Rao, Reddy, R.D.V. and Chakrabarty, A.S. (1990).Outbreaks and new records. India. Interception of peanut stripe virus in groundnut seeds imported from Myanmar. *FAO Plant Protection Bulletin,* 38 (1), 48-49.

Prasada Rao, Reddy, R.D.V.., Chakrabarty, A.S., and Reddy, A.S. (1988a). First report of peanut stripe virus from India. *Plant Diseases,* 72(10), 912.

Prasada Rao, Reddy, R.D.V.., Chakrabarty, A.S., and Reddy, A.S. (1989a). Peanut stripe virus occurence in India. *Indian Phytopathology,* 42(4), 487-491.

Prasada Rao, Reddy, R. D. V., Chakrabarty, A.S., Sastry, S.K., Ramnath and Moss, J.P. (1989b). Peanut stripe virus research in India. In *Second Coordinators' Meeting on Peanut Stripe Virus,* 1-4 August 1989, ICRISAT, Patancheru, India. pp. 8-9.

Rajeswari, R. and Murant, A. F. (1988). Purification and particle properties of groundnut rosette assistor virus and production of specific antiserum. *Annals of Applied Biology.,* 112, 403-414.

Rajeswari, R., Iizuka, N., Nolt, B. L. and Reddy, D. V. R. (1983). Purification, sereology, and physiochemical properties of a peanut mottle virus isolate from India. *Plant Pathology.,* 32, 197-205.

Ratna, A.S., A.S.Rao, A.S.Reddy, B.L.Nolt, D.V.R.Rddy, M.Vijayalakshmi and McDonald,D. (1991). Studies on transmission of Indian peanut clump virus diseases by Polymyxa graminis. Annals of Applied Biology, 118, 71-78.

Rechcigl, N.A., Tolin, S.A., Grayson, R.L. and Hooper, G.R. (1989). Ultrastructural comparison of peanut infected with stripe and blotch variants of peanut stripe virus. *Phytopathology,* 79(2), 156-161.

Reddy, D.V.R. (1991). Groundnut viruses and virus diseases: distribution, identification and control. Review of Plant Pathology, 70 (9), 665-678.

Reddy, D.V.R., Demski, J.W. and HORN, N. (1988 a). Identification of peanut stripe virus. p. 14. In First Coordinators' Meeting on Peanut Stripe Virus, 9-12 June 1987, held at Malang, Indonesia, ICRISAT, Patancheru, India.

Reddy, D.V.R., Wongkaew, S. and Santos, R. (1985). Peanut mottle and peanut stripe virus diseases in Thailand and the Philippines. Plant Disease, 69, 1101.

Reddy, D.V.R., Wightman, J.A., Bashear, R.J., Highland, B., Black,M., Sreenivasan,P., Dwivedi, S.L., Demski, J.W., McDonald, D., Smith,Jr. J.W. and Smith, D.H. (1991). Budnecrosis : a disease of groundut caused by tomato spotted wilt virus. ICRISAT, Patancheru, A.P, India. Information Bulletin no.31.

Reddy, D. V. R. (1980). International aspect of groundnut virus research. In R. W. Gibbons and J. V. Mertin eds. *Proc. International Workshop Groundnut at ICRISAT*. Patancheru, A. P. India. pp 203-210.

Reddy, D. V. R., Amin, P. W., Mcdonald, D. and Ghanekar, A. M. (1983). Epidemiology and control of groundnut bud necrosis and other disease of legume crops in India caused by tomato spotted wilt virus. In: *Plant Diseasesease Epidemiology* (eds, R. T. Plumb and J. M. Tresh) Oxford, U. K., Blackwell, pp 93-102.

Reddy, D. V. R., Buiel, A. A. M., Satyanarayana, T., Dwivedi, S. L., Reddy, A. S., Vijayalaxmi, K., Ranga Rao, G. V., Naidu, R. A. and Wightman, J. A. (1995). Peanut Bud Necrosis Disease - An Overview, In recent studies on Peanut Bud Necrosis disease, Proceedings of a meeting, 20 Mar, 1995, ICRISAT, Asia Centre, India, (eds. Buiel, A. A. M., Parleviliet, J. E. and Lenne, J. M. eds.) Patancheru 502 324, Andhra Predesh, India: International Crop Research Institute for Semi- Arid Tropics; and P. O. Box 386, 6700 A. J. Wasgeningen, The Netherlands: Department of Plant Breeding, Agriculture University of Wasgeningen. pp 3-7.

Reddy, D. V. R., Murant, A. F.Duncan, G. H., Ansa, O. A., Demski, J. W. and Kuhn, C. W. (1985a). Viruses associated with chlorotic rosette and green rosette diseases of groundnut in Nigeria. *Annal of Applied Biology.*, 107, 57-64

Reddy, D.D.R., Ranga Rao, G. V. and Sanower, T. G. (1993). Bio-ecology and management of groundnut leaf miner, *Aproaerema modicella.* Group Discussion on Integrated Pest management Srategies in Oilseeds in India, held at Panjab Agricultural University, Ludhiana from 23-24 December, 1993.

Reddy, D.V.R. (1984). In Compendium of Peanut Diseases (Eds. Porter, D.M., Smith, D.H. Rodrigner-Kabana R.) pp 73.

Reddy, D.V.R. (1988). Virus diseases. In Groundnut Monograph. (Ed. P.S. Reddy). Indian Council of Agricultural Research, New Delhi. pp 508-525.

Reddy, D.V.R. (1984). In Compendium of Peanut Diseases (Eds. Porter, D.M., Smith, D.H. Rodrigner-Kabana R.) 73p.

Reddy, D.V.R., Demski, J.W., Wongkaew, S. and Middleton, K.J. (1988 b).Occurrence of isolates of peanut stripe virus. p. 15. In First Coordinators' Meeting on Peanut Stripe Virus, 9-12 June 1987 held at Malang, Indonesia, ICRISAT, Patancheru, India.

Reddy, D.V.R., Nambiar, P.T.C., Rajeswari, R., Mohan, V.K, Anjaiah, V. and Mc Donald, D. (1988 c). Potential of Enzyme-linked Immunosorbent Assay for Detecting viruses, fungi, bacteria, mycoplasma-like organisms, Mycotoxins, and Hormones, pp. 43-49, ICRISAT 1988. Biotechnology in tropial crop improvement.

Reddy, D.V.R., Lizuka, N., Ghanekar, A.M., Murthy, V.K., Kuhn, C.W., Gibbons, R.W. Chohan, J.S. (1978). Plant Disease Reporter 62: 978-982.

Reddy, D.V.R., Rajeswari, R. (1984). In Compendium of Peanut Diseases (Eds. Porter, D.M., Smith, D.H., Rodrignez Kabana, R.) 73p.

Reddy, M., D.V.R.Reddy, Appa Rao, A. (1968). Plant Disease Reporter 52 :494-495.

SALEH, N. and TANTERA, D. M. (1988). Peanut stripe virus disease in Indonesia. pp. 3-4. In First Coordinators' Meeting on Peanut Stripe Virus, 9-12 June 1987, held at Malang, Indonesia, ICRISAT, Patancheru, India.

Saleh, N., Horn, N.M., Reddy, D. V. R. and Middleton, K.J. (1989a). Peanut stripe virus in Indonesia. *Netherlands Journal of Plant Pathology*, 95 (2), 123-127.

Saleh, N., Middleton, K.J., Baliadi, Y., Horn, N. and Reddy, D.V.R. (1989b). Research on peanut stripe virus in Indonesia. In *Second Coordinators' Meeting on Peanut Stripe Virus*, 1-4 Aug. 1989, ICRISAT, Patancheru, India. pp. 9-10.

Sdoodee, R. and Teakle, D. S. (1987). Transmission of tobacco stripe virus by *Thrips tabaci*: a new method of plant virus transmission. *Plant Pathology*. 51, 755-758

SENBOKU, T. (1989). Research on peanut stripe virus in Japan p.11. In Second Coordinators' Meeting on Peanut Stripe Virus, 1-4 Aug. 1989, CRISAT, Patancheru, India.

Singh, A. B. and Srivastava, S. K. (1995). Status and control strategy of peanut bud necrosis disease in Uttar Predesh, In Recent studies on peanut bud necrotic disease: Proceedings of a meeting, 20 March, 1995, ICRISAT Asia Center, India (Buiel, A. A. M., Parleviliet, J. E. and Lenne, J. M. eds.) Patancheru 502 0324, Andhra Predesh, India: International Crop Research Institute for Semi- Arid Tropics; and P. O. Box 386, 6700 A. J. Wasgeningen, The Netherlands: Department of Plant Breeding, Agriculture University of Wasgeningen. pp 65-68

Singh, B. R. and Gupta, S. P. (1989). Prevalence and losses in groundnut due to bud necrosis disease and peanut mottle disease in Uttar Predesh., *Proc. 10th Conven. Indian Virology Society and National Symposium on Epidemiology of viral diseases*, CPRI, Simla, OCT. 1989, pp 36.

Singh,V., M.P.Ghewade and Reddy.P.S. (1993). Peanut Stripe Virus disease-Present status and its management. *International Journal Pest Management*,39,422-430.

Sreenivasulu, P. and Demski, J.W. (1988). Transmission of peanut mottle and peanut stripe viruses by Aphis craccivora and Myzus persicae. *Plant Diseases*, 72 (8), 722-723.

Sreenivasulu, P., Kuhn, C.W., Naidu, R.A., Demski, J.W., Reddy, D.V.R. and Nayudu, M.V. (1991). Viruses infecting peanuts : Taxonomy, identification, and disease management. *Research Bulletin 406,* The Georgia Agricultural Experiment Stations, The University of Georgia College of Agricultural and Environmental Sciences, p. 34.

Subrahmanyam, P., Reddy, L.J., Faris, D.G., Dan, N.T. and Hein, N.X. (1990). Diseases of groundnut in Vietnam. *International Arachis Newsletter,* 7, 9-10.

Sudarshan, M.R., Reddy, D.V.R. and Reddy, A. S. (1990). Methods for the detection of seed borne viruses in groundnut (*Arachis hypogaea* L.) In *Proceedings National Seminar on Advances in Seed Science and Technology,* pp. 203-209.

Sudhakar Rao,A., R.D.V.J.P.Rao, and P.S.Reddy.(1980). A whitefly transmitted yellow mosaic disease on groundnut (*Arachis hypogaea* L.), Current Science,49,160.

Theuri, J. M., Bock, K. R. and Woods, R. D. (1987). Distribution, host range and some properties of a virus diseases of sunflower in Kenya. *Tropical Pest Management.* 33, 202-207.

Thouvenel, J. C., Dollet, M. and Fauquet, C. (1976). Some properties of peanut clump, a newly discovered virus. *Annals of Applied Biology.* 84, 311-320.

Tolin, S.A. and Rechcigl, N.A. (1985). Occurrence of peanut stripe virus in Virginia (Abstr.) *Phytopathology,* 75, 629.

Tsai S. Y., Hwang G.H. and Dhing T. (1978). Some insect viruses discovered in China. *Actes Entomol.* Sinica., 21, 101-102.

Vijayalakshmi, K. (1994). Transmission and ecology of *Thrips palmi* Karny, the vector of peanut bud necrosis virus. Ph. D. thesis. Andra Pradesh Agric. University, Rajendra Nagar, Hyderabad, India. pp 99.

Wakman, W., Pakki, S. and Hasanuddin, A. (1989). Yield loss of groundnut due to peanut stripe virus, p.11, In *Second Coordinators' Meeting on Peanut Stripe Virus,* 1-4 Aug. 1989, ICRISAT, Patancheru, India.

Warwick, D. and Demski, J.W. (1988). Susceptibility and resistance of soybeans to peanut stripe virus. *Plant Diseases,* 72, 19-21.

Waterston, J. (1922). On chaicidoidea, mainly bred at Dehra Dun, U.P. from pests of Sal, Toon.

Wongkaew, S. and Kantrong, S. (1989b). Purification and serology of peanut stripe virus. In *Proceedings of the Seventh Thailand National Groundnut Meeting* for 1987, Pattaya, Chonburi, Thailand, 16-18 March 1988. Khon Kaen, Thailand: Khon Kaen University,Faculty of Agriculture. pp. 213-219.

Wongkaew, S. (1989). Groundnut virus research in Thailand. pp. 12-13, In Second Coordinators' Meeting on Peanut Stripe Virus, 1-4 Aug. 1989, ICRISAT, Patancheru, India.

Wongkaew, S. and Dollet, M. (1990). Comparison of peanut stripe virus isolates using symptomatology on particular hosts and serology. Oleagineux, 45 (6), 267-278.

Wongkaew, S. and Dollet, M. (1989). Comparison of peanut stripe virus isolates using symptomatology on particular hosts and serology. pp.16-17, In Second Coordinators' Meeting on Peanut Stripe Virus, 1-4 g. 1989, ICRISAT, Patancheru, India.

Wongkaew, S. and Kantrong, S. (1989 a). Yield loss assessment for peanut stripe virus in groundnuts. pp. 206-209, In Proccedings of the Seventh Thailand National Groundnut Meeting for 1987, Pattaya, Chonburi, Thailand, 16-18 March 1988. Khon Kaen, Thailand : Khon Kaen University, Faculty of Agriculture.

WONGKAEW, S. (1986). Peanut stripe and other viruses in Thailand. pp. 86-90. In Proceedings of Peanut CRSP Workshop, 19-21 Aug. 1986, held at Khon Kaen University, Khon Kaen, Thailand. (H. T. Stalker, A. Waranyuwat, and B.B. Shew, Ed.) 106 pp.

Wongkaew, S., Kantrong, S. and Choopanya, D. (1988). Peanut stripe virus in Thailand. p. 8, In First Coordinators' Meeting on Peanut Stripe Virus 9-12 June 1987, held at Malang, Indonesia, ICRISAT, Patancheru, India.

Wynne, J.C. (1986). Effect of peanut stripe virus on peanut breeding. Proceedings of American Peanut Research and Education Society, 18, 78.

Xu, Z. and Barnett, O. W. (1984). Identification of a cucumber mosaic virus strain from naturally infected peanuts in China. *Plant Disease*. 68: 386-389.

XU, Z. (1988). Research on peanut stripe virus disease in The People's Republic of China. In *First Coordinators' Meeting on Peanut Stripe Virus*, 9-12 June 1987, held at Malang, Indonesia. ICRISAT, Patancheru, India. pp. 6-7.

XU, Z., CHEN, K., ZHANG, Z. and CHEN, J. (1989 a). Screening groundnut genotypes for seed transmission of peanut stripe virus. International Arachis Newsletter, 6, 18-19.

XU, Z. (1988). Research on peanut stripe virus disease in The People's Republic of China. pp. 6-7. In First Coordinators' Meeting on Peanut Stripe Virus, 9-12 June 1987, held at Malang, Indonesia. ICRISAT, Patancheru, India.

XU, Z., CHEN K., ZHANG, Z. and CHEN, J. (1991). Seed transmission of peanut stripe virus in peanut. Plant Disease, 75, 723-726.

Xu, Z., Yu, Z., Lui, J. and Barnette, O. W. (1983). A virus causing peanut mild mottle in Hubei Province, China. *Plant Diseases*. 67, 1029-1032.

XU, Z., ZHANG, Z., CHEN, K. and CHEN, J. (1989 b). Groundnut virus research in China. pp. 7-8. In Second Coordinators' Meeting on Peanut Stripe Virus, 1-4 Aug. 1989, ICRISAT, Patancheru, India.

XU, Z.Y., CAI, Z.A., YU, S.L. and QIU, W.F. (1984). The types of peanut virus diseases and serologicl diagnosis of the viruses in Northern China. Oil Crops of China, 3, 48-54.

XU, Z.Y. (1985). Detection of PMMV and CMV-CA in peanut seeds. Oil Crops of China, 11, 56-58.

Yadav, L. S. and Yadav, P. R. (1983). Pest complex of Cowpea (*Vigna sinensis* Savi). in Haryana, Bulletin of Entomology.

Yang, C. and Liu, H. (1966). Biological observations on *Stomopteryx subsecivella* Zell. in Den Bei District, Kwantung. (In Chinese). *Acta Entomol. Sin.*, 15, 39-46.

□□□

Botanicals

K. SAHAYARAJ AND V. NANDAGOPAL

18.1. Introduction

The indiscriminate use of synthetic insecticides for the control of insect pests in groundnut has led to several environmental problems such as serious health hazards to human-being and animals, development of resistance by the insect pests, destruction of natural enemies and pesticide residues. This growing awareness towards the environment has made it imperative to search for alternatives to the chemicals. Since ancient times, neem (*Azadirachta indica* A. Juss.) has been known for its insecticidal and anti-microbial properties. The leaves, bark, fruit and seed are known to possess toxic principles with insecticidal and anti-microbial activities. But use of neem in the management of insect pests of groundnut is limited. Since, neem and its products are abundantly available in India, efforts were made to use them for the management of insect pests (Ghewande *et al.*,1993; Nandagopal, 1992; Nandagopal *et al.*, 1992; Sahayaraj and Paulraj, 1998c).

18.2. Neem as Insecticide

Aqueous extracts of neem leaf, bark and seed, neem oil and neem-based formulations have been tested under laboratory/field conditions for the management of insect pests of groundnut. The neem based components of the IPM found promising under the laboratory and insectary conditions were also studied in combination with synthetic insecticides, trap crops and plant products under field condition. Besides, experiments were also conducted on the efficacy of neem products singly and in combination with other components. Reviews of Saxena (1989) and Schmutterer (1990) revealed that not much work has been done on the use of neem products in the control of insect pests in groundnut. Neem and its products affect biology, feeding, sex communication and repel insect pests; the mode of action is mainly as a contact insecticide and to some extent as feeding detterent. The compatibility of crude neem oil with other pesticides is essential to use such mixtures in the groundnut ecosystem. Various concentrations of crude neem oil either alone or blended with synthetic pesticides were used at vegetative, pegging and crop maturity. Crude neem oil at 2% and in combination with phosphamidon, endosulfan, carbendazim or with mancozeb did not cause phytotoxicity in groundnut foliage but high doses were phyto-toxic. Hence, crude neem oil (2%) blended with compatible synthetic pesticides may be used in integrated pest management. The mode of action has been investigated (Jacobson, 1986; Kraus *et al.*,1985; Saxena, 1989). The neem products are primarily feeding poisons for nymphs/larvae and adults of polyphagous insects and non-toxic natural enemies of pests (Schemutterer, 1990). Unlike synthetic insecticides, the neem products have the advantage of not inducing resistance in insects (Saxena, 1989) and may be included in the integrated pest management (Schemutterer, 1990). The efficacy of botanicals (*Vitex negundo* (5%), *Argemone mexicana* (5%), *Calotropis gigantea* (5%), neem seed kernel extract (NSKE 5%), and pongamia oil) with entomopathogenic fungi *N. rileyi* in managing *Spodoptera litura* in groundnut (cv. Dh-3-30) ecosystem. Both 0.05% monocrotophos and untreated control were maintained for comparison. In 1996, *N. rileyi* at 2x1011 conidia per ha combined with either *V. negundo*, NSKE or A. mexicana were the most effective treatments, next to monocrotophos spray. NSKE and *V. negundo* combined with N. rileyi recorded higher yields than *N. rileyi* alone and the untreated control. In 1997, two sprays of *N. rileyi* combined with NSKE, with *V. negundo* or with *A. mexicana* at 15-day intervals were the next best treatments to monocrotophos. No difference was observed in pod yields between combination treatments and chemical control in both years. NSKE + *N. rileyi* and *V. negundo* + *N. rileyi* were as effective as the monocrotophos spray. Although the highest benefit:cost (B:C) ratio was obtained in *N. rileyi* spray alone (15:1), it failed to afford complete or desired level of protection against the defoliator. Among the treatments, NSKE + *N. rileyi, V. negundo* + *N. rileyi* and *A. mexicana* + *N. rileyi* were the most cost-effective, with B:C ratios of 12.28:1, 14.22:1 and 14.01:1, respectively. NSKE and *V. negundo* enhanced the bioefficacy of *N. rileyi* against S. litura in groundnut, resulting in high B:C ratio and yields comparable to that of monocrotophos, aside from being safer to environment (Patil *et al.*, 2002).

18.2.1. Major insect pests

18.2.1.1. Aphids, *A. craccivora*

The extent of losses in groundnut crop caused by aphids direct damage (feeding) and by transmitting virus diseases are well documented (Amin, 1988; Reddy, 1988). Some sources of resistance to aphids have been reported (Brar, 1981; Brar and Sandhu, 1975). Crude neem oil and neem seed extract at 2% concentration has not affected the biology of the aphids, but, it may prolong the duration of second instar from 3 days to 5.5 days. However, When *Oncopeltus fasciatus* nymphs were treated with pure azadirachtin, they could not develop into adults even for five months but, then produced mature eggs (Dorn *et al.*,1987). Crude neem oil and seed extract could not affect the fecundity, pre-oviposition, oviposition and life-cycle of the aphid (Nandagopal, 1992; Bakhetia and Sandhu, 1977; Behura,1956; Dorge *et al.*,1966).

The mortality of aphid varied significantly with various concentrations of crude neem oil with high concentrations causing quick mortality. Complete mortality of both nymphs and adults occurred after 6-7 h of exposure to the lowest concentration (0.2%) of crude neem oil (Nandagopal, 1992). Devakumar *et al.* (1986) observed that the LD_{50} of the safflower aphid, *Dactynotus carthami* as 0.24%. Neem products had no influence on the population build up of aphid on groundnut plants.

18.2.1.2. Leafhoppers, *Empoasca kerri, Balclutha hortensis*

Of the 16 species of leafhoppers occurring in groundnut (Amin,1988), only two namely *E. kerri, B. hortensis* are harmful to the crop and are a major constituent of leafhopper complex in Saurashtra, Gujarat. Leafhoppers cause yield loss of about 584 kg/ha (Anon,1988). Investigations showed that crude neem oil-based plant products may control leafhoppers. Neem oil, neem leaf extract and neem leaf decoction at 5% gave 87 to 91% cumulative mortality over 48h of exposure of *B. hortensis* (Nandagopal *et al.*,1990). Synthetic insecticides and plant products were also studied for the management of this insect. The toxicity of cypermethrin and dimethoate persisted for 10 days and was least in monocrotophos and neem oil (12-23 h). Thus, crude neem oil needs to be applied frequently.

18.2.1.3. *Thrips, Scirtothrips dorsalis* Hood

Neem product such as neem seed kernel 5%, crude neem oil 3%, Nimbicidine 3 ml/lit, neem gold 3 ml/ lit, neem azol 2 ml/lit and Achook 3gm/ lit when sprayed reduced the thrips population in groundnut ranging from 25 to 50% compared to control (Senguttuvan, 1999).

18.2.1.4. Leafminer, *Aproaerema modicella*

Among the defoliating insects of groundnut, the leaf miner results in 23 to 89% loss in pod yield (Mohammad, 1981). Neem-based insecticides namely, Jawan and Margoside OK and CK were compared with crude neem oil or its blending with hostathion. The immediate toxicity of crude neem oil at lowest concentration of 2% either alone or with hostathion deterred 100 % oviposition of leafminer and the residual toxicity persisted up to 15 days, while other treatments had no residual effect. The field study also confirmed these results. Thus, crude neem oil proved the most effective in deterring the oviposition of leaf miner. Crude neem oil suppressed the oviposition had ovicidal property but had no influence on larval and pupal mortality of leafminer. There are varied reports on the activity of the crude neem oil on various insect-pests of crops, the oligophagous insects were more sensitive to neem oil than the polyphagous pests (Simmonds and Blaney,1984), leafminer, being oligophagous, responded favourably (Mohammad,1981). Suppression in the hatching of eggs of leafminer was probably due to the choking effect of oil. Hence, to control this insect, neem oil may be used to prevent its oviposition. The investigation on the parasite emergence from the neem oil treated leafminer larvae indicated its safety as there was normal emergence of parasites. Among the various plant products, crude neem oil and neem seed extract were found to be encouraging either alone or in blending with endosulfan (0.04%) and phosphamidon (0.02%) for leafminer management in groundnut. Neem seed extract 5% + 0.025 % phosphamidon had lowest (1.55 %) intensity of leafminer when compared to control (16.9 %), followed by 2% crude neem oil and gave maximum of pod yield (2068 kg/ha). While, crude neem oil 2% + dimethoate 0.04% although gave pod yield of 1404 kg/ha, but caused yellowing and scorching of foliage in groundnut.

In field experiments, with soybean and castor as trap crops, neem seed extract 2% with two synthetic insecticides, namely, 0.02% phosphamidon and 0.04% dimethoate at two and three spray schedules were evaluated. The results indicated that leafhoppers and leafminer were effectively controlled by three sprays of neem seed extract blended with dimethoate in the presence of castor and soybean. The pod yield of groundnut was 950 kg/ha with Rs.7233 monetary value under castor as a trap crop with two rounds of spray. Since, crude neem oil deterred oviposition in leafminer, its mixture with various synthetic insecticides was investigated. The crude neem oil in combination with the mixture of 0.02% phosphamidon and 0.04% endosulfan efficiently controlled insect population in groundnut. The crude neem oil suppressed the oviposition, whereas endosulfan and phosphamidon controlled sucking pests and defoliators, respectively. The neem oil based formulation (Margoside CK and OK) either alone or in blending with monocrotophos resulted in higher pod yield (ranging 3038 to 3229 kg/ha) compared to control (2771 kg/ha) in the AICRPG trial (Anon, 1990). This is in conformity with the findings of Chandramohan and Sivasubramaniam (1987) who reported significant higher pod yield (2383 kg/ha) when 2% neem oil was sprayed at 35 and 55 days after sowing.

The crude neem oil toxicity persisted for 12 days against leafminer. The longer persistence against leafminer might be attributed to the shiny coating on the leaflets which might have prevented the orientation of the oviposition females. Since the residual effect was more than 12 days against leafminer three sprays coinciding with emergence of each brood could be sufficient to contain this pest below the economic injury levels. Different parts of neem tree particularly seeds, contain the triterpenoid azadirachtin (Zanno *et al.*,1975) that has anti-feedant effect and disturb metamorphosis in Lepidoptera, Coleoptera, Heteroptera and Diptera (Meisner *et al.*, 1987; Schemuterer, 1990; Schemuterer *et al.*, 1983; Steets,1976; Steffens and Schemuterer,1982).

18.2.1.5. Gram pod borer (*Helicoverpa armigera*), red hairy caterpillar (*Amsacta albistriga*) and tobacco caterpillar (*Spodoptera litura*)

Patel (1968) evaluated efficacy of neem seed extract, neem leaf extract at 10% and neem oil at 2% on the hairy caterpillar *A. albistriga*. Crude neem oil (2%) spray increased the per hectare pod yields significantly than control. Neem kernel suspension (1%) and neem seed extract inhibited the growth of *H. armigera* and *S. litura* larvae (Jotwani and Srivastava,1981; Koul,1985; Rao and Subramanya,1987). Neem leaf extract even at 5% conc. resulted 65% larval mortality in *H. armigera*. The neem based insecticides (Margoside OK and CK) suppressed the *S. litura* by reducing the foliage damage in groundnut and gave higher pod yield (Anon,1990). When 5% neem oil was sprayed on the eggs of *S. litura*, a mean mortality of 36% was observed. Similarly there was a least foliage damage of 27% of the groundnut crop. Almost similar result was obtained when the concentration of the neem oil was raised to 10% and 15% (Ghelani, 1989). Sahayraj, (1998) exposed the groundnut shoot treated with 10% crude water extract of neem leaves to the final instar larvae of *S.litura*. The food consumption and the approximate digestibility was higher in neem leaf extract. The reduction in the body weight was 53% resulting about 25% deformity of adults emerged.

18.2.1.6. Termites (*Microtermes obesi* and *Odonototermes* spp.)

Application of neem cake mulch to a depth of 2.5 cm of the soil resulted low level of pod damage (2%) by termites compared to 36 % in the control (Gold *et al.*, 1989).

Euborellia stali: Application of 250 kg of neem seed cake resulted in lowest pod damage of 4.9% compared to 9% pod damages in untreated control plot (Senguttuvan and Dhanakodi, 1997).

18.2.1.7. Storage insect pests

1. *Caryedon serratus* : Neem seed kernel powder, neem leaf powder at 25 gm/kg of groundnut pod resulted in total protection for a period of 6 month against *C. serratus* (Kumari *et al.*, 1998).

2. *Corcyra cephalonica* : Neem leaf powder and neem oil were found most effective against *C. cephalonica* in stored groundnut. (Senguttan *et al.*, 1995). Azadirachtin when tried against the last instar larvae of *C. cephalonica* affected further development. Similarly the larval to pupal

and pupal to adult development was also disturbed due to interference with the morphgenetic hormone pool size (Sharma, 1992). Neem leaf powder, neem kernel powder and neem oil were found to be most effective against *C. cephalonica* based on the damage caused to the storage groundnut on the basis of % dry matter loss (Senguttuvan *et al.*,1995). Deformed adult with a prolonged period of development were recorded when 0.03 % neem oil was mixed with the food. Similarly growth inhibition, developmental disturbance and increased mortality were also recorded (Sudipta-Chanda and Sanjib Chakroborty., 1998). In a peculiar response the de-odorizes neem attracted *C. cephalonica* larvae at 0.5 % concentration (Behal, 1998). When neem derivatives like Fortune Aza 0,15 %, when exposed at 75 to 100 ppm the mortality and oviposition apart from pupal and adult duration were reduced drastically (Murugasen *et al.*,1998). There was no deleterious effect of the aqueous suspension and ethonolic extract of neem seed kernels at 0.3, 0.6, 1.2, 2.5 and 5 % concentration on the parasitoid *Bracon hebator* for oviposition, larval and pupal development (Raghuraman and Singh, 1998).

18.3. Integrated pest management

Various components of IPM such as, plant products, trap crops, pheromones and mixture of insecticides with neem oil have been studied under field conditions. Pheromones, trap crops and crude neem oil (2%) mixed with synthetic insecticides phosphamidon (0.02%) or endosulfan (0.04%) were tried either alone or in suitable combinations. The pesticide mixture (crude neem oil 2% + endosulfan 0.04% or phosphamidon 0.02%) was sprayed on main crop, trap crop or on both the crops in the presence of trap crops and pheromone. Leafhoppers, leaf miner and thrips populations were drastically suppressed throughout the growth period when insecticides mixture including neem oil was sprayed. The use of pheromones alone, is not likely to give complete control of all the larval pests. The role of insecticides mixture proved efficient in the management of these pests. The trap crops also proved useful in trapping the insects as they lowered the pest population considerably. The judicious integration of these IPM components may solve the insect pests problem in groundnut without causing undesirable side effects.

Though neem products proved less effective in their residual toxicity for sucking pests in groundnut, they were found to possess excellent oviposition deterrent and ovicidal action for the leafminer (Nandagopal,1992). The commercial formulations of neem products insecticides are listed in the Table 13.1. Though synthetic insecticides (phosphomidon, endosulfan and monocrotophos) were inefficient for the control of thrips (Wightman and Amin,1988), their blending with neem products decreased the dose of insecticides to 50%. This blend can efficiently solve the pest problem in groundnut. A dynamic IPM for leafminer was developed (Dudley *et al.*,1989). The neem based IPM was tested over years and developed (Nandagopal, 1992).

18.4. Other plant products

18.4.1. A. modicella

The combined effect of *Bacillus thuringiensis* subsp. kurstaki (Dipel) and certain botanical insecticides on growth, feeding and nutritional efficiency of *Helicoverpa armigera* was investigated in the laboratory at 27+or-1 degrees C, LD 14:10 and 75% RH. The antifeedant activity of neem oil and *Vitex negundo* leaf extract was greater than that of Pongamia oil. However, growth was severely inhibited by Pongamia oil. Mortality was significantly increased by Pongamia oil with *B.' thuringiensis* subsp. *kurstaki.* at different concentrations considerably inhibited the food consumption of *H. armigera*. Approximate digestibility was decreased at higher concentrations of *Bt.* The efficiency of conversion of ingested and digested food was significantly reduced and the profiles of digestive enzymes were decreased by *Bt.* treatment. A significantly greater reduction in growth rate was observed when *Bt.* was mixed with Pongamia oil than with the *Vitex negundo* extracts or neem oil. The complete inhibition of nutritional efficiency measures and digestive enzymes in the midgut of *H. armigera* suggested that *Bt.* and plant extracts greatly affect the gut homeostasis and suppress the feeding physiology (Murugan and Babu, 1998).

18.4.2. S. litura

For *S. litura* , *Pongamia* glabra (Ghelani, 1989) *V. negundo* (Sahayraj, 1998), *Zingiber officinale* (Sahayraj and Paulraj, 1998) and *C. giganta* was found most toxic. The pupae formed were deformed in nature.

18.4.3. H. armigera

Pongamia glabra, Vitex negundo and *Calotrophis giganta* were found equally effective against *H. armigera* causing more than 50 % mortality at 6 % concentration (Sahayraj and Paulraj, 2001) and the survivors became deformed pupae and adults **(Plate 1, 2, 3 & 4)**.

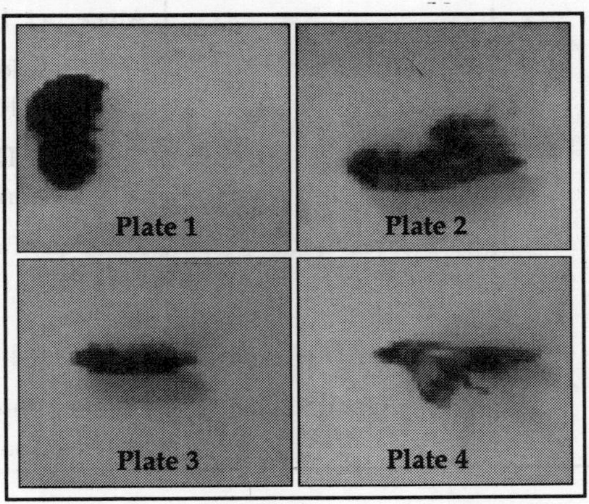
Plate 1 Plate 2 Plate 3 Plate 4

18.4.4. C. serratus

Lantana camare leaf powder at 25 gm / mg of pods was found to be effective oviposition deterrant for *C. serrata* but the effect was lasted for a month (Kumari *et al.*, 1998).

18.5. Conclusion and future areas of research

To reduce the yield losses in groundnut caused by insect pests neem and its products could be a viable alternative to the synthetic pesticides. Integrated pest management includes a judicious blend of a multi-pronged approach that will reduce the incidence or delay the build up of the intensity of insect pests. The effects of the insecticidal and antimicrobial toxic principles of different plant parts of neem (*Azadirachta indica* A. Juss.) on the insect pests are well established. The commercial availabile list of the neem based pesticides is given in Table 14.1 (Puri *et al.*, 1997).

Table 18.1. Commercial formulations of neem available in India

S.No.	Trade Name	Formulation	Sources
1	Achook	0.5% Azadiradion, Nimbocinol and Epinimbocinol	Godrej Agrovet Ltd., Pirojshanagar Estern Express Highway, Vikhroli, Mumbai-400079 M/s Bahar Agrochemicals Ltd., E/24. MIDC, Lote Parshuram Road, Raynagiri-451722. (Maharashtra)
2	Aphidin	Soapy humus extract	Eco-Max Agrosystem Ltd., Good Value Marketing Company Mumbai-400020
3	Aza	0.15% Azadirachtin	M/s SOM-IPM System India Ltd., Flat No.-12, Kempson Apartment Ameerpet, Hydrabad-500016
4	Azadit	0.3% Azadirachtin	M/s Pesticide India Ltd., Udaisagar Road, Udaipur-313001.
5	Biosol	Oil	M/s A. V. Thomas Pvt. Ltd.,Anna Salai, Chennai-600032
6	Bioneem	0.03% Azadirachtin	M/S Ajay Bio-Tech Laboratories Pvt Ltd.,A-12/5, Kubera Park, Kanchawa, Pune-411040 (Maharashtra)
7	Biopest	0.15% Azadirachtin	Bio-Ved Research Society 103/42, M. L. N. Road, Allahabad-211002

Contd...

Table 18.1. Contd...

8	Eco-neem	0.3% Azadirachtin	M/S P. J. Margo Pvt. Ltd., 344/8, IV Main, Sdashivnagar Banglore- 560080
9	Field marshal	0.3% EC Azadirachtin	Khetiwadi Corner,Shiyabave, Vadodara-390001 (Gujarat)
10.	Fortuna aza	0.15% Azadirachtin	Fortune Bio-Tech Pvt. Ltd., 14, Ishaq colony, 108, Bazzar Road, Secunderabad-500015. (A. P.)
11.	Jai neem (developed by iari)	0.03% Azadirachtin	Jai Chemicals, 14/1, Matura Road,Faridabad-121003
12.	Jawan	0.03% Azadirachtin	Mc DA Agro Pvt. Ltd., Mumbai-400001.
13.	Jeevan crop protector	0.05% Azadirachtin	Mc DA Agro Pvt. Ltd., Mumbai-400001.
14.	Juerken	0.03% Azadirachtin	Madurai Chemicals & Agro Industries Ltd., Rajaji Road, Madurai – 625104 (Tamilnadu)
15.	Limonool	75% EC 0.3% Azadirachtin	M/S Bio- Multi Tech Pvt. Ltd., Ganganagar, Banglore-500065 (Karnataka)
16.	Margocideck	20% EC (Azadirachtin 0.15% w/w)	M/s Monofix Agro Product Ltd. V. G. Limbikai Building, Gokule Road Hubli-580030 (Karnataka)
17.	Margosan-0	80% EC Azadirachtin	M/s Monofix Agro Product Ltd. V. G. Limbikai Building, Gokule Road Hubli-580030 (Karnataka)
18.	Neemark	0.03% Azadirachtin	M/s West Cost Herbochem Pvt. Ltd. 105/B,Industrial Suburb, II Stage, Rajajinagar, III Cross Rd. Banglore-22
19.	Neemazal-f	5% Azadirachtin	Pesticide Division, E. I. D, Parry (India) Ltd., "Dare House", P. B. No.-12, Chennai-600001
20.	Neemgold	0.15% Azadirachtin	Southern Petrochemical Industries

Contd...

Table 18.1. Contd...

			Corporation Ltd, Bio-Tech Division, Agro Ind., Complex, 97, Mount Road, Chennai-600032
21.	Neem guard	0.03% Azadirachtin	M/S Ghadra Chemicals Pvt. Ltd., 48, Hill Road, Mumbai-400050
22.	Neem rich	0.03% Azadirachtin	National Chemical Labortory,CSIR Laboratory, Pune-414001
23.	Neemolin	0.15% Azadirachtin	M/s Khatau Agro Tech., Khatau House, Mogul Lane, Mahim, Mumbai-400016
24	Neempourn	0.15% Azadirachtin	M/s Prabhakar Oil Mills, Trichy-632217. (T. N.)
25	Neemosan	0.03% Azadirachtin	M/s Agronule Indutries, Trichy-632217. (T. N.)
26.	Neemox	80% EC Azadirachtin	Prakash Farm Chemicals, Gulmohar Cross, Road No.-9, Vile Parle West, Mumbai-400049
27.	Neemata	0.15% Azadirachtin	M/S A. J. Chemicals, Kisan brother Pvt. Ltd. New Cloth Market,Ahmedabad-380002 (Gujarat)
28.	Neethrin	Oil 0.03% Azadirachtin	Amitul Agro Chem Pvt. Ltd., P. O. New Snehapuri Colony, Gorakpur-273001 (U. P.)
29.	Nimba	Wettable powder 0.15% Azadirachtin	Division of Entomology, Indian Agriculture Research Institute, Pusa, New Delhi-110012
30.	Nimbcidine	0.03% Azadirachtin	M/s T. Stanes & Co. Ltd., 8/23-24, Race Course Road, Coimbatore-614018 (Tamilnadu)
31.	Nimbosol	0.15% Azadirachtin	Victoria Laboratories, Mahatma Gandhi Road, Salem-632004 (Tamilnadu)
32.	Nimbasal	0.15% Azadirachtin	M/s Nimba Food & Agro Chem. Pvt. Ltd. Agro Chemical Division, 14-A/10, Western Extension, Pusa Road, New Delhi-110055

Contd...

Table 18.1. Contd...

33.	Nimlin	0.03% Azadirachtin	Sunline Agro. Chemicals Pvt. Ltd. P. B. No.-73, Sakri Road, Dhulia (Maharashtra)
34.	Neem base	20% EC 20 WDPAzadirachtin	Agrochemical Division, Indian Agriculture Research Institute, Pusa, New Delhi-110012
35.	Neem hit	0.15% Azadirachtin	M/s Sky Lark Agro Chemicals, W-92, M.I.D.C, Phase-II, Dombivali (East), Thane-412204 (Maharashtra)
36.	Neem oil	0.03% EC Azadirachtin	Six Agro Research Laboratories, Dhanraj Apartment, Mulund (West) Mumbai-400080
37.	Neem plus	0.15% Azadirachtin	M/s B. D. Keathen & co., Bandra (East), Mumbai-400029
38.	Neem top	0.03% Azadirachtin	Sri Krishna Company, Race Coure Road, Coimbatore-641018 (T. N.)
39.	Neemol	0.03% Azadirachtin	Ramson Laboratories, 29-14-53, Prakasm Road, Vijayawada-520002. (A. P.)
40	Nim-76	0.05% Azadirachtin	Defence Institute of Physiology & Applied Sciences, Delhi Cantonment., New Delhi-110022.
41.	Neem oil	2% Azadirachtin	National Organic Chemical Industries Limited, 951/A, Appasaheb Marathi Marg, Prabhadevi, Mumbai-400025.
42.	Neemax	Seed Powder 300 ppm Azadirachtin	Eco-Max Agrosystem Ltd., Good Value Marketing Company, Industrial Assurance building, Church Gate, Mumbai-400020
43	Neemocide	0.03% Azadirachtin	Vinayak Fats & Protein Pvt. Ltd., Agri. in-put Division, 206, Bentinck Chambers, 37 A, Bentinck Street, Calcutta-700069
44.	Neemactine	0.15% Azadirachtin	M/s Wockhardt Limited, Ready Money Terrace167, Dr. Annie

Contd...

Table 18.1. Contd...

			Besant Road, Worli, Mumbai-400018.
45.	Nimbo bas	0.03% Azadirachtin	Basarass Bio-Control Research Laboratory, 3/204, Main Road, Eraiyur, Pennadanam, R.S., S. A. Vallar District-606111 (T. N.)
46.	Nimbo bas	0.15% Azadirachtin	Basarass Bio-Control Research Laboratory, 3/204, Main Road, Eraiyur, Pennadanam, R.S., S. A. Vallar District-606111 (T. N.)
47.	Nimbitor (za – 199)	0.03% Azadirachtin	M/s Zandu Pharmaceutical Works Limited., 70, Ghokhale Road (South), Mumbai-400025
48.	Phytowin	0.03% Azadirachtin	Phyto Product Ltd., Thirutharipoondi, S.A. District, Tamilnadu.
49.	Rakshak	0.15% Azadirachtin	M/s Murkumbi Bio Agro Pvt. Ltd., BC-105, Pov Lock Road, Cantonment, Belgaum-590001 (Karnataka)
50.	Replin 555	93% EC Azadirachtin	Micro Chemicals Limited, Rajajo Road, Madurai-625104 (T. N.)
51.	RD- 9 replin	100% EC Azadirachtin, Pongamia, Anona	M/s I.T.C. Limited, I.L.T.D. Division, IBD-6-3-1110, Amrutha Hall, Begumpet, Hydrabad-500016 (A.P.)
52.	Shaktiman	0.03% Azadirachtin, Nimbim	Krishna Bio-Tech Pvt. Ltd., 8-B, National Highway, Post Veraval, Rajkot District, 360002 (Gujarat)
53.	Sukrina	0.03% Azadirachtin, Meliontriol	M/s Constar Chemical Ltd Adayar,Chennai-600 020

References

Amin, P.W. (1988). Insect and mite pests and their control. In *Groundnut* (Ed. P.S. Reddy). pp. 393-452. Indian Council of Agricultural Research, New Delhi.

Anonymous, (1990). Annual Progress Report, All India Coordinated Research project on Oilseeds. Hyderabad.

Anonymous, (1988). Annual report *International Crop Rresearch Institute for the Semi-Arid Tropics (ICRISAT).* Patancheru, Andhra Pradesh, India.

Anonymous, (1990). Annual Progress Report (1989 to 1990). of AICRP on Rodents Control. (Junagadh: Gujarat Agricultural).

Anonymous, (1988). Annual Report, National Research Centre for Groundnut, Junagadh, Gujarat, 75 pp.

Bakhetia, D.R.C. and Sidhu, A.S. (1977). Biology and seasonal activity of the groundnut aphid *Aphis craccivora* Koch. *Journal of Research of Punjab Agricultural University* 14, 299-303.

Behal-S. R. (1998). Effect of some plant oils on the olfactory response of the larvae of rice-moth *Corcyra cephalonica* Stainton. *Annals-of-Plant-Protection-Sciences*, 6: (2), 146-150.

Behura, B.K. (1956). Observations on the biology of Aphis *craccivora* Koch (Aphididae: Homoptera). *Proceedings of the 43rd Indian Science Congress* Vol. III: 295. Indian National Science Association: New Delhi.

Brar, K.S. and Sandhu, B.K. (1975). A note on the evaluation of resistance among different cultivars against aphid and grey weevil. *Science and culture* 41, 445-448.

Brar, K.S. (1981). Note on the comparative susceptibility of groundnut varieties to *Aphis craccivora* Koch. *Madras Agricultural Journal* 68, 207-208.

Chandramohan, N and Sivasubramanian, P. (1987). Evaluation of neem products against leafminer *Aproaerema modicella* (D.). *Neem newsletter*, 4(4), 44.

Devakumar, C., Saxena, V.S. and Mukerjee, S.K. (1986). Evaluation of neem (*Azadirachta indica* A. Juss). limonoids and azadirachtin against safflower aphid (*Dactynotus carthami* H.R.L.). *Indian Journal of Entomology* 61, 299-302.

Dorge, S.K., Dalaya, V.P. and Kaul, O.B. (1966). Studies on bionomics and control of groundnut aphid (*Aphis craccivora* Koch). *Labdev Journal Science Technology*, 4, 165-167.

Dorn, A., Rademacher. J. M. and Sehn, E. (1987). Effect of azadirachtin on reproductive organs and fertility in the large milk weed bug *Oncopeltus fasciatus*. In Natural Pesticides from the Neem Tree and Other Tropical Plants (Eds. H. Schmutterer, and K.R.S.Ascher). pp. 273-288. Proceedings of the 3rd *International Neem Conference, Nairobi*, (1986). Eschborn: G T Z. 703 pp.

Dudley, N.J., Muller, R.A. and Wightman, J. A. (1989). Application of dynamic programming for guiding IPM on groundnut leafminer in India. *Crop Protection.*, 8, 349 - 357.

Ghelani, (1989). M. Sc. thesis, submitted to Gujarat Agriculture University (GAU, Campus) Junagadh, Gujarat.

Ghewande, M.P., Desai, S., Prem Narayan and Ingle, A.P. (1993). Integrated management of foliar diseases of groundnut (*Arachis hypogaea* L). in India. *International Journal of Pest Management* 39, 375-378.

Gold, C.S., Wightman, J.A. and Pimbert, M. (1989). Mulching effects on termite scarification of drying groundnut pods. *International Arachis Newsletter*, 6, 22-23.

Jacobson, M. (1986). The neem tree: natural resistance par excellence. In Natural Resistance of Plants to Insects : Role of Allelochemicals (Eds. M.B. Green, and P.A.Hedin). *American Chemical Society Symposium Series* No.296, 220-223. ACS, Washinton, DC.

Jotwani, M.G. and Srivastava, K.P. (1981). Neem insecticides of the future II. Protection against field pests. *Pesticides* 15(11), 40-47.

Koul, D. (1985). Adiradichtin inetraction with development of *Spodoptera litura* Fab. *Indian Journal of Experimental Biology*, 23, 160-63.

Kraus, W., Bokel, M., Klent, A. and Pohil, H. (1985). The structure of azadirachtin and 22, 23- dihydro-23, b- methoxyazadirachtin. *Tetrahedron letter.*, 6, 6435-38.

Kumari,D.A., Kumar,S.T. and Reddy,V.S. (1998), Management of the groundnut bruchid, *Caryedon serratus* (Olivier). with some plant products. *Pest Management and Economic Zoology*, 6(2), 127-131.

Meisner, J., Mclamed-Majdar, V., Yathom, S and Ascher, K.A.S. (1987). The influence of neem on the European corn borer (*Ostrinia nibulalis*). and the serpentine leafminer (*Lirimyza trifolii*). In Natural *Pesticides from Neem Tree and Other Tropical Plants.* pp. 461- 477.

Mohammad, A. (1981). The groundnut leafminer (*Stomopteryx subsecivella* Zeller). (Lepidoptera : Gelechidae). A review of world literature. Occasional paper 3, Groundnut Improvement Program, International Crops Research Institute for Semi Arid Tropics (ICRISAT), Patancheru, A.P., India. Pp 21.

Murugesan, S., Balu, A., Durairaj, S., Pankajam, S. and Sunitha, B.(1998). Insecticidal performance of a neem product in control of two major seed pests of forestry tree species. *Entomon*. 23(3), 197-201.

Murugan, K. and Babu, R. (1998). Impact of certain plant products and *Bacillus thuringiensis* Berliner subspecies kurstaki on the growth and feeding physiology of *Helicoverpa armigera* Hubner (Insecta: Lepidoptera: Noctuidae). Journal of Scientific and Industrial Research. 57(10/11): 757-765.

Nandagopal, V. (1992). Studies on Integrated pest management in Groundnut in Saurashtra, Ph. D. thesis, Saurashtra University, Rajkot, pp 246.

Nandagopal, V. (1990). Present status of *Helicoverpa (=Heliothis) armigera* (Hub.). on groundnut in major growing area in the country. *In* Proceeding of the First National Workshop on *Heliothis* Management: current status and future strategies, 30-31, August, 1990, Division of Entomology, Directorate of Pulses Research, Kanpur. Pp.151-159.

Nandagopal, V., Jaroli, R.K., Ramkumar and Reddy, P.S. (1990). Neem products: possible insecticide on the groundnut jassid *Balclutha hortensis* Lindb. *International Arachis Newsletter*, 8, 22.

Patel, H.K. (1968). Neem seed suspension-A sure deterrent to hairy caterpillar *Amsacta moorei* B. *Madras Agriculture Journal*, 55, 509-510.

Puri, S.N., K.S.Murthy and O.P.Sharma.1997. Resource Inventory for *IPM-I, NCIPM Information Bulletin No.3,* National Centre for Integrated Pest Management (ICAR),New Delhi, pp 64.

Patil, R.K., Lingappa, S., Hiremath, I.G; Hiremath,-R.V. (2003). Efficacy of botanicals with *Nomuraea rileyi* (Farlow) Samson in groundnut ecosystem.Biological-control-of-lepidopteran- pests- Proceedings- of- the- Symposium- of- Biological- Control- of-Lepidopteran-Pests,-July-17-18,-2002,-Bangalore,-India. 2003; 313-317.

Raguraman, S. and Singh, R. P (1998). Behavioral and physiological effects of neem (*Azadirachta indica*). seed kernel extracts on larval parasitoid, *Bracon hebetor*. *Journal of Chemical Ecology*. 24(7), 1241-1250.

Rao, P.T and Subramanya, B. (1987). Effect of azadirachtin on *Achaea janata* Linn. and *Spodoptera litura* F. (Noctuidae: Lepidoptera). *Journal Entomological Research*., 11, 166-169.

Reddy, D. V. R. (1988). Groundnut virus diseases in India. in: Groundnut. Indian Council of Agricultural Resaerch, New Delhi, pp 509-525.

Reddy, P.S. (1988). Genetics, breeding and varieties. In Groundnut, (ed. P.S.Reddy) pp. 508-525.

Sahayaraj, K. (1998), Antifeedant effect of some plant extracts on the Asian armyworm, *Spodoptera litura* (Fabricius). *Current Science*. 74: 6, 523-525.

Sahayaraj, K. and Paulraj, M.G. (1998a). Screening the relative toxicity of some plant extracts to *Spodoptera litura* Fab. (Insecta: Lepidoptera: Noctuidae). of groundnut. *Fresenius Environment Bulletin*, 7(9&10), 557-560.

Sahayaraj, K. and Paulraj, M.G. (1998b). Relative toxicity of some plant extracts on groundnut leafminer *Aproaerema modicella*. *International Arachis Newsletter*., 18, 19-27.

Sahayaraj, K. and Paulraj, M.G. (1998c). Effect of neem leaf extract on *Amsacta albistriga* walker larvae of groundnut. *Insect Environment*., 4(2), 42-43.

Sahayaraj, K. and Paulraj, M.G. (2000). Impact of some plant products on the behavior of *Tribolium castaneum* Herbst in groundnut kernels. *International Arachis Newsletter*20 (1). (in press).

Saxena,R.C. (1989). Insecticides from neem. In Insecticides of Plant Origin. (Eds. J.T. Arnasan, J.R. Philogene and P. Morand). *American Chemical Society symposium series*., 387, 110-135, ACS, Washington, DC.

Schemutterer, H. (1990). Properties and potential of natural pesticides from the neem tree *Azadirachta indica*. *Annual Review of Entomology*, 35,271-297.

Schmutterer, H., Saxena, R.C. and Vonder Heyde, J. (1983). Morphogenetic effects of partially purified fractions and methonolic extracts of neem seeds on the *Mythimna seperata* (Walker). and *Cnaphalocrosis medinalis* (Guenee). *Zeitschrift fuer Zoologie und angewandte* Entomology. 95, 230-237.

Senguttuvan, T. (1999). Effect of plant products against thrips in groundnut. *International Arachis Newsletter*., 19, 36 – 39.

Senguttuvan, T. and Dhanakodi, C.V. (1997b). Management of pod borer in groundnut through manipulation of cultural practices in alfisols. *Journal of Oilseed Research*., 14(2), 269-273.

Senguttuvan, T. and Dhanakodi. C.V. (1997). Control of pod borers of groundnut by soil application of insecticides in alfisols. *Journal of Oilseed Research.*, 14, 274-277.

Senguttuvan, T., Kareem, A.A. and Rajendran, R. (1995). Effects of plant products and edible oils against rice moth *Corcyra cephalonica* stainton in stored groundnut. *Journal of Stored Product Research,* 31(3), 207-210.

Sharma,G.K. (1992). Growth-inhibiting activity of azadirachtin on *Corcyra cephalonica.* Phytoparasitica., 20.

Simmonds, M.S. and Blaney, W.M. (1984). Some neurological effect of azadirachtin on lepidopteran larvae and their feeding response. In Natural Pesticides from the Neem Tree and Other Tropical Plants. pp. 163-180. *Proceedings of the 2nd International Neem Conference.* Rauischholzhauen. 1983. GTZ: Eschborn .

Steets. R. (1976). Die wirkung von Rohextracten aus den Meliaceen *Azadirachta indica* und Melia azadirachta. *Zeitschrift fuer Zoologie und angewandte Entomologie.,* 77, 306-312.

Steffens, R and Schmutterer, H. (1982). The effect of a crude methonolic neem (*Azadirachta indica*). seed extract on metamorphosis and quality of adults of the Mediterranean fruit fly *Ceratitis capitata* (Diptera:Tephrididae). *Zeitschrift fuer Zoologie und angewandte Entomologie.,* 14, 98-103.

Sudipta, Chanda and Sanjib, Chakravorty. (1998). Effect of stress on heart beat and post embryonic development in *Corcyra cephalonica* larvae. *Indian-Journal-of-Experimental-Biology,* 36: 8, 796-799.

Wightman, J.A and Amin, P.W. (1988). Groundnut pests and their control in the Semi-Arid Tropics. *Tropical Pest Management.,* 34: 218-226.

Zanno, P.R., Miura, I., Nakanishi, K. and Elder, D.L. (1975). Structure of the insect phagorepellent azadirachtin. *Journal American Chemical Society,* 97: 1975-1977.

□□□

19

Biocontrol

S. P. Singh, D. N. Yadav, T. V. Prasad and K. Gunathilakaraj

19.1. Introduction

Unlike cereals and commercial crops, groundnut is a low input crop. As far as protection of this crop from insect pests and diseases is concerned, technology involving minimum inputs will be the most appropriate. Biological suppression, *i.e.,* conservation and augmentation of biotic agents, offers such opportunity. The efforts made in India to quantify the biodiversity of natural enemies of groundnut pests and its utilization are enumerated pestwise.

19.2.Groundnut leaf miner, *Aproaerema modicella* (Deventer) (Lepidoptera, Gelechiidae)

Groundnut leaf miner, *A. modicella* is the key pest of groundnut in many parts of India. It also develops on soyabean and lucerne. It is particularly serious in Karnataka, Maharashtra, Andhra Pradesh and Tamil Nadu.The natural enemies recorded from *A. modicella* are presented in chapter 3. Over 40 species of hymenopterous parasitoids attack *A. modicella* eggs, larvae and pupae on groundnut and soybean.

In Andhra Pradesh, larval parasitism around Hyderabad, reaches up to 90% (Murthy, 1989). *Goniozus* sp. was most abundant in the post-rainy season (December-April), whereas, *Stenomesius japonicus* (Ashmead) and *Apanteles* sp. were abundant in the rainy season. The percentage parasitism was greatest in September-November and January-March around Tirupati, Andhra Pradesh (Srinivasan and Rao, 1987). The average parasitism of a decade was 34 and 40 per cent in rainy and post-rainy seasons, respectively (Ranga Rao *et al.*, 1995). Parasitism in sprayed areas was quite high. Recent findings indicate that in addition to *S. japonicus* and *Goniozus* spp., *Sympiesis dolichogaster* (Ashmead) was the most abundant parasitoid. Viral and fungal pathogens killed up to 30 per cent of the larvae of *A. modicella* (Shanower *et al.*, 1992). Larvae of carabid predator, *Chlaenius* sp. were found inside mines of *A.*

modicella during the rainy season at Patancheru, Andhra Pradesh in groundnut and soyabean leaves (Shanower and Ranga Rao, 1990).

Out of the eight species of parasitoids recorded in Gujarat (Anand), *Apanteles* sp., *Goniozus* sp. and *S. japonicus* were the most important, with maximum rates of parasitism of 38.21, 50.00 and 79.16 per cent, respectively. The parasitoids were very effective and did not allow the pest to build up requiring chemical control (Yadav *et. al.*, 1987).

In Karnataka (Raichur), 16 species of hymenopterous parasitoids parasitise *A. modicella.* Parasitism reaches 66.22 and 90.57 per cent in the 3rd week of August and the 2nd week of September, respectively. The eulophids, *S. japonicus, Sympiesis* sp. and *Tetrastichus* sp. are most active in August and the braconid *Chelonus* sp. in September (Shekharappa *et al.*, 1990).

In Maharashtra, *Goniozus* sp., *Brachymeria* sp., *S. japonicus, Bracon gelechiae* Ashmead, *Chelonus curvimaculatus* Cameron and *Apanteles* sp. have been recorded in third week of August reaching peak (83%) in mid-September (Khan and Raodeo, 1978). Out of seven parasitoids recorded in Maharashtra, four (*Dolichogenidea* sp., *Sympiesis dolichogaster, Eurytoma, Anagyrus* sp.) were not reported earlier. The parasitoids caused 16.4, 25.0 and 26.2 per cent mortality of the pest in three generations, respectively (Shetgar *et al.*, 1993).

In Tamil Nadu, over a dozen species of hymenopterous parasitoids have been recorded (Muthiah, 1991). Among the natural enemies listed (table 1), rearing techniques are available for *Apanteles* spp., *Brachymeria* spp., *Bracon* spp., *Copidosoma* sp., *Elasmus* spp., *Goniozus* spp. and *Apertochrysa* sp. (Singh, 1994). Parasitoids such as *Avga choaspes* Nixon, *S. japonicus* and *S. dolichogaster* are fairly abundant in the field and efforts should be made to conserve them.

Population of coccinellid predators, *viz.*, *Cheilomenes sexmaculata* (F.) and *Coccinella septempunctata* Linnaeus showed peak activity during August and the populations of leaf miner, jassids and thrips remained low during this period, at Junagadh in Gujarat (Upadhyay and Vyas, 1987). Fifty three species, belonging to 34 genera and 14 families of spiders have also been found in groundnut fields. Theridiidae, Lycosidae and Oxyopidae were the dominant families. *Coleosoma blandum* Pickard-Cambridge, *Pardosa birmanica* Simon and *Oxyopes shweta* Tikader were common and their population reached a peak in October (Patel and Pillai, 1988; Peter and Biswas, 1990).

Intercropping of groundnut with cowpea (*Vigna unguiculata* L. Walp.) or black gram (*V. mungo*) at a ratio of 3:1 gave a significantly lower incidence of the pest and higher cost-benefit ratio than pure groundnut plots (Logiswaran and Mohanasundaram, 1985). Intercropping with pearl millet (*Pennisetum glaucum*) significantly reduced pest incidence and increased coccinellid numbers per plant compared with groundnut monocultures. *Coccinella* sp. and *C. sexmaculata* were dominant predators and *Chelonus* sp. the dominant parasitoid (Kennedy *et al.*, 1990). Intercropping with pearl millet at a ratio of 4:1 is recommended to reduce the occurrence of leaf miner and increase natural enemy activitiy (Baskaran and Thangavelu, 1990). Application of granular insecticides (isofenphos 5G at 2 or 5 kg a.i./ha, carbosulfan 10G at 4 kg ai/ha and phorate 10G at 4 kg ai/ha) during *kharif* on rainfed groundnut had no harmful effect on the natural enemies of *A. modicella* (Rajagopal and Gowda, 1992).

Not much work has been done on the augmentation of biotic agents of *A. modicella* but there is ample scope for mass production of egg-larval parasitoid *Chelonus* sp. as well as egg parasitoids. As reported from Gujarat, conservation of biotic agents alone can provide sufficient suppression of the pest. Once parasitism reaches 50 per cent all deliberate actions for the management of this pest could be suspended to encourage the build up of biotic agents.

19.3. White grubs, *Holotrichia consanguinea* (Blanchard), *H. serrata* (Fabricius) (Coleoptera, Scarabaeidae)

H. consanguinea is common in the light soils of Rajasthan, Punjab, Haryana and Gujarat, and few pockets in the southern states. *H. serrata* on the other hand is widely distributed but more commonly in the heavier soils of southern India. *Maladera* sp. is a serious pest in canal command areas of western Rajasthan. Unlike other species it has more than one generation per growing season. *L. lepidophora* is common in areas around Sholapur in Maharashtra.

Among the predators of beetles, birds - black drango, *Dicrurus adsimilis* (Bechstein), blue jay *Coracias benghalensis*(Linnaeus), common quail *Coturnix coturnix* (Linnaeus), house sparrow *Passer domesticus* (Linnaeus) and barn owl *Tyto alba* (Scopoli); insectivorous bat *Scotophilus* sp.; toad *Bufo* sp.; hedge hog *Hemiechinus auritus collaris* Grey and lizard *Varanus* sp. may feed on beetles (chapter 8). The grubs are picked up by crows *Corvus splendens* Vieillot, mynah *Acridotheres tristis* (Linnaeus) and egrets *Bubulcus ibis* (Linnaeus) which follow the plough. Two cross ploughings in the presence of birds may reduce 70 per cent of the grubs present in the field (Yadav, 1991). Such birds need encouragement and protection.

Carabid, *Scarites indus* Olivier burrows in soil and feeds on grubs. A mite of genus *Sancassania* is an efficient predator of eggs.

The scoliids *Campsomeris collaris* (F.) and *Scolia aureipennis* Lepeletier parasitised the grubs. Some micro organisms are pathogenic to the grubs as well as beetles. The milky disease was recorded in 1977 (Patel *et al.*, 1977).

Bacillus popilliae var. *holotrichiae* was effective in small-scale laboratory tests against grubs of *H. consanguinea* and *H. serrata* (Vyas *et al.*, 1991a,b). The application of *B. popilliae* resulted in 72.2 per cent kill in 40 days of the grubs of *H. consanguinea* in groundnut when applied as spot treatment @ 5 x

10^{13} spores/ha. It was also infective to *H. serrata* and *Maladera nathani* (Frey) and few other species of white grubs (Vyas *et al.*, 1986).

Entomofungus *Metarhizium anisopliae* (Metchnikoff) Sorokin may cause disease in all the three species of grubs. *M. anisopliae* and *Beauveria brongniartii* Sacc.Petch. caused 54.6 per cent and 45.5 per cent kill in the grub of *H. consanguinea* when applied @ 10^{15} conidia/ha. *M. anisopliae* is compatible with soil insecticides like phorate, carbofuran, sevidol and others. A spore dust having the combination of *B. popilliae* and *B. lentimorbus* was found to be more pathogenic to the dust containing the spores of *B. popilliae* alone (Sharma *et al*, 1980).

B. bassiana (Balsamo) Vuillemin has also been found to be pathogenic to the grubs. *B. brongniartii* is effective against the root grubs *H. serrata* when applied @ 10" spores/ton of soil along with 50 g of HCH 50% wp/m². The fungus mixed with farm yard manure or pig manure before application gave better results (Jayaramaiah, 1981). Phorate granules applied in soil did not inhibit growth and sporulation of *B. brongniartii* (Vyas *et al.*, 1990). Entomophilic nematode, *Rhabditis* sp. is effective in killing the eggs of white grubs and *Steinernema glaseri* (Steiner) and *Heterorhabditis* sp. parasitise the grubs.

19.4. Termites, *Microtermes* spp., and *Odontotermes* spp. (Isoptera, Termitidae).

Termites (*Microtermes* spp. and *Odontotermes* spp.) are serious pests of groundnut, particularly in red and sandy soils. Very little work seems to have been done on biological suppression of termites. The reduviid predator *Acanthaspis pedestris* Stal has been identified as a potential biological suppression agent of the groundnut infesting termite *Odontotermes assmuthi* Holmgren (Sahayaraj, 1995b).

A carabid beetle, *Omphra pilosa* Klug fed exclusively on termites. The adult beetle consumed 15 termite workers and 2.1 alates per day. Larvae of the predator stored termites in their tunnels. Both larvae and adults were fairly abundant in groundnut ecosystem and lived under weed stubble that was infested with termites (Kumar and Rajagopal, 1990).

19.5. Hairy caterpillars, *Amsacta albistriga* (Walker), *A. moorei* (Butler), *Spilosoma obliqua* (Walker) (Arctiidae : Lepidoptera)

The red hairy caterpillars, *A. albistriga* found in southern India, and *A. mooeri* in the northern states create serious problems in groundnut cultivation in certain areas. *S. obliqua* is an important pest of groundnut mainly in the northern states. Its damage to groundnut has been recorded in Andhra Pradesh, Assam, Bengal, Bihar, Haryana, Madhya Pradesh, Maharashtra,

Punjab and Uttar Pradesh. The natural enemies recorded from hairy caterpillars are listed in chapter 6.

• The egg parasitoid *Telenomus manolus* Nixon parasitised up to 94.3 per cent eggs of *A. albistriga,* but only 3.8 - 59.4 per cent adult parasitoids were able to emerge from the parasitised eggs (Sundaramoorthy *et al.,* 1976). It is considered as a potential bioagent for the management of this pest. *Trichogramma brasiliensis* (Ashmead) was tested against the eggs of *A. moorei.* In the laboratory, the average per cent parasitisation was 98.4 and the average number of parasitoids developed per host egg was 1.90. In the field, the per cent parasitisation ranged from 1.9 to 34.8 in different locations in Rajasthan. Hence, *T. brasiliensis* could be effectively used against *A. moorei* (Tuhan *et al.,* 1987). Thirty per cent larvae of *A. moorei* were parasitised by *Apanteles obliqua* Wilkinson and *Meteorus* sp. (Siddiqi, 1981). Gunathilagaraj and Sundara Babu (1987) suggested that tachinid parasitoids, *Palexorista solennis* (Wlk.) and *Exorista xanthaspis* Wiedemann could be effectively used against *A. albistriga* . A predatory pentatomid, *Eocanthecona furcellata* (Wolff) often feeds on the larvae of *A. albistriga* (David and Basheer, 1961).

A. albistriga was found to be infected with nuclear polyhedrosis virus (Aa NPV) and in laboratory tests, a concentration of 10^8 POBs per ml provided 86.4 per cent mortality (Jayaraj *et al.,* 1976). AaNPV @4.3 x 10^{10} PIBs proved effective in the field trials conducted at Madurai (Chandramohan and Kumaraswamy, 1979). The virus was found to persist in the field for many years destroying larvae years after application by having been transmitted from generation to generation through infected eggs, persistence in the soil and through the dead larval bodies found in the field (Rabindra and Balasubramanian, 1980). The gregarious and migrating behaviour of the larvae also influenced the spread of the virus disease (Jayaraj, 1989). AaNPV could be produced by collecting large number of moths on light traps, obtaining eggs in the laboratory, infesting the larvae with AaNPV, harvesting and storing AaNPV for use in the next season. Biocontrol potential of native entomopathogenic nematodes was tried in controlling red hairy caterpillar, *Amsacta albistriga* (Baskaran *et al.,* 1993).

A bio-intensive integrated pest management action plan for *Amsacta* spp. was suggested by Singh and Bakthavatsalam (1994) which includes deep summer ploughing to expose the pupae to predators and sunlight and collection and destruction of pupae; setting up of bonfires at community level between 7 and 11 pm for killing of emerging adults; collection and destruction of the eggs which are usually deposited on the undersurface of the leaves in masses; inundative release of *Trichogramma chilonis* or *T. brasiliensis* immediately after adult emergence @ 50,000 adults per hectare per release, repeat the release one week after the first release in case adult emergence is seen; handpick the young larvae and kill them by putting in bonfires; foliar sprays of AaNPV 4.3 x 10^{10} POBs per ml on early stage larvae or *Bacillus thuringiensis* var. *kurstaki* at 2g/l; use dust formulations for controlling the young larvae in case AaNPV is not available; spray 4 per cent neem kernel suspension or malathion against mature larvae and avoid spraying insecticides such as carbaryl, chlorpyriphos, dichlorvos, endosulfan, ethion, fenitrothion, phosalone and quinalphos as these insecticides are very toxic to adult parasitoids.

19.6. Tobacco caterpillar, *Spodoptera litura* (Fabricius) (Noctuidae, Lepidoptera)

Tobacco caterpillar, as the name suggests, is a serious pest of tobacco. But in recent years it has shifted to groundnut, particularly in parts of Karnataka and coastal areas of Andhra Pradesh and Tamil Nadu in the post-rainy season.

Over 30 species of parasitoids, 31 predators including 9 species of birds, and 15 entomopathogens have been recorded on different crop hosts of *S. litura* in India (Singh and Jalali, 1997). The natural enemies frequently recorded were *Chelonus formosanus* Sonan, *C. carbonator* Marshall, *C. heliopae* Gupta, *Cotesia ruficrus* (Haliday), *Ichneumon* sp., *Peribaea orbata* (Wiedemann), *Exorista xanthaspis* (Wiedemann), *Cyrtopeltis tenuis* Reuter, *Harpactor costalis* Stal, *Rhynocoris fuscipes* (Fabricius), *Apanteles* sp. nr. *colemani* Viereck, *Euplectrus gopimohani* Mani, *Dicrurus adsimilis* (Bechstein), *Bubulcus ibis*(L.) and nuclear polyhedrosis virus. *Rhynocoris marginatus* (F.) preferred 3-day-old *S. litura* larvae (Sahayaraj, 1994,1999). Hence it was considered as a potential biocontrol agent for *S. litura* on groundnut. Similarly *Rhynocoris kumari* Ambrose and Livingstone preferred *S. litura* larvae among the lepidopteran pests of groundnut tested (Sahayaraj and Sivakumar, 1995). The egg parasitism varies from 5-30 per cent in different states and seasons. The larval parasitism reaches 35 per cent (Ranga Rao, 1996 personal communication).

In Karnataka, *Metarhizium anisopliae* infection began to appear in the 2nd fortnight of June, was high in mid August and fell to nil in November (Siddaramaiah *et al*., 1986). Application of nuclear polyhedrosis virus (NPV) of *Spodoptera litura* @ 250-500 larval equivalents in the evening hours has proved quite effective. Combined use of *S. litura* and *Helicoverpa armigera* NPVs at 250 larval equivalents with crude sugar (jaggery/gur) at 2.5 kg/ha was effective in checking the population of both the pests (Dhandapani *et al*., 1993).

19.7. Gram pod borer, *Helicoverpa armigera* (Hubner) (Noctuidae, Lepidoptera)

The gram pod borer, *H. armigera* is a polyphagous pest. It is a serious pest of groundnut in coastal Andhra Pradesh, Tamil Nadu and Karnataka. The overall losses caused by this pest on different crops exceeds Rs.1000 crores per annum.

In Gujarat, about 23 per cent of the 3rd to 6th instar larvae were killed by nuclear polyhedrosis virus. A few pupae died as a result of parasitism by the tachinid *Senometopia illota* (Curran) (Koshiya and Patel, 1987). Inundative release of *Trichogramma chilonis* @ 2,00,000/week was effective for the suppression of *H. armigera* on summer groundnut only during first four weeks of March. But in April other measures including sprays of *Ha* NPV have to be adopted (Anon., 1993) (chapter 5). Parasitoid *Bracon gelechiae* and *Campoletis chlorideae* Uchida, an unidentified staphylinid, and the migratory bird, rosy pastor, *Sturnus roseus* (Linnaeus) were common in released plots and checked the pest population effectively (Anon., 1988).*Rhynocoris marginatus* consumed more small size *H. armigera* larvae than medium or large size larvae (Sahayaraj, 1995a).

The potential of the parasitoids *Eucelatoria bryani* ; *Apanteles* sp. and nemotode (*Ovomermis albicans*); predators such as spiders, preying mantis, robber fly, dragonfly, lace wings, reduvids, lady bird beetles, mud wasp, king crow and diseases for the biological suppression of this pest may be thoroughly explored.

19.8. Other lepidopterans

A nuclear polyhedrosis virus was found to infect the larvae of *Chrysodeixis chalcites* (Esp.) (Noctuidae) feeding on groundnut (Rabindra *et al.*, 1975). *Anarsia ephippias* (Meyr.) (Gelechiidae) feeds on the shoots and webs the leaves of groundnut in Punjab. The larvae were parasitised by *Bracon gelechiae* and the pupae by *Brachymeria* sp. (Bakhetia, 1977). *Scopula albiflava* (Warren) (Geometridae) larva feeding on groundnut was parasitised up to 10 per cent by *Microgaster* sp. (Kalra *et al.*, 1989).

19.9. Aphids, *Aphis craccivora* Koch (Aphididae, Homoptera)

A. craccivora is a widely distributed pest of groundnut and other leguminous crops throughout India. It could create problems in young crops particularly those under moisture stress. It also transmits diseases like rosette in Africa, stripe in Asia and mottle in Asia and Africa.

Coccinellids *Cheilomenes sexmaculata* and *Coccinella septempunctata*, syrphid, *Ischiodon scutellaris* and chrysopid, *Chrysoperla carnea* have been recorded as common predators (Patel *et al.*, 1976). In some parts of Gujarat, *Hippodamia variegata* (Goeze), *C. septempunctata* and *C. sexmaculata* were the common predators of aphids. A positive correlation was observed between the aphid index and the population of active stages of

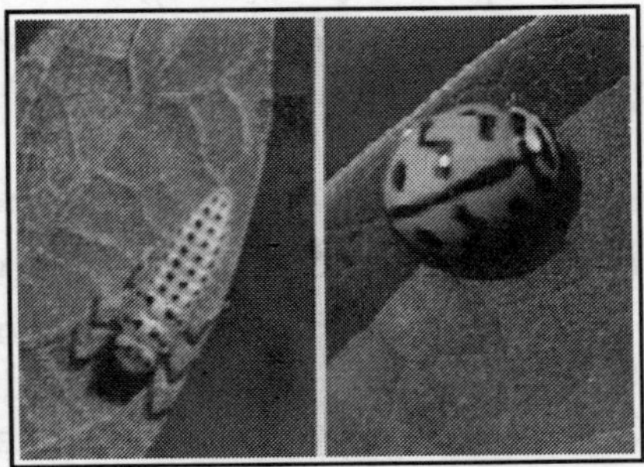

Plate 1. Lady bird beetle

the predators during March, while in April the aphid population decreased with increased abundance of coccinellids followed by decrease in aphid as well as predator population in 1st week of May. Application of insecticides is unnecessary if coccinellids are present in the groundnut crop in summer (Butani and Bharodia, 1984). *Coccinella transversalis* Fabr (*C. repanda*) is attracted to ultraviolet light (Veeresh, 1974). The regulation of light traps according to the activity of predators or parasitoids will help in conserving these beneficials. *C. septempunctata* larva consumes as many as 467 nymphs of *A. craccivora* during its development time of 7 days, while a single adult consumed up to 143 nymphs per day (Talati and Butani, 1979). *C. sexmaculata* larva consumed 474 aphids during the 6 days of larval development and 16-49 days of adult life (Patel and Vyas, 1984). Among the pesticides recommended, thiometon 0.03 or 0.05 per cent was found to be least harmful to the coccinellids, followed by oxydemeton methyl (Upadhyay and Vyas, 1986).

There is scope for manipulating the population of *C. sexmaculata* through augmentative releases because mass rearing technique for this predator is available. Synchronisation of the releases of this predator with the appearance of aphid in the field results in substantial reduction of aphid populations (Anon., 1987). Inundative releases of *C. sexmaculata* @ 1,50,000/ha gives complete control of *A. craccivora* in groundnut when 50 per cent of the plants are having medium to heavy infestation (Anon., 1988).

19.10. Thrips, *Caliothrips indicus* Bagnall, *Scirtothrips dorsalis* Hood, *Thrips palmi* Karny, *Frankliniella schultzei* (Trybom) (Thripidae : Thysanoptera)

Thrips live in flowers and folded leaflets of groundnut. *T. palmi* transmits peanut bud necrosis virus, which results in the death of the attacked plant. Natural enemies recorded from thrips includes mainly predators such as *Franklinothrips megalops* (Trybom)(Thysanoptera, Aeolothripidae), *Scolothrips indicus* Pries. (Thysanoptera, Thripidae), *Orius* sp. and *O. maxidentex* Ghauri (Heteroptera, Anthocoridae) feeding on adults (Amin, 1987). Spiders (Araneae), *Laius malleifer* Champion, *Coccinella septempunctata*, *Orius* sp. and *Cheilomenes sexmaculata* effectively controlled thrips infesting groundnuts (Singh *et al.,* 1991) (chapter 11).

19.11. Jassids or leaf hoppers, *Empoasca kerri* Pruthi, *Balclutha hortensis* Lindb. (Cicadellidae, Homoptera:)

E. kerri is the most common species of jassid recorded on groundnut. *Balclutha hortensis* is often recorded in Gujarat and some other parts of western India. The leaf hoppers have the potential of causing 'hopper burn' in severe case of infestation.

Among the natural enemies of *E. kerri, Orius maxidentex* Ghauri (Hemiptera, Anthocoridae) from Andhra Pradesh (Amin, 1987) and *Crossopalpus* sp. (Diptera, Empididae) from Gujarat (Nandagopal, 1988) were frequently recorded. *Brumoides suturalis, Cheilomenes sexmaculata, Coccinella septempunctata* and *Hippodamia variegata* were recorded as predators elsewhere (Singh *et al.*, 1991). Fluctuations in the number of cicadellids were accompanied by a corresponding increase or decrease in the number of *Brumoides* sp. (Singh *et al.*, 1991). The general predators recorded in groundnut ecosystem include *Stenus pictus* Motschulsky (Coleoptera, Staphylinidae), *Chrysoperla carnea* (Steph.)(Neuroptera, Chrysopidae), *Monomorium indicum* Forel (Hymenoptera : Formicidae), *Laius malleifer* Champion (Coleoptera, Melyridae), *Geocoris* sp. (Hemiptera, Lygaeidae), *Polistes olivaceus* (De Geer) (Hymenoptera, Vespidae), *Orius* sp. (Hemiptera, Anthocoridae), *Cryptocephalus* sp. (Coleoptera, Chrysomelidae), *Cheilomenes sexmaculata* (Fabricius), *Hippodamia variegata* (Goeze), *Coccinella septempunctata* Linnaeus, *Brumoides suturalis* (Fabricius) (Coleoptera, Coccinellidae), *Campylomma* sp. (Heteroptera, Miridae) and some unidentified spiders (Araneae, Salticidae) (Singh et *al.*, 1990).

19.12. Future needs

- Exploratory surveys should be intensified to identify new natural enemies.
- New biotic agents may be imported or introduced to combat accidentally introduced pests like serpentine leaf miner.
- Bio-ecology of all the pests and natural enemies should be studied to acquire better understanding for initiating appropriate actions for the management of different pests.
- Inoculative/inundative release application methods of natural enemies have to be standardised for different pests.
- Mass multiplication techniques have to be evolved for predators, parasitoids and entomopathogens.
- Use of pheromones/kairomones to time the release and increasing the efficiency of natural enemies has to be standardised.
- Interaction with resistant/tolerant cultivars, traps and trap crops should be studied to improve the overall pest management efficiency.
- Conservation methods such as selective use of pesticides based on phenology of predators and parasitoids, spot treatment, habitat modification, growing plants on bunds or border, adoption of ant control methods and others may be standardised.

- Simple BIPM system evaluation techniques may be evolved so that the end user-the farmer- is convinced of the system.
- Need based large scale predator/parasitoid/pathogen multiplication units should be established.

References

Amin, P.W . (1987). Insect pests of groundnut in India and their management. In: *Plant Protection in Field Crops,* pp. 219-233. Veerabhadra Rao, M. and Sithananthan, S. (Eds). Plant Protection Association of India, Rajendranagar, Hyderabad, India.

Anonymous, (1987). Annual Report All India Coordinated Research Project on Biological Control of Crop Pests and Weeds, Indian Council of Agricultural Research, IIHR, Bangalore, India. Tech. Doc. No. 20, 1-236.

Anonymous, (1988). Annual Report. All India Coordinated Research Project on Biological Control of Crop Pests and Weeds, Biological Control Centre,(ICAR- NCIPM)., Bangalore, India. *Tech. Doc.* No. 23, 1-260.

Anonymous, (1993). Annual Report (1st April, 1992 to 31st March, 1993). All India Co-ordinated Research Project on Biological Control of Crop Pests and Weeds. Biological Control Centre (ICAR-NCIPM)., Bangalore, India. Tech. Doc. No. 40, 1-313.

Bakhetia, D.R.C. (1977). *Anarsia ephippias* (Meyrick).(Lepidoptera : Gelechiidae). damaging the groundnut crop in the Punjab . *Journal of Research. PAU* 14 (2),232-233.

Baskaran, R. K. M and Thangavelu, S. (1990) Non-conventional method to minimise the incidence of groundnut leaf miner. Groundnut-News., 2: 2, 6.

Baskaran, R.K.M., Sivakumar, C.V. and Venugopal, M.S. (1993). Biocontrol potential of native entomopathogenic nematodes in controlling red hairy caterpillar, *Amsacta albistriga* (Lepidoptera:Arctiidae). of groundnut (*Arachis hypogaea*). *Indian ournal of Agricultual Science,* 64(9), 655-657.

Butani, P.C. and Bharodia, R.K. (1984). Rotation of groundnut aphid population with its natural enemies, lady bird beetles. *Gujarat Agricultual University Research Journal,* 9(2), 72-74.

Chandramohan, N. and Kumaraswamy, T. (1979). Comparative efficacy of chemicals and nuclear polyhedrosis virus in the control of red hairy caterpillar *Amsacta albistriga* (Wlk.). *Science & Culture,* 45(5), 202-204.

David, B.V. and Basheer, M. (1961). Mass occurrence of the predatory stink bug *Cantheconidia (Canthecona) furcellata* (Wolff). on *Amsacta albistriga* Walk. in south India. *Journal of Bombay Natural History Society,* 58(3), 817-819.

Dhandapani, N., Sundara Babu, P.C., Sathiah, N. and Jayaraj, S. (1993). Combined use of nuclear polyhedrosis viruses of *Spodoptera litura* (Fb.) and *Heliothis armigera* (Hb.). on groundnut (*Arachis hypogaea* L.), *J. Applied Entomol,* 116 (5), 523-526.

Gunathilagaraj, K. and Sunderbabu, P.C. (1987). *Amsacta albistriga* Wlk. on groundnut and other crops. *FAO Plant Prot. Bull.* 35(2), 63-64.

Jayaraj, S. (1989). Pest control: Advances in biological means. *Survey of Indian Agriculture* Special Issue Published by The Hindu. Madras, India, pp. 181-187.

Jayaraj, S., Sundaramurthy, V.T. and Mahadevan, N.R. (1976). Laboratory studies on the nuclear polyhedrosis of red hairy caterpillar of groundnut *Amsacta albistriga* (Walker). Symposium on Plant Protection Research and Development. Part B-Herbicides, Pesticide degradation and Microbiology, and Integrated Pest Control. *Madras* , 63,567-569.

Jayaramiah, M. (1981). Studies on the entomogenous fungus, *Beauveria brongniartii* Sacc. Petch. in relation to white grub, *Holotrichia serrata* Fab. and silkworm, *Bombyx mori* L. and possibilities of its use in the management of the white grub. Ph.D. thesis (unpublished)., University of Agricultural Sciences, Bangalore.

Karla, V.K. , Singh Harvir, Ram, H., and Yadava, T.P. (1989). *Scopula albiflava* Warren (Geometridae,Lepidoptera). damaging groundnut crop in Haryana. *Haryana Agricultural University Journal of Research, 19*(1), 74-75.

Kennedy, F.J.S., Rajamanickam, K. and Raveendran, T.S. (1990). Effect of intercropping on insect pests of groundnut and their natural enemies. *Journal of Biological Control,* 4 (1), 63-64.

Khan, M.I. and Raodeo, A.K. (1978). Importance of larval parasites in the control of *Stomopteryx subsecivella* Zeller *Journal of Maharashtra Agricultural University,* 3, 261-263.

Koshiya, D.J. and Patel, H.K., (1987). Life table and population dynamics of *Heliothis armigera* Hubn. on groundnut. *G.A.U. Research Journal*, 12, 11-14.

Kumar, P. and Rajagopal, D. (1990). Carabid beetle, *Omphra pilosa* Klug. (Coleoptera: Carabidae). a potential predator on termites. *J. Biol. Control* 4(2), 105-108.

Logiswaran, G. and Mohanasundram. M.(1985). Effect of intercropping, spacing and mulching in the control of groundnut leafminer, *Aproaerema modicella* Deventer (Gelechiidae:Lepidoptera). *Madras Agricultural Journal*. 72,695-700.

Murthy, S.N. (1989). Biological suppression of pulse and oilseed pests, pp. 84-106. Proceedings of the Seminar-cum-Seventh Workshop on Biological Control of Crop Pests & Weeds. AICRP on BCCP & W, Biological Control Centre (ICAR-NCIPM). Bangalore. Tech. Doc. No. 26, pp 191.

Muthiah, C. (1991). Some new parasites of groundnut leaf miner (*Aproaerema modicella* Dev.). *Indian Journal of Plant Protection,* 19(1),105.

Nandagopal, V. (1988). *Crossopalpus* sp. (Empididae: Diptera).- an efficient predator of jassids of groundnut. *International Arachis Newsletter,* (4),19-20.

Patel, A.G. and Vyas, H.N. (1984). Studies on predatory capacity of lady bird beetle, *Menochilus sexmaculata* Fabricius against *Aphis craccivora* Koch under laboratory condition. *Pesticides,* 18(11), 8-9.

Patel, B.H. and Pillai, G.K. (1988). Studies on the spider fauna of groundnut fields in Gujarat, Indian *Journal of Biological Control,* 2(2), 83-88.

Patel, R. C., Yadav, D.N. and Patel, J.R. (1976). Natural control of groundnut aphid, *Aphis craccivora* Koch in Central Gujarat. *Current Scienceence,* 45(1), 34-35.

Patel, R.M., Patel, C.B. and Vyas, H.N. (1977). Record and some observations on local milky disease in white grubs (*Holotrichia* sp. near *consanguinea* Blanch.). in India. *Indian Journal of Entomology,* publ. 1978, 39 (2), 81-182.

Peter, C. and Biswas, B.K. (1990). A note on the spider fauna associated with various crops and their potential prey. *Current Research,* 19(2), 61-62.

Rabindra, R.J. and Balasubramanian, M. (1980). Observations on an induced epizootic of nuclear polyhedrosis virus of *Amsacta albistriga* Wlk. *Current Scienceence.* 49, 279.

Rabindra, R.J., Paul, A.V.N. David, B.V. and Subramanium, T.R. (1975). On the nuclear polyhedrosis of *Plusia chalcites* Esp. (Lepidoptera:Noctuidae). *Current Scienceence.* 44 (8), 273-274.

Rajagopal, D and Gowda, V. (1992). Efficacy of granular insecticides in the management of groundnut leaf miner *Aproaerema modicella* Deventer (Lepidoptera: Gelechiidae). *Tropical Pest Management* 38(1), 82-84.

Ranga Rao, G.V., Wightman, J.A., Arjuna Rao, P., Mahesh Babu, V., Sitharama Rao, G and Rameshwar Rao, V. (1995). Integrated pest management of groundnut insect pests,. Paper presented at *National Symposium of Plant Protection and Environment.* Breeze Hotel, Madras, Tamil Nadu from February 2-4, 1995). 26 pp.

Sahayaraj, K. and Sivakumar, K. (1995). Groundnut pest and pest stage preference of a reduviid predator, *Rhynocoris kumari* Ambrose and Livingstone (Heteroptera : Reduviidae). *Fresenius Environmental Bulletin.* 4(5), 263-269.

Sahayaraj, K. (1995a). Bioefficiency and prey size suitability of *Rhynocoris marginatus* Fabricius to *Helicoverpa armigera* Hubner on groundnut (Insecta: Heteroptera: Reduviidae). *Fresenius Environmental Bulletin.* 4(5), 270-278.

Sahayaraj, K. (1995b). Developmental stages and biocontrol potential of a reduviid predator, *Acanthaspis pedestris* against termites on groundnut . *International Arachis Newsletter,* (15), 57-59.

Sahayaraj, K. (1994). Biocontrol potential evaluation of the reduviid predator *Rhynocoris marginatus* (Fabricius). to the serious groundnut pest *Spodoptera litura* (Fabricius). by functional response study. *Fresenius Environmental Bulletinetin,* 3(9), 546-550.

Sahayaraj-K (1999). Field evaluation of the predator, *Rhynocoris marginatus* (Fab.) on two groundnut defoliators. *International Arachis Newsletter.* 1999, No. 19, 41-42.

Shanower, T.G. and Ranga Rao, G.V. (1990). *Chlaenius sp.* (Col.:Carabidae): a new predator of groundnut leafminer larvae. *International Arachis Newsletter.,* 8, 19- 20.

Shanower, T.,G., Wightman, J.A., Gutierrez, A.P.,and Ranga Rao, G.V. (1992). Larval parasitoids and pathogens of the groundnut leaf miner, *Aproaerema odicella* (Lep.: Gelechiidae)., in India. *Entomophaga.,* 37(3), 419-427.

Sharma,S.K., Vikram Singh and J. Vijayvergli. (1980). Ad-hoc Project on Studies on White Grub and its Control. Annual Report University of Udaipur, Agri.Research Station, Durgapura, Jaipur.

Shekharappa, Somasekhar and Patil B.V. (1990). Parasitoid complex of the groundnut leaf miner, *Aproaerema modicella* Deventer (Lepidoptera:Gelechiidae). *Journal of Biological Control.*, 4 (1), 55-56.

Shetgar, S.S., Bilapate, G.G., Patil, V.V.and Londhe, G.M. (1993). A note on the natural enemies of Bihar hairy caterpillar, *Spilosoma obliqua* (Walker). *Indian Journal of Entomology.*, 52 (1), 158.

Siddaramaiah, A.L., Narendrappa, T. and Panchabhavi, K.S. (1986). Occurrence and severity of *Metarrhizium anisopliae* on *Spodoptera litura* Fb. in groundnut field. *Current Research,* 15 (5-6), 48-49.

Siddiqi, J. I. (1981). Record of *Apanteles obliquae* Wlkn. and *Meteorus* sp. (Hymenoptera: Braconidae) as parasites of the hairy caterpillar *Amsacta moorei* Butl. (Lepidoptera: Arctiidae). *Journal of Bombay Natural History Society.*, 78(1), 181-182.

Singh, S.P. (1994). Technology for production of natural enemies. Project Directorate of Biological Control, Bangalore, India. *Technical Bulletin* (4), pp 221.

Singh, S.P. and Bakthavatsalam, N. (1994). Management of *Amsacta* spp. (Lepidoptera: Arctiidae).: the present status. Project Directorate of Biological Control, Bangalore, India. Technical Document (42)., 19-21.

Singh, S.P. and Jalali, S.K, (1997)(Dhaliwal, G. S. (ed.)., Arora, R. (ed.), Randhawa, N. S. (ed.) and Dhawan, A. K.) Impact of pesticides on natural enemies of agricultural pests. Ecological agriculture and sustainable development: Volume 2. *Proceedings of International Conference on Ecological Agriculture: Towards Sustainable Development.* Chandigarh, India, 15-17 November, 1998, 162-175.

Singh, T.V.K., Singh, K.M. and Singh, R.N. (1990). Groundnut - *Archis hypopaea* Linn., I.Pestcomplex. *Indian Journal of Entomology,* 52 (3), 482-492.

Singh, T.V.K., Singh, K.M. and Singh, R.N. (1991). Influence of intercropping. III. Natural enemy complex in groundnut. *Indian Journal of Entomologyomol.,* 53 (3), 333-368.

Srinivasan, S. and Rao, D.V.S. (1987). New report of parasites of groundnut leaf webber, *Aproaerema modicella* Deventer (Lepidoptera: Gelechiidae). *Entomon.*, 12 (2), 117-119.

Sundaramurthy, V.T. , Jayaraj, S. and Swamiappan, M. (1976). *Telenomus manolus* Nixon, a potential egg parasitoid of groundnut red hairy caterpillar *Amsacta albistriga* Walker. *Current Sciences.* 45, 21.

Talati, G.M. and Butani, P.G. (1979). Predatory capacity of *Coccinella septempunctata* on groundnut aphid. *Indian Journal of Plant Protection.*, 7(1), 107.

Tuhan, N.C., Pawar, A.D. and Singh, B. (1987). Preliminary studies on *Trichogramma brasiliensis* Ashmead against the red hairy caterpillar, *Amsacta moorei* Butler. *Journal of Biological control,* 1(1), 72-73.

Upadhyay, V. R. and Vyas, H. N. (1987). Integrated pest control of major insect pests of groundnut in Gujarat in relation to pest population levels and coccinellid predatory populations. *National Symposium on Integrated Pest Control-Progress and Perspectives* (Oct. 15 - 17,1987). (abstracts). *Association for Advancement of Entomology,* Trivandrum, Kerala, India, p.6.

Upadhyay, V.R. and Vyas, H.N. (1986). Comparative toxicity of certain insecticides to predatory coccinellids associated with sucking pests of groundnut in Saurashtra. *Indian Journal Plant Protection.*, 13(2), 91-93

Veeresh, G.K. (1974). Attraction of coccinellid predator *Coccinella repanda* Thunb. var. *transversalis* Fabr. (Coccinelidae: Coleoptera). to black light. *Current Research.*, 3(9), 101-102.

Vyas, H.G., Joshi, D.P., Patel, K.C., Yadav, D.N. and Dodia, J.F. (1986). Natural incidence of milky disease of white grubs in Gujarat. *Indian Journal of agricultural Sciences.*, 56(3), 213-214.

Vyas, R.V. , Yadav, D.N. and Patel, R.J. (1990). Compatibility of *Beauveria brongniartii* with some pesticides used in groundnut pest management. *Annals of Biology.*, Ludhiana, 6(1), 21-26.

Vyas, R.V., Yadav, D.N. and Patel, R.J. (1991a). Efficacy of *Bacillus popilliae* var. *holotrichiae* against white grub (*Holotrichia consanguinea*). *Indian Journal of agricultural Sciences,* 61(1), 80-81.

Vyas, R.V., Yadav, D.N. and Patel, R.J. (1991b). Studies on mass production of *Bacillus popilliae* var. *holotrichiae,* a promising pathogen of white grub in Gujarat. *Gujarat Agriculture University Research Journal.,* 17(1), 30-37.

Yadav, D.N. , Patel, R.R. and Patel, R.C. (1987). Natural enemies of groundnut leaf-miner, *Aproaerema modicella* Deventer (Lepidoptera : Gelechiidae) and their impact on its infestation at Anand (Gujarat). *Gujarat agriculture University research Journal.*, 13 (1), 13-16.

Yadava, C.P.S. (1991). Whitegrub management in groundnut. PCC, AICRP (Whitegrubs)., Agricultural Research Station, Durgapura, Jaipur. *Technical Bulletin. No.*1, pp14.

□□□

Sex Pheromones

V. NANDAGOPAL,
G. V. RANGA RAO AND T. V. PRASAD

20.1. Introduction

A variety of stimuli are being used by communicating individuals, the chemical communication is important for the reproductive behaviour of many species of insect group. Pheromone was originally defined as " substances which are secreted to the outside by an individual and received by a second individual of the same species, in which they release a specific reaction. This definition clearly indicates that a chemical or a mixture of chemicals that is secreted to the outside by an individual and triggers an immediate response of sexual behaviour in the receiving individual has been called as sex pheromone. The pheromone based behaviour has been demonstrated in a wide range of orders of insects, e.g., Lepidoptera, Hymenoptera, Coleoptera, Diptera, Homoptera, Hemiptera, Orthoptera, Isoptera, Neuroptera and Siphonaptera (Jacobson,1972). In majority of the Lepidopteran insects, females secrete sex pheromone and some cases, the males produce the sex pheromones (Yoshio Tamaki,1985). Among the Lepidopteran insects inflicting heavy damage to groundnut crop, the groundnut leaf miner (GLM) (*Aproaerema modicella*), (Gelechiidae: Lepidoptera); the red hairy caterpillars (RHC) *Amsacta albistriga* Wlk. and *Amsacta moorei* Btlr., the tobacco caterpillar (*Spodoptera litura* (F.)), the gram pod borer (*Helicoverpa armigera* (Hub.)) are most important yield barriers in India. The less important insects are Bihar hairy caterpillar *Spilosoma obliqua* W. and the castor semi - looper *Achaea janata* Linn. In the past, these pest were monitored using light traps. Since effective monitoring plays a key roll in the pest management programmes, in recent years, considerable importance has been given to strengthen the monitoring system. Lepidopteran are known to use sex pheromones, in their mating process. Except the leaf miner, these insects cause damage to foliage by direct feeding, while the Leaf miner mines the leaflets and feed the tissues leaving the lower and upper epidermis intact. Starting from the first instar to the emergence of moths, the larvae and pupae of leaf miner live under the protective covering. The tobacco caterpillar, the gram pod borer and hairy caterpillars hide under the dried foliage, crevices, clods etc. The dense hairs on the red hairy caterpillars protect them from insecticides. The larvae of these insects pupate inside the soil, while the leaf miner pupate inside the folded leaflets, hence the suppression of these insects becomes difficult by synthetic insecticides and warrants a different management strategy to bring the population below the economic threshold levels (ETL). Among the modern tactics employed in the insect management, the use of sex pheromones offers scope for proper monitoring of the activity of these defoliators and also in reducing the fertility potential of the wild females, thereby bringing down the fertile egg production. This method can also be integrated with other methods of insect management, especially with the use of organic pesticides such as neem products. Studies on sex pheromones have shown that a remarkable scope exists in the use of these pheromones as one of a viable components of IPM.

20.2. Occurrence

The GLM is endemic in South India and also been recognized as potential pest of Gujarat and Maharashtra. The tobacco caterpillar (*S. litura*) has been assuming importance in groundnut ecosystem since the last decade. It also appears occasionally in Saurashtra of Gujarat, partially in Andhra Pradesh, Karnataka and Tamil Nadu (Nair,1986; Amin,1988). At present the gram pod borer, *H. armigera* is considered to be one of the major insects of groundnut (Koshiya and Patel, 1987; Nandagopal,1990). The RHC *A. albistriga* regularly occurs in *Kharif* (rainy) seasons in Tamil Nadu, Andhra Pradesh and Karnataka and *A. moorie* is essentially a North Indian pest infesting on several crops and weeds. The other species *S. obliqua* is reported to occur in parts of Northern India, namely Punjab, Delhi, Bihar, Assam, M.P. and Maharashtra. The yield loss caused by these insect pests is discussed elsewhere in this book.

20.3. Sexual activity

All these moths are generally nocturnal in habit, however, leaf miner can be seen flying and mating during the day time also. Mating has been observed usually in the early hours (0300 hr.) of the day in *S. litura* and between 0200 hr. and 0600 hr. in the case of leaf miner. Under laboratory conditions, day-old males mate with the virgin females under artificial light (Nandagopal and Reddy, 1990). Prior to mating, the virgin females assume a "calling position" by raising the abdominal tips and release the pheromones during this process. As a result, a large number of the males are attracted and fly excitedly around these virgin females resulting in the mating. This chemical communication can be taken advantage of for the monitoring of their activity (Anon,1992) and for mass trapping as a component in IPM (Nandagopal, 1992).

20.4. Identification of the sex pheromones

The biochemical compositions of the sex pheromones have been identified in case of *S. litura* (Tamaki *et al.*, 1973 and Yushima *et al.*, 1974), *H. armigera* (Piccardia *et al.*, 1979 and 1980), *A. modicella* (Hall *et al.*, 1993) and *A. albistriga, Achaea janata* and *S. obligua* (Persoon *et al.*,1993; Yadav *et al.*,1998) (Table 20. 1).

Table 20.1. Chemical composition of the sex pheromones of major insect pests groundnut

Species	chemicals identified	reference
S. litura	1. (Z,E)-9,11-tetra decadienyl acetate 2. (Z,E)-9,12-tetra decadienyl acetate **effective ratio of 10:1**	Tamaki *et al.*, 1973 Yashima *et al.*, 1974
H. armigera	1. (Z) - 11 - hexadecenal 2. (Z) - 9 - hexadecenal **effective ratio of 88.6:2.9**	Picardia *et al.*, 1977 Nesbitt *et al.*, 1979, 1980
A. modicella	1. (Z) - 7,9- decadienyl acetate 2. (E) - 7 - decanyl acetate 3. (Z) - 7 - decanyl acetate **effective ratio of 100:20:14**	Hall *et al.*, 1993
A. albistriga	1.Octadecadienal 2.Z9, Z12- Octadecadienal 3.Z9, Z12, Z15-Octadecatrienal 4.Z3,Z6,Z9-heneicosatrienal **effective ratio of 1:1:11.5:24**	Persoons *et al.*, 1993
S. obligua	1.(Z3,Z6)- cis-9.10-epoxy-3, 6 heneicosadiene 2.(Z3,Z6)cis-9,10-epoxy-1,3,6 henecosatriene 3.Z9,Z12,Z15-Octadecatrienal 4.Z9,Z6,Z9-henecosatriene **effective ratio of 1:0.5:1:3:2.6**	Persoons *et al.*, 1993 Yadav *et al.*, 1998
Achaea janata	1.Heneicosane 2.Z6,Z9-Heneicosadiene 3.Z3,Z6,Z9-Heneicosatriene 4.Z9,Z12-Octadecadienal **effective ratio of 1:1:65:1**	Persoons *et al.*, 1993

20.5. Current status

The activities of the pheromone technologies include identification of presence of pheromones, identification of components of pheromones, synthesis of pheromone, supply system, monitoring using pheromone, mass trapping using pheromones, communication distruption using pheromone, regional networking of the working group, training and extension activities in relation with pheromone technology. Wightman and Ranga Rao (1993) has given a fair account of the such of these activities for different species of the insects attacking oilseeds in India which has been modified (Table 20. 2).

Table 20.2. Activities for different species of the insects attacking oilseeds in India

Species	presence	idetification	synthesis	supply	monitoring	mass trapping	distruption	NETWORK	
Apraerema modicella	1	1	1	*	1	?	*	*	*
Spodoptera litura	1	1	1	1	1	1	*	1	*
Helicoverpa armigera	1	1	1	1	1	1	*	1	*
Amsacta albistriga	1	1	1	1	1	1	?	*	*
Spilosoma obligua	1	1	1	1	1	1	?	*	*
Achaea janata	1	1	1	1	1	1	?	*	*
Caryedon serratus	*	*	*	*	*	*	*	*	*
Holotrichea consanguinea, H.serrata	*	*	*	*	*	*	*	*	*

1=Considerable progress made; *= Considerable effort needed; ?=Possible application

20.6. Pheromones as a monitoring tool

Monitoring

The occurrence of the major insect pests is difficult to forecast unless it is noticed in a damaging proportion. Hence monitoring of the insect pests using various gadgets becomes imperative. In the monitoring programme it is very important to recommend the required number of the traps to be used per unit area. Based on the trapping efficiency of the trap **(Plate 1)**, the number of insects trapped, the management options have to be decided. Careful consideration is to be given to the availability of the host plant, the mode of pupation, occurrence of natural enemies of the target insects etc. Lot of research has to be done before recommending the number of traps to be used/unit area. In India, monitoring of the major insect pests of

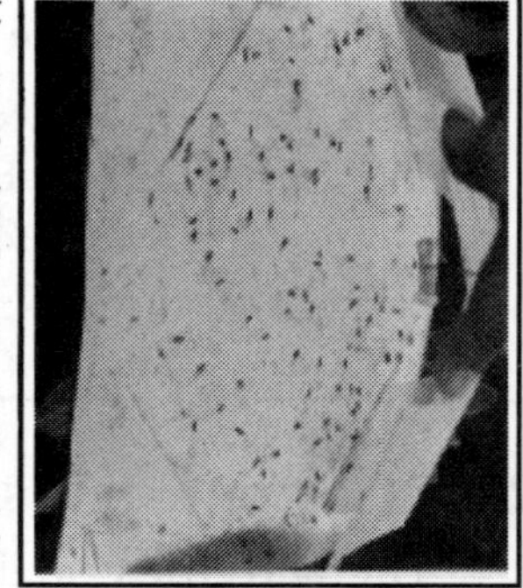

Plate 1. Male moths

groundnut such as *H. armigera*, *S. litura* were undertaken as early as 1985. Intensive work has been done on these insects spreading over Karnataka, Haryana, M.P, T. Nadu, A.P., Maharashtra, Himachal Pradesh, Punjab and Gujarat (Table 20.3).

Table 20.3. Monitoring of defoliators of groundnut

Species	Area	Reference
H. armigera (Hub.)	23 locations in India	Srivastav *et al.*, 1990
	Karnataka	Patil *et al.*, 1992, Ballal *et al.*, 1998; Patil and Kulkarni,1997; Naik *et al.*, 1993; Nandihalli *et al.*, 1990
	Haryana	Sinha and Jain, 1992; Sinha and Malhotra, 1993
	M.P.	Bhaduria and Bhaduria,1998; Anju Bhatnagar *et al.*, 1998; Gupta *et al.*, 1996; Lal *et al.*, 1985
	T, Nadu	Loganathan *et al.*, 1999; Loganathan and Uthamasami,1998
	A.P.	Venkataiah and Subbaratnam,1992; Prasad *et al.*, 1993
	Maharashtra	Dhanorkar and Puri,1993; Tadas *et al.*, 1994
	Himachal Pradesh	Verma and Sankhyan,1993
	Punjab	Chhabra and Kooner,1992
	Gujarat	Chari *et al.*, 1985.
	T Nadu	Sundaramurthy *et al* 1985
S. litura (Fabricius)	Karnataka	Nandihalli *et al.*, 1989; 1990;
	U.P.	Singh and Sachan, 1993,1991
	T. Nadu	Muthukrishnan and Balasubramanian,1992; Balasubramanian *et al.*, 1985.
	A.P.	Karuppuchamy *et al.*, 1992
	W.Bengal	Senapati *et al.*, 1990; Das *et al.*, 1990
	Orissa	Mishra and Sontakke, 1992
	A.P.	Gupta,M; Pawar,A D.1989.
	Haryana	Singh,J P; Lather,B P S. 1989
Aproaerema modicella	Gujarat	Nandagopal and Soni,1993
(Dev.)	M.P.	Das, 1999

The pheromone lure of the *S. litura* has been in use as a monitoring tool in the All India Coordinated Research Project on Oilseeds (AICORPO) since 1985 using sleeve trap **(Fig:1)**. Initially it was started at three centres and further it was extended to another five centers representing all the groundnut growing zones in India (Fig 1). The monitoring activity from 1985 to 1997 indicated that *S. litura* occurs throughout the year with a highest population of 2307 in a week. The peak population density was recorded between July and October, which is the main groundnut growing season in the rainy season. Ranga Rao and Wightman (1994) consolidated the trap catches data from 1986 to 1992 of *S. litura* and found that a continuous existence of this insect in South India, which could be due to prevalence of more favourable climatic conditions, throughout the year compared to north India where insects go hibernation due to extreme cold. Consistently high density of population of *S. litura* was recorded during the rainy season at Chiplima, Dharwad, Jalgaon and Junagadh. The other centres such as Ludhiana, Latur, Vriddachalam and Akola are also appeared to be endemic for this pest. At ICRISAT (Hydrabad), average of eight years data indicated that there was a decline in the moth catch (62/trap/week) from the first week of May till the last week of June. Similarly, at Gwalior, Madhya Pradesh apart from the reduction in the summer months, there seemed to be second spell of decline during the last week of November to third week of February. Almost similar situation was there at Hissar, Haryana. In the case of Ludhiana, Punjab), the declining trend was extended upto last week of March. In Padapai, Tamil Nadu the population density was considerably less. At Jalgaon, Maharashtra similar situation existed. The high population density at Dharwad helped in the screening of the advanced breeding lines against this insect and a considerable number of resistant lines were identified (Anon, 1994).

The sex pheromone of the gram pod borer (*H. armigera*) was used as monitoring tool only in few centers namely Palem, Latur, Vriddachalam and Hyderabad (ICRISAT). The population, in the groundnut ecosystem was not so much as to be considered as a major insect, which collates with the views of Amin (1988). Though much time and effort has been devoted to the development of pheromone traps for *H. armigera* (Pawar *et al.*,1988). Unfortunately, the efforts has not been successful because the moths are not caught with the same efficiency as those for *S. litura*, the reasons for instance, the blend of pheromones, the trap design or the olfactory competition from the environment, they could not able to predict egg or larval densities from moth catches in such a way that they could be employed for pest-forecasting (Dent,1986). In contrast to the above observations, Srivastava *et al.* (1992) have calculated regression equations by consolidating larval counts for the whole ICRISAT farm (1300 ha) and taking the total weekly pheromone trap catch. It was encouraging that this exercise gave significant results to predict egg densities in a field complex of up to 5 ha from one or two traps on a daily basis.

The hairy caterpillars (e.g. *Amsacta* and *Spilosoma* spp.) are major pests of dry land agriculture in South Asian countries. There is usually only one generation a year, if conditions are favourable, however, *A. albistriga* completes three generations within a short period (Dr. S. Jayaraj, personal communi-cation), but in bad years this alone is devastating, irrespective of the crops, over wide areas. Clearly, an early warning of flight activity could be helpful to farmers. There is a need to explore further on the biology of the *Amsacta.* Patel *et al.,* 1987) recorded that moths of *A. moorei* emerged within 3 days of the first rain (first week of July), in large numbers (4334, 7000,and 2420 in a ultra violet light trap and on the fourth day the population reduced to 68 indicating the moth emerging pattern during 1981. Subsequently, in 2 years, in the same period the population density was too low ranging 79 to 158 in a UV light trap. Further no work on this insect has been reported. Since, only very recently the pheromone system in *A. moorei* and *S. oblique, A. janata* have been identified, the work on trap design, effective dose, traps per unit area have to be worked out.

In the case of the leaf miner, the monitoring was initiated from 1992 onwards (Hall *et al.,*1993). Six centers were identified based on the occurrence of the leaf miner, which included Junagadh (NRCG) in Gujarat, Latur, Akola and Jalgaon in Maharashtra, Vriddachalam in Tamil nadu and Hyderabad (ICRISAT) in Andhra Pradesh. At Junagadh, the density of the leaf miner moths was 137 and 175 during kharif 1992 and 1993 respectively. The population was meager (maximum of 152/trap/week) in other centers such as Latur, Akola and Jalgaon and Vriddhachalam. In Hyderabad (ICRISAT), the population pressure was extremely high in water trap developed by ICRISAT. During 1990-91 light trap catches was to a maximum of 34034/ trap/week, indicating the severity of this insect problem (Anon,1991). During 1992-93, there were 4-5 generation in rainy and 6 generations in post-rainy season. The moth population exceeded 3000 /trap/week during February-March (Reddy *et al.,* 1993). Similarly during 1993-94, the moth catch in pheromone traps touched 7312/trap/week, in 1994-95 the population was 2500, in 1995-96, it was just below 1500 numbers and in 1996-97 it was 2200 moth/week/trap (Ranga Rao and Wightman,1994).

20.7. Pheromones as a mass trapping tool

This is a direct method of managing the insects. Based on the number of insects trapped, a decision can be taken to continue the recommended number of traps/unit area for mass trapping or remove all the traps and use only for monitoring. Because, if the anticipated insects are going to occur only below ET levels, there is no need for continuing the traps meant for mass trapping which will lead to unnecessary waste of resources. If the population is of moderate level, mass trapping followed by application of bio-pesticides would be sufficient to contain the insect before reaching economically important level. If the insect is occurring in a endemic way, necessarily, we have to think of going for too many methods of available tactics to manage the insects. Pheromone as a mass trapping tool have also been utilized in *S. litura,* , *H. armigera, A. modicella* (Table 20.4).

Table 20.4. Mass trapping of defoliators of groundnut

S. litura (Fabricius)	U.P.	Singh and Sachan,1991
	A.P.	Krishnaiah, 1986; Lalith kumari *et al.*, 1992
	T. Nadu	Dhandapani, 1985.
H. armigera (Hub.)	A.P.	Lalith kumari *et al.*, 1992
	Haryana	Pawar *et al.*, 1984.
	Gujarat	Patel *et al* 1985.
Aproaerema modicella (Dev.)	Gujarat	Nandagopal,1992;

20.8. Pheromones as a mating disruption tool

In a very few cases mating disruption technique was successful in India. The communication disruption as a tool has been tried in *S. litura, A. modicella* in India (Table 20.5) with limited success. Based on the information's available, communication disruption may be successful in those of the crops where dense foliage gives such a atmosphere where the wind velocity is reduced considerably and the pheromone persist for few hours *Aproaerema modicella* (Dev.) in groundnut intercropped with pigeon pea (groundnut based cropping system (Anon, 2002) *S. litura* (Fabricius) (Noctuidae: Lepidoptera) in groundnut (Srinivas and Rao, 1993; Rao *et al.*, 1989).

Table 20.5. Mating Disruption

S. litura (Fabricius)	AP.	Srinivas and Rao, 1993; Rao *et al.*, 1989
Aproaerema modicella (Dev.)	Tamil Nadu	Anon,2002

20.9. Trap design

There are attempt made to design the trap for efficient trapping of the target insect in groundnut ecosystem (Table 20.6). Limited information is available on the development of traps for various insects in Indian agriculture. There has been efficient trap design for groundnut leaf miner (Nandagopal and Soni,1993) in groundnut ecosystem, for *H armigera* (Pawar *et al.*,1988 and Nandagopal *et al.*,2003), for *S. litura,* (Prasad *et al.*, 1985; Krishnananda *and* Satyanarayana, 1985). The efficiency of these traps were between 60 to

90 %. The trap and the lures used for *H. armigera* attracted the *Pectinophora* males.

Table 20.6. Trap design

H. armigera (Hub.) (Noctuidae:Lepidoptera)	Pawar *et al.*, 1988 ; Nandagopal *et al.*, 2003
Aproaerema modicella (Dev.) (Gelechiidae:Lepidoptera)	Nandagopal and Soni, 1993 Nandagopal *et al.*, 2006
Spodoptera litura	Ranga Rao *et al.*, 1993

At Junagadh, a square glue trap (David's Trap), developed at NRCG (Nandagopal and Soni, 1993) was used, while in other centres water trap designed at ICRISAT was used. A simple study was conducted to see if the male of GLM trapped are potential or spent by collecting the males approaching the trap in day time in a test tube and allowing them in ratios to mate with freshly emerged virgin females. The result indicated that there was fertilized eggs as there was emergence of larvae and damage to the foliage. Different types of traps were designed and tried in the trapping of GLM. The result indicated that a square glue trap designed with 22 cm side and with 4 entry holes (Nandagopal and Soni,1993) was found to be efficient in trapping the males.

20.10. Interaction effects of pheromones

Pheromone lures of *H. armigera* and *S. litura,* when tied together and tried in groundnut, showed that the lure of the latter species suppressed the attraction of the former species as there were only 4.2 males of *H. armigera* trapped compared to *S. litura* (97.6 males) (Pawar and Srivastava, 1988). Similarly, the height of the trap affected the catch of *H. armigera* (Pawar *et al.*,1988), *S. litura* (Ranga Rao *et al.*, 1991) and *A. modicella* (Nandagopal, 1992). Different shapes of the trap also affected the catch of male moths of leaf miner. When the David's Trap was placed on the foliage of groundnut, a maximum of 90 moths/trap/ day was recorded. While, the same trap when placed about 15 cm above the soil surface affected the catch. Nandagopal *et al.* (2003) studied the compatibility of sex pheromones of *S. litura, H. armigera* and *Aproaerema modicella* (Deventer) in cotton and groundnut ecosystem and concluded that there is so much variation in trap catches having a pheromone lures singly and in combination. The moth catches of *S. litura* and *H. armigera* were lower in number when the pheromone trap baited with combination of spodolure, helilure and leaf miner lure. The moth catches/ trap were also lower in number when the pheromone traps were baited with combination lures of *S. litura* and *H. armigera* compared to the moth catches in pheromone traps baited with spodolure and helilure alone. Hence the present investigation revealed that the trapping of *H. armigera* and *S. litura* in the same trap is not advisable. Similar results were also obtained by Pawar and Srivastava (1988). This may be due to the masking effect of the sex pheromone of *S. litura* on sex pheromone of *H. armigera* and vice-versa. Some of the lures in combination had similar chemical compounds which attracted different species of insects. Similarly Nandagopal *et al.* (2003) reported that trap which are having helilure that attract *Pectinophora gossypiella* (Saunder) males moth because of having some similar chemical components. Two structurally related compounds or pheromone "Mimics" (Z)-9-tetradecenyl formate (III) and (Z)-13-hexadecenyl formate (IV) were tested and found that both are structurally related to the natural pheromones of striped rice borer, *Chilo suppressalis* (Walker) (Beevor *et al.*, 1977).

Table 20.7. Moth catches of *Spodoptera litura* (F.) and *Helicoverpa armigera* (Hub.) in cotton ecosystem.

Lure combination	Total	Mean	Per cent reduction of moth catches
Spdoptera male moth catches			
Trap1 (S+H)	387.0	29.8	41.27
Trap 2 (S+L)	368.0	28.3	44.16
Trap 3 (S+H+L)	282.0	21.7	57.21
Trap 4 (S)	659.0	50.7	
H. armigera male moth catches			
Trap1 (S+H)	131.0	10.1	69.01
Trap 2 (H+L)	264.0	20.3	41.97
Trap 3 (S+H+L)	175.0	13.5	61.54
Trap 4 (H)	455.0	35.0	

S = spodolure ; H = helilure ; L = Leaf miner lure
Leaf miner incidence was not found.

20.11. Problems in adoption

Even after its introduction about 25 years ago, pheromone technology in Indian Agriculture is still in its infancy. This may be attributed to defects in manufacture of the lures and traps, difficulties in installation of traps at farmers level and insect related issues. An analysis of these issues has been made and suggestions are given to overcome the same. Then only there be actual fillip to the novel, eco-friendly pheromone technology in Indian agriculture.

1. Defects in manufacture

- Poor quality rubber band which tied around the neck of the trap, can't stand for a month time in the field condition allowing plastic sleeve to fall apart

- Poor quality rubber band tied to bottom of the sleeve, may escape of trapped insects
- The improper fixed trap tilting on one side allowing visiting insects to escape
- Most of the available trap in the market do not have handle. So that it can be tied with single stick, warranting two pieces of stick tied to a "t" shaped two pieces bamboo sticks
- Poor quality lures or delivery system allowing quick decimation of the pheromone lasting for a week or two
- Optimization of the gap between the lure and the mouth, to avoid of the trapped insect
- Improper pheromone impregnation in the lure
- Use of poor quality dispenser as delivery system
- The traps should be designed to either by using talc powder or a funnel etc.
- Poor quality plastic sleeve which if run for one month becoming brittle and tear off allowing escape of the trapped males
- Different species requires a definite type of trap

For example:Pyrallids (eg.,sugarcane borer like Early shoot borer, inter-node borer etc) is trapped in water trap only (David et al., 19985; Dr.Mukunthan and Dr. Easwaramoorthi, 2002, Sugarcnae Breeding Institute, Coimbatore- personnal communication).

Noctuids (eg., gram pod borer, tobacco caterpillar) are efficiently trapped in sleeve trap, while Gellechids (groundnut leaf miner, cottom pink bollworm) are trapped in sticky trap (Nandagopal and Soni,1993).

2. *Installation of the trap and fixing of lure related issues*

- The fixing of the lures in traps are not proper
- The bottom of the sleeve are not toed with the stick
- Very thin or damaged bamboo sticks can not sustain the mass of trap leading to tilting and thus escape of the insects attracted.
- Farmers can not afford to store the lures in the deep fridge and lure stored openly last for a few days only

3. *Insects related issues*

3.1 Pheromone specificity

Usually it is said that pheromones are species specific, but actually it is not so for all the species. There are some exceptions such as,

- Cis-8-Dodecenyl acetate attract 40 insect species (Kydonieus and Beroza, 1982)
- Is amyl Acetate- Shares two order
  1) Hymenoptera—*Aphis mellifera*
  2) Lepidoptera—*Manduca sexta*
- Cis-14-methyle.8.hexadecenal- is shared by many *Trogoderma* species
- Cis-9, trans-12-Tetra Decadienyl acetate- Shares two different genus
  1) Pyralidae—*Plodia interpunctella*
  2) Noctuidae—*Spodoptera litura*

The sub-species of the main species of the sugarcane inter-node borer shares the same pheromone. The species *Chilo sacchariphagus* is a insect of West Indies, Trinidad and Cuba etc, while the sub-species *Chilo sacchariphagus indicus* is occurring in India (David et al., 1981;David et al., 1985; Brinda F. Nesbitt et all., 1986).

The pheromone components for both the species and the sub-species are having the same two components

1. (Z)-13-Octadecanyl acetate
2. (Z)-13-Octadecenol

3.2. Variability in attraction for same species across the location

The ratio of blend of pheromone compounds for same species may vary for biotype, The species of *Adoxophyes orana* and *Clepsis spectrans* in the Netherland share the two compound and (Z)-9 and (Z)-11-Tetra decanyle acetate in 9:1 and 1:3 ratio, respectively. The biotype *Ostrinia nubilalis* in Iowa, Z/E-tetradecanyle acetate is used in the ratio of 97:3 (Klun and Bindley, 1970) whereas in New York a ratio of 4:94 is more effective (Kochousky et al.,1975). Interestingly in Pennsylvania, both 97:3 and 2:98 ratio blends (Carde,et al., 1975) work.

In India similar observation in relation to groundnut leaf miner *Aproaerema modicella* has been recorded. In Tindivaram, Tamil Nadu one component (Z) - 7 - decanyl acetate or a blend of (Z) - 7,9- decadienyl acetate (1), (E) - 7 - decanyl acetate (2), (Z) - 7 - decanyl acetate(3) in the ratio of 10:2:1.4 were equally effective against GLM (Hall et al., 1993). However, at Junagadh either the complete blend of the pheromone or either (Z) - 7 or (E) - 7 - decanyl acetate in combination with with (Z) - 7,9- decadienyl acetate was only found to be effective. Similar result were also obtained in Hyderabad, A.P. excepting that (E) 9 and (Z) 7,9 combination which was not so effective.

3.3. One or two related components of the blends may be attractive to the other species

eg. *Helicoverpa armigera* pheromone lures attracting the pink boll worm *Pectiniphora gossypiella*

Lure of : *Helicoverpa armigera*

H.armigera moths caught: 1 to 34 males /trap/ days

Components of *Helicoverpa armigera:*

1. (Z) –11-hexadecenyl acetate
2. (Z)-9- hexadecenal

P. gossypiella moths caught:: 1 to 19 males /trap/ days

Components of *Pectiniphora gossypiella*

1. (Z,Z)-7,11-hexadecadienyl acetate
2. (E,Z)-7,11-hexadecadienyl acetate

20.12. Steps to overcome/improve the pheromone technology

For scientists

- There should be thorough experimentations on the trap designs to be used for various species of insects as they differ in their trapability. This could be possible only on the observation of the males movements of both to-wards and more specifically return activity during its peak activity periods.
- Proper ratios of the various components of the pheromone for various agro-climatic zone for a particulars species should be examined
- There should be experimentations on the use of a particular components for more than one species of the insects so that the multiplicity of the traps in a unit of cropped area can be avoided
- Durability of the traps/sleeves and delivery system of the pheromones should be evaluated thoroughly before recommendation to large scale manufacturing by the industries
- There should be development of standard trapping system for each insect encompassing trapability distance, trap density/unit area, minimum load of the pheromone components, placement of the trap, longevity of the lures and traps
- The correct delivery system should be devised for each of the insect
- Experimentation on the minimum distance between any two traps or source of lures should be found out to minimise the number of traps/ lures per unit area

- In communication techniques, the reduction in attractioncy in comparison with control should be more than 95%

For Industries:

- Proper quality of the materials for the manufacturing the delivery system /traps should be used for longer period in the field conditions
- Proper fixing facilities should be provided in the trap itself so that the farmers can use it properly instantly; i.e., inverted 'L" handle would be more suitable to tie the trap in a single bamboo stick, instead of 'T' shaped bamboo stick
- High purity of the pheromone components should be used for actual blend of the pheromone (above 97%)
- Proper quantity of the pheromone should be loaded in to the delivery system
- The cost factor is very important for the resources poor farmers
- Supply of the pheromone delivery system /traps and lures should be nearest to the farmers
- Instructions with flow chart in pictorial form for proper installation of the traps and lures should be supplied along with the pheromone traps

For extension agency:

- There should be regular training for the extension agencies on the installation and the methods adopted on the monitoring, mass trapping and communication disruption of the insects using the pheromones
- The extension agency should explain the farmers about the proper instillations of the traps and the actual number of traps/unit area
- Proper advise to farmers on the timing of use of the pheromone is necessary

For farmers:

- Fresh lures should be purchased every time and lures should not be kept in room temperature, instead, if there is facilities like refrigerators, lures in a plastic bag should be kept in side the deep fridge or lures in a plastic bag should be kept in side the container with weight in the pack filled with water. See that no water should enter the pockets
- Install the actual number of traps to be used per unite area
- The traps should be installed upright, so that the lure is in correct position and not likely to fall due to wind
- The tying of the sleeve both at the top and at the bottom should be proper to avoid opening of the sleeve and escape of the trapped insects

- Do not allow the trapped insects inside the sleeve for many days as it may facilitate for slow movement and escape of few of the insects
- At least once in a week the trap should be monitored for insects and if found, they should be killed by crushing inside the sleeve itself and , the killed insects can be removed by opening the bottom of the sleeve
- Use strong sticks so that it can withstand the weight of the moths trapped and the strong winds
- While moving inside the field, see that any trap is loosened from its position and sleeve is detached, if found it should be rectified
- If the crop is nearing harvest collect the traps carefully with out damage the fittings and the sleeves and store them properly so that the same can be used for other pest where similar trap is recommended or in the next season
- If the neibouring farmers are also having crops harbouring the same species, should also be encouraged to go for pheromone trapping-through community approach
- Replace the lures at least once in 25 to 30 days or as given in the instructions on the pocket

20.13. Discussion

The analysis of the trap catches of these species of defoliators explained the potential of the sex pheromones, which can be tapped to compliment as an adjunct in IPM. Intensive studies in the case of *A. modicella* and *S. litura* indicated that they differed in their population fluctuations. Also, the GLM registered four clear cut broods both in *kharif* and in summer on the groundnut crop in Junagadh conditions, While it was 5 to 6 generation at Hyderabad. The *S. litura* and *H. armigera* recorded overlapping broods. In the case of leaf miner, a low initial population will be useful in identifying the initial foci of infestation which is normally not so easily detectable under field conditions. The high density of the moths in southern India warrants a serious approaches in its management. The results of monitoring of *S. litura* trial taken up over nine years all over India, indicated that it may be useful for field application under farmers conditions. In an another study (Ranga Rao *et al.*, 1991) revealed that light trap catches of the *S. litura* gave a wrong picture of the bionomics of this species compared to the pheromone catches data. Similar observation were made in the case of GLM trapping in the pheromone trap and the light trap, the later giving the wrong picture. The mobility of the GLM was 50m and having uneven distribution even in the same field, its mass trapping with a specific number of traps/ unit area might help in the economic and eco-friendly management of the GLM. Further detailed information on less expensive and durable dispensers of the lure, optimum number of traps/unit area, its role as an adjunct in the IPM programme are to be explored.

Farmers have been using pheromone traps against *H. armigera* in pigeon pea as well as cotton. Among 78 fields surveyed (pigeon pea as intercrop), the installation of the traps, the looping of the rubber bands around the sleeve, sleeve fitting around the funnel, fixing of lures were defective. Very rarely traps were in order. Even though, the lures were fresh and effective to attract males, but the attracted males were found to be escaped due to the above installation problems of pheromone traps.

20.14. Future thrust

- Since over 80% of the groundnut cultivation is rain dependent and the withdrawal of rains during the active growth stage of the crop is regular, an efficient mass trapping programme using dry funnel for the lepidopteran pest would be useful.
- For GLM, owing to its size and the slow flying activity, a cheap sticky trap would be economical and useful for monitoring and mass trapping.
- Mating disruption for GLM would be useful as the occurrence of the broods in both the seasons (rainy and post-rainy) is clear and there is no overlapping broods reported.
- Screening the available dispensers such as rubber septa, plastic laminate, hollow fibre, polyethelene tube, PVC resins, sprayable polymer beads, sprayable hollow fibres, sprayable micro-encapsulations etc.
- The microencapsulated pheromone offers the advantage of both utilising the conventional pesticides application equipments and of distributing pheromones in a relatively uniform manner. The other option is to use micro-dispersible technique, hollow fibre dispensers applied with a modified spray procedure, appears to obviate the difficulty of maintaining the required emission rate, since longevity of emission depends on the length of the fibre.
- It would be useful to use pheromone traps for mass trapping, against poly phagus insects, over large area in villages, especially where poly crop is normal in Indian villages.
- It is important to ascertain that the trapped males are potent or spent males.
- Collaboration with Natural Resources Institute, London and IICT at Hyderabad, India in developing data - base for pheromone and weather.
- Training of field level workers for proper use of pheromone trap.
- Periodic workshops to exchange information.

- Collaborative approach with commercial and scientific groups for proper improvement in the trap designs.

References

Abdul Kareem, A., and Subramaniam,T.R. (1976). Antifeeding effect of two organotin compounds on *Stomopteryx subsecivella* Zell. (Lepidoptera). *Madras agri. J.*, 63, 354-357.

Abdul Kareem, A., S. Sadakathulla and Subramaniam. T.R.(1974). Control surulpoochi on groundnut. *Farm and Factory* 8(4), 35-36.

Amin, P.W., (1988). Insect and mite pests and their control. In Groundnut (P.S.Reddy, *Ed.*),. Indian Council of Agricultural Research, New Delhi, India, pp. 393-452.

Anonymous, (1989).Annual Reort, National Research Center for Groundnut (Indian Council of Agricultural Reseach), Junagadh, Gujarat. pp75.

Anonymous, (1985 to 1998). Annual progress report, All India Coordinated Research Project on Groundnut (ICAR), National Research Center for Groundnut, Junagadh-362 001, Gujarat, India.

Anonymous, (1984 –1994). Annual Progress Report, All India Coordinated Research project on Oilseeds. Hyderabad.

Anonymous, (1989). Annual Progress Report (1987-1988 and 1988-1989). of AICRP on Rodents Control. (Junagadh: Gujarat Agricultural).

Anonymous, (1989). *Groundnut.* Publications and Informations Directorate, CSIR, New Delhi.55 pp.

Anonymous, (1989). Twenty-seventh meeting of the Research Council held on Dec 22, 1989 at TNAU, Coimbatore. *Agenda notes.* Directorate of Research, TNAU, Coimbatore. 137p.

Anonymous, (1989-1997).Annual Progress Reports (1988-89 to 1996-97) of AICRP on Rodent Control. Gujarat Agricultural University, Junagadh.

Anonymous, (1990). Annual Progress Report (1989-1990). of AICRP on Rodents Control. (Junagadh: Gujarat Agricultural).

Anonymous, (1991). *Annual progress report Groundnut 1990.* 38th annual Kharif Oilseed research worker's group meeting. All India Co-ordinated Research Project on Oilseeds. pp 16-18.

Anonymous, (1991). Annual Progress Report (1990-1991). of AICRP on Rodents Control. (Junagadh: Gujarat Agricultural).

Anonymous, (1994). Annual Report. Project Directorate of Biological Control, Banglore, India, 143 pp.

Anonymous, (1994). Elite groundnut germplasm lines ICGV 86252, ICGV 86393, ICGV 86455, ICGV 86462. *Plant-Material-Description.* 1994, No. 47, 4 pp. International Crops Research Institute for the Semi-Arid Tropics. ICRISAT, Patancheru, Andhra Pradesh 502 324, India.

Anonymous, (1990). Rodent control measures in Gujarat-Report of Expert Team,(Ahmedabad: Directorate of Agriculture, Gujarat State.

Anonymous, (1994). Annual Research Report AICRP biological control, Gujarat Agricultual Universisty, Anand, Guajarat.

Anonymous, (2002). Final Research Report AP Cess Fund Scheme on leaf miner pheromone in Groundnut, submitted to ICAR,New Delhi, pp.

Anuj Bhatnagar, Saxena, R. R., Bhatnagar, A. (1998). Monitoring of *Helicoverpa armigera* through light and pheromone traps and seasonal activity of their natural enemies at Bastar plateau zone, India. *Annals,of,Plant,Protection Sciences* 6: 142-145.

Arida, G.S., Dyck,V.A. and Beevor, P.S. (1978). Sex pheromone in insect pest control. Saturday Seminar, 4th March, 1978.

Balasubramanian, G., Chelliah, S., Balasubramanian M, and Regupathy, A. (1985). Studies on light trap and pheromone trap catches of *Spodoptera litura* (abricius); Jayaraj S (ed.). *Behavioral,and,Physiological,Approaches,in,Pest,Management*. 78-84pp.

Ballal, C. R., Singh, S. P., Reddy, P. P., Kumar, N. K. K. and Verghese,A. (1998). Effect of pesticide applications on pheromone trap catches and distribution pattern of *Helicoverpa armigera* (Hubner) (Lepidoptera: Noctuidae) in pigeonpea ecosystem. Advances in IPM for horticultural crops. Proceedings of the First National Symposium on Pest Management in Horticultural Crops: environmental implications and thrusts, Bangalore, India, 15,17 October 1997. 124-128pp.

Barber, C.S.(1906). The pests of groundnut in Madras. *Agricultural Journal of India.*, 1, 170.

Beevor, P.S., Hall, D.R., Nesbitt, B.F., Dyck, V.A., Arida, G.S., Lippold, P.S. and Oloumi-Sadeghi, H. (1977). Field trials of the synthetic sex pheromones of the striped rice borer, *Chilo suppressalis* (Walker) (Lepidoptera : Pyralidae), and of related compounds. *Bulletin Enomoligical Research* 67: 439-447.

Bhadauria, N. K. S. and Bhadauria, N. S. (1998). Seasonal population fluctuation of gram pod borer, *Heliothis armigera* (Hub.). *Advances in Plant Sciences*.11: 283-285.

Bindra, O.S and Kittur,S.U. (1961). Biology and control of the red hairy caterpillar (*Amsacta moorei* B.). in Madhya Pradesh. *Vikram Science*,. 5, 62-71.

Brinda F. Nesbitt,P.S.Beever,A.Cork, D.R.Hall, H.David and V.Nandagopal.(1986). The female sex pheromone of four components of the sugarcane stalk borer *Chilo auricillus* Ddgn. and the field tests. Journal Chemical Ecology.12:11377-1388.

Carde,R.T., Kochansky,J., Stimmel,J.F., Wheeler,A.G.Jr. and Roelofs,W.L.(1975). Sex pheromone of the European corn borer (*Ostrinia nibilalis*): cis and trans-responding ,males in Pennsylvania. Environmetal Entomology. 4, 413-414.

Channabasavanna, G.P. (1951). 'The surulpuchi' (leaf miner). of groundnut, *Mysore Agricultural Journal.*, 26, 67-68.

Channabasavanna, G.P. (1954). Insect pest of some cash crops- nature of damage and control. *Mysore Agricultural College Year Book*. pp 89.

Channabasavanna, G.P. (1957). The groundnut leaf miner and its control. *Mysore Agricultural College Year Book* (1956-57). Pages 123-124.

Chari, M. S., Patel, A. R., Rao, B. S., Bharpoda, T. M. and Patel, N. M. (1985). Population studies on tobacco capsule borer *Heliothis armigera* Huber. *Tobacco Research.* 11: 98-104.

Cherian, M.C. and Basheer, M. (1942). Studies on *Stomopteryx nerteria* Meyr., a pest of groundnut in the Madras Presidency. *Madras Agricultural Journal,* 30, 379-381.

Chhabra, K. S., Kooner, B. S. (1992). Assessment of *Heliothis armigera* (Hubner) population through pheromone trap at Ludhiana, India. *Zeitschrift,fur,Angewandte,Zoologie.* 79: 309-317.

Das, B. K., Senapati, A. K., Panda, M. (1990). Seasonal occurrence of *Spodoptera litura* in West Bengal as reflected by pheromone trap catches. *Environment and Ecology.*8: 825-829.

Das, S. B. (1999). Seasonal activity of groundnut leaf miner, *Aproaerema modicella* in West Nimer Valley, Madhya Pradesh. *Insect,Environment.* 5: 69.

David R.Hall, Peter S. Beevor, Derek, G., Campion, David, J., Chamberlain, Alan Cork, Rosemary White, Aurelio Almestre, Thomas J. Henneberry, V.Nandagopal, John A. Whightman and Ranga Rao,G.V. (1993). Identification and sysnthesis of new pheromones.IOBC wprs Bulletin 16: 1-8.

David, H., S.Easwaramurthi, B.P. Nesbitt and Nandagopal, V. (1981). Sex attraction studies in sugarcane internode borer *Chilo sacchariphagus indicus* (K) with synthetic pheromone. Pestology.5:5-11.

David. H, B.P. Nesbitt S. Easwaramurthi, and Nandagopal, V.(1985). Application of sex pheromone in sugarcane pest management. Proceedings of the Indian Academy of Science. 94:333-340.

Dent, R. David.(1985). Population studies of *Helicoverpa armigera* through the analysis and interpretation of light and pheromone trap catches and larval counts. Pulse Entomology departmental progress report 18 . International Crop Research Institute for Semi Arid Tropics , Patencheru,A.P. 502 324, India. Pp 154.

Dent,D.R. and Pawar,C.S. (1988). The influence of moon light and weather on catches of *Helicoverpa armigera* Hub. (Lepidoptera :Noctuidae) in light and pheromone traps. Bulletin of Entomological Research,78, 365-377.

Dhandapani, N. (1985). Trapping of *Spodoptera litura* (Fb) moths using sex pheromone. *Madras Agricultural Journal.* 72: 236-238.

Dhanorkar, B. K and Puri, S. N. (1993). Monitoring of cotton bollworms using sex pheromone traps in Marathwada. *Journal of Cotton Research and Development.* 7(1): 171-173.

Fletcher, T.B. (1914). Some South Indian Insect and other Animals (of Importance Considered Especially from an Economic Point of View). Govt. Press, Madras. 565p.

Gahukar, R.T. (1992). Groundnut Entomology. Retrospect and Prospect. *Agricultural Zoological Reviews*, 5, 139-199.

Gupta, M. and Pawar, A. D. (1989). Monitoring *Spodoptera litura* (F.) in vegetable crops through synthetic sex pheromone traps. *Indian Journal of Plant Protection*. 17: 117-118.

Gupta, M. P., Gupta, D. P. and Shrivastava, K. K. (1996). Population dynamics of cotton bollworms in Madhya Pradesh. *Annals of Entomology*. 14(1): 61-66.

Gupta, R. L., Pandey, N. D., Ram, S., and Sukhani, T. R. (1966). Bionomics of *Amsacta moorei* Buller (Lepidoptera : Arctiidae). *Allahabad Farme,r* 40(2), 65-70.

Hall, D.R., Peter S. Beevor, Derek G. Campion, David J. Chamberlain, Alan Cork, Rosemary White, Aurelio Almetre, Thomus J. Henneberry, Nandagopal,V., Wightman,J. and Ranga Rao,G.V. (1994). Identification and syntheisi of new pheromones. *IOBC Wprs Bulletin*, 16, 1-10.

Kapadia, M.N., Bharodia, R.K. and Vora, V.J. (1982). Biology and estimation of incidence of groundnut leafminer, *Stomopteryx subsecivella* Zell). (Gelechiidae: Lepidopter). *GAU. Research. Journal*, 8, 37-39.

Klun,J.A and Bindley,T.A. (1970). cis Tetradecenyl acetate, a sex pheromone of the European corn borer. Journal of Economic Entomology. 63,779-780.

Kochousky,J.P., Carde,R.T., Libherr,J., Roelofs,W.L. (1975). Sex pheromone of the European corn borer (*Ostrinia nibilalis*)

Koshiya, D.J. and Patel, H.K. (1987). Life table and population dynamics of *Heliothis armigera* Hubn. on groundnut. *G.A.U. Research Journal*, 12, 11-14.

Krishnaiah, K. (1986). Studies on the use of pheromones for the control of *Spodoptera litura* Fab. on black gram grown in rice fallows. *Indian Journal of Plant Protection*. 14: 43-46.

Krishnananda, N. and Kaiwar, S.R. (1965). A preliminary study on the control of leafminer of groundnut. *Indian Oilseeds Journal*, 9, 180-185.

Krishnananda, N. and Satyanarayana, S. V. V. (1985). Effect of height of pheromone trap on the capture of *Spodoptera litura* moths in tobacco nurseries. *Phytoparasitica*. 13: 59-62.

Kumari, V.L.L. and Reddy, D.D.R. (1992) Evaluation of pheromone trap designs for trapping *Spodoptera litura* and *Helicoverpa armigera* and their reproductive behaviour. *Indian Journal of Plant Protection*. 20: 18-23.

Kydonieu, A.F. and Beroza,M. (1982).Insect suppression with controlled release pheromone system, Vol I CRC Press, Inc., Boca Raton, Florida.

Lal, S S., Sachan, J. N. and Yadava, C. P. (1985). Sex pheromone trap , a novel tool for monitoring gram pod borer populations. *Plant Protection Bulletin India*. 37: 3-5.

Lal,S.S.,Varma,B.K., Sanyal,P.K. and Sagar, R.D. (1974). A new approach to the control of groundnut leafminer, *Stomopteryx nerteria* M. by seed treatment with carbofuran. *Indian Journal of Plant Protection*, 2, 111-116.

Lefroy, H. M. (1906). Hairy caterpillar pest of crops. *Agricultural Journal of India.*, 1, 187-91.

Lewin, H.D., Saroja, R., Leela David, A., and Padmanabhan, M.D. (1973). Chemical control of groundnut leafminer (*Stomopteryx subsecivella* Zell). *Agriculture and Agro-Industries Journal*, 6, 12-16.

Loganathan, M. and Uthamasamy, S. (1998). Efficacy of a sex pheromone formulation for monitoring *Heliothis armigera* Hubner moths on cotton. *Journal of Entomological Research*. 22: 35-38.

Loganathan, M., Sasikumar, M. and Uthamasamy, S. (1999). Assessment of duration of pheromone dispersion for monitoring *Heliothis armigera* (Hub.) on cotton. *Journal of Entomological Research*.23: 61-64.

Logiswaran, G. and Mohanasundram. M.(1985). Effect of intercropping, spacing and mulching in the control of groundnut leafminer, *Aproaerema modicella* Deventer (Gelechiidae:Lepidoptera). *Madras Agricultural Journal*. 72,695-700.

Mathur, A. C.(1962). Food Plant spectrum of *Dicrisia oblique* Walk (Arctiidae: Lepidoptera). *Indian Journal of Entomology*, 24, 286-287.

Maxwell-Lefroy, H. (1907). The groundnut leaf miner. *Agricultural Journal of India.*, 2, 292.

Mishra, P. R. and Sontakke, B. K. (1992). Monitoring of *Spodoptera litura* F. by pheromone trap catches. *Indian Journal of Plant Protection*. 20: 188-190.

Mukundan, M. (1964). Achievements of campaign for the control of the groundnut red hairy caterpillar, *Amsacta albistriga* M. in Madurai division (Madras State). *Madras Agricultural Journal.*, 51, 46-49.

Muthukrishnan, N. and Balasubramanian, M. (1992). Relationship between the pheromone trap catches of *Pectinophora gossypiella* and *Spodoptera litura* their field infestation and larval population. *Madras Agricultural Journal*. 79: 321-326.

Naik, M., Lingappa, S. and Patil, B. V. (1993). A note on the efficiency of different pheromone traps in monitoring *Helicoverpa (Heliothis) armigera* (Hb.) males. *Karnataka Journal of Agricultural Sciences*. 6: 64-66.

Nair, M.R.G.K. (1975). Insects and mites of crops in India. ICAR, New Delhi,pp 404.

Nair, M.R.G.K. (1986). Insects and mites of crops in India. Indian council of Agricultural Research, New Delhi, India.

Nandagopal, V. (1992). Studies on the integrated pest management in groundnut in Saurashtra. Ph.D. thesis submitted to the Saurashtra University, Rajkot, pp. 247.

Nandagopal, V. and Reddy, P.S. (1990). Report of sex pheromone in groundnut leaf miner *Aproaerema modicella* Dev. *International Arachis Newsletter*, 7, 25-26.

Nandagopal, V. and Soni, V. C. (1993). Evaluation of pheromone traps for groundnut leaf miner (*Aproaerema modicella*). *Indian Journal of Agricultural Sciences*. 63: 64-67.

Nandagopal, V. and Soni,V.C. (1993). Evaluation of pheromone traps for groundnut leaf miner (*Aproaerema modicella* Dev.). *Indian Journal of .Agricultural .Science*, 63, 64-67.

Nandagopal, V., 1990. Present status of *Helicoverpa (=Heliothis) armigera* (Hub.) on groundnut in major growing area in the country. *In* Proceeding of the First National

Workshop on Heliothis Management: current status and future strategies, 30-31, August, 1990, Division of Entomology, Directorate of Pulses Research, Kanpur. pp.151-159.

Nandagopal, V., Pradip Fulmali, Gedia, M.V. and Rajiv Rathod (2003). Development of an efficient pheromone trap for catching Helicoverpa armigera in pigeonpea ecosystem in the field. International Pest Control (London) 45: 32-36.

Nandagopal.V. and V.C.Soni (1993). Evaluation of pheromone traps for the groundnut leaf miner (*Aproaerema modiceela* Deventer). Indian Journal of Agricultural Science. 63: 64-677.

Nandihalli, B. S., Patil, B. V., Somasekar, S. and Hugar, P. (1989). Influence of weather parameters on the population dynamics of *Spodoptera litura* (Fb.) in pheromone and light traps. *Karnataka Journal of Agricultural Sciences*. 2: 62-67.

Nandihalli, B. S., Patil, B. V., Somasekar, S. and Hugar, P. (1990). A comparative assessment of pheromone and light trap catches of cotton pests. *Journal of the Indian Society for Cotton Improvement*. 15: 148-150.

Nesbitt, B.F., Beever,P.S., Hall, D.R., and R. Lesta. (1980). (Z).-9- hexadecenal, a minor component of the female sex pheromone of *Helicoverpa armigera* Hubner). (lepidoptera:Noctuidae). *Entomologia Experimenta Applicata,* 27, 306-308.

Nesbitt, B.F., Beever, P.S., Hall, D.R., and Lesta, R. (1979). Female sex pheromone components of the cotton boll worm *Heliothis armigera. Journal of Insect Physiology,* 25, 535-541.

New York. Journal of chemical Ecology 1, 225-231.

Pachori , R., Rathore, V. S., Rawat, R. R., Jakhmola, S. S. and Khatri, A. K. (1980). Economic threshold of *Diacrisia oblique* wlk on groundnut. *Indian Journal of Agriculture Sciences* 50, 69-72 .

Palanisamy, M.S. and Ramchandran,T.K. (1978). Chemical control of groundnut leafminer *Stomopteryx subsecivella* Zell. (Gelechiidae: lepidoptera). *Pesticides,* 12, 24-26.

Patel, J. S. (1943). Indian Central Jute Committee. Annual report. Agri. Res. Scheme, 1942-43, Calcutta, pp. 51.

Patel, J. S.(1944). Indian central Jute Committee, Calcutta, India. Annual report of the Agriculture research scheme for the year 1942-43, pp 43-44.

Patel, H. K. and Patel, V. C. (1965). Life history, epidemiology and seasonal history of Gujarat hairy caterpillar (*Amsacta moorei* Butler). Bansilal and Amritlala Coll. Mag. 16 and 17:44-56.

Patel, R. C., Yadav, D. N., Joshi, A. D. and Patel, A. K. Regupathy,A (ed.); Jayaraj,S.(1985). Impact of mass trapping of *Heliothis armigera* and *Spodoptera litura* from cotton with sex pheromones. *Bhavioural and Physiological Approaches in Pest Management*. 120-122.

Patil, B. V., and Kulkarni, N. S. (1997). Monitoring of *Helicoverpa armigera* (Hubner) with light and pheromone traps. *Indian Journal of Plant Protection*. 25: 13-19.

Patil, B. V., Nandihalli, B. S., Hugar, P. and Somasekhar. (1992). Influence of weather parameters on pheromone trap catches of cotton bollworms. *Karnataka Journal of Agricultural Sciences*. 5: 346-350.

Pawar, A. D., Prasad, J., Raj,Singh and Singh, R. (1984). Field evaluation of traps and sex pheromones of *Heliothis armigera* Hb. (Noctuidae: Lepidoptera) and effect of releases of parasites on male moth catches in pheromone traps. *Plant,Protection,Bulletin,,* 36: 29-30.

Pawar, A. D., Tuhan, N. C. and Parry, M. (1984). Management of codling moth *Cydia pomonella* (L.) (Lepidoptera Olethreutidae) at Khalsi (Ladakh), Jammu and Kashmir (India). *Pesticides*. 18: 39-42.

Pawar, C. S. and Srivastava, C. P. (1988). Interaction between Sex Pheromones of *Spodoptera litura* Fab. and *Heliothis armigera* (Hub). *International Arachis News letter* 4 : 22.

Pawar, C. S., Sithanantham, S., Bhatnagar, V. S. Srivastava, C. P. and Reed, W. (1988). The development of sex pheromone trapping of *Heliothis armigera* at ICRISAT, India. *Tropical Pest Management*. 34: 115- 120.

Pawar, C..S. and Srivastav, C.P. (1988). Interaction of pheromones of *Spodoptera litura* Fab. and *Heliothis armigera* (Hub.). *International Arachis Newsletter*, 4, 22.

Pawar, C.S., S.Sithananthamam., V.S. Bhatnagar., Srivastav, C.P. and Reed,W. (1988). The development of sex pheromone trapping of *Heliothis armigera* at ICRISAT. *Tropical Pest Management*, 34, 39-43.

Persoons, C. J. , Vos, J.D., Yadav, J.S. and Prasad,A.R., Sighomony,S.,Jyothi,K.N. and Prasuna,A.L.(1993). Indo-Dutch Co-operation on pheromone of Indian Agricultural insectpests.In pheromone technology in Europe and the developing countries.Working group:Use of pheromone and other semiochemicals in integrated control held on 10-14,May 1993,Natural Resources Institute, Chathan,England. *IOBC Wprs Bulletin*, 16, 136-140.

Piccardia, P., Capizzi, A., Cassani,G., Spinelli, P., Arura, E and Massado,P. (1977). A sex pheromone component of the old world boll worm *Heliothis armigera*. Insect *Physiology*, 23, 1443-1445.

Prasad, N. K. K., Mallikarjunaiah, H., Nandihalli, B. S. (1985). Efficiency of different types of traps in monitoring Spodoptera litura (F.) populations with synthetic sex pheromones. In Regupathy, A (ed.) and Jayaraj, S. (ed.) Behavioural and physiological approaches in pest management. 53-55pp.

Prasad, V. D., Rambabu, L and Reddy, G. P. V. (1993). An action threshold for *Helicoverpa armigera* Hb. based on pheromone trap catches in cotton. *Indian Journal of Plant Protection*. 21: 17-18.

Rai, K. (1976). Pests of oilseed crops in India and their control. Indian Council of Agricultural Research (ICAR)., New Delhi. 121 Pages.

Rajput, S.G., Dalaya, V.P., and Awate, B.G. (1985). Field evaluation of synthetic pyrethroids and other insecticides against groundnut leafminer (*Aproaerema modicella*, D.). *Pesticides*, 19, 34-35.

Ramasivan, T., Krishnamurthy, K. and Pingale, S.V. (1968). Studies on preservation of food grains in rural storages. Part-II. Storage of grains in villages near Hapur. *Bulletin of Grain. Technology*., 6(2), 69-75.

Ramaswamy, K. A. and Kuppuswamy, S and Santhanaraman, T. (1968). Ridding groundnut of red hairy caterpillar. Indian Farming 18, 21-22.

Ramaswamy, K. A. and Kuppuswamy, S. (1973). Some observation on the biology of the red hairy caterpillar (*Amsacta* spp.). on groundnut. *Madras Agricultural Journal.*, 60(7), 637-9.

Ranga Rao,G.V. and Wightman, J.A., (1994). First Annual Rabi/Summer Groundnut Research Workers' Group Meeting held at Indian Institute of Technology, Kharagpur, West Bengal from 12- 15 September, 1994.

Ranga Rao, D.V., Singh.K.N., Wightman,J.A and Ranga Rao,G.V. (1991). Economic status of neem cake mulch for termite control in groundnut. *International Arachis Newsletter.*, 9, 12-13.

Rao, S. V., Azam, K. M. and Ahmed, K. (1989). Mating disruption through behavioural manipulation of *Spodoptera litura* Fabricius. *Current Science*. 58: 933-935.

Reddy, D.D.R., Ranga Rao, G.V. and Sanower, T.G. (1993). Bio-ecology and management of groundnut leaf miner ,*Aproaerema modicella*. Group Discussion on Integrated Pest management Srategies in Oilseeds in India, held at Panjab Agricultural University, Ludhiana from 23-24 December, 1993.

Sandhu, G.S. (1977). Some insects recorded as pests of fodder crops at Ludhiana (Punjab). *Punjab Agriculture University Research Journal*. 14, 449-459.

Sangappa, H.K. and Ali, T.M.M. (1977). Evaluation of some newer insecticides in the control of the groundnut leaf miner, *Stomopteryx subsecivella* Zell. (Gelechiidae:Lepidoptera). *The Mysore Journal of Agricultural Sciences.*, 11(4), 559- 561.

Senapati,AK.,Das,BK and Panda, M.(1990). Influence of weather factors on the catches of male moths of tobacco caterpillar *Spodoptera litura* F. in the pheromone trap. *Environment and Ecology.*, 8: 758-760.

Singh, J. P. and Lather, B. P. S. (1989). Monitoring of pink bollworm moths and larvae. *Indian Journal of Plant Protection*. 17: 199- 204.

Singh, K. N and Sachan, G. C. (1993). Assessment of the use of sex pheromone traps in the management of *Spodoptera litura* F. *Indian Journal of Plant Protection*. 21: 7-13.

Singh, K. N. and Sachan, G. C. (1991). Relationship between pheromone trap catches of male *Spodoptera litura* and oviposition in soybean and sugarbeet at Pantnagar, Uttar Pradesh, India. *Journal of Plant Protection in the Tropics*.8: 121-124.

Singh, K. N. and Sachan,G. C. (1993). *Spodoptera litura* male moth catches in pheromone traps and their relationship with oviposition in groundnut field at Pantnagar, India. *Insect Science and its Application*. 14: 11-14.

Singh, S. and Singh, J. (1956) On the control of Kutra *Amsacta moorei* Butler (Arctidae: Lepidoptera). Indian Journal of Horticulture, 13, 107.

Singh,KN and Sachan,GC. (1991). Monitoring of *Spodoptera litura* F. populations with pheromone baited traps at Pantnagar, India. *Journal,of,Plant,Protection,in,the,Tropics*, 8: 97-101.

Singh,O.P. and Gangrade, G.A.(1974). Biology of *Diacrosia oblique* Walker (Lepidoptera: Arctiidae) on soybean and effect of loss of chlorophyll on pod and gram. *Jawaharlal Nehru Krishi Viswa Vidyalaya Research Journal* 8(2):86-91.

Sinha, M. M., Yadav, R. P. and Kumar, A. (1975). Outbreak of the Bihar hairy caterpillar, *Diacrisia oblique* Walker in North Bihar. *Entomologist Newsletter,* 5, 47.

Sinha, S. N and Jain, D. K. (1992). Field studies in pheromone trap catches of *Helicoverpa armigera* Hubn.: I. Monitoring seasonal abundance and larval infestation on chickpea and the effect of climate on trap catches. *Indian Journal of Plant Protection.* 20: 149-157.

Sinha, S. N. and Mehrotra, K. N. (1993). Factors influencing pheromone trap catches of *Helicoverpa armigera* Hubner. *Indian Journal of Plant Protection.* 21: 1-6.

Siva Rao, D.V., Swamy, M.T. and Raju, A.P.(1977). Outbreak of *Amsacta albistriga* Walker on groundnut in Andhra Pradesh. *Entomologists' Newsletter* 7: 3.

Sivasubramanian, P. and Palaniswamy, G.A. (1983). Studies on the chemical control of groundnut leafminer, *Aproaerema modicella* Dev. *Madras Agricultural Journal.* 70, 485-486.

Srinivas, K. and Rao, P.A. (1999). Management of *Spodoptera litura* (F.) infesting groundnut by mating disruption technique with synthetic sex pheromone. *Journal of Entomological Research,* 23(2): 115-119.

Srivastava, A. S. and Goel, B. K. (1962). Bionomics and control of red hairy caterpillar *Amsacta moorei* Butl. *Proceedings fo the national Academy of Sciences,* India B 32, 97-100.

Srivastava, C. P., Pimbert, M. P. and Reed, W. (1990). Spatial and temporal changes in the abundance of *Helicoverpa (=Heliothis) armigera* (Hubner) in India. *Insect Science and its Application.* 11: 869-876.

Srivastava,C.P., Pimbert,M.P. and Reed,W. (1992)Monitoring of *Helicoverpa* (= *Heliothis*) *armigera* (Hubner) moths with light and pheromone traps in India. Insect-Science-and-its-Application., 13: 2, 205-210.

Sundaramurthy, V. T. (1985). Abundance of adult males of *Heliothis armigera* (Hb) in poly crop system as affected by certain abiotic factors. In Regupathy, A (ed.) and Jayaraj, S (ed) *Behavioral,and,physiological,approaches,in,pest,management.* 197-199pp.

Tadas, P. L., Sarnaik, D. N., Kene, H. K. and Satpute, U. S. (1994). Effect of weather parameters on monitoring of cotton bollworms with pheromone trap. *PKV Research Journal.* 18: 87-90.

Tamaki, Y., Noguchi, H and Yashima, T. (1973). Sex pheromones of the *Spodoptera litura* (Lepidoptera:Noctuidae): Isolation, identification and synthesis. *Applied Entomolgy and Zoology,* 8, 200-203.

Trehan, K. N. and Talgeri, G. M. (1947). Combatting the hairy caterpillar, *Amsacta moorei* in Bombay province. *Indian Farming.,* 17, 560-564.

Trehan,K.N. and Bagal,S.R. (1957). Life history, bionomics and control of sweet potato weevil (*Cylas formicarius* N.) with short notes on the pests of sweet potato in Bombay State. *Indian Journal of Entomology* 19: 245-52.

Usman, S. and Puttarudraiah, M. (1955). A list of insects of Mysore including mites. Department of Agriculture Mysore State- Entomology Series Bulletin 16. pp 194 .

Venkataiah, M and Subbaratnam, G. V. (1992). Influence of weather factors on the abundance of *Helicoverpa armigera* (Hubner). *Indian Journal of Ecology*. 19: 55-60.

Venkataraman, A., Abraham, E. V. and Sreenivasan, D. (1970). A note on trial of two insecticide for the control of red hairy caterpillar *Amsacta* spp on groundnut in Tanjavur District. *Madras Agricultural Journal.*, 57, 478-9.

Verma, A. K and Sankhyan, S. (1993). Pheromone monitoring of *Heliothis armigera* (Hubner) and relationship of moth activity with larval infestation on important cash crops in mid-hills of *Himachal Pradesh. Pest Management and Economic Zoology*. 1: 43-49.

Vittal, S.M. and Saroja, R. (1965). Insecticidal control of 'surulpoochi' *Stomopteryx subsecivella* (Zell.) on groundnut. *Madras Agricultural Journal.*, 52, 275-279.

Vittal, S.M., Azeez Basha, A., and Saroja, R. (1964). A note on the control of 'surulpoochi' (*Stomopteryx subsecivella* Zell.) on groundnut. *Madras Agricultural Journal.*, 51, 475-476.

Wightman, J.A. and Ranga Rao,G.V. (Comps). (1997). *Spodoptera litura* in India: *Proceedings of the National Scientist Forum on Spodoptera litura* (F.)., 2-4 April,1996,ICRISAT Asia Centre, Patancheru, 502 324 and 94, Victoria Street, London SL1 E8JL,UK: Overseas Development Administration, pp 158.

Wightman, J.A and G.V. Ranga Rao (1993). A Groundnut Insect Identification Handbook for India. Information Bulletin No. 39 ICRISAT. Patancheru, Andhra Pradesh, India, pp. 28–37.

Yadav, J. S., Madabhushi,V., Valli. and Prasad. A.R. (1998). Total synthesis of Enantiomers of (3Z,6Z).-cis-9.10-epox 1,3,6-Henicosatriene- the Pheromonal Component of *Diacrisia obliqua. Tetrahedron*, 54,7551-7562.

Yadav,C.P.S., Bhatnagar, K.N. and Srivastava, B. P. (1966). Incidence of *Amsacta moorei* Butler in semi-arid zones of Rajasthan. *Indian Journal of Entomology.*, 28, 274.

Yashima, T., Tamaki, T., Kamano, S. and Oyama, M., (1974). Field evaluation of a synthetic sex pheromone "litlure" as an attractant for males of *Spodoptera litura* (F.). (Lepidoptera :Noctuidae). *Applied Entomology and Zoology*, 9, 147-152.

□□□

Chemical Control

G. LOGISWARAN AND R. VISHNU PRIYA

21.1. Introduction

Groundnut crop is affected by more than 100 species of insects of which some cause damage regularly whereas others are only occasionally important. Some of the insects injurious to groundnut, enjoy a world wide cosmopolitan distribution, while insect pests which are of minor importance in one area enjoys a major status in another geographical situation. Considering the chemical control of insect pests on groundnut, these insect pests can conveniently be categorized as those affecting the above ground parts and below ground parts. Those attacking the below ground parts needs more attention as they kill the entire plant reducing the number of plants per unit area which is an important factor for deciding the groundnut yield. The weather factors also play a major role in increasing or reducing the insect population mainly by influencing the bio-ecology of the pest. Hence chemical control measures must be considered in relation to biology and ecology of the pests. The economics and the expected return should also be considered while advising costly insecticidal treatments.

This chapter provides information on the work done on chemical control measures with respect to groundnut pests. Some of the insecticides recommended or reported effective have either been subsequently banned in India or permitted for restricted use. Hence the readers are requested to exercise caution while making recommendations.

21.2. Foliage feeding insects

21.2.1. Sucking pests

Sucking pests like aphids *Aphis craccivora* Koch ,thrips, *Scirtothrips dorsalis* Hood , *Caliothrips indicus* (Bagnell), *Frankliniella schultzei* (Trybom) and *Thrips palmi* Karny and jassids *Empoasca kerri* and *Balclutha hortensis* Lindb. are commonly observed in all groundnut growing areas. These insect pests may occur alone at times or in combination as pest complex, all occurring at the same time, causing considerable damage to groundnut by sucking the sap.

The yield losses due to damage caused by jassids, thrips and aphids were worked out as 9, 17 and 16 per cent decrease in pod yield respectively (AICORPO, 1982). Therefore need arises to check the damage caused by these sucking pests. These sucking pests attacking groundnut crop are not monophagous and have a wide host range. The alternate hosts play an important role in the carry over of these pests from season to season. The insecticides recommended for the control of these pests on other crops will also afford effective control on groundnut. Monocrotophos 0.05 % was found to be effective against the sucking pest complex by many workers (Shetgar, *et. al.*, 1989; Patil, *et. al.*, 1992; Ghorpade and Thakur, 1989).

21.2.1.a.Control of Jassids

Studies regarding insecticidal control on jassids indicated that emulsifiable concentrates and dust formulations of quinalphos, phosalone, endosulfan, carbaryl and monocrotophos were effective (Saboo and Puri, 1979). Patel and Vora (1981) reported that dimethoate, phosphamidon, methyl-o-demeton and vamidothion are effective. Feakin (1973) recommended carbaryl 1.12 kg ai/ha; dichlorvos 350 g to 500 g ai/ha; monocrotophos 150g to 200 g ai/ha ; phosphamidon 150 to 200 g ai/ha and malathion 1.12 kg ai/ha. Soil application of aldicarb (Temik) @ 2.25 kg ai/ha at 15 and 30 days after sowing gave maximum protection for 60 to 80 days (Bhushan, *et. al.* 1988). Spraying dimethoate at 200 to 250 ml ai/ha was effective (Wightman and Ranga Rao, 1993). Peter and Sundararajan (1991) reported that monocrotophos 250 g a.i./ha , pyraclofos 500 g a.i./ha and synthetic pyrethroid deltamethrin 15 g a.i./ha were effective against jassids. Insecticides like deltamethrin > dimethoate > monocrotophos were also effective (Kennedy *et. al.*, 1992).

21.2.1.b. Control of Thrips

Groundnut thrips can be controlled effectively by spraying dimethoate, demeton methyl and monocrotophos (Singh and Vaishampayan, 1967; Panchabhavi and Thimmiah, 1973; Senapathi and Patnaik, 1973, 1980). Wightman and Ranga Rao, (1993) suggested to spray dimethoate @ 200 to 250 ml ai/ha when more than five thrips per terminal leaf is present that too up to 30 days after seedling emergence. Acharya and Koshiya (1991) also reported that dimethoate 0.03% was highly effective against thrips. However, Feakin (1973) listed the following insecticides for use against thrips damage. Dichlorvos 350 to 500 g ai/ha; Dimethoate 1.12 kg ai/ha; Malathion 1.12 kg ai/ha; Monocrotophos 150 to 200 g ai/ha; Phorate 1.12 kg ai/ha and Phosphamidon 150 to 200 g ai/ha.

21.2.1.c.Control of Aphids

Sporadic outbreaks of this pest have been observed in recent years due to long period of drought stress. Organo-phosphorus compounds like mesopestox

0.03%, parathion 0.025% (Basheer, 1958); methyl demeton 0.05% (Nagaraja Rao, 1959) were found effective. Vasantharaj David *et al.* (1965) reported that metasystox 0.1%; trithion 0.06% and thiometon 0.08% are effective. Pot culture studies conducted by Sundara Babu (1969) revealed that menazon 0.03%; carbophenothion (trithion) 0.06%, demeton methyl 0.05% and thiometon 0.08% are effective against aphids. Thakkar *et al.* (1981) reported methyl demeton @ 400 ml/ha to be more effective than dimethoate, phosphamidon or melathion at the same dosage rates. However, all these insecticides are highly toxic to predators and parasites, which are abundantly seen in aphid colonies. Feakin (1973) has also listed the following chemicals effective against aphids; Demeton 140 to 180 ml ai/ha; Demeton- S- methyl 72 to 96 ml ai/ha; Dichlorvos 350 to 500 gm ai/ha; Dimethoate 450 ml ai/ha; Endosulfan 480 ml ai/ha; Malathion dust 388 to 580 g ai/ha; Menazon 294 g ai/ha; Monocrotophos 150 to 200 g ai/ha; Phorate 1 to 1.5 kg ai/ha; Phosphamidon 150 to 200 g ai/ha and Thiometon 130 ml ai/ha. Recently, Jena and Kuila (1997) have reported the following insecticides in descending order to be effective against aphids, dimethoate > endosulfan > carbaryl > quinalphos > dichlorvos > chlorpyriphos > malathion.

21.2.2. Leaf miner

Leafminer, *Aproaerema modicella* Dev. is a regular pest on groundnut wherever the crop is grown. It is presently a major pest in Tamilnadu, Andhra Pradesh, Karnataka, Maharashtra, and Gujarat States in India. Soybean is another principal host crop for this oligophagous pest although green gram (*Phaseolus aureus*), pigeon pea and Lucerne (were reported to be host crops to a smaller extent. The pest has a short life cycle of 21 to 25 days and occurs both in *kharif* and *rabi* groundnut. Concerted research works on this pest established various methods of integrated pest management encompassing physical, cultural, insecticidal and biological aspects. However, insecticidal control dominates among all. A very wide range of chemicals including chlorinated hydrocarbons, terpenes, cyclodienes, organophosphates, carbamates and synthetic pyrethroids were reported effective. Besides, the efficacy of tin compounds as antifeedants, moult inhibitory action of benzyl phenyl urea compounds and efficacy of various plant products were also reported. Nair (1975) reported that most organophosphates with systemic and contact action are quiet effective against this pest. Among the cyclodiene compounds, endosulfan 35 EC at 500 ml to 1000ml/ha was found effective (Gopal,1988; Gopal *et al.*, 1980; Palaniswami *et al.*,1978 a, 1978 b; Logiswaran and Madhava Rao, 1982 b; Radhakrishnan *et al.*, 1976; Vasudevan *et al.*, 1980; Jean and Kuila, 1997; Muthiah and Hussain, 1991). Many organophosporus insecticides as well as synthetic pyrethroids were reported effective (Table 21.1). Among carbamates, carbaryl 0.02% (Anon, 1983), bendiocarb (Garvax 80 WP) 0.5 kg ai/ha (Logiswaran and Madhava Rao, 1982 b) and carbosulfan 10 G at 4 kg ai/ha (Rajagopal and Gowda, 1992) were reported to be effective. Among the insecticides of animal origin, cartap (Padan 50 EC) at 1 kg ai/ha (Jesudasan *et al.*, 1989) was reported to be effective. Antifeeding effects of TPTA (Brestan 60% WP) and TPTH (Du Ter 20% WP) both at concentrations of 0.05% were reported to be effective by Abdul Kareem (1976) and Abdul Kareem and Subramaniam (1976). Jesudasan *et al.* (1986) reported the efficacy of the moult inhibitor, diflubenzuron 0.02% in controlling this pest. Chandrasekaran *et al.* (1992) reported that Dimilin 25 WP (diflubenzuron) at 1.6 Kg ai/ha was effective. SIR 8514 is another moult inhibitor which was reported to be effective at 100 g ai/ha (Natarajan, 1982; Lakshmanan, 1982). Many plant products have also been reported to be effective on this pest, the details of which were furnished in Table 21.2.

21.2.3.Defoliators

Polyphagous pests such as tobacco caterpillar, *Spodoptera litura* (F.), gram pod borer, *Helicoverpa armigera* Hub., red hairy caterpillar, *Amsacta albistriga* W., *Amsacta moorei* Butler, Bihar hairy caterpillar, *Spilosoma oblique* W. are considered as important defoliators of groundnut crop in India. The larvae of these pests feed gregariously on leaves and as the larvae grow they disperse, march from fields to fields and devour the entire foliage, flowers and growing points. When large number of caterpillars infest the crop, bare stems of the plants are seen in the fields resulting in considerable loss in yield. Extensive studies on bio-ecology of these pests have been carried out as reviewed by Amin (1983) resulting in the development of integrated pest management strategies covering all possible tactics such as cultural, mechanical, biological, legal and chemical methods. The farmers mostly rely on insecticidal strategy since the incidences of the pests are not invariably noticed at early stages of attack. The insecticides used against one defoliator will also control other defoliators. The chemicals recommended for defoliators are given in Table 21.3.

21.3. Soil inhabiting and root feeding insects

These insects cause considerable damage to the groundnut crop. When the infestation occurs at the early stage of the crop, the entire seedling or plant dies out. If the infestation occurs at the late stage of the crop, pegs are damaged resulting in reduced yield. Hence the losses inflicted are considerable. Among the soil inhabiting pests, termites enjoys a wide distribution of which *Odontotermes obesus* is the predominant species. Their damage includes scarification of pods leading to fungal entry and penetration and hollowing of the tap root. Generally the spilled insecticides from the spray schedules adopted for other foliage feeders depress the termite population from one season to the next making specific management strategies unnecessary. However chemical insecticides have to be used whenever the termites are known to be regular in an area. In such areas soil application of chlorinated hydrocarbons is recommended (Rawat *et al.*, 1970; Feakin, 1973). A mixture of aldicarb and chlorpyriphos has been reported effective against termites (Dicko and Lynch, 1995). Carbofuran 1 kg/ha reduced pod scarification by termites (Sithanantham and Kannaiyan, 1991). Application of aldicarb at sowing and chlorpyriphos at 50 and 100 days after reduced the damage by termites to 10 per cent and increased the yield of groundnut (Lynch *et al.*, 1991).

Several species of white grubs, *Holotrichia* attacks groundnut especially in Northern States of India. The grubs feed on rootlets and roots causing the death of the plant. The adult beetles gather in large numbers on many trees and shrubs which serve as hosts for carry over and perpetuation of this pest. Hence the chemical treatment is necessary both for tackling the grub and adult stage of the pest. Aldrin 2% dust applied in soil at 125 kg/ha was found effective (Srivatsava and Khan, 1963). Aldrin as dust, granule or emulsifiable concentrate formulations applied to soil at 3.4 kg a.i./ha was reported effective (Feakin, 1973). Soil application of phorate granules at 1.1 and 1.6 kg a.i./ha (Kulshreshta *et al.*, 1964) 2.5 kg a.i./ha (Feakin, 1973; Gandhale *et al.*, 1982) 1.5 kg a.i./ha (Yadava and Yadava, 1973), 1.0 kg a.i./ha (Bakhetia, 1982) were effective. Bakhetia (1982) reported the efficacy of soil applied carbofuran, isofenphos and fensulfothion granules at 1 kg a.i./ha. The efficacy of soil application of 5% diazinon granules at 25 kg/ha was reported by Patil *et al.*,1981. Foliar sprayings with methyl parathion at 6 to 7.5 kg a.i./ha in 675 lit. of water was found to provide effective control of first and early second stages of white grubs (Wang,1994).

The earwigs or pod borers bore into the pods and feeds on kernels. The holes are plugged with excreta, sand particles and discoloured pulp. The symptoms are similar to those caused by millipedes, wireworms and termites. In the earlier periods soil application of chlorinated hydrocarbons (Padmanabhan *et al.*, 1973; Melamed-Madjar and Shalomo, 1970) was recommended. Recently Senguttuvan and Dhanakodi (1997) conducted insecticidal trials against pod borer complex *viz.*, earwigs, wireworms and *Dorylus* sp and reported that soil application of carbofuran 3G or phorate 10G at 10 kg/ha or quinolphos 1.5D at 25 kg/ha at the time of sowing were highly effective. Some of the insects such as groundnut root borer, *Sphenoptera perotetti* (Guer) pod sucking bugs *Elasmolomus sordidus* (Fab.), lesser corn stalk borer, *Elasmopalpus lignosellus* are considered as minor pests but many attain major importance in certain localized areas. Soil application of insecticides recommended for other soil inhabiting pests like white grubs will reduce the population of these minor pests also.

21.4. Storage pests

Groundnuts are stored both as unshelled nuts and also as kernels. In both the cases insect damage is experienced although kernels are easily damaged than the unshelled nuts. Proper drying and storage conditions minimizes insect attack. The important insects that damage the groundnuts in storage are the bruchid *Caryedon serratus* Ol., the red flour beetle, *Tribolium castaneum* Herbst., merchant grain beetle, *Oryzaephilus mercator (Fauvel)*, khapra beetle, *Trogoderma granarium* Everts, almond moth *Ephestia cautella* Walker and the rice moth *Corcyra cephalonica* Stainton. All these are secondary pests and are not capable of infesting sound pods. They can infest kernels or once the pods are damaged due to other causes, the kernels are infested. Besides the above, cadelle beetle *Tenebroides mauritanicus* (L.), lesser grain borer *Rhizopertha dominica* Fabr., Indian meal moth, *Plodia interpunctella* become serious at times. The best approach is to use insecticides as needed after adopting all post-harvest pest management practices such as proper drying and hygienic storage. Bisell and Du Pree (1946) reported that bins and store houses should be clean and should be sprayed with malathion EC at 225 g/ 100 m^2. For spraying on absorbent surfaces such as brick or cement-faced wall, a wettable powder is preferable to an emulsion because the solid particles are filtered out on the surface while the water is absorbed into the wall. Feakin (1973) reported that fenitrothion 25 WP applied at 166g per 5 lit. water per 100 m^2 is useful. Vilogen *et al.*, (1981) reported that when fumigated groundnuts are stored in small bag stacks and when these bags were sprayed with phoxim and iodofenphos the lepidopterans *viz., E.cautella* and *P. interpunctella* were controlled. Feakin (1973) reported that in the case of unshelled nuts stored in large stacks, a surface application of 0.5% lindane dust (500 g per 10 m) is effective and for seed materials a dose of 170 g per 45 kg of nuts is recommended.

Whenever infested groundnuts have to be taken into store room, fumigation with methyl bromide at 16 to 24 g/ m^3 for an exposure period of 24 h at 25^{0}C or above is effective (Feakin, 1973). Groundnut retains higher residues of inorganic bromind than cereals and not more than three treatments should be given. He also recommended twenty five tablets (each weighing 0.6 g) of Aluminium phosphide per ton of groundnut with fumigation period of not less than 5 days is effective. Further, Leesch *et al.* (1979) reported that methyl bromide fumigation reduced the germination of seeds in direct proportion to the dosages applied and number of fumigations while phosphin did not adversely affect the germination.

Table 21.1. Organophosphorous insecticides and synthetic pyrethoid reported effective against groundnut leaf miner, *Aproaerema modicella* Dev.

Insecticides	Dose	References
Methyl Parathion 50 EC	0.05%	Radhakrishnan *et al.* (1976)
Methyl Parathion 2% D	25 Kg/ha.	Gopal (1988)
Quinalphos 1.5% D	25 Kg/ha.	Gopal (1988)
Quinalphos 25 EC Dicrotophos Dimethoate 20 EC	4 l/ha 0.5 Kg ai/ha 0.6 Kg ai/ha	Abdul Kareem and Subramaniam (1976)
Dichlorvos 100 EC	0.05%	Radhakrishnan *et al.* (1976) Lewin *et al.* (1973)
Dichlorvos 76 EC	–	Somasekhar *et al.* (1991)
Dichlorvos 100 EC	0.1%	Sadakathulla *et al.* (1976)
Dichlorvos 100 EC Fenvalerate 20 EC	0.15% 0.67%	Mutiah *et al.* (1991)
Chlorpyriphos 20 EC	0.05%	Sadakathulla *et al.* (1976)
Monocrotophos 40 EC	1.6 l/ha	Abdul Kareem *et al.* (1974)
Monocrotophos 36EC	750 ml/ha	Sivasubramanian and Palanisamy (1982) (1983) Vasudevan *et al.* (1980)
Monocrotophos 36 EC	0.05%	Anon (1989), Gopal (1988) Logiswaran and Madhava Rao (1982 b) Shetgar *et al.* (1989) (1992) Patil *et al.* (1992) Ghorpade and Thakur (1989)
Monocrotophos 36 EC	1 l/ha.	Misra (1995)
Monocrotophos 36 SL Pyraclofos Deltamethrin	250 g ai/ha 500 g ai/ha 15 g ai/ha	Peter and Sundararajan (1991)
Quinalphos 25 EC	0.05%	Radhakrishnan *et al.* (1976)
Quinalphos 25 EC Permethrin Decamethrin	0.05% 0.01% 0.01%	Anon (1983)
Cypermethrin 25 EC Quinalphos 25 EC Fenvalerate 20 EC	0.01% 0.05 % 0.01%	Chandramohan and.Manoharan (1986)
Cypermethrin 25 EC Deltamethrin 2.8 EC	0.004% 0.001%	Gopal (1988)

Contd...

Table 21.1 Contd...

Quinalphos 25 EC Phenthoate 50 EC	750 ml/ha 750 ml/ha	Gopal *et al.* (1980)
Quinalphos 25 EC Phosphamidon 100 EC Permethrin 25 EC	750 ml/ha 375 ml/ha 600 ml/ha	Paneerselvam *et al.* (1990)
Phosalone 35 EC	0.05%	Sadakathulla *et al.* (1976)
Fenvalerate 20 EC	250 g/ha	Radhakrishnan *et al.* (1982)
Phosphamidon 100 EC Methyl demeton 25 EC Formothion 20 EC	0.05% 0.05% 0.075%	Logiswaran and Madhava Rao (1982 a)
Isofenphos 40 SD Isofenphos 5 G Isofenphos 50 EC	2 g/kg seeds 0.5 Kg ai/ha 750 ml/ha	Logiswaran and Madhava Rao (1984)
Isofenphos 5 G Phorate 10 G	2-5 Kgai/ha 4 Kg ai/ha	Rajagopal and Gowda (1992)
Quinalphos 25 EC Decamethrin Permethrin,Cypermethrin Fenvalerate	0.05% 0.001% 0.01% 0.01%	Panchabhavi (1992)
Fenvalerate Dust Quinalphos Dust	31.25 kg/ha 25 Kg/ha	Muthiah (1993)
Cypermethrin Permethrin	100 g ai/ha 100 g ai/ha	Chary *et al.* (1990)
Monocrotophos Dichotophos,Phosphamidon Dichlorvos	300 g ai/ha 300 g ai/ha 500 gai/ha	Feakin (1973)
Phosalone 35 EC Chlorpyriphos 20 EC	750 ml/ha 2 l/ha	Chandrasekaran *et al.* (1992)

Table 21.2. Plant products reported effective against groundnut leafminer, *Aproaerema modicella* Dev.

Plant products	Dose	Reference
Neem oil	2%	Chandramohan (1988)
Neem seed kernel extract Neem cake extract Neem leaves extract	10% 10% 10%	Chandramohan and Sivasubramanian (1987)
Neem oil Neem seed kernel extract	5% 5%	Gopal (1988)
Margosan (neem extract)	5 l/ha	Gopal *et al.* (1980)

Table 21.3. Recommendations for chemical control of groundnut defoliators

Defoliators	Insecticide & Dosage	References
Tobacco caterpillar (TC)	Diazinon 300 g ai/ha.	Feakin (1973)
Red hairy caterpillar (RHC)	Endosulfan 1 Kg ai/ha	
American boll worm (ABW)	Endosulfan 644 ml ai/ha	
Dichlorvos 350-500 g ai/ha	Feakin (1973)	
RHC	Dichlorvos 300-500 g ai/ha	
	Dicrotophos 150-200 g ai/ha	Feakin (1973)
	Monocrotophos 200-300 g ai/ha	
	Phosphamidon 200-300 g ai/ha	
RHC	Parathion 0.025%	Kuppusamy *et al.* (1965)
RHC	Phoxim (Valexon) 1000 ml/ha	
ULV formulation	Paramasivam *et al.,* (1973)	
RHC	Parathion 0.05%	Bindra and Kittur (1961)
RHC	Parathion 1.1 Kg ai/ha	Srivatsava and Misra (1968)
	Dichlorvos 350-500 g ai/ha	
TC	Carbaryl 840 g ai/ha,	Feakin (1973)
	Dichlorvos 2.5 Kg ai/ha	
TC	Monocrotophos, Quinalphos, Endosulfan, Methamidophos, fenitrothion and Chlordimeform (all at 0.5%)	Ayyanna *et al.* (1982)
TC	Endosulfan 0.03 or 0.06%,	Jagadish (1978)
	Trichlorfon 0.03 or 0.06%,	
	Carbaryl 0.05 or 0.1%,	
	Methomyl 0.025 or 0.05%	
TC	Monocrotophos 0.07 %	Muthiah and Hussain (1991)
TC and ABW	Chlorpyriphos 200 g ai/ha	Dhandapani *et al. (1993)*
TC	Dichlorvos 0.15 %	Muthiah *et al.* (1991)
TC	Quinalphos 0.5 %	Patil *et al.* (1997)
TC	Cypermethrin 0.016 %,	Singh (1998)
	Deltamethrin 0.003 %	
	Endosulfan 0.05 %,	
	Phosphamidon 0.05 %	
	Fluvalinate 0.05 %	
TC	Brestan (TPTA) 0.05%	Abdul Kareem (1976)
	Du Ter (TPTH) 0.05%	Abdul Kareem and Subramaniam (1976)
TC	SIR 8514 (Moult inhibitor) 100 g ai/ha	Natarajan (1982)

References

Abdul Kareem, A. (1976). A study on the antifeeding effects of two organotin compounds TPTA and TPTH on *Spodoptera litura* F., *Pericallia ricini* F., and *Stomopteryx subsecivella* Zell . *Pesticides*, 10: 57.

Abdul Kareem, A., and Subramaniam,T.R. (1976). Antifeeding effect of two organotin compounds on *Stomopteryx subsecivella* Zell. (Lepidoptera). *Madras agri. J.*, 63, 354-357.

Abdul Kareem, A., S. Sadakathulla and Subramaniam. T.R.(1974). Control surulpoochi on groundnut. *Farm and Factory* 8: (4), 35-36.

Acharya, M.F and D. J. Koshiya. (1991). Efficacy of different insecticides against thrips, *Caliothrips indicus* Bagnell. (Thysanoptera: Thripidae). on groundnut. *Gujarat Agricultural University Research Journal*. 16: (2), 87-90.

AICORPO. (1982). Annual kharif oilseeds workshop. *All India Coordinated Research Project on Oilseeds* (AICORPO). May 7-11, 1982, Bangalore.

Amin, P.W. (1983). Major field and storage insect pests of groundnut in India and their control. Paper presented at the Summer Institute on *'Recent Advances in Groundnut Productivity'* held at Junagadh, 9-30, April 1983.

Amin, P.W. (1983). Major field insect pests of groundnut in India and associated crop losses. Special Issue, Indian *Journal of Entomol*. 2: 337-344.

Anonymous, (1989).Annual Reort, National Research Center for Groundnut (Indian Council of Agricultural Reseach), Junagadh, Gujarat. pp75.

Anonymous, (1983). Annual Report 1983-84. International Crops Research Institute for the Semi-Arid Tropics, Patancheru, Andhra Pradesh, India.

Anonymous. (1983). Entomological Research in Tamilnadu. Annual Report 1982-83. TNAU Coimbatore 235p.

Anonymous. (1989). Annual Progress Report (1987-1988 and 1988-1989). of AICRP on Rodents Control. (Junagadh: Gujarat Agricultural).

Anonymous. (1989). *Groundnut*. Publications and Informations Directorate, CSIR, New Delhi.55 pp.

Anonymous. (1989). Twenty-seventh meeting of the Research Council held on Dec 22, 1989 at TNAU, Coimbatore. *Agenda notes*. Directorate of Research, TNAU, Coimbatore. 137p.

Ayyanna, T., P. Arjunrao, G.V. Subbaratnam, B.H.K. Rao and K.L.Narayana. (1982). Chemical control of *Spodoptera litura* Fab. on groundnut crop. *Pesticides*; 16, 19-20.

Bakhetia, D.R.C. (1982). Studies on the white grub, *Holotrichia consanguinea* Blanchand in the Punjab- Control and yield losses due to white grub and collar rot of groundnut. *Journal of Soi. Biolology* and *Ecology*, 2: 40-46.

Basheer, M. (1958). Observations on *Aphis craccivora* infesting *Gliricidia maculata* and its easy method of control. *Indian Journal of Entomology*, 20: 66–67.

Bhushan, V.S., K.Ramasubbiah, V.V.N. Nath and K.M. Azam.(1988). Bio-efficiency of different doses of aldicarb (Temik) against certain groundnut pests. *Pesticides* 22: (4), 23-26.

Bindra, O.S and Kittur,S.U. (1961). Biology and control of the red hairy caterpillar (*Amsacta moorei* B.). in Madhya Pradesh. *Vikram Science*,. 5: 62-71.

Bisell, T.L and M. Du Pree. (1946). Insects in shelled peanuts in relation to storage and bagging. *Journal of Econonmic Entomology*, 39: 550-552.

Chandramohan, N and Sivasubramanian, P. (1987). Evaluation of neem products against leafminer Aproaerema modicella (D.). Neem newsletter 4(4):44.

Chandramohan, N and T.Manoharan. (1986). Insecticidal evaluation against groundnut leafminer, *Aproaerema modicella* (D.). *Pestology*, 10: (3), 12-13.

Chandramohan, N. (1988). Evaluation of neem products against groundnut leafminer, *Aproaerema modicella* D. *Neem Newletter*, 8: (11), 3.

Chandrasekaran, J.,Balasubramanian, G., Balasubramanian,M. and Sivaprakasam,N. (1992). Evaluation of insecticides and moult inhibitor in the control of groundnut leafminer , *Aproaerema modicella* Dev. *Madras Agricutlral Journal*, 79: 142-145.

Chary, M.S., K. Ramasubbiah and V.S. Bhushan. (1990). Effect of certain modern insecticides on the yield of groundnut. *Indian Journal of Entomology*, 52: (3), 478-481.

Dhandapani, N., Sundara Babu, P.C., Sathiah, N. and Jayaraj, S. (1993). Combined use of nuclear polyhedrosis viruses of *Spodoptera litura* (Fb.). and *Heliothis armigera* (Hb.). on groundnut (*Arachis hypogaea* L.), *J. Applied Entomol*, 116 : (5), 523-526.

Dicko, O. and R.E. Lynch. (1995). Effectiveness of soil applied insecticides against foliage and pod feeding arthropod pests in Burkina Faso. *International Arachis Newsletter* 15: 60-62.

Feakin, S.D. (1973). Pest control in groundnuts. *PANS mannual* No.2, 3rd edn., Centre for Oversees Pest Research, London, pp. 155-175.

Feakin, S.D (ed.).(1973). Pest control in groundnuts. *PANS Manual No.2*. Overseas Development Natural Resourses Institute London U.K 197 p.

Gandhala, D.N., L.M.Naik and S.V. Khandge. (1982). Soil application of various insecticidal dusts for the control of white grubs (*Holotrichia serrata* F.). on groundnut (*Arachis hypogaea* L.) *Pesticides*, 16: (8), 21-22.

Ghorpade, S.A and Thakur,S.G. (1989). Assessment of need based monocrotophos applications for the control of major pests of groundnut. *Pesticides*, 23:(11), 19-20.

Gopal, S. (1988). Studies on *Aproaerema modicella* Deventer (Gelechiidae: Lepidoptera). on Soybean (*Glycine max*.L). M. Sc. (Ag). Thesis. Tamilnadu Agric. Univ. Madurai.

Gopal, S., S. Radhakrishnan, M.D. Padmanabhan, G.Logiswaran and Madhava Rao,S. (1980). Biological efficacy of phenthoate 50 E.C on the control of leafminer, *Biloba (Stomopteryx subsecivella)*. infesting groundnut. *Farm Science*, 7: (9-10), 30-32.

Jagadish, A. (1978). Evaluation of insecticides against aphids (*Myzus persicae* Sulzer). and caterpillars (*Spodoptera litura* F.). on tobacco in Karnataka State (India). *Phytoparasitica* 6: 45-50.

Jena, B.C. and Kuila, B. (1997). Effect of insecticides on population of aphids and their natural enemies in groundnut. *Madras Agricultural Journal*, 84: (2), 101-102.

Jena, B.C. and Kuila, B.(1996). Jassid infestation in groundnut and its chemical control. *Indian Journal of Entomology* 58: (4), 378–381.

Jesudasan, R.W.A., D.A. Samraj and C.J. Priyakumari. (1989). Field evaluation of a new insecticide against the groundnut leafwebber. *Pesticide,*. 23: (2), 26-27.

Jesudasan, R.W.A., Kandasamy, C. and Ravikumar,S. (1986). Use of diflubenzuron for the control of groundnut leaf webber, *Aproaerema modicella* D. (Gelechiidae: Lepidoptera). *Pestolog,* 10: (1), 19.20.

Kennedy, F.J.S., K Rajamanickam, R.Balagurunathan, and Lourduraj,A.C. (1992). Yield loss due to insect pests and their control in groundnut. *Groundnut News*.4: (2), 7.

Kulshreshtha, J. P., Vember,R.N. and Kulkarni,L.C. (1964). Success with groundnuts begins with seed treatment. *Indian Farming* 14:11-27.

Kuppusamy, S., Ramasamy,K.A. and Basheer,M. (1965). Recent insecticidal trials to control red hairy caterpillar on groundnut in Madras State. *Indian Oilseeds Journal,* 9: 195-200.

Lakshmanan, L. (1982). Studies on the efficacy of Cyfloxylate (Baythroid 50 SL). in the control of major pests of cotton. M. Sc. (Ag) Thesis. Tamilnadu Agriculture University, Coimbatore.

Leesch, J.G., Gillenwater,H.B.,Davis,R. and Wilson Jr.R. (1979). Phosphine and methyl bromide fumigation of shelled peanuts. *Peanut Science,* 6: 18-26.

Lewin, H.D., Saroja, R., Leela David, A., and Padmanabhan, M.D. (1973). Chemical control of groundnut leafminer (*Stomopteryx subsecivella* Zell). *Agriculture and Agro-Industries Journal,* 6: 12-16.

Logiswaran ,G and Madhava Rao, S. (1982b). Efficacy of two synthetic pyrethroids and a carbamate insecticide on the pests infesting groundnut. *Madras Agricultural Journal,* 69: 416-420.

Logiswaran, G. and Madhava Rao, S. (1984). Studies on the efficacy of the insecticide isofenphos in the control of leafminer infesting groundnut. *Pesticides,* 18:(8), 42-46.

Logiswaran, G. and S. Madhava Rao. (1982a). Control of leafminer, *Aproaerema modicella* Deventer (*Stompteryx subsecivella* Zell.). infesting groundnut with certain systemic insecticides. *Madras Agricultural Journal,* 69: 129-131.

Lynch, R.E., I.O. Dicko., S.A. Some and A.P. Ouedraogo. (1991). Effect of harvest date on termite damage, yield and aflatoxin contamination in groundnut in Burkina Faso. *International Arachis Newsletter* 10: 24-25.

Melamed – Madjar, V. and Shalomo, T. (1970). Tests with alternative insecticides in the control of earwigs in peanuts. *Int. Pest Control,* 12: (5), 29-3.

Misra, H.P. (1995). Optimium spray schedule of monocrotophos for the control of groundnut leafminer. *Current Research, University of Agricultural Science, Bangalore,* 24: (3), 49-51.

Muthiah, C. (1993). Bioefficacy of insecticidal dusts for the control of groundnut leaf miner *Indian Journal of Agricultural Science,* 63:(4), 251-253.

Muthiah, C. and Hussain. H.J.(1991). Bioefficacy of insecticides against groundnut pests. *Groundnut News,* 3,6.

Muthiah, C., V.Valluvaparidasan and H.S.J.Husain. (1991). Combined effect of insecticides and fungicides on groundnut pests. *Groundnut News,* 3: (1), 7-8.

Nagaraja Rao, K.R. (1959). The bean aphid control with systemic pesticides. *Madras Agricultural Journal,* 46, 421–425.

Nair, M.R.G.K. (1975). Insects and mites of crops in India. ICAR, New Delhi, pp 404.

Natarajan, N. (1982). Studies on the biological effects of moult inhibitor SIR 8514 on *Spodoptera litura* Fabricus and field efficacy against pests of cotton and groundnut. M.Sc (Ag). Thesis. Tamilnadu Agric. Univ., Coimbatore.

Padmanabhan, M.D., Lewin,H.D. and Saroja, R. (1973). Soil treatment to control groundnut pod borer (*Euborellia stali* Dohn.). *Pesticides,* 7:(8), 34-35.

Palaniswami, M.S and Ramachandran, T.K. (1978a). Effect of certain insecticides against the groundnut leaf miner *(Stomopteryx subsecivella* Zell.) in Tamilnadu *Agricultural and Agro-Industry Journal,* 11: 37-38.

Palaniswami, M.S and Ramachandran, T.K. (1978b). Chemical control of groundnut leaf miner. *Pesticides,* 12:(7), 24-26.

Panchabavi, K.S and Thimmaiah. G. (1973). A note on the evaluation of groundnut varieties for resistance to thrips, (*Heliothrips indicus* Bagnell). *Mysore Agricultural Journal,* 7: 334–335.

Panchabhavi, K.S, (1992). A note on field screening of insecticides against groundnut pests. *Karnataka Journal of Agricultural Science,* 5:(3), 271-273.

Paneerselvam, K., Logiswaran,G. and Gopal, S. (1990). Time of insecticidal application for control of groundnut leafminer (*Aproaerema modicella* D.) *Indian Journal of Agricultural Science,* 60: 86-87.

Paramasivan, V., Letchoumanae,S. and Sreeramulu,O.S. (1973). Preliminary studies on the insecticidal control of the groundnut red hairy caterpillar (*Amsacta* spp.). *Madras Agricultural Journal,* 7: 36-38.

Patel,B.R. and Vora,V.J. (1981). Efficacy of different insecticides for the control of groundnut jassid, *Empoasca kerri* Pruthi. *Pesticide,* 15: 33-34.

Patil, B.R., Hasabe, B.K. and Agri, D.S. (1981). Chemical control of white grub, *Holotrichia serrata* Fabr. infesting groundnut , paddy and sorghum. *Indian Journal of Entomology,* 43: 232-236.

Patil, N.M., Ghule, B.D., Dhumal, V.S. and Deokar,A.B. (1992). Critical time and frequency of insecticidal sprays for the control of groundnut sucking pests. *Journal of Maharashtra Agricultural Universities,* 1992, 17:(3), 477-478.

Patil, R.K., Mahabaleshwar Hegde, Kulkarni, K.A. and Hegde, M. (1997). Efficacy of insecticides against *Spodoptera litura* on groundnut. *Karnataka Journal of Agricultural Science,* 674-677.

Peter, C. and Sundararajan,R.(1991). Chemical control of major insect pests of groundnut in Tamil Nadu. *Journal of Insect Science,* 4:(1), 64–66.

Radhakrishnan, S., Mohanan, M. and Vittal. S.M. (1982). A note on the efficacy of Fenvalerate 25 EC against groundnut surulpoochi. *Pesticides.,* 16: (9), 34-35.

Radhakrishnan, S., Saroja, R., Venkatanarayanan, D. and Viswanathan, A.R (1976). On the control of groundnut leaf webber. *Pesticides* 10: (10), 41-42.

Rajagopal, D and Gowda, V. (1992). Efficacy of granular insecticides in the management of groundnut leaf miner *Aproaerema modicella* Deventer (Lepidoptera: Gelechiidae). *Tropical Pest Management* 38: (1), 82-84.

Rawat, S.R., Jakhmola, S.S and Sohuch, R.H. (1970). Assessment of losses of hybrid sorghum (S.H. 1). to earhead caterpillar and comparison of insecticidal control. *PANS.* 16, 367-379.

Saboo, K.C and Puri, S.N. (1979). Effect of insecticides on incidence of sucking pests and yield of groundnut, *Arachis hypogaea* Linn. *Indian Journal of Entomology.,* 40, 311–315.

Sadakathulla, S., A. Abdul Kareem, P.M .Srinivasan and S. Jayaraj. (1976). Chemical conrol of groundnut leaf webber, *Stomopteryx subsecivella* Zell. (Gelechiidae: Lepidoptera). in Tamilnadu. *Pesticides.,* 10: 30-31.

Senapathi B and Patnaik N C 1973 Occurrence of *Scirtothirps dorsalis* Hood on groundnut and preliminary evaluation of loss in pod yield. Journal of Research of the Orissa University of Agriculture and Technology 3(1,2), 110–113.

Senapathi, B. and Patnaik, N.C. (1980). Studies of ecology of *Scirtothrips dorsalis* Hood. on groundnut. *Journal of Research of the Orissa University of Agriculture and Technology.* 10: (1), 17-20.

Senguttuvan, T. and Dhanakodi. C.V. (1997). Control of pod borers of groundnut by soil application of insecticides in alfisols. *Journal of Oilseed Research.,* 14, 274-277.

Shetgar, S.S., Bilapate, G.G., Puri, S. N., Patil, V.V.and Londhe, G.M. (1992). Losses in yield due to foliage pests of groundnut. *Journal of Maharashtra Agricultural University.,* 17: (2), 304-305.

Shetgar, S.S., Bilapate, G.G., Puri, S. N., Patil, V.V.and Londhe, G.M. (1989). Bio-efficacy of different insecticides against foliage pests of groundnut. *Pestology* 13: (11), 12–17.

Shetgar, S.S., Bilapate, G.G., Puri, S. N., Patil, V.V.and Londhe, G.M. (1992). Losses in yield due to foliage pests of groundnut. *Journal of Maharashtra Agricultural University.,* 17: (2), 304-305.

Singh, A.K. (1998). Insecticidal management of tobacco caterpillar, *Spodoptera litura* (Fab.). infesting groundnut. *Paras Nath Shashpa.,* 5: 203-206.

Singh, Z and Vaishampayan, S.M. (1967). Results of trials on the chemical control of *Caliothrips indicus* (Bagnell). in pea. *Indian Journal of Entomology.* 29: 219–220.

Sithanantham, S and J. Kannaiyan. (1991). Effects of soil application of two insecticides on groundnut yield and quality in Zambia. *International Arachis Newsletter.,* 9: 18-19.

Sivagami, V., Mohansundaram, M, and Subba Rao, P. V. (1975). Record of parasites and predators of some south Indian crop pests. *Indian Journal of Entomologyomol.,* 37: 100-101.

Sivasubramanian, P. and G.A. Palanisamy. (1982). Control of groundnut leaf miner. *TNAU Newsletter.,* 12: (3), 4.

Sivasubramanian, P. and Palaniswamy, G.A. (1983). Studies on the chemical control of groundnut leafminer, *Aproaerema modicella* Dev. *Madras Agricultural Journal.* 70: 485-486.

Somasekhar, Shekharappa, Patil, B.V. and Patil, S.A. (1991) Evaluation of insecticides for control of groundnut-leafminer (*Aproaerema modicella*) (Lepidoptera: Gelechiidae). Indian-Journal-of-Agricultural-Sciences., 61: 11, 850-852.

Srivatsava, A.S and Khan, R.M. (1963). Bionomics and control of *Holotrichia insularis* (Melolonthidae : Coleoptera).. *Indian Journal of Entomology.,* 25, 347-354.

Srivatsava, B.P and R..K.Mishra. (1968). Studies on the control of the red hairy caterpillar, *Amsacta moorei* Butler. *Indian Journal of Entomology.,* 30: 125-129.

Sundara Babu, P.C. (1969). Trials against the aphid *Aphis craccivora* Koch. on groundnut with some modern synthetic insecticides. *Madras Agricultural Journal,* 56: (7), 452-454.

Thakkar, V.K., Talati, G.M. and Vyas, H.N. (1981). Comparative efficacy of some insecticides against aphid (*Aphis craccivora* Koch.). infesting groundnut in relation to their toxicity to coccinellid predator. *Indian Journal of Plant Protection.,* 9: 132-136.

Vasantharaj David. B., S.Vijayaraghavan and K.A.Ramaswamy. (1965). Field trials for control of groundnut aphid. *Indian Oilseeds Journal,* 9: 188–191.

Vasudevan, G., G. Logiswaran, V. Kannan and S. Madhava Rao. (1980). Chemical control of *Biloba (Stomopteryx). subsecivella* Zell. and aphids, *Aphis craccivora* infesting groundnut. *Farm. Science,* 2: (5-6), 12-18.

Vilogen, J.H., P.R Beer, D.M. Toit, Du Tonder, van. H.J. (1981). Contact insecticides for the control of stored product pests. Series No.5. The protection against reinfestation of maize and groundnuts in bag stacks. Technical Communication, Department of Agriculture and Fisheries, South Africa No. 168, 10pp.

Wang, K. (1994). A new technique for controlling autumn white grub in peanut. *Bulletin of Agriculture Science and Technology.,* 3: 27.

Wightman, J.A and G.V. Ranga Rao (1993). A Groundnut Insect Identification Handbook for India. Information Bulletin No. 39 ICRISAT. Patancheru, Andhra Pradesh, India, pp. 28–37.

Yadava, CP.S and S.R.S. Yadava. (1973). Field trials with certain commonly used soil insecticides against white grub, *Holotichia consanguinea* Blanchand infesting groundnut. *Indian Journal of Entomology.,* 35: 329-332.

□□□

Integrated Pest Management

V. NANDAGOPAL AND M. P. GHEWANDE

22.1. Introduction

Many insect and non-insect pests are known to be associated with groundnut crop. But all are not economically important, instead only one or two would be economically important at a time in any given situation. Different aspects like distribution, losses, symptomatology, control measures and their integration wherever possible for effective management of major pests have been discussed. Advanced breeding lines possessing resistance to major insect pests are also provided. Bio-control agents like NPV against *Spodoptera* and thuricides for red hairy caterpillar, have been successfully used. Intercrops such as pearl millet and soybean against thrips, leaf hoppers and leaf miner, castor against leaf hoppers and *Spodoptera* are useful as traps/repellent crops for reducing the incidence of these pests. Approaches like pheromones for gram pod borer, tobacco caterpillar and leaf miner, and use of neem products and bio-pesticides against major insect pests would go a long way in reducing the economic burden of groundnut farmers. The best plant protection scenario is to integrate all the possible pest management measures, even it were to be one or two components with minimum inputs with out affecting yields and environment.

22.2. Biotic stress and their influence in groundnut

Biotic factors like insect pests, diseases and weeds have been recognized as the major limiting factors in groundnut production. The groundnut crop suffers from a variety of insect pests, pathogens and weeds, to an extent of 2380 million rupees per annum in India (Balasubramanian,1986), afterwards no attempt were made approximate the possible economic loss in groundnut. The relative importance of pests in different groundnut growing zones are variable depending upon the local cultivation practices, the environment and other varieties and the cropping patterns. Some pests are widely distributed and cause significant yield reduction while others are restricted in distribution and are not considered to be economically important. However, there would always be shifts expected and experienced in pests situation. Therefore, timely pest surveillance is of greater importance. For this purpose well trained personnel should periodically visit farmers fields, study the symptoms/injury and evaluate the damage and wherever necessary, offer advice appropriate management strategies.

Though chemical pesticides have played an important role in increasing groundnut production in the past, their indiscriminate use for over two decades has led to several problems such as development of resistance in pest to pesticides, pesticides residues, destruction of beneficial insects and the out break of secondary pests. Therefore, the strategy should aim at economically feasible and environmentally safe Integrated Pest Management (IPM). In Holistic Integrated Pest Management (HIPM), a series of steps are followed to keep the population level of the pests below the economic threshold level and demands a judicious use of pesticides, resistant/tolerant varieties, biological suppression of the pests population, use of pheromones, timely spray of pesticides etc. to form part of the plant production umbrella. Special attention should be given to those pests which occur consistently and cause heavy losses. With economic considerations of input versus out put ratios, it would be most appropriate and feasible to develop IPM Programme and implement it for the effective management of pests in groundnut.

22.3. Major pests of groundnut

a. Foliar insects and mite pests

The foliage feeding insects which cause economic loss mostly belong to the order Lepidoptera. These include the leaf miner, *Aproaerema modicella* Dev. (Gelechidae), the red hairy caterpillars, *Amsacta albistriga* Wlk and *A. moorei* Bult. (Arctiidae); Bihar hairy caterpillar, *Spilosoma obliqua* Wlk. (Arctiidae); tobacco caterpillar, *Spodoptera litura* (F); (Notcuidae) and gram pod borer *Helicoverpa armigera* Hb (Noctuidae). Among the sucking insects, the leaf hoppers, *Balclutha hortensis* Lindb. and *Empoasca kerri* Pruthi; aphids *Aphis craccivora* Koch, and H setaria); mites, the two spotted white spider mite, *Tetranychus hypogaea* and two spotted red spider mite, *Tetranychus cinnabarinus* are important. Symptoms of damage caused by various insects is given by Amin (1988).

b. Subterranean pests

Among the subterranean insects, *Holotrichia consanguinea* Blanch is a major pest causing severe damage to groundnut crops in Rajasthan, Gujarat, Haryana, Punjab, Uttar Pradesh, Andhra Pradesh and Bihar. Another species, *H. serrata* occurs in southern parts of the country like Karnataka and Tamil Nadu. *Odontotermes obesus* (Rambur), *Odontotermes brunneus* (Hagen)and *Trinervitermes biformis* (Wasmann) and *Microtermes* sp. are important termites affecting groundnut.

c. Rodents and birds

The rodents and birds are of minor importance as pests of groundnut under field and storage conditions. Three species of rats are associated with groundnut *viz., Mus booduga, Rattus meltoda* and *Talera indica*. Under field conditions the rats cause a clear runway with legs impression seen from channel to field or from field bunds to field. Holes with loose soil near the entry could be seen. Pieces of groundnut shells and cut branches are also seen in the corners of house, storage go-downs or near storage bins. The birds are generally not harmful to groundnut except crows (*Corvus splendens*), pigeons and black ibis (*Psaubis papillasa*). The former two birds generally pick up the sown seeds of groundnut leading to reduction in plant population and the occurrence of gaps in the fields. The latter one removes the matured pods from plant and feed on the kernel. Otherwise all these birds are helpful by way of predating the insects like white grubs, grasshopper and grown up lepidopteron larvae.

Detailed list of insect pests and non-insect pest have been given in Chapter1.

d. Diseases

i. Foliar Fungal Diseases

Foliar fungal diseases, take heavy toll on the crop in congenial environmental conditions. The magnitude of yield losses caused by these diseases is very high and ranged from 10% to 70%, all over the groundnut growing zones (Ghewande,1983; Ghewande *et al.*, 1983; Ghewande, 1985,1987, 1989a,1989b, 1990, 1990a; Ghewande and Desai,1992; Ghewande *et al.*, 1993; Ghewande *et al.*, 1993a; Moraes and Godoy,1985; Subramanium and Mcdonald,1983). Early leaf spot is caused by *Cercospora arachidicola* Hori (Perfect stage-*Mycosphaerella arachidis* Deighton) and late leaf spot is caused by *Phaeoisariopsis personata* (Berk & Curt) V.Arx (Perfect stage-*Mycosphaerella berkeleyi* Jenkins). Temperatures in the range of 25° - 30° C and high relative humidity favour infection and disease development. The pathogens may survive on volunteer groundnut plants and infected crop debris. Long distance distribution of the pathogens may be by airborne conidia, by movement of the infected crop debris, or by movement of pods or seeds that are surface-contaminated with conidia. Early and late leaf spot pathogens are also soil-borne in nature (Moraes and Godoy,1985). Groundnut rust is caused by *Puccinia arachidis* Speg. Rust is known to perpetuate, spread and produce severe disease outbreaks by means of its uredinospores. Groundnut crop or volunteer crops are available in one or the other parts of the groundnut zones enabling the survival of uredinial stage round the year. Long distance dissemination of rust may occur via air-borne urediniospores, by the movement of infected crop debris, or by the movement of pods or seeds surface contaminated with urediniospores. An optimum temperature of 20° C, the presence of moisture on the leaf surface, and high relative humidity, favour infection and subsequently disease development. Spread of the disease within the field is facilitated by wind movement, by rain splash and also by insects (Subramanium and Mcdonald,1987).

ii. Seed and soil-borne diseases

Seed and soil-borne diseases may develop from fungi already established in the seeds before sowing or may result from direct invasions of seeds or seedlings by soil inhabiting fungi. Among the seed and soil-borne diseases, collar rot/crown rot/seedling blight (*Aspergillus niger* Van Tieghem, *A. pulverulentus* (McAlp) Thom), stem rot/Sclerotium wilt (*Sclelrotium rolfsii* Sacc) dry wilt or dry root rot (*Macrophomina phaseolina* (Tassi) Goid = *Rhizoctonia bataticola* (Taub.) Butler) have been recognized as the major diseases. These diseases cause severe seedling mortality resulting in patchy crop stand mostly in sandy loam soils and reduce the yields by 25-50 per cent (Ghewande,1985). Collar rot is serious in the sandy loam and medium black soils of Punjab, Tamil Nadu, Uttar Pradesh, Rajasthan and Haryana. Dry root rot and stem rot are sporadic in occurrence. Dry root rot is serious in Rajasthan, Uttar Pradesh, Tamil Nadu, Andhra Pradesh and Maharashtra. Stem rot is becoming serious in Tamil Nadu, Andhra Pradesh, Karnataka, Madhya Pradesh and Gujarat. The collar rot disease spreads by the spores and mycelium adhering to the seeds pods from season to season. In addition collar rot pathogens are saprophytic, common soil dwellers in soil humus and plant debris and capable of initial infection. The main carry over from season to season is in plant debris rotting in the soil and on the seeds harvested from a infected field. *A. niger* can tolerate low soil-moisture and optimum development at temperature between 30° C and 35° C. Dry root rot has wide host range. The pathogen. *M. phaseolina* is facultative saprophyte and a soil dweller. Infected soil, plant debris, and pods serve as sources of inoculums. It can be obtained from kernel shell surface and shell tissue. The optimum temperature for seedling infection is 29° - 35° C. For pods invasion the temperature range between 26° - 32° C is preferable. The dissemination of the fungus is by sclerotia via plant debris and soil, and also infected pods, shell and kernel (Garren,1964). The Sclerotium wilt has also a wide host range and spreads through infected soil, rain drops, sclerotia and in infected farm yard manre.

iii. Virus diseases

Economically important virus diseases are bud necrosis diseases (PBND), peanut mottle virus (PMV) and clump virus (PCV). Recently peanut stripe virus (PStV) disease has been reported on groundnut from India (Reddy *et al.*,1983). Peanut Stem Necrosis diseases has been reported recently (Dr.S.Desai personal communication). Bud necrosis disease is widely distributed in India and has been recognized as one of the most important virus diseases of groundnut in Andhra Pradesh, Uttar Pradesh and Madhya Pradesh and of late in Punjab, Haryana, Rajasthan, Maharashtra, Northern Karnataka and Tamil Nadu on rainy season crop. It has also been reported on rabi/summer (post rainy season crop) crop of Mehaboob Nagar districts of Andhra Pradesh and Northern and Vidarbha region of Maharashtra. The incidence of the disease was observed to be as high as 92% on spanish bunch varieties of groundnut during summer, 1985 in Jamnagar district of Saurashtra region of Gujarat (Ghewande,1987). Bud necrosis disease has been reported to cause yield losses up to 50 per cent in early infected crop of groundnut in India (Ghewande,1983). The first sign of infection of PBND appears as chlorotic speckling or ring spots on the terminal leaf followed by necrosis of the terminal bud, stunted and shoots bear small leaves that are distorted and mottled leading to death of the plant. In late infections, pod size is reduced. Seeds are often shrivelled and have mottled and discoloured testa (Reddy *et al,* 1991). The virus is sap transmissible. Three species of thrips, *Scirtothrips dorsalis* Hood, *Frankliniella schultzei* Trybom and *Thrips palmi* Karny have been shown to transmit the disease of which *T. palmi* is the efficient vector. Crops like green gram, black gram, pea, soybeans, beans, groundnut, tomatoes and pea are highly susceptible to both virus and vector apart from ornamental and weed plants. Major spread of the disease is associated with thrips builds up again during January and February and as such crops sown in December-January suffer the maximum (Amin,1988).

Peanut mottle virus (PMV) is reported to occur on rabi/summer groundnut mainly in the states of Andhra Pradesh, Maharashtra and Gujarat. In PMV, newly formed leaves show mild mottling and vein clearing, whereas older leaves show upward curling and interveinal depression with dark green islands. Seeds from infected pods are discoloured. PMV is reported to be seed transmitted in a rage from 0.1 to 3.5 per cent depending on the type of cultivar. The disease can cause up to 30 per cent loss in yield (Kuhn and Demski,1984). PMV is transmitted in a non-persistent manner by aphids, *A. craccivora* and *Myzus persicae* (Kuhn *et al.*,1978). It occurs in nature on several important legume crops and weeds.

Peanut clump virus (PCV) has been reported from Punjab, Gujarat, Andhra Pradesh and Utter Pradesh on crops grown in sandy soils. It has also shown its presence in certain pockets of Rajasthan, particularly around Jaipur. Yield losses up to 60% have been recorded in late infected plants. The virus has a wide host range and several weeds commonly occurring in groundnut fields are also infected by the virus. Indian PCV isolates are transmitted through seed (Ratna *et al.,* 1991). The virus has also been reported to be soil-borne and possibly transmitted by fungus *Polymyxa graminis.*

Peanut Stripe Virus (PStV) was observed in a few groundnut research centres of All India Coordinated Research Project on Groundnut (AICORPG) located in Andhra Pradesh, Karnataka, Tamil Nadu, Maharashtra and Gujarat, during 1987 to 1992. It was also noticed in some farmers fields around

Junagadh district of Saurashtra region of Gujarat during kharif, 1991 and summer, 1992 (Vijendra Singh *et al.*, 1993). The typical symptom of the PStV is the development of discontinuous dark green banding along with lateral leaflet veins. The other variations of PStV symptoms include blotches, chlorotic rings surrounding the blotching of round leaflets, oak leaf pattern and vein clearing, a mosaic pattern of dark green islands and stunting of plants due to primary infection. PStV is reported to be transmitted through seed and several species of aphids, vectors. It has a wide host range. No authentic reports of yield losses due to PStV are available from India.

e. Nematode Diseases

Although more than 90 species of plant-parasitic nematodes have been reported in association with groundnut crop (Sharma *et al.*,1990), only a few species are known to be economically important (Varaprasad and Sharma,1990) Pod and root-knot disease, caused by four species of *Meloidogyne* viz., *M. arenaria* (Neal) Chitwood, *M. hapla* Chitwood, *M. incognita* Kofoid & White and *M. javanica* (Treub) Chiitwood is the most important disease of groundnut. It is almost distributed throughout the groundnut growing areas of the world. *M. arenaria* and *M. javanica* are more common in the Warmer regions, and *M. hapla* in the cooler regions and *M. incognita* in the Mediterranean region (Sharma and Mcdonald,1990). *M. arenaria* is the most wide spread of the groundnut root-knot nematodes. Estimates of economic losses caused by pod and root-knot disease are not available so far, however, this disease is considered to be important in India (Sharma *et al.*, 1990). Groundnut plants infected with root-knot nematode commonly develop enlarged roots and pegs which develop into galls of various sizes. *M.hapla* produces small galls but extensive root proliferation. Roots often form a dense mat in severe infection. Pods also become infected and develop knobs, protuberances, or small warts. Development of the root system is commonly much reduced. Plants infected with root-knot nematodes may show various degrees of stunting, chlorosis and wilting. Pod and root lesion disease caused by *Pratylenchus brachyrus* (Godfrey). *Filipjevs Schstekh* was recently reported in India from one of the major groundnut growing areas of Andra Pradesh (Sharma,1988), but it is distributed in other states in India is not known. *Cricoemella ornata* (Raski) de Grisse & Loof has also been reported to be associated with groundnut in India (Sharma,1988). Pod and root lesion disease commonly called as Kalahasti malady caused by *Tylechorhycus brevilieatus* Williams is also important disease which damages groundnut crop in Nellore and Chittur districts of Andhra Pradesh. The disease caused by *P. brachyrus* is characterized by the production of small brown flecks where the nematode enters and increase in size with time. *T. brevilieatus* infection causes brownish black discolouration of the pod surface. Small brownish yellow lesions appear on pegs, and young developing pods. Affected plants are stunted and greener than foliage (Reddy *et al.*, 1984).

f. Weeds

Weeds are the regular menace in most part of groundnut growing area in India during rainy seasons. They interfere with growth and development of groundnut apart from acting as alternate host for many insects and diseases. Controlling the weeds at appropriate time is essential to realize good yield of the groundnut. The critical periods for weed competition in groundnut was between 3 to 6 weeks after planting. Hence the weed control practice should be taken up during early part of crop phase.

g. Major weeds

The weeds occurring in groundnut ecosystem represent atleast 15 families of plant species. They can be classified broadly into monocot and dicot weeds. Monocots: *Brachiaria eruciformis* (Graminae)); *Cynotis* sp. *Commelina bengalensis* L.(Commelinaceae); *Cynodon dactylon* L. Pers, *Dactloctenium aegyptium* (L.) P. Beaue; *Diotaria sanguinalis* L.Scop.; *Echnochloa colonum* (Linn) Link; *E. cruegalli* (L.) Beanv.; *Eleusine indica* (L) Gaertn; *Eragarostis tenella* (L) Stapf, *Paspalum distichum; Setaria glauca* (Graminae). Dicots: *Abutilon indicum* (Malvaceaea); *Acalypha indica* (Papavaeraceae); *Alvsicarpus* sp. (Papilionaeae); *Acalypha indica* (Euphorbiaceae); *Celosia argaentea* L. (Amaranthaceaea); *Argaemone mexicana* (Tiliaceaea); *Digera arvensis* Forsk (Amaranthaceaae); *Euphorbia parviflora, E. microphylla* (Eupborbiaceaea); *Leucas aspera* (Lebiatae); *Portulaca oleraceae* L. (Portulacecaea); *Vernonia cinerae* (Composite) and sedges: *Cyparus rotundus* L.; *C. defformis; C. compressus* (Cyperaceae) (Devidayal *et al.*,1987).

22.4. Methods of pests control

In general, the pests control programme involves use of pesticides mostly and very rarely farmers go in for other alternative methods such as cultural, physical, biological etc. The losses due to pests are of high magnitude all over the world. Hence, comprehensive pest management programme is most desirable. Various pest control methods adopted in India and elsewhere are discussed below:

22.4.1. *Insect pests*

a. Host plant resistance

Greater importance is given in India and elsewhere to the identification of resistant varieties to insect pests as the resistant lines generally support less population and lesser damage due to the feeding of insects. A groundnut line, ICGV 50 was found to be highly resistant to leaf miner and tobacco caterpillar. Two cultivars, ICGV 86031 and ICGV 86699 developed at ICRISAST were found to be highly resistant to leaf miner. A variety Girnar 1 developed by the NRCG during 1988 was found to possess high level of resistance to leaf hopper (*B. hortensis*) (Anon.,1989). To develop a resistant variety for all the insect pests occurring in a locality is rather difficult as the feeding behaviour of various insects differs considerably.

b. Cultural practices

Cultural methods involve all aspects of pest management except use of insecticides, natural enemies and resistant varieties. Considering yield as well as the low level of resistance to leaf miner, August 1st or 2nd week appeared to be the best sowing time for higher yields in TMV 7. A significant negative correlation was observed between incidence of leaf miner and maximum - minimum temperature, while positive correlation existed with relative humidity and maximum-minimum temperature (Logiswaran *et al.*,1982). Groundnut either as pure crop with 30 x 10 cm or 20 x 10 cm spacing alone with straw as mulch gave the maximum control of leaf miner as well as increased the extent of larval parasitization (Logiswaran and Mohanasundaram,1985). As an intercrop, cowpea with groundnut turn out to be the best combination with least larval population. Interestingly, groundnut with paddy straw mulch resulted in least weed number/unit area. Nandagopal (1992) reported the low incidence of leaf miner, when bajra was intercropped with groundnut and 74% reduction in egg laying by leaf miner on the foliage treated with 10% concentration of bajra leaf water extract. A highly significant positive correlation was observed between three generations of leaf miner and percent leaflets attacked and larvae per plant. The population of leaf miner moth catch was high in the month of October followed by August and September (Logiswaran and Mohanasundaram,1985, Srinivasan *et al.*,1986) reported high leaf miner population at high plant populations levels irrespective of varieties or level of nitrogen applied. This observation was further confirmed as there was highly significant positive correlation between the plant density and leaf miner incidence (r=0.588). In case of thrips and damage by BND, the incidence was reduced when crop was sown with the onset of rainy season. Closer spacing also lessens the impact of diseases (Reddy *et al.*,1983). Intercrop or multi crop groundnut with cereals reduce the incidence by 10% (Reddy and Wightman,1989).

c. Biological control

Several parasites, predators and pathogens have been reported on different insect pests of groundnut. A brief account on the natural enemies reported on the major pests of the groundnut in India is presented here under:

c.1.Aphids

Many parasites and predators are reported to cause mortality in aphids (Amin,1988, Bakhetia and Siddhu,1976; Upadhyay and Vyas,1987). Among the predators, *Menochilus sexmaculatus* Fab., *Coccinella repanda* Thumbeg. var. *transversalis* Fab., *C. 8 maculata, C.septempunnctata ; Brumus suturalis* Fab and syrphid, *Ischiodon scutellans* Fab., I.javan are important. Some of the bugs such as *Anthocoris* sp., and green lacewings *Chrysoperla cornea* were also reported to play a dominant role in the suppression of aphids. The peak activity of coccinellids appeared from 6th week to 10th week after sowing at Anand during 1979-1981 (Upadhyay and Vyas,1987). Among coccinellids, *M.sexmaculatus* dominated with more than 50% mean population followed by *C.septumpunctata* and *B.suturalis.*

c.2.Leafhoppers

Not much work has been done on the role of parasites and predators in the control of leafhoppers in groundnut. Some lygaeid bugs and spiders were observed predating on them, but they did not appear to cause large reduction (Amin,1988). A predatory fly *Crassopalpus* sp. (Empididae: Diptera) was found predating on *E. kerri, B. hortensis* and *Exitianus taeniaticeps* (Kirchbaum) and it was reported feeding 27 leafhoppers in 10 h when field condition was simulated (Nandagopal,1988).

c.3.Thrips

There were only two reports on the natural enemies of thrips, Muralidharan and Ananathakrishnan (1978) have reported a bug *Carayonocoris indicus* feeding on *F. schultzei.* Three species belonging to Heteroptera and four belonging to Salticidae (Spiders) were recorded predating on *C. indicus* in Junagadh (Nandagopal, 1992).

c.4. Leaf miner

Though 28 parasites have been recorded from South India, Maharashtra and Gujarat (Mohammad,1981; Shanower *et al.* 1992) none was put to use under field conditions. Kothai (1974) reported a peak (23.5%) in the parasitism of leaf miner during October, Khan and Raodeo (1978) reported high parasitism during August to September from Maharashtra. Yadav *et al.* (1987) reported a mean parasitism of 50 per cent during March-May by Goniozus sp. and 79.2 per cent by *Stenomesius japonicus.* Oblisami *et al.* (1969) reported a fungus *Aspergillus flavus* on the larvae of leaf miner. Occurrence of a mermithed nematode from larvae of leaf miner was also reported (Kothai,1974). A series of experiments at ICRISAT centre has shown that larval parasitism can reach to 90 per cent due to this control pressure and carry over of larval population from one generation to the next was not observed (Wightman, and Amin, 1988).

d. Chemical Control

Synthetic insecticides are inevitable in the management of insect pests in agriculture and they continue to dominate as major insect control agents. Despite increased cost, handling hazards, farmers are inclined to go in for synthetic insecticides, as the quick knock-down effect of these synthetic insecticides is convincing to them. A lot of information has been generated on the use of synthetic insecticides by various workers in India and abroad and only cost effective chemicals (Alexander Jesudasan *et al.*, 1989; Amin, 1988,

Das,1988; Ghorpade and Thakur,1989; Jayaraj and Rabindra, 1985; Joshi *et al.*, 1978; Kothai,1974; Mohammad,1981; Mohanasundram,1986; Muthaih and Hussain,1991; Nandagopal,1992; Upadyay and Vyas,1989; Wightman and Amin,1988) are presented here under: Following insecticides were found to be effective against various insect pests. Against aphids Metasystox 0.1%; trithion 0.06%; thiometon 0.08%; menazone 0.03%, demetan-o-methyl 0.05%, 25% EC 400 ml/ha, against Leaf hoppers Dimethoate 400 ml/ha monocrotophos 0.05%, 0.07%; 0.5 kg/ha phosalone 4%D; phenthoate 50 EC 750 ml/ha; dichlorovos 76% EC 65, 62 ml/ha; phosphamidon 375 ml/ha, demeton-o-methyl 0.3%; thiometon 0.03%., against thrips Carbofuron seed dress 4%; monocrotophos 0.075%, 0.05%, 0.04%; phosphamidon 0.02%; dimethoate 0.05%; demeton-o-methyl 0.03%., against Leaf miner, Dimethoate 0.03%, 0.05%, dichlorovas 0.05%, 0.1%; chloropyriphos EC 0.05%; carbofuron EC 0.025 and 0.05%, WP 0.05%; Carbaryl WP 0.2% and 10% D; BHC WP 0.05%, 0.1% monocrotophos 180 g a.i./ha, EC 1.6, 1kg/ha, 0.04%, 0.05%, 0.07%; phosalone EC 0.25 and 0.05%; quinolphos EC 0.025% and 0.05%; endosulfan 0.5 kg a.i./ha; ripcard 50 g a.i./ha were found to be effective.

e. Other methods

The use of one or two methods at a time may not help in controlling the majority of the insects occurring over time and location. Hence new approaches are being examined which can be included in the existing pest management practices. These includes, moult inhibitors, pheromones, plant origin insecticides, trap cropping etc,.

e.1.Moult inhibitors

The cuticular chitin in insects is degraded along with protein at the time of moulting by chitinase enzyme. If synthesis of chitin is prevented the newly formed cuticle will be devoid of chitin. The insect will be unable to withstand the muscular contraction during the moulting which may prevent the casting of excuvium and finally leading to the death of insect (Balasubramanian,1986). These moult inhibitors include diflubenzuron, penfluron, triflumuron and apart from 11 coded components.

e.2.Sex pheromones

The sex pheromones in insects are the chemicals excreted outside which are species specific in attraction and in some species some components could attract subspecies of a species (David *et al.*, 1985). Sex pheromones have the potential role in IPM in groundnutand have beensuccessfully used in the field for the management of *H. armigera* and *S. litura* (Anon.,1974-1991; Pawar *et al.*, 1988). The presence of sex pheromones in leaf miner was reported recently (Nandagopal and Reddy, 1990). Though these sex pheromones are under field use, no economics or its role was reported in reducing the larval population in the field. Sex pheromones for *H. armigera* and *S.litura* are being explored in groundnut through the AICRPG (Anon.,1974-1991). With regard to *H. armigera* in groundnut, a mean of 3.9 males were trapped at 1 m height of sex lure and then there was decrease with increase in height of the trap (Pawar *et al.,* 1988). In an another study when the pheromone of *S.litura* was mixed with *H. armigera,* there was 200 per cent decrease in catch of this moth as compared to trap kept without mixing. Under AICORPO the adult male moths collected using sex pheromone revealed that in Ludhiana the insects continued to be present all through the year (Anon.,1974-1991). Pest population based on three years data (1986, 1987, 1988), the maximum population occurred during April (340), May (649), June (581), September (246), and October (308) while in Jalgaon last week of July (125) and first week of August (108) were the peak periods (Anon.,1974-1991). The population of leaf miner at Hyderabad occurred through out the year recording a maximum of 1662, 1344 and 1190 males/week between March 3rd week to 4th week of February, 1989, respectively (our un -published data). Pawar and Srivastava (1988) reported that even though the sex pheromone of *S.litura* was mixed with pheromone of *H. armigera,* no reduction in the moth catch in the former was observed, however there was drastic reduction in *H. armigera* catch.

e.3.Neem products

Efficiency of neem products in insect control was reviewed by Saxena (1989) and Schemutterer (1990). According to these reviews not much work has been done on the use of neem products in the control of insect pests in groundnut. Neem oil, neem leaf extract and neem leaf decoction at 5% gave 87 to 91% cumulative mortality over 48 hrs of exposure of the leaf hopper *B. hortensis* (Nandagopal *et al.,* 1991). Under the AICORPG system, the first year trial with neem based products like Margoside CK 0.1% and aqueous leaf extract resulted significantly higher pod yield at Vridhachalam and Junagadh centres apart from reducing the leaf miner damage (Anon,1974-1991). Chandramohan and Sivasubramanaian (1987) reported a yield of 2383 kg pods/ha in neem oil treated plot which was on par with monocrotophos. Patel (1968) evaluated efficacy of neem seed extract, neem leaf extract at 10% and neem oil 2% in comparison with monocrotophos against *Amsacta albistriga.* Though monocrotophos gave the highest pod yield, crude neem oil gave 1542 kg/ha pod which was significantly higher compared to control. In an another study both neem kernel suspension (1%) and neem seed extract inhibited the growth of *H. armigera* and *S. litura* larvae (Koul,1985). The neem based insecticides (Margoside OK and CK) suppressed the *S.litura* damage and also gave higher pod yield (Anon.,1974-1991). Growth regulating hormones for *S.litura* have been recorded from 43 plants species belonging to 27 families. Among them Indian neem (*Azadirachta indica*), *Pongamia glabra* and *Parthenium hysterophorus* are important. Crude neem oil has been successfully utilised in the management of gram pod borer, hairy caterpillar and leaf miner in groundnut. Ovicidal property of crude neem oil was demonstrated in leaf miner (Nandagopal,1992).

22.4.2. *Control of Rodents and Birds*

22.4.2.1. Storage (Rodents)

Trapping with 10 g of mixture of Zinc phosphide (3 to 5 kg) + popped rice (1 kg) + groundnut oil (5 g)+ sugar (5 g)} was found effective in the management of rodents.

The bramodiolon is the single doze anti-coagulant available at 0.005% concentration. This is marketed in the name of "moosh moosh". Each cake can be divided into 4 pieces and each piece can be placed inside a live borrow and the mouth of the borrow should be plastered with soil.

22.4.2.2. Birds control

The current practice of birds control includes destroying the roosting sites near crop areas, normal scaring through drumming, birds scaring through guns, use of colour balloons and bright lighting ribbons (Sivanarayan,1985).

22.4.3. *Diseases*

22.4.3.1. Host plant resistance

Developing diseases resistant cultivars is one of the best means of reducing crop losses from diseases. The disease resistance breeding programme requires proven stable and easily available resistant sources for economically important diseases. Such sources have been identified both in the primary gene pool and in the wild relatives of *A. hypogaea*. Efforts are underway in many parts of the world to incorporate the genes responsible for resistance into cultivars with agronomic and quality characters suitable to different environments. Hundreds of land races, advanced breeding lines, cultivars and wild species have been screened for resistance against foliar fungal diseases by several workers, and a large number of resistant sources have been identified (Ghewande *et al.*, 1992). Sources of multiple resistance especially for late leaf spot and rust are available from ICRISAT. These are: EC 76446 (292), NCAc 17133 (RF), PI 259747, PI 350680, PI 389595 and USA 63. Recently five multiple disease resistant germplasm accessions, namely NCAc 17149, NCAc 927, NCAc 17133(RF), PI 393646 and PI 341879 have been identified to be resistant to early leaf spot, late leaf spot, rust and Alternaria leaf spot. Genetic resistance to foliar diseases caused by *P. personata* and *C. arachidicola* has usually been associated with low yields and late maturity (Reddy,1988). It is not difficult to obtain lines with moderate levels of resistance and produce reasonably high yields with minimal or no fungicidal applications (Huang Bing Chao, 1991). Several groundnut research centres of the AICRPG, the National

Research Centre for Groundnut (NRCG) and ICRISAT have started disease resistance breeding programmes to evolve genetically resistant genotypes of groundnut for minimizing losses caused by diseases. As a result of this effort, for the first time in the country, multiple foliar disease resistant groundnut varieties *viz.*, Girnar 1 from NRCG, ALR 1,DH 8TG 3, MH 16, ICG 37 TAG 24, ICGV 87160 and ICGV 86590 from India and Southern Runner and Carduipayne from America have been released for general cultivation. In general, it is very difficult to get promising sources of resistance to many of the seed and soil-borne pathogens such as *S. rolfsii, M. phaseolina* and *A. niger* as they are not very specialized pathogens and they have a very wide spectrum of host attacking mechanism (Shew *et al.*, 1987; Garren,1964). However, a few tolerant genotypes have been identified to these pathogens Chohan *et al.*, (1970) identified one highly resistant genotype, EC 21115 (U4-47-7) to collar rot. J 11 a spanish bunch released variety and other genotypes like C No.45-23, PI 350680, NCAc 17157, B 201 and RAU 31-3 have also been reported to be resistant to collar rot (Anon., 1974-1991; Chahal *et al.*, 1974). Resistant sources like TMV 1, TMV 8 (Sharma,1982), B 30 and B 31 (Sakhuja and Sethi,1985) for root rot (*M. phaseolina*) have been reported. Genotypes, T 17, T 11-11, EC 1682, RB 4, T 25, T 9, NC 2, NC 18016, NCAc 17941 AX Florigiant have been reported to be resistant to stem rot pathogen (*S. rolfsii*) (Garren,1964;Sharma,1982; Rodrigujez-Kabna *et al.*, 1991).

Among released varieties, Kadiri 3 (R 33-1) and ICGS 11 are tolerant to bud necrosis disease. These varieties should be used in endemic areas. At ICRISAT, ICGV 86029, ICGV 86030, 86031, ICGV 86032, ICGV 86033 and ICGV 86538 were identified to possess field tolerance to bud necrosis with good agronomic traits (Reddy *et al.*, 1991). *Arachis chacoense, A. glabrata, A. skp* (PI 262848) and *A. pusilla,* were found to be immune to bud necrosis (Reddy *et al.*, 1983; 1991). while *A.chacoense* and *A.pussilla* were found to be resistant to peanut mottle virus (Reddy,1988). Two genotypes viz., EC 76446 and NCAc 17133 RF have not shown any seed transmission of peanut mottle virus (Ghewande,1983) and PI 261945 and PI 261946 were tolerant to PMV (Reddy,1988). A. sp. ICG 30036 was reported to be tolerant to peanut clump virus. Genotypes, PI 468141, PI 468142, PI 468144, PI 468345, PI 468174, PI 468363, PI 468366 were reported to be resistant to PStV (Culver *et al.*,1987). ICG Nos. 4983, 8215, 8958, 8973, 11558, 11560, 11562, and 12168 were found to be resistant to PStV. Out of these, ICG 11558 was immune (Prasada Rao *et al.*, 1988; Vijendra Singh *et al.*, 1993). Four genotypes JSP 1, ICGMS 2, NC 4X and CGC four have been reported to be resistant to root knot disease caused by *M.arenaria* (56). Sakhuja and Sethi (1985) identified G201, Japtin 220-15, local-256, local Ambali and 4018 as resistant to *M.javanica.* PI 1290606, PI 1295233 and PI 1365553, have been found resistant to pod and root lesion disease caused by *P.brachyurus* (Starr,1984; Varaprasad and Sharma,1990).

22.4.3.2. Cultural control

Removal of volunteer groundnut plants, removal or burial of infected crop debris, crop rotation etc. are important cultural operations in reducing the leaf spots and rust diseases (Subramanium *et al.,* 1989). Early sowing of the crop reduces incidence of rust and leaf spots (Ghewande *et al.,*1993; Ghewande and Desai,1992). Sowing date also influenced the severity of the early leaf spot (Subramanian and Hildebram, 1993) Foliar disease severity could also be reduced by altering plant population (Ghewande *et al.,* 1986; Ghewande *et al.,* 1993). Leaf spots and rust being mainly air borne diseases, spread quickly where there is continuity of host plants over large areas. Therefore, various economically profitable crop combinations were tried to check the spread of these diseases at the NRCG and elsewhere.

Muthiah *et al.* (1991) reported that the incidence of late leaf spot in groundnut + cowpea was minimum followed by groundnut + pearl millet and groundnut + black gram and the rust incidence was minimum in groundnut + sorghum. Deep planting of seed should be avoided as etiolated hypocotyls are prone to infection by collar rot pathogen. Deep ploughing of fields and rotation of groundnut with gram and wheat are helpful in reducing the collar rot disease incidence. Chohan (1974) reported control of collar rot by resorting to early sowing in June. Groundnut planted with a spacing of 30 cm between the rows had lowest root rot incidence than when planted with a row spacing of 45 cm or 60 cm (Sharma, *et al.,* 1990). The lower incidence of collar rot and stem rot in early sown crop and closer plant spacing has also been reported (Ghewande,1983). Cultural practices such as deep covering or burial of organic matter before planting, non-durting cultivation by avoiding movement of soil up around the base of plants and prevention of development of organic debris are useful in reducing the incidence of stem rot. Cotton is a suitable rotation crop for management of *S. rolfsii* (Rodriguez *et al.,* 1991).

Plant density can be increased to decrease the proportion of infected plants of bud necrosis (Reddy *et al.,* 1991). Groundnut intercropped with pearl millet and early sowing showed less incidence of bud necrosis disease as compared with sole crop at ICRISAT (Ghewande, 1983). Cultivation of groundnut away from soybean, cowpea and navy beans could reduce peanut mottle virus considerably as in these crops the virus is seed transmitted and serve as foci of inoculums (Garren, 1964). Groundnut crop grown during *rabi* escapes the clump disease (Delfosse *et al.,* 1996). PStV could be managed by adopting measures like use of virus free seed for sowing, sowing of suitable trap crops around groundnut fields, practice of suitable mixed cropping and intercropping to reduce secondary infection due to aphids, adopting crop rotations of groundnut with non-legumes, roughing of PStV infected plants and burning them, removal of weed hosts of PStV and avoiding of growing crops like soybean, cluster bean, French bean, green gram, cowpea and sesame in the nearby areas of groundnut crop since these are hosts of PStV (Vijendra

Singh *et al.*, 1993). Crops like maize, sorghum, pearl millet and rice may be rotated with groundnut to reduce the incidence of root-knot namatodes. Sesamum, castor and cotton are good crops to include in rotations for reducing the incidence of *M. arenaria* (Rodriguez *et al.*, 1991). Nematode problems in groundnut can be effectively managed through deep ploughing during summer months, increase the use of F.Y.M., oilcakes and other organic amendments and soil solarization (D.J.Patel, personal communication).

22.4.3.3. Biological control

Mycoparasites, *Dicyma pulvinata* (= *Hansfordia pulvinata*) and *Verticillum lecanii* were found to parasitize the early and late leaf spot pathogens (Ghewande *et al.*, 1991a). *Acronomium persicinum, A.salmonaum, Darluca filum, Eudarluca caricis, Tuberculina costaricana, Penicillium islandicum* and *V. lecanii* were found parasitizing groundnut rust (Ghewande *et al.*, 1983; 1991a). *P. islandicum* and *V. lecanii* and their culture filtrates were found to be effective in controlling both late leaf spot and rust diseases under field condition (Ghewande *et al.*, 1991a). Harvey and Bailey (1983) reported that *Bacillus thuringiensis* and *Pseudomonas ceparia* provided upto 70 per cent control of early and late leaf spot in small scale field tests. In 2 of 3 field tests, significant reductions in the severity of early leaf spot was observed due to *Bacillus cerus* (strain 304) on chitin amended leaves (Kokalis and Burelle et al , 1992). Some soil fungi, actinomycetes and bacteria antagonistic to collar rot pathogen have been reported (Chogan,1974). *Trichoderma harzianum, T. auareoviride, T. viride, T. polysporum, Bacillus* spp. are reported to be antagonistic to *S. rolfsii, M. phaseolina* and *Rhizoctonia solani.* For example, it is estimated that about 140 kg of Trichoderma granules are required to suppress *S. rolfsii* in 1 ha area (Beckman and Rodriguz-Kabana,1975). Both *T. viride* and *T. harzianum* were found to be capable of reducing the sclerotial population of *M. phaseolina* (Sharma,1982). Seed treatment with spores and mycelium of *T. polysporum* protects the seeds from invasion by *M. phaseolina* (Chohan,1974). Elad *et al.* (1982) have proved the beneficial effects of *T. harzianum* against *S. rolfsii* and *R. solani. T. harzianum* was found to be antagonistic to *A. niger* (Lastin *et al.*, 1989). *Gliocladium virens* was found to be effective against *S. rolfsii* (Papavizas and Cillins,1990).

22.4.3 4.Chemical control

Several systemic and non-systemic fungicides were tried for reducing the severity early leaf spot, late leaf spot and rust in India and elsewhere (Salako and Adu,1990; Culbreath and Breneman,1992). However, no single fungicide was found to be very effective against all the foliar diseases. In India, leaf spots and rust were effectively managed when three sprays of carbendazim (0.05%) +mancozeb (0.2%) were given at 15 days interval commencing from the first appearance of symptoms (Anon.,1974-1991). Vidhyasekharan (1981) reported that chlorothalonil was effective against leaf spots and rust. Similarly, positive effects of chlorothalonil in controlling leafspots were reported (Breneman and Sumner,1990). Sekhawat *et al.* (1985) reported that the spray schedule consisting two sprays of mancozeb (0.2%) at 35 and 70 days after germination and one spary of carbendazim (0.025%) at 50 days after germination was most economical. Vyas *et al.* (1986) also recommended that carbendazim (0.075%) + mancozeb (0.15%) mixture should be sprayed around August 15 and second and third spray should be repeated depending on the intensity of both leaf spot and rust. The rust and leaf spots were substantially reduced due to the application of carbendazim (0.025%) + tridemorph (0.4%) five times at fortnightly intervals commencing 35 days after sowing during summer season (Ghewande,1983). Application of 1.06 kg a.i./ha of tridomorph-carbendazim-maneb resulted in the most effective-foliar diseases control and the highest pod yield (Shekhawat *et al.*, 1985). Gorbet *et al.* (1990) reported that five rounds of spray, of carbendazim (0.05%) + mancozeb (0.2%), starting from 30 days after sowing with 12 days intervals was found to have maximum control of both late leaf spot and rust diseases which resulted in 57% increased in dry pod yield.

Seed treatment with either thiram @ 3g/kg seed or carbandazim @ 2g/kg seed is effective in controlling seed and soil-borne diseases (Ghewande, 1989a). Seed treatments of peanut with Vitavax-captan at 3g/kg seed and the spraying of crop with benlate at a rate of 2.5g/L caused a significant reduction in crown rot incidence (Nofal *et al.* 1990). Good control of pre-emergence root rot caused by *M. phaseolina* was achieved by seed dressing with captafol (Shanmugam and Govindaswami,1973). Soil drenching with carboxin has been reported to be effective against *S. rolfsii* (Amma and Shanmugam,1974). Flutolanil was found effective against *S. rolfsii* and *R. solani* (Csinol, 1987). Diniconazole was effective against *R. solani* and *S. rolfsii* (Csinos *et al.* 1987). To avoid re-infestation of thrips transmitting bud necrosis disease, an application of insecticides with strong residual effect is required (Reddy and Wightman, 1989). Granular formulations of thiofanox, aldicarb, acephate and disulfotan gave significant control of thrips when applied at the time of sowing. Subsequently the crop may be applied with systematic insecticides at regular intervals (Reddy *et al.* 1991). Similarly soil application of 1,2-dibromide, 3-chloropropane, carbofuran and temik reduce peanut clump disease effectively, but they are not economical (Delfosse *et al.*, 1996). Aphids and other species of PStV vectors should be controlled through commonly used systemic insecticides to avoid secondary transmission of PStV (Vijendra Singh *et al.* 1993). Application of granular nematicides like nemacur 10 g, carbofuran 3g, thimet 10g, mocap 10g etc. @ 3kg a.i./ha is useful in reducing the incidence of nematodes in groundnut. Seed treatments with 3 and 6% aldicarb, sulfone and carbofuran reduce the damage by *M. arenaria* (Patel *et al.* 1986).

22.4.3.5. Other methods of control

Use of chemicals in the control of diseases has had created a lot of environmental problems. Therefore, hunt for renewable sources of alternative chemotherapeutants is highly desirable. Green plants appear to be the reservoirs of effective chemotherapeutants and can provide renewable sources of useful biofungicides. For examples, aqueous leaf extracts (2%) of neem and mehandi (*Lawsonia inermis*) found to be effective and economical in controlling leaf spots and rust diseases (Ghewande, 1989). Neem oil (1%) and nerium leaf extract (10%) reduced the incidence of ring mosaic, late leaf spot and rust (Ganapathi and Narayanaswami,1990). Usman *et al.,* (1991) reported that foliar application of neem kernel extract (3%) reduced the late leaf spot and rust diseases. Ghewande *et al.* (1993) reported that the application of 3 sprays of 5% aqueous neem leaf extract reduced early leaf spot by 14% and late leaf spot by 19%. The aqueous neem leaf extract was found to be efficacious in controlling both late leaf spot and rust diseases and in increasing pod yield at hot spot locations in Tamil Nadu, Karnataka, Maharashtra and Orissa (Anon.,1974-1991). Seed treatment with neem seed extract can check the infection of *S. rolfsii* and *R. bataticola* pathogens (Venkatarao and Alice,1991). Soil solarisation using a thin transparent plastic sheet eliminates soil-borne pathogens. This environmentally safe and efficient cultural practice obviates use of soil pesticides to disinfect the fields before sowing.

22.4.4. Weeds

23.4.4.1. Mechanical/cultural control

The most prevalent methods of weed control in groundnut are hand weeding and harrowing. Experiments conducted through out the country revealed that 2-3 hand weeding or harrowing upto 45 days are optimum for groundnut. First weeding should be done at about 20-25 days after sowing and second a fortnight later.

22.4.4.2.Chemical weed control

Many weedicides like Nitrofen, Fluchloralin, Alachlor have been recommended to control the weeds in groundnut crop throughout the country (Anon.,1974-1991). Pendimithalin has also been recommended by All India Coordinated Research Project on Oilseeds in groundnut. Alachlor is applied primarily as a pre-emergence treatment at the rate of 1.5 to 2 kg a.i./ha. This is effective in controlling annual grasses and Acalypha, Amaranthes and *E. species*. Dephenemid is a water soluble wettable powder applied at the planting time @ 2.25 to 3.40 kg a.i./ha for the control of annual and perennial grasses and partial control of broad leaved weeds. It persists for 3 to 6 months in soil. Trifluralin is available as both EC and granules, for pre-emergence application @ 0.5 kg a.i./ha. Fluchloralin is a selective pre-emergence weedicide controlling grasses and broad leaved weeds. Immediate irrigation is necessary after soil application @ 1.25 to 1.5 kg a.i./ha. Pendimithalin is a selective, pre-emergence weedicide controlling annual grasses and certain broad leaved weeds. Application of this weedicide @ 1 kg a.i./ha controls most of the grassy weeds in groundnut crop. Fusilade is a post emergence weedicide formulated as EC and is very effective against annual and perennial grasses when applied @ 0.25 to 1 kg a.i./ha.

22.5. Integrated pest management

Majority of these methods discussed so far have not been fully effective from economics point of view and various control measures need to be suitably integrated to cover a broad spectrum of pests. The approach, therefore, should be an integrated protection of the crop rather than integrated management of individual pest. Unfortunately, serious efforts were not made in the past to integrate the various pests control measures and as such research on integrated pest management in groundnut is still in its infancy. However, there are a few research efforts where the pest control measures have been integrated and it appears that there is a scope for an integrated pest management in groundnut (Ghewande and Nandagopal, 1997).

22.5.1. *Insect pests*

The strategy of IPM is problem oriented and varies depending upon the pest situation and climatic factors. So a strategy developed for a particular situation need not be necessarily relevant to other situations. Combination of different components of IPM may also vary. In groundnut, integrated management of individual pests like red hairy caterpillar, tobacco caterpillar (Jayaraj and Rabindra,1985; 1985a) and white grub has been given: A comprehensive scheme for the integrated management of pests namely leaf miner, aphid, leaf hoppers, thrips, gram pod borer and tobacco caterpillar is also proposed (Nandagopal,1992).

22.5.1.1. Integrated management of red hairy caterpillar, *A. albistiga*

The management strategies include organizing campaigns to hand pick and destroy pupae after summer ploughing on receipt of monsoon showers, growing cowpea as an intercrop to attract moths to lay eggs, setting up light traps or bonfires immediately after receipt of rain, collecting and destroying egg masses, monitoring the cowpea leaves for lace-like appearance due to gregarious larval feeding, spraying of NPV at 250 larval equivalent/ha (Jayaraj and Rabindra,1985a). handpicking and destroying grown up caterpillars, digging a trench around the field to avoid further movements and applying any one of the insecticides namely BHC 10%, quinophos 1.5%, phosalone 4%, endosulfan 0.05%, carbaryl 10%, 25 kg/ha each, spray of endosulfan 35 EC, quinophos 25 EC and phosalone 35 EC 750 ml/ha and chlorpyriphos at 625 ml/ha.

22.5.1.2. Integrated management of tobacco caterpillar, *S. litura*

To manage the tobacco caterpillar, growing castor/sunflower as border or intercrops to serve as trap crop, monitoring the emergence of adult moth through light or pheromone traps, mass trapping with pheromone traps, collecting and destroying egg masses is from trap crops and groundnut and baiting with rice bran 5 kg + molasses or jaggery 1 kg + DDT 50% WP or carbaryl 50% WP 0.5 kg, to cover 0.4 ha of area are being recommended (Jayaraj and Rabindra,1985).

22.5.1.3. Integrated management of white grub

Spraying host trees of the adult beetles such as ber, neem, khejari and jular with 2% carbaryl or 0.05% monocrotophos during the evening hours just after the start of the monsoon, collecting the adult beetle from trees, treating soil with phorate 10 G at 25 kg/ha, treating seeds with chlorpyriphos 20 EC at 1 liter/80 kg of seed or quinophos 25 EC at 2 liter/80 kg of seed summer ploughing to expose the grubs to predatory birds and flooding the field to impound water for at least 6 days to kill the grub population are the management strategies for the white grubs (Yadav,1985).

22.5.1.4. Integrated management of leaf miner, gram pod borer, tobacco caterpillar, leaf hoppers, thrips and aphids

Integrated management of leaf miner, leaf hoppers, thrips and aphids. Sow groundnut early and synchronously in the rainfed and irrigated seasons, in areas where leaf miner is a major problem, sow spreading and semi-spreading varieties instead of bunch varieties which are highly susceptible, sow cowpea as an intercrop with groundnut (1:4) to minimise the leaf miner infestation, wherever possible, depending on the availability, mulch the soil with straw within 10 days after germination to reduce leaf miner incidence, avoid water stress in irrigated crop to reduce leaf miner infestation, maintain the fields and bunds free from weeds, pull out and destroy plants with bud necrosis or bud blight as thrips act as vectors of the disease. Set up light traps up to midnight to monitor, attract and kill leaf miner moths and synchronize insecticidal application within a week after brood emergence, monitor the crop from the second week onwards and assess the leaf miner larval population. If two or more larvae are observed per plant, apply any one of the following insecticides: phosalone 4%, D, endosulfan 4% D or carbaryl 10% D or fenitrothion 2% D @ 25 kg/ha or phosalone 35% EC, endosulfan 35% EC or monocrotophos 36% WSC, quinolphos 25% EC or phenthoate 50% EC at 750 ml/ha, dichlorvos 76% SC 625 ml or phosphamidom 375 ml/ha.

There has been a report on IPM scheme (Nandagopal,1992) which includes components such as trap crops (soybean, pearl millet, castor), Pheromones (for gram pod borer, tobacco caterpillar and leaf miner), spray application of neem oil (2%) + phosphamidon (0.02%) + endosulfan (0.04%)

at every appearance of leaf miner brood. Sow 1 row of soybean after every 2 rows of groundnut, sow 3 rows of pearl millet all around the field followed by a row of castor, spray insecticide mixture at 30 to 35 days after germination followed by 2 sprays of the same on the appearance of second and third broods of leaf miner respectively. Place pheromones traps for all the 3 larval pests at the rate of 8 each/ha.

22.5.2.Integrated management of diseases

Collar rot causes very high plant mortality at the seedling stage in the groundnut crop during June-July, An integration of date of sowing (10th or 20th June), and seed treatment with thiram @ 5 g/ha kernel reduced the plant mortality and gave higher returns at Ludhiana and Samrala in Punjab (Anon,1974-1991). At Aliyarnagar and Coimbatore in Tamil Nadu adoption of practices such as closer spacing, advancement of date of sowing by two weeks, spray of insecticides like monocrotophos and orthene, spray of antiviral principles like sorghum (JVP) and coconut (CAVP), removal of bud necrosis infected plants particularly at the early stages of crop development and removal of weeds reduced the incidence of bud necrosis disease (Anon.,1974-1991). It is reported that when low rates of insecticides are applied for the control of an insect complex in groundnut, NC 6 a resistant variety yielded 16% more yields than a susceptible Florigiant variety of groundnut (Campbell *et al.*, 1977). Gorbet *et al.* (1990) have successfully used host resistance along with fungicides for the control of leaf spots. Combining Southern Runner, a moderately resistant variety to late leaf spot, *S. rolfsii* and tomato spotted wilt virus with applications of Chlorothalonil can provide late leaf spot control (Vidhyasekharan, 1981).

At NRCG, various control measures were integrated for the management of early and late leaf spot during *kharif* 1988 to 1990. Three years data showed that most effective early and late leaf spot control was achieved when groundnut was intercropped with pigeon pea and sprayed twice (55 and 70 DAS) with fungicide mixture (carbendazim 0.05% + mancozeb 0.2%) or groundnut intercropped and received one spray each of neem leaf extract (2%), fungicide mixture, and cell-free filtrate of *P.islandicum*. These treatments and the intercrops with neem leaf extract at 55 and 70 DAS and with inoculum *P.islandicum* at 55 and 70 DAS gave the highest monetary returns over three years (Ghewande *et al.,* 1989a).

22.5.3. Integrated management of weeds

So far no attempt has been made to integrate all the feasible methods of weed control in groundnut. Any single weed control method either chemical or cultural is not completely effective in managing the weed problem. For effective weed management all the feasible methods have to be used harmoniously. The known practice such as row spacing, tillage and crop rotation are worth to include in any weed control strategy. Generally, application of mixture of two or more herbicides is effective against different species of the weeds compared to application of single weedicide. For example, mixture of Nitrofen + fluchloralin @ 0.2 kg a.i. and 0.5 kg a.i./ha respectively controls various weeds in rainfed crop. The pre-emergence application of mixture of alachlor and Nitrofen is more effective in controlling weeds. Pre-sowing application of fluchloralin followed pre- emergence application of Nitrofen @ 0.7 to 1 kg a.i./ha. control weeds effectively. So the approach should be sequential application of two different herbicides which will be more effective. Pre-emergence application of fluchloralin followed by one or two hand weeding at 25 to 30 and 50 to 60 days after sowing is the most economical. Integration of herbicide and hand weeding at 30 DAS gave consistent and effective weed control and higher pod yields over two years (Ramkrishna, 1991).

22.6. Holistic IPM in Groundnut

A holistic IPM was developed in groundnut. Three years rainy season experiments revealed that by following IPM, under varied rainfall patterns, it is not only the productivity that is sustained but also the requirements for a farm family and their cattle are met. The need based use of crude neem oil with pesticides to the minimum along with use of intercrops such as pearl millet, castor, soybean and pigeon pea increases the net profit apart from acting as a security crop against the failure of groundnut crop due to some reason or the other.

Need based spray of crude neem oil (CNO) 2% with insecticide, phosphamidon 0.02% could effectively control the population build up of leafhopper and thrips. Intercropping of pearl millet reduces the insect pests when grown with groundnut (Gangwar *et al.*, 1994; Kennady *et al.*, 1990.) This is in conformity with the finding of Nandagopal (1992) and Nandagopal *et al.*, (1995). Male moths of *S. litura*, *H. armigera* and *A. modicella* trapped are given in Fig.1. Number of moths trapped were 1690 *S. litura*, 1520 *H. armigera*, and 2610 *A. modicella* during rainy season of 1993. Fungicide mixture (mancozeb 0.025% + carbandazim 0.05%) could effectively control the ELS and LLS of groundnut. Application of 2% aqueous neem leaf extract spray at 40 DAS, followed by spray of fungicide mixture 55 DAS, and application of culture filtrate of *P. islandicum* (50%) (1ml culture filtrate in 2 ml of water) at 70DAS also effectively controlled the diseases (ELS, LLS and rust). The low per cent leaf damage due to GLM in the unsprayed plot during *kharif* 95 was due to smothering of weeds over the groundnut crop. The weed intensity was significantly suppressed under control wherever Flochloralin at 1.5 kg ./ ha was applied along with one hand weeding and one inter culture operation.

It was obvious that, wherever groundnut was intercropped with pigeon pea, the monetary return was higher mostly because of additional returns obtained from pigeon pea. In the good year i.e. when rainfall was optimum, the CBR was (1:1.93) in the treatment.

It can be concluded that the biotic stresses were managed more effectively, when the IPM package *i.e.* (seed treatment with carbendazin + trap crops (pearl millet, castor –around plot and soybean and pigeonpea – middle of plot) + insecticide mixture (neem oil 2 % + phosphamidon 0.02% or endosulfan 0.04%) + pheromone trap (for *H. armigera, S. litura* and *A. modicella*) (trap crop (red gram) + 2% neem leaf extract spray(40 DAS), Mancozeb 0.25% + carbendezin 0.05% (55 DAS), culture filtrate of *P. ilandicum* (70 DAS). (weedicide: fluchloralin 1.5kg a.i./ha + one inter-culturing (35 DAE) + one hand weeding (30 DAE) was fallowed. The IPM module along with pigeonpea, castor, pearl millet and soyabean contributed for higher CBR as growing of intercrops was always profitable in groundnut (Devi Dayal and Reddy, 1991, 1991 a). Also intercrop system always is less likely to fail and more stable than mono-cropping (Rao and Willey, 1980). In Indian situation, where failure of either beginning, mid season or end season rainfall is frequent, the inter crop such as pigeon pea and castor may support as security crop to the farmers. Lingappa (1990) reviewed the effect of intercropping on the pest management in crops in India and expressed that reduced pest damage is possible in intercropping system. For an Indian farmer to whom intercropping is the norm, and so may not be an impossible task to persuade the farmer to take up intercropping (Anon., 1997).

22.7. Future prospects of IPM

It is evident from the literature that there is a great scope for integration of suitable and compatible pest management options and thus develop cost effective and highly efficient package for the integrated pest management in groundnut. The components of IPM such as use of multiple resistant varieties mass trapping of lepidopetran pests, using sex pheromones, use of biocontrol agents like NPV, thuricides, *V. lecanii, P. islandicum,* use of moulting inhibitors, use of suitable intercrops, use of plant products of neem origin, soil solarization, alellopathy for weeds and judicious use of pesticides should be integrated suitably wherever possible keeping in view of the environmental conditions, cultivation practices, cost and social factors since these components have proved to be highly successful. The interdisciplinary approach in IPM groundnut programmes involving field scouting of insect pests, plant diseases, nematodes and weeds should be given high priority. Though such a system is in operation on a limited scale in India it needs to be strengthened and implemented in a large scale for economically important pests.

References

Alexander Jesudasan, R.W.,D.Arulsamaraj and Joyce Priyakumari.(1989). Field evaluation of new insecticides against the groundnut leaf miner, *Pesticides*, 22, 26-27.

Amin, P.W. (1988). Insect and mite pests and their control. In Groundnut (ed. P.S.Reddy) pp.393-452.

Amma, P.A. and Shanmugam. N.(1974). Studies on the root rot of groundnut caused by *Sclerotium rolfsii* Sacc. *Madras Agric. J.* 61,973-978.

Anonymous, (1974 to 1991). Progress Report, Annual Kharif Oilseeds Work shop, All India Coordinated Research Project on Oilseeds Research (AICORPO), Directorate of Oilseeds Research, Hydrabad.

Anonymous,(1989).Annual Reort, National Research Center for Groundnut (Indian Council of Agricultural Reseach), Junagadh, Gujarat pp75.

Anonymous, (1997). Annual Report, National Research Center for Groundnut (Indian Council of Agricultural Reseach), Junagadh, Gujarat.pp,69.

Backman, P.A. and Rodriguez-kabana.(1975). A system for the growth and delivery of bilogical control agents to the soil. *Phytopathology*, 65,819-821.

Bakhetia, D.R.C. and Sidhu. A.S.(1976). Biology and seasonal activity of the groundnut aphid *Aphis craccivora* Koch. *J. Res. Punjab Agric. Univ.*, 14,299-303.

Balasubramanian, M. (1986).Integrated Pest and Disease Management. The need of the day In Pest and Disease Management (ed. S.Jayaraj) pp. 252.

Brenneman, T.B. and Sumner, D.R. (1990). Effects of tractor traffic and chlorothalonil applied via ground sprays or Center Pivot irrigation system on peanut and pod yield. *Pl. Dis.* 74,277-279.

Campbell, W.V., Wyne, J.C. and Emery, D.A. (1977). Control of an insect complex with a resistant variety Nc 6. Proceedings of APRES. 9, 69.

Chahal, A.S., Dutt, S. and Chogan, J.S. (1974). Varietal resistance in groundnut (*Arachis hypogaea* L.) to collar rot (*Aspergillus niger*) in the Punjab state, *J.Res.Punjab Agric. Univ.*, 11,200-203.

Chandramohan, N. and Sivasubramaian, P. (1987). Evaluation of neem product against leaf miner *Aproaerema modicella* (D.). *Neem Newsl.*, 4,44.

Chohan, J.S. (1974). Recent Advances in diseases of groundnut in India. In Current Trends in Plant Pathology Lucknow, Utter Pradesh, India, Unversity Press. pp. 171-184.

Chohan, J.S., Sandhu, G.S. and Dalal, J.L. (1970). Screening of world collection of groundnut (*Arachis hypogaea* L.) germplasm for resistance to collar rot disease caused by *Aspergillus niger* Van Tieghem in Punjab. *Indian J. Agric.Sci.*, 40, 546-551.

Csinos, A.S., Littrell, R.H. and Kvien, C.S. (1987). Activity of diniconazole on foliar and soil-borne diseases of peanut. *Appl. Agric.Res.*, 2, 113-116.

Culbreath, A. K. and Bresnneman, T.B. (1992). Combining center pivot irrigation, application of chlorothalonil with moderately resistant cultivars for control of late leaf spot in groundnut. *Plant Diseases*, 76, 29-32.

Culver, J.N., Sherwood, J.L. and Melouk, H.A. (1987). Resistance to peanut stripe virus in Arachis germplasm. *Plant Dis.*, 71,1080- 1082.

Das, B.B. (1988). Efficacy of some synthetic pyrethroids and conventional insecticides spray against groundnut pest complex. *Pesticides*, 22,10-12.

David, H., Nesbitt, B.F., Easwaramoorthy, S. and Nandagopal, V. (1985). Application of sex pheromone in sugarcane pest management. Proceedings of the Indian Academy of Sciences, Animal Sciences. 94, 333-339.

Delfosse, P., Legreve, A., Devi, P. S., Reddy, A. S., and Maraite, H. (1999). Indian peanut clump virus is highly infectious on graminaceous plants. *Zeitschrift-fur-Pflanzenkrankheiten-und-Pflanzenschutz*. 106(4), 418-424.

Demski, J.M. and Kuhn, C.W. (1983). Peanut mottle. In Peanut Disease Compendium (eds. D.M.Porter, D.H.Smith and R.Rodriguez-Kabana) PP: 45- 46, St. Paul minnesota. *The U.S.American Phytopathological Society*.

Devidayal and Reddy, P. S. (1991). Effect of intercropping sunflower (*Helianthus annus*) in different planting pattern of groundnut (*Arachis hypogaea*) on yeild and economics in Sautashtra region of Gujarat. Ind. J. Agric. Sci., 61(7), 490-2.

Devidayal, Basu, M. S. and Reddy. P.S. (1987). Weed control in groundnut Technologies for better crops, (Bulletin No.32), ICAR,New Delhi, pp 16.

Elad, Y., Hadar, Y., Chat, I. and Hemis, Y. (1982). Prevention with *Trichoderma harzianum* Rifai of reinfestation by *Sclerotium rolfsii* Sacc. and *Rhizoctonia solani* Kuhn of soil fumigated with methyl bromide and improvement of disease control in tomatoes and peanuts. *Crop Prot.*, 1,199-211.

Ganapathy, T. and Narayanasamy, P. (1990). Effect of Plant Products on the incidence of major disease of groundnut. *Int. Arachis Newsl.*,7, 20-21.

Gangwar, S.K., Singh, Y.P., Patel, C. S. (1994). Influence of intercropping on infestation by insect pests of crops at medium-high altitude of Meghalaya. Indian Journal of Agricultural Sciences, 64: 2, 137-140.

Garren, K.H. (1964). Inoculum potential and differences among peanuts in susceptibility to *Sclerotium rolfsii*. *Phytopathology*, 54,279-281.

Ghewande, M.P..and Nandagopal, V. (1997). Integrated Pest Management in Groundnut (*Arachis hypogaea* L.) in India. Integrated Pest Management Reviews 2(1):1-15.

Ghewande, M.P. (1989a). Management of foliar diseases of groundnut (*Arachis hypogaea*) using plant extracts. *Inadian journal of Agricultural science*, 59, 133-134.

Ghewande, M.P. (1987). Virus diseases of groundnut and their management. In symposium on Diseases of Oilseed and Pulse Crops and their control. (Eds. S.C.Vyas and V.N.Shroff), Jawaharlal Nehru Krishi Vishwa Vidhyalaya, Indore, (M.P.) pp.62-67.

Ghewande, M.P. (1989 b). Biological control of late leaf spot (*Phaeoisariopsis personata*) of groundnut (*Arachis hypogaea* L.) *Indian J. Agric. Sci.*, 59,189-190.

Ghewande, M.P. and Desai,S. (1992). Integrated Management of Groundnut Diseases. In *Advances in Plant Disease Management* (Eds. Vyas *et al.*) (In press).

Ghewande, M.P. (1985). Seed and seedling diseases of groundnut deserve attention for better productivity. *Seeds and Farms*, 11,9-12 & 19.

Ghewande, M.P., Desai, S. Prem Narayan, Ingle, A.P. (1993). Integrated Management of foliar diseases of groundnut (*Arachis hypogaea* L.) in India. *Int. J. Pest Mgmt.*, 39, 375-378.

Ghewande, M.P., Desai, S., Prem Narayan and Kamble, S.D. (1992). Some sources of resistance to early leaf spot of groundnut (*Arachis hypogaea* L.) in India. *Trop. Agric.*(Trinidad), 69,284-286.

Ghewande, M.P. (1983a). Effect of cultural practices on the disease (budnecrosis, collar rot, stem rot) incidence and yield of groundnut. *Indian Bot.Reptr*, 2, 176-177.

Ghewande, M.P. (1989a). Management of foliar diseases of groundnut (*Arachis hypogaea*) using plant extracts. *Indian J. Agric. Sci.*, 59, 133-134.

Ghewande, M.P. (1990a). Diseases of groundnut and their management. *J. Oilseeds Res.*, 7,78-97.

Ghewande, M.P., A.K.SUkla and Pande. R.N. (1986). Management of foliar disease of groundnut through agronomic practics. *Indian Bot. Reptr*, 5,179-181.

Ghewande, M.P., Desai, S., Prem Narayan and Ingle, A.P. (1991a). Evaluation of fungicides and plant products to manage foliar diseases of groundnut (*Arachis hypogaea* L.). Proc. *4th Ann. zonal meet. Indian Phytopathology soc.*, 1, 2.

Ghewande, M.P., Shukla, A.K., Pandey, R.N.and Misra, D.P. (1983). Losses in groundnut yields due to leafspots and rust at different intensity levels. *Indian J. Mycology and Pl.Path.*, 13,125- 127.

Ghorpade, S.A.,and Thakur.S.G. (1989). Assessment of need based monocrotophos application for the control of major pest of groundnut. *Pesticides*, 11,19-20.

Gorbet, D. W., Knauft, D. A. and Shokes, F. M. (1990).response of peanut genotype with differential level of leafspot resistance to fungicides treatment. *Crop Science*, 30, 529-33.

Harvey Spurr and Bailey. (1983). Biological control of peanut leaf spot. Proceedings of American Peanut Research and Education Society APRES. 15, 93.

Huang Bingchao, (1991). Review for groundnut pests and diseases and IPM implementation in Guangdong Province of South China. Poster Abstract: Second International groundnut work shop, 25-29 November, 1991.Interational Research Institute for the Smi-Arid tropics, Patancheru, Adhra Pradesh, India. ICRISAT, 1980-1993. Annual Report, Internatioanl Crops Research Institute for Semi-Arid Crops, Patancheru,Andhra Pradesh, India.

Jayaraj, S. and Rabindra. R.J. (1985). Management of red hairy caterpillar *Amsacta* spp.(Lepidoptera:Arctiidae) on groundnut. In Pest and Diseases Managemet - Oilseeds, Pulses, Millets and Cotton (ed.S.Jayaraj) pp. 38-43.

Jayaraj, S. and Rabindra. R.J. (1985a). Integrated management of *Spodoptera litura* with special reference to NPV. In Pest and Diseases Managemet -Oilseeds,Pulses,Millets and Cotton (ed.S.Jayaraj) pp.30-37.

Joshi, B.G., Rama Prasad. G., and Satiyanaryana.S.V.V. (1978). Relative hydroxide as antifeedant against tobacco caterpillar *Spodoptera litura* F. in the nursery. *Indian J. Agric. Sci.*, 48,19-22.

Kennedy, F.J.S., Rajamanickam, K. and Raveendran, T.S. (1990). Effect of intercropping on insect pests of groundnut and their natural enemies. *Journal of Biological Control*, 4 (1), 63-64.

Khan, M.I., and Raodeo, A.K. (1978). Importance of larval parasites in the control of *Stomopteryx subsecivella* Zeller Journal of Maharashtra Agricultural University 3, 261-263.

Khan, M.I. and Raodeo. A.K.(1987). Importance of larval parasites in the control of *Stomopteryx subecivella* Zeller. *J. Maharashtra Agric. Univ.*, 3,261-263.

Kokalis-Burelle, N., Backman, P.A., Rodriguez-Kabana,R. and Ploper, L.D. (1992). Potential for biocontrol of early leaf spot of peanut using *Bacillus cereus* and chitin as foliar amendments. *Biol. Cont.*2,321-328.

Kothai, K. (1974). Bionomics and chemical control of the groundnut leaf miner *Stomopteryx subsecivella* (Zeller) (lepidoptea: Gelechiidae), Msc(Ag) thesis, Univrsity of Agricultural Sciences, Bangalore, India, pp.78.

Koul, D. (1985). Adiradichtin inetraction with development of *Spodoptera litura* Fab. *Ind. J. of Expt. Biology*, 23, 160-63.

Kuhn, C.W., Paguio, D.R. and Adams, D.B. (1978). Tolerance in peanuts to peanut mottle virus. *Pl.Dis. Reptr.*, 62,365-368.

Kuhn, C.W. and Demski, J.W. (1984). Peanut mottle. In Compendium of Peanut Diseases (Eds, D.M.Porter., D.H.Smith., R. Rodriguez-Kabana).*The American Phytopathological Society*, pp.73.

Lastin, S.M., Seif El- Nast, H.I., El-Nagar, M.A.A. and Nofal, M.A. (1989). Biological control of *Aspergillus niger* the causal organism of peanut crown rot by *Trichoderma harizanum*. *Ann. Agric. Sci.*, 34,795-803.

Lingappa,S.(1990).Insect pest management under intercrop system,pp 197-206 *in* Intercropping Principles and Practices (eds. M.M.Hosmani,B.M. Chittpur and S.M.Hiramth). University of Agricultural Science, Dharward.

Logiswaran, G, Madhava Rao, G.vasudevan and V.Kannan. (1982). Influence of time of sowing and weather factors on the infestation of leaf miner *Aproaerema modicella* Deventer and yiekd in rainfed groundnut. *Madras Agricultural Journal*. 69,359-363.

Logiswaran, G. and Mohanasundram. M.(1985). Effect of intercropping, spacing and mulching in the control of groundnut leafminer, *Aproaerema modicella* Deventer (Gelechiidae:Lepidoptera). *Madras Agricultural Journal*. 72,695-700.

Mohammad, A. (1981). The groundnut leaf miner (*Stomopteryx subsecivella* Zeller) (Lepidoptera: Gelechiidae). A Review of world literature. Occasional paper 3, Groundnut Improvement Program, Interational Crops Research Institute for Semi- Arid Tropics (ICRISAST), Patancheru, A.P., India. pp.21.

Mohansundram, M. (1986). Insect pest management in groundnut. In Pest and Disease Management- Oilseeds, Pulses, Millets and Cotton (ed. S.Jayaraj). Tamil Nadu Agril. Univ. Coimbatore.

Moraes, S.A. and Godoy, I.J. (1985). Evaluation of resistance to *Cercosporidium personatum* In *Arachis hypogaea* genotypes. *Summa Phytopathologica*, 11: 140-151.

Muthiah, C. and Hussain. H.J.(1991). Bioefficacy of insecticides against groundnut pests. *Groundnut News*, 3,6.

Muthiah, C., Senthivel, T., Venkatkrishnan, J. and Sivram, M.R. (1991). Effect of intercropping on incidence of pest and disease in groundnut (*Arachis hypogaea*). *Indian J.Agric. Sci.*, 61,152-153.

Nandagopal,V., V.C.Soni, David R.Hall, M.V.Gedia. 1995. Effects of certain components of IPM on the damage and yields in groundnut. Rsearch presented in the National Seminor on " IPM in Agriculture" held at the college of Agriculture, Nagpur on 29 and 30 th of December,1995.

Nandagopal.V.(1992). First record of insect pest and predator of thrips and jassids in groundnut. International Arachis Newsletter.11:26.

Nandagopal, V. (1988). *Crossopalpus* sp. an efficicent predator of jassids of groundnut. *Int.Arachis Newsl.*, 4,18-19.

Nandagopal, V. (1992). Studies on the Integrated Pest Management in Groundnuut in Saurashtra. Ph.D.Thesis,, Saurashtra University, Rajkot. pp. 246.

Nandagopal, V., R.K.Jaroli, Ramkumar and P.S.Reddy. (1991). Neem products: possible insecticide on the groundnut jassid *Balclutha hortensis* Lindb. *Int. Arachis Newsl.*, 8,22.

Nandagopal,, V. and P.S. Reddy. (1990). Report of sex pheromone in groundnut leaf miner. *Int.Arachis Newsl.*, 7,25-26.

Nandagopal. V. (1992a). First record of insect pest and predators of thrips and jassids in groundnut. *Int. Arachis Newsl.*,11, 26.

Nofal, M.A., Seif EL-Nasr, H.I., Diab, M.M., EL-Nagar, M.A.A., and EL-Said, S.A.I. (1990). Effect of the systemic fungicides benlate and vitavax - capltan on *Asplergillus* crown rot incidence of peanut plants. *Ann. Agric.Sci.*,35, 407-415.

Oblisami, G., K.Ramamurthy and G.Rangaswami. (1969). Studies on the pathology of some crop pests of South India. *Mysore J. Agric. Sci.*, 3,86-98.

Papavizas, G.C. and Collins, D.J. (1990). Influence of *Glioccadium virens* on germination and infectivity of sclerotia of *Sclerotium rolfsii*. *Phytopathology*, 80: 627-630.

Patel, H.K.(1968).Neem seed suspension-a sure deterrent to hairy caterpillar (Amsacta moorie.Madras Agricultural Jornal. 55(11)509-510.

Patel, H.R., M.V.Vaishna, and I.U.Dhruj.(1986). Efficacy of aldicarb sulfone and carbofuron flowable seed treatment on plant growth and against Meliodogyne arenaria on groundnut. Pesticides,20, 29-31.

Pawar, C.S. and C.P.Srivastava. (1988). Interaction between sex pheromones of *Spodoptera litura* Fab. and *Heliothis armigera* (Hub.). *Int. Arachis Newsl.*, 4,22.

Pawar,C.S., S.Sithanantham, V.S. Bhatnagar, C.P.Srivastava and W. Reed. (1988). The development of sex pheromone trapping of *Heliothis armigera* at ICRISAT, India. *Trop. Pest Mgmt.*, 34,39-43.

Prasada Rao, R.D.V.J., Chakrabarty, S.K., and Reddy, A.S.(1988). First report of peanut stripe virus from India. *Plant Dis.*, 72,912.

Ramkrishna, A. (1991). Integrated weed management - An efficient method of groundnut weed control. Poster Astracts,, Second International Groundnut Workshop 25-29 November1991, ICRISAT,Patancheru,Andhra Pradesh 502 324,India pp.633.

Rao, M.R and Willey, R.W. (1980). Preliminary studies of intercropping combinations based on pigeonpea or sorghum. *Experimental Agriculture.*, 16, 1-11.

Ratna, A. S. A. S., Rao, A. S., Reddy, B. L., Nolt, D. V. R., Reddy, Vijayalakshmi,M. and McDonald,D.(1991). Studies on transmission of Indian peanut clump virus diseases by *Polymyxa graminis*. Annals of Applied Biology, 118, 71-78.

Reddy, D.V.R., Wightman, J.A., Bashear, R.J., Highland, B., Black,M., Sreenivasan,P., Dwivedi, S.L., Demski, J.W., McDonald, D., Smith,Jr. J.W. and Smith,D.H. (1991). Budnecrosis : a disease of groundut caused by tomato spotted wilt virus. Information Bulletin no.31. ICRISAT, Patanchru,A.P, India.

Reddy, P.S. (1988). Genetics, breeding and varieties. In Groundnut, (ed. P.S.Reddy) pp. 508-525.

Reddy, D.V.R., Amin, P.W., MccDonald, D. and Ghanekar, A.M. (1983). Epidemiology and control of groundnut bud necrosis and other diseases of legume crops in India caused by tomato spotted wilt virus. *In Plant Disease Eplidemiology* (eds. R.T.Plumb and J.M.Tresh) Oxxford, U.K., Blackwell. pp. 93-102.

Reddy, D.D.R., Subrahmanyam, P., Shanker Reddy, G.H., Raja Reddy. C. and Siva Rao, D.V. (1984). A nematode disease of peanut caused by *Tylenchorhynchus brevilineatus*. *Plant Dis.*, 68,528-529.

Reddy, D.V.R. and Wihgtman, J.A. (1989). Tomato spotted wilt virus: Thrips transmission in disease vector research Vol. 5 (ed. K.F.Harris) pp. 203-219.

Rodriguez-Kabana,R., Roberson,D.G., Wells,L., Weaver,C.F. and King,P.S. (1991). Cotton as a rotation crop for the management of *Meloidogyne arenaria* and *Sclerotium rolfsii* in peanut. *Suppl. J. Nematology*, 23,652-657.

Rodriguez-kabana, R., Roberson, D. J., Wells, L., Weaver, C. F., and King, P. S.(1991). Cotton as a rotation crop for the management of *Meloidogyne arenaria* and *Sclerotium rolfsii* in peanut. Suppl. J. Nematology, 23, 652-657.

Sakhuja, P.K. and Sethi, C.L. (1985). Screening of groundnut germplasm for resistance to root-kot nematode *Meloidogyne javanica. Indian J. Nematology*, 15,129-130.

Salako, E.A. and Adu, J.K. (1990). Effect of foliar fungicides on foliar disease and pod, haulm and plant crude protein yield of groundnut. *Trop. Pest Mgmt.*,36,380-383.

Saxena,R.C. (1989). Insecticides from neem. In Insecticides of Plant Origin. (Eds. J.T. Arnasan, J.R. Philogene and P. Morand). American Chemical Society symposium series 387:110-135, ACS, Washington, DC.

Schemutterer, H. (1990). Properties and potential of natural pesticides from the neem tree *Azadirachta indica. Annual Rev. Ent.*, 35,271-297.

Shekhawat, P. S., Patel, V. N., and Patel, J. G.(1985). Economical fungicides spray schedule for control of tikka and rust disease of rainfed groundnut. Indian J.Mycology Pl. Path. 17, 11-16.

Shanmugam, N. and Govindasamy,G.V. (1973). Control of Marcrophomina root rot of Groundnut. *Madras Agric. J.* 60,500-503.

Shanower,T. G., Wightman, J. A., Gutierrez, A. P. and Ranga Rao, G. V. (1992). Larval parasitoids and pathogens of the groundnut leaf miner, *Aproaerema modicella* (Lep.: Gelechiidae)., in India. Entomophaga 37(3), 419-427.

Sharma, S.B., Subrahmanyan, P. and Starr, E. (1990). Plant parasitic nematodes associated with groundnut in Niger. *Trop. Pest Mgmt.*, 36,71-72.

Sharma,R.C. (1982). Studies on Groundnut root rot caused by *Macrophomina phaseolina* (Tessi) Goid (=*Rhizoctina batatiola* (Taub) Butler. Ph.D thesis, IARI,New Delhi.

Sharma,S.B. and McDoald,D. (1990). Global status of nematode problem of groundnut, pigeopea, chickpea, sorghum and pearlmillet and suggestions for future work. *Crop Prot.*, 9,453-458.

Sharma,S.B. (1988). Survey for nematode diseases of Groundnut in some coastal and Rayalasama districts of Adhra Predesh, India. *Int. Arachis Newsl.*,3,11-13.

Shekhawat, P.S., Patel, V.N. and Patel, J. G. (1985). Economical fungicides spray schedule for control of tikka and rust disease of rainfed groundnut. *Indian J.Mycology Pl. Path.*, 17,11-16.

Shew,B.B., J.C. Wynne, and M.K.Beute. (1987). Field, microplot, and green house evaluations of resistance to *Sclerotium rolfsii* in peanut. *Plant Dis.*,71,188-191.

Shivanarayan, N. (1985). Bird pests of national importance. In Integtrated Pest annd Disease Management (ed. S.Jayaraj) pp.285-289.

Srinivasan, S.D.V., Siva Rao. and Sundrarajan, M. (1986). Influence of crop mannagement practices on incidences of leaf miner (*Aproaerema modicella* dev.) in groundnut. *J. Agric. Res. Andhra Pradesh Agric. Univ.*,33,381-382.

Starr, J.L. (1984). Expression of resistance in peanut, *Arachis hypogaea* to *Pratylenchus brachyrus*: Impact on screening for resistance. *J. Nematl.*, 16,404-406.

Subrahmanyan, P. and McDonald, D. (1983). Rust disease of groundnut.Information Bulletin no.13, Patanncheru, AP, India, International Crops Research Innstitute for the Semi-Arid Tropics.

Subrahmanyan, P., Rao,V.R., McDonald, D., Moss, J.P. and Gibbons, R.W. (1989). Origin of resistance to rust and late leaf spot in peanut (*Arachis hypogaea* L.). *Econ. Bot.*, 43, 444 - 455.

Subrahmanyan, P. and McDonald, D. (1987). Groundnut rust disease: Epidemiology and control. ICRISAT (International Crops Research Innstitute for the Semi-Arid Tropics), 1987, Groundnut rust disease. Proceedings of Discussion Group Meeting.24-28, September, 1984, ICRISAT, center, Patancheru, AP, India.

Subramanyam ,P. and Hildebrand, G. L.(1993). Current research on the management of early leaf spot of groundnut at SADC/ ICRISAT Groundnut Project. In: Proceedings of the third ICRISAT Regional Groundnut Meeting for west Africa, Summary Ouagadougou, Burkina Faso, 14-17 Sept. 1992. pp 49-50.

Upadhyay, V.R, and Vyas, H.N. (1987). Integrated pest control of major pest of groundnut in Gujarat in relation to population level and Coccinellid predators population. *Proc. Natn. Symp. Integrated Pest Cont.*- Prog pros., 15- 17 October,1987, Trivandrum, Kerala.

Usman, K.M., Durairaj, P., Karunanithi,K. and Thangavelu,S. (1991). Effect of neem products on the control of foliar disease of groundnut. Poster Abstract, *Second International Groundnut Workshop*, 25- November, 1991. ICRISAT, Patancheru, AP, India.

Varaprasad, K.S. and Sharma, S.B. (1990). First report of the lesion nematode ,*Pratylenchus bachyurus* on groundnut in India. *Indian J. Pl. Prot.*, 18,140.

Venkatarao, A. and Alice, D. (1991). Neem extract checks plant diseases. *Capart Press clippigs*, 7: 23.

Vidhyasekaran, P. (1981). Control of rust and tikka disease of groundnut. *Indian Phytopathology*, 34:20-23.

Vijendra Singh,V., M.P.Ghewade and Reddy, P.S. (1993). Peanut Stripe Virus disease-Present status and its management. *International Journal Pest Management*, 39, 422-430.

Vyas, S. C., Prasad, K. V. V., Shastry, P. P., and Shukla, B. N. (1986). Field control of leaf spots and rust of groundnut by fungicides. Indian Phytopath. 39, 447-448.

Wightman, J.A. and Amin, P.W. (1988). Groundnut pests and their control in the Semi-Arid Tropics. *Trop. Pest Mgmt.*, 34, 218-226.

Yadav, C.P.C. (1985). White grubs and their control (Bulletin). Indian Council of Agricultural Research, Krishi Bhavan, New Delhi, pp.15.

Yadav, D.N., Patel, R.R. and Patel, R.C. (1987). Natural enemies of groundnut leaf miner *Aproaerema modicella* Deventer (Lepidoptera : Gelechiidae) and their impact in its infestation at Anand (Gujarat). *Gujarat Agric.Univ. Res. J.*, 13,13-16.

□□□

Breeding for Insects Resistance

P. MANIVEL, M. SAMDUR
AND V. NANDAGOPAL

23.1. Introduction

The changing scenario of cropping pattern and multiple cropping systems in different seasons has aggravated the insect pest problems. Most pests are accidentally introduced, arriving on our shores without their natural enemies. Insect and disease are important constraints to groundnut production and breeding resistant cultivars is one of the primary aims of groundnut improvement programmes worldwide. Pest resistance is advantageous in developing countries because there is no cost to farmers and resistance cultivars are easily adopted and disseminated unlike "knowledge based" technologies. Also concern for the environment is becoming an important public policy issue in developing countries, and so breeding resistant lines is a burning topic in crop improvement. Among the various alternatives, growing varieties with insect resistance, offers the most economical way of reducing losses. This is not a new concept, but has assumed importance, because, of increased dependence on host plant resistance in IPM systems. In this chapter the insect resistance breeding aspects in groundnut are discussed.

23.2. Definitions about resistance and mechanism of resistance

23.2.1. Definitions

Snelling (1941) defined, "Plant resistance as those characters which enables a plant to avoid, tolerate or recover from the attack of insects under conditions that could cause greater injury to other plants of same species". Painter (1951, 1958) gave slightly different definition, as "Plant resistant as relative amount of heritable qualities possessed by a plant which influence the ultimate degree of damage done by the insects". However, Beck (1965) restricted plant resistance to 'the collective heritable characteristics by which a plant species, races, clones as individual can successfully utilize the plant as a suitable host. Beck's definition narrows the spectrum of insect plant interactions to successful use of plant as host by the insect, but it excluded the plants ability to recover or repair losses after injury occurred (Horber, 1980). According to Fowden *et al.* (1972) Resistance is the ability of certain variety to produce a larger yield of good quality than other varieties at the same initial level of infestation and other similar environmental conditions". Gallum and Khush (1980) defined, "Resistance is the ability of host plant to reduce the infestation or damage or both by an insect.

23.2.2. *Types of resistance*

McColloch (1923) reported that plant resistance to insects could be classified into two general categories: (1) Natural and (2) Artificial. Natural resistance is that form shown by native plants or acquired by cultivated ones.

23.2.2.1. Classification of resistance

Some of the more conventional classification of resistance includes those of Painter (1951, 1958), Horber (1980) and Wiseman (1985). Early classification of resistance in plants to insects has included such terms as 'physicochemical' and 'physiological' resistance. Physicochemical resistance dealt with the integument of the plant, presence of hairs or presence of alkaloids, essential oils, etc. Physiological resistance was based on vigour, quick recovery or seasonal adaptation.

Mumford (1931) classified on the basis of 'Epiphylaxis' or 'Endophylaxis'. Epiphylaxis related to external protection agencies, whereas Endophylaxis was used for describing the internal protection afforded by biochemical qualities of the plant. Even though he did not include tolerance of plant as part of the terms proposed, he did include it with plants overall resistance to describe the cultivar's ability to survive or pull through severe insect attack. For several years, many workers described this new science in terms of resistance or tolerance but did not include both terms under the same umbrella 'resistance'.

Immunity: An immune plant or cultivar is one which a specific insect will not damage or use under any known conditions. An immune plant is a non-host (Horber, 1980).

High resistance: A cultivar with high resistance one which possesses attributes that result in small or minor damage by a specific insect under a given set of conditions.

Moderate resistance: A moderate or intermediate level of resistance is one that may result from any of at least three situations: (a) a mixture of phenotypically high and low resistant plants; (b) plants homozygous for genes which under a given environmental condition produce an intermediate level of injury; and (c) a single clone which is heterozygous for incomplete dominance for high resistance.

Low resistance: A low level of plant resistance indicates attributes possessed by the cultivar that result in less damage or infestation by an insect than the average for the crop.

Susceptibility: a susceptible cultivar is one on which average or more than average damage is inflicted by an insect pest species.

Van der Plank (1963, 1968) classified resistance as vertical or horizontal and Gallun and Khush (1980) encouraged using of the following terms in plant-insect studies. Vertical or specific resistance implies that a series of different cultivars of the same crop infested with a series of different insect biotypes of the same species show a differential interaction. Some cultivars are classed as resistant and others are susceptible when infested with the same insect biotype. Biotype specific resistance is another term used to describe vertical or specific resistance, which is usually controlled by major gene(s). Vertical resistance is usually less stable than horizontal resistance. Horizontal or general resistance implies that the level of resistance offered by a particular host cultivar is similar against all or several insect biotypes. Horizontal or general resistance is controlled by several genes and is usually considered to be stable and permanent. The mode of inheritance of resistance to insect pest will largely determine which method of breeding and selection to be used. It is therefore of great importance for a plant breeder to discover at the start of the breeding programme whether the resistance is controlled by a few or many genes, and whether the influence of the cytoplasm is likely to be significant.

(a) *Monogenic resistance:* There are many advantages in handling Monogenic resistance, which is controlled by a single gene, particularly if it is dominant. Difference between resistance and susceptible plants are clear-cut, and segregation for resistance occurs in simple ratios.

(b) *Oligogenic resistance:* It is control by a small number of major genes each of which has a large effects. Such resistance can involve either a complex or different resistance mechanism, each of which is conditioned by a single gene, or a single mechanism which is controlled by several genes.

(c) *Polygenic resistance:* The term is used for types of resistance, which are known to be controlled by many genes. Very few examples of polygenic controlled resistance have so far been analysed. Nevertheless, resistance that is conditioned by many genes has usually been more durable and has race-specific that resistance that is controlled by only a few genes.

(d) *Cytoplasmic inheritance of resistance:* The resistance is controlled by cytoplasm.

A few other terms that needs definitions are:

Adult plant or mature plant resistance: Resistance observed mainly in maturing plants. It may involve mostly horizontal resistance.

Field resistance: Resistance observed under field conditions. It may involve seedling resistance as well as mature plant resistance.

Multiple resistance: Resistance to many biotic and abiotic stresses like insects, nematodes, disease, drought, cold, etc. In some cases, the environment may alter expression of resistance. Such phenomena related to resistance (ecological resistance) are based on non-heritable traits (Painter, 1951).

Pseudo-resistance: Apparent resistance resulting from transitory characters in potentially susceptible host plants. Three types may be distinguished, host evasion, induced resistance and escape.

Host evasion: Under some circumstances a host may pass through the most susceptible stage quickly or when insect number are reduced. Some cultivars evade insect injury by maturing early.

Induced resistance: This term may be used for the temporarily increased resistance resulting form some condition of plant or environment such as a change in the amount of water or soil fertility.

Escape: Escape refers to the lack of infestation of, or injury to, the plant because of such transitory circumstances as incomplete infestation.

23.3. Mechanisms of resistance

23.3.1. Genetic Mechanism of resistance

The genetic mechanisms responsible for resistance of plant to pest attack are not very well known. Painter (1951) divided plant resistance mechanism into three categories.

Non-preference: This term denotes a group of plant characters and insect response that lead an insect away from a plant or plant part for oviposition, for food, or for shelter or for any combination of the three. Painter (1968) delineated non-preference into two distinct actions of choice among cultivars: (1) a choice to oviposit or feed when several cultivars are grown, or (2) a choice to oviposit, stabilize, or feed when only one cultivar is present. Owens (1975) further described these two uses of non-preference as relative or absolute. Non-preference is often projected as a property of the plant, which is not incongruous with the process to be described. For this reason, Kogan and Ortman (1978) proposed to substitute 'antixenosis' for the term non-preference. It is a parallel term to 'antibiosis' and coonveys the idea that the plant is avoided as a 'bad host'.

Antibiosis: In which a plant is resistance by exerting an adverse influence on the growth and survival of insects. The influence may be in many forms such as death of the neonate larvae or nymph, small or low weight, abnormal longevity, low food reserves, death in the prepupal stage and reduced fecundity.

Tolerance: A tolerant plant can grow and reproduce or repair injury to a marked degree in spite of supporting a density of insects approximately equal to that damaging a susceptible cultivar.

The basic trial of the mechanisms of resistance proposed by Painter (1951) is usually delineated by specifically designated experiments to demonstrate independently these components. Normally, resistant cultivars will possess combinations and or varying levels of the resistance mechanisms. Thus a cultivar that is non-preferred does not require the same level of antibiosis or tolerance that a more preferred cultivar must possess. Therefore, different cultivar may possess the same level of resistance and or levels of resistance components. Hence, it is highly possible to increase the cultivar's resistance level when only one mechanism is present, simply by the addition of another mechanism of resistance.

23.4. Methodology for Field Screening of Groundnut against major insects

1. The specific screening method

Identification of the resistant /tolerant lines to be used for breeding programme is an important activity leading to use of donor parents for specific traits. A simple scheme has been given by Amin (1987) has been given in fig.1.

23.4.1. Screening methods for major insects:

23.4.1. a. Defoliators

23.4.1. a.1. Leaf miner (Aproaerema modicella Dev.):

Screening for leaf miner resistance is done through two means viz., 1. Counting the eggs or larvae and 2. Assessment of dried foliage. The former is time consuming processes and may followed when limited genotypes are being screened. The later one would be highly feasible. For this purpose, a visual scoring on the percent dried foliage on different lines is practiced using the following 0-9 scale.

0-9 scale

Rating	% Dried leaf area
0	No damage
1	1-10
2	11-20
3	21-30
4	31-40
5	41-50
6	51-60
7	61-70
8	71-80
9	81-100

Fig. 1. 0-9 Scale.

For scoring, the sample from a line of 4m could be any one of the sample sizes may be followed based on the severity of the incidence of leaf miner. Twenty compound leaves per plot if the leaf miner incidence by visual scoring is above 6 (in 0 to 9 scale), Forty compound leaves per plot if the leaf miner incidence by visual scoring is between 3 to 6 (in 0 to 9 scale) , Sixty compound leaves per plot if the leaf miner incidence by visual scoring is below 3 (in 0 to 9 scale), Score again the samples leaves in the laboratory for further precision.

Based on the severity of the pest occurrence, it is suggested to take at least two observations on each brood of the insect to have better understanding on the consistency of resistance. Due to sporadic nature of this pest in many locations, total dependence on natural population delay the process of screening substantially. Hence an alternative artificial screening method has been proposed by ICRISAT. In this method the leaf miner cultures have to be maintained in groundnut or soybean plants in a small cages (100 L x 60 W x 55 H (m) under glasshouse environment. Thirty pairs of moths from insectory should be released per cage to induce optimum damage on the test line. By this way in the above sized cage one can accommodate 15 pots of 12 cm diameter. The data on Ovipositional preference and larvae growth should be recorded. In this method if necessary modifications or improvements may be made to suit different situations.

23.4.1.a.2. Tobacco caterpillar (Spodoptera litura) and gram pod borer (Helicoverpa armigera)

Sample size and plot size may be followed as above given for leaf miner. At NRCG a simple laboratory screening method "Choice test" for resistance to *Spodoptera* and *Helicoverpa* was developed (NRCG 1992-93). In this method, The fully opened 3rd - 5th leaves from the top should be collected and arranged in a petridish (20 cm diameter x 2 cm depth or plastic container (18 cm diameter x 5.5 cm depth) touching the petiole on the inner surface of the container and release 25 or 50 or 100 larvae of 3rd or 4th or 5th instars, based on the availability of the larvae in the laboratory, at the centre of the container and should be allowed to feed for 24 hours. The larvae feeding on each leaf can be recorded and noted as "larvae oriented" and the area of leaf (in cm2) eaten by the larvae should be recorded. The genotypes categorized based on the means of percentage of leaf area eaten for third and fifth instar and fitted in zero to nine scale (Fig. 2).

For *Spodoptera*, ICRISAT has developed a simple and effective artificial screening technique. Laboratory reared 4th instar larvae from the insectory should be released to the test rows planted in the field. Normally the test material should be planted in replicated design and the area has to be surrounded by a 15 cm aluminum barrier (or similar kind) to arrest the escape of larvae from the experimental area. In general screening should be done whsen the crop is 30-40 days after emergence (DAE) with 2 larvae per plant

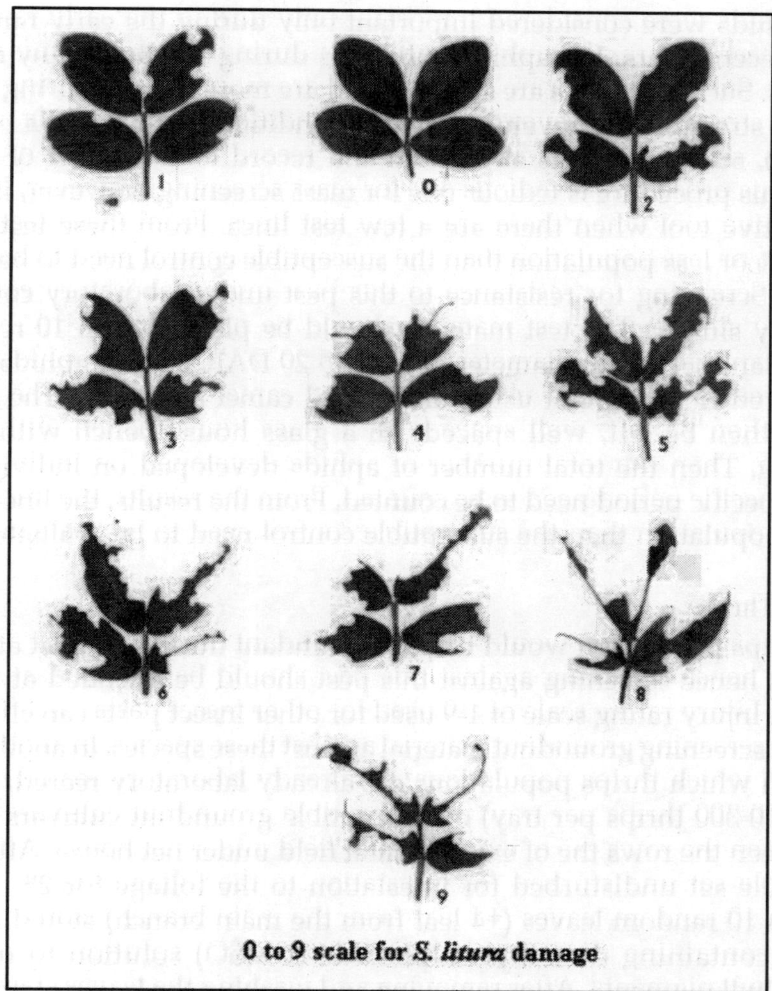

0 to 9 scale for *S. litura* damage

Fig. 2

and it may cause 50% foliage damage in susceptible. If screening is done after 50DAE, about 5 larvae per plant should be released to get > 50% defoliation. Another set of test material should be planted outside the barrier in order to have a pest free comparison. Ten randomly selected plants should be sampled from the central rows and the total leaf area shou'ld be measured using leaf area meter to assess the loss of leaf area eaten and their relative performance among the test line. Bird-proof net should be provided to avoid predation.

23.4.2. Sucking pests

23.4.2.1. Aphids

Nymphs and adults suck sap from the tender growing shoots, flowers, and pegs. They cause stunting and distortion of the foliage and stems. In the past, aphids were considered important only during the early rainy season, but in recent years, the aphids outbreaks during the post rainy season are frequent. Such outbreaks are sporadic and are more intense during periods of drought stress. Under severe infestation conditions (>100 aphids per plant at 30 DAE), select 10 plants at random and record total number of aphids on them. This procedure is tedious one for mass screening; however, it would be an effective tool when there are a few test lines. From these tests the lines with 50% or less population than the susceptible control need to be evaluated further. Screening for resistance to this pest under laboratory conditions is relatively simple. The test material should be planted with 10 replicates of single plant per 10 cm diameter pot. At 15 20 DAE, 2 adult aphids should be transferred to each plant using fine tipped camel hairbrush. The test plants should then be left, well spaced, on a glass house bench with adequate watering. Then the total number of aphids developed on individual plant over a specific period need to be counted. From the results, the lines with 50% or less population than the susceptible control need to be evaluated further.

23.4.2.2. Thrips

Thrips population would be more abundant during August and January months; hence screening against this pest should be attended at these peak periods. Injury rating scale of 1-9 used for other insect pests can effectively be used for screening groundnut material against these species. In another method a tray in which thrips populations are already laboratory reared/multiplied (with 250-300 thrips per tray) on susceptible groundnut cultivars are placed in between the rows the of experimental field under net house. After keeping the whole set undisturbed for infestation to the foliage for 25- to 30 days collect a 10 random leaves (+4 leaf from the main branch) stored in a plastic bottles containing dimethyl sulphoxide (DMSO) solution to remove the chlorophyll pigments. After removing and washing the leaves stain the leaves with 0.50% fushion acid, followed by washing in running water. Count the thrips eggs under dissection microscope as suggested by Nandagopal and Soni (1994). After finding the resistant sources in order to have data on the thrips population, 10 randomly selected terminals should be collected in 10% alcohol per sample. These field-collected samples can be easily sorted out to different species under microscope.

23.4.2.3. Jassids or leafhoppers

Jassids are very mobile insects and hence it is difficult to take absolute count of pest population in any given area. The most reasonable and accurate way of screening lines against jassids is by observing the jassid injury. During the peak population periods scoring should be done on a 1-9 scale at least twice with 15 days interval. Once the selection is made the nymphal observations on the top three terminal leaves will provide information on the relative abundance of the pest. Amin and Mohammed 1980 gave a screening method based on visual rating 1-9 scale on percentage of yellowing recorded on the third leaf from top in the primaries on 50-60 days after sowing. Leafhopper resistance has been associated with density of trichomes (Campbell *et al.*, 1976; Amin and Mohammed, 1980). Hence, the density of trichomes per mm2 on petiole and midrib should be counted using Sudan IV dye or similar dye. Another simple method is to collect jassids with 3 sweeps/5m row using a sweep net of known efficiency and count the number. To count the eggs deposited by the jassids, collect a 10 random leaves (+4 leaf from the main branch) stored in a plastic bottles containing dimethyl sulphoxide (DMSO) solution to remove the chlorophyll pigments. After removing and washing the leaves stain the leaves with 0.50% fushion acid, followed by washing in running water. Count the thrips eggs under dissection microscope as suggested by Nandagopal and Soni (1994).

23.4.3.1. Soil insects

Several kinds of insects feed on roots and pods. Insects that feed on roots are always dangerous to a crop because the crop is subjected to sudden plant mortality. In the case of pod borers the damage the damage is usually detected when the crop is harvested and it is not always easy to work out which insect caused the damage unless guided by the appropriate identification keys, especially when the pods are rotting. For scoring against soil insects, one should take up general observation on plant mortality once in the early part of the season and the later before harvest. Observations on total number of plants at the time of emergence and plant mortality at different stages need to be scored and the mean damages should be recorded as percent plant mortality. Close observations on dead plants will reveal the causal agents responsible for the mortality. Generally, white grubs, termites and jewel beetle kill the plants while the damage caused by earwigs, red ants, wireworms and false wireworms is confined to only pods.

23.4.4. Nematodes

Worldwide more than 90 plant parasitic nematodes associated with groundnut have been reported of which about 37 species have been reported in India (Sharma, 1990). Screening method for nematode resistance described by Jones and Nirula (1963) is given below:

Step 1: Growing plants in pots (preferably 15 cm diameter and one plant per pot) filled with sterilized soil and farmyard manure (9:1).

Step 2: One day after germination inoculation with 2000 second stage juveniles per plant.

Step 3: Removal of plants 60 days after nematode inoculation, washing of roots with water and asses the root galling (%) on 0-5 scale ie.

0 = No root galls (highly resistant)
1 = 1to 20% root galls (resistant)
2 = 21 to 40% root galls (moderately resistant)
3 = 41 to 60% root galls (moderately susceptible)
4 = 61 to 80% root galls (susceptible)
5 = > 80 root galls (highly susceptible)

Step 4: Roots and soil samples should be analyzed and final nematode population has to be recorded

Step 5: Plant growth parameters have to be recorded

Step 6: conclusions has to be drawn based on the analysis

23.4.5. Storage pests

Basic procedures for resistance screening are, the assessment of the inherent susceptibility of crops to attack by storage insects usually involves the measurement of parameters such as length of development period, mortality of juvenile stages, amount of food consumed, and oviposotion rate. To ensure that any differences in the values of these parameters obtained in a screening trial reflect genuine differences between genotypes, both the grain and the insects used in the experiment should be as uniform as is practical. Before insects are placed on samples of the genotypes to be examined, the grain should be pre-conditioned to the experimental temperature and humidity levels for a period of at least two weeks to ensure that the moisture contents of the different genotypes are as similar as possible. The moisture content of a sub sample of each genotype should be determined (as described above) at the start of the experiment. If mass loss is being assessed, then the moisture content should also be determined at the end of the experiment so that the loss can be expressed as dry mass of grain.

Ideally the experiment should be performed in a room or chamber with controlled temperature and humidity. If this is not possible then the range of temperature and humidity should be recorded. This is essential when reporting the length of the insects' development period. Cultures maintained to supply insects for screening should be of relatively constant density so that the size of the adults, and therefore their likely egg-laying capacity is comparable. If the screening technique involves placing adults on the experimental material then they should be of identical age and reproductive status (mated or virgin). Whatever stage of the life cycle is used (adult, larva or egg) a range of insect population densities per unit of grain mass should be examined in a preliminary experiment to determine the density at which differences between genotypes are most obvious and to check for any interaction between density and genotype.

Some important tips for field screening (Up dated from Ranga Rao and Whitman, 1997).

Before starting any evaluation for host plant resistance, it is essential to obtain clear-cut information of the biology, seasonality and damage potential of individual pest species in that area.

One of the most important requirements in any host plant resistance screening is obtaining adequate pest pressure. Methods should be developed to obtain infestation that gives maximum variation between resistant and susceptible lines. On the other hand exceedingly severe infestation is not desired because lines with moderate levels of resistance may be eliminated.

Since insect populations not only vary from season to season be also from field to field in the same filed it is always desirable to have uniform populations wherever possible by releasing artificial population reared under laboratory conditions.

Select testing spots where intensity is regular and the escapes are minimum.

Field must be absolute even character. Areas that become water logged or drought stressed should be avoided, together with patches where nutrient variation and/or salinity are apparent. Select plots of homogenous fertility status. Variation in plant growth resulting from soil fertility may influence the pest populations.

Avoid plots close to tall buildings; trees, crops or wind breaks to eliminate any disturbance to pest movement.

Always conduct screening exercise away form insecticidal sprayed plots.

When the screening is meant for insect pest, always take precaution to control other diseases that affect the observations such as fungicide application to control foliar diseases.

To assess the damage by visual scoring against insect pest follow 1-9 scale where 1 = no damage and 9 = 81-100% foliage destroyed.

A well-trained entomologist / person / scientist should do Scoring. Never allow different people to score the same trials.

Scoring must be taken up as many times as possible, but normally 2-3 observations during peak infestation periods.

Scoring should be initiated only when the susceptible checks have adequate infestation. Otherwise the whole trial should be cancelled.

Scoring against particular pest in a trial should be completed in a single stretch or if the trial could not be completed on the same day the whole set need to be scored again the next day to avoid variation.

Sun direction plays a critical role while taking visual scores, hence care should be taken one should not stand facing eh sun while scoring.

While scoring, if any other good character (observation) is noticed make note in the register and follow up further.

When the scoring is meant for some key pest always keep record of other minor insects that visit the crop during the season.

Under epidemic situation, if the whole trial has maximum damage of the key pest at some stage of the crop, do not discard the trial; take further observations to see any other good character in that material.

To maintain purity in the lines for further evaluation remove off-types during harvest and maintain them separately if they are of some interest.

Borders should be avoided while taking observations.

When artificially reared populations are released on to the field trials or laboratory conditions excess care must be taken for their establishment. For example, if *Spodoptera* larvae are released, release them in the evening time, or in case of *Helicoverpa* put bird nets to avoid predation.

Keep records of meteorological data for the whole period.

Always consult a statistition before setting up the trial and discuss thoroughly about the pest distribution, sampling methods etc.

23.5. Genetics of resistance for various insect pests

Achievements in breeding for resistance to insect pests in groundnut have been limited and that to for a few sucking insects only. Though moderate degrees of resistance to the sucking insects like aphids, jassids, thrips and leaf miner are available in the cultivated germplasm but the transfer of these into high yielding backgrounds has not been very successful due to lack of genetic information either on the host resistance or on the virulence of the pest. Resistance to *Heliothis* and *Spodoptera* is practically limited to the wild species only and so the results of breeding for resistance to the above are yet to come. White grubs being a complex group of polyphagous species it may not be possible to breed for resistance to them in the near future at least by conventional means. Generalised protection through Bt or similar genes may be helpful.

Methods of biometrical genetics have to be applied to study the nature of inheritance. The segregation from crosses with horizontal resistance are not distributed into discrete classes but remain continuous from the level of one parent to another. The genetic studies for resistance to various insect pest are very much limited as compared to genetic studies for resistance to various groundnut diseases. The major reasons are, Providing controlled condition is difficult

Ensuring the sufficient pest population or pressure by artificial mean is comparatively difficult than the diseases pressure

The scientist working in insect resistance in groundnut in India are very few. Non availability of required facility for insect resistance screening. Sporadic occurrence of insect pests from year to year etc.,

The available genetic studies related to different insect pest are discussed here

23.5. 1. Leaf Hopper (Empoasca kerri Pruthi)

It has been reported that resistance to leaf hopper is conformed by the presence of thick epidermis, long trichomes and with a high percentage of straight trichomes on leaves (Campbell *et al.*, 1976; Dwivedi *et al.*, 1986). Leaf trichomes were longer on leaf hopper resistant cultures than the susceptible ones. A high percentage (86-99%) of trichomes on resistant lines was straight trichomes. Leafhopper susceptible groundnuts possessed about 50% curved trichomes. A study of segregating generation from crosses with commercial cultivars indicated that resistance to leaf hopper is probably controlled by a few major genes (Norden, 1980).

Grade	Genotype	Frequency
1	--	0
2	r1r1r2r2r3r3	1
3	R1r1r2r2r3r3	6
4	R1R1r2r2r3r3	15
5	R1R1R2r2r3r3	20
6	R1R1R2R2r3r3	15
7	R1R1R2R2R3r3	6
8	R1R1R2R2R3R3	1
9	--	0

Vindhya varman *et al.* (1989) crossed the resistant donor NCAc 2230 with susceptible Virginia bunch ALR 1 and Valencia ALR 33. The F_1 showed an intermediate resistance reaction (4.5 and 5.0 on a 1-9 scale) bud had tendency towards susceptibility. They also had a trichomes number that was intermediate (21 and 19) between the parents (30 trichomes/petiole in NCAc 2230, 13 in ALR 1 and 10 in ALR 33). The trichomes number showed a significant negative correlation (r=-0.99**) with percent yellowing in the F_2 generation. The susceptible and resistance plants segregated in a ratio of 63:1, indicating that the resistance is controlled by recessive genes (Table below). The frequency distribution of resistance grade of F_2 plants followed a normal distribution. Hence, they concluded that three recessive genes with additive effects govern leafhopper resistance. The genotypes proposed for the F_2 progeny scoring different grades of reaction (on a 1-9 scale) to leaf hopper damage are as follow

The basis of resistance to jassids was investigated at the ICRISAT. The resistance was found to be associated with hairiness of leaves and not with tannin (AICORPO groundnut Annual Report Kharif, 1983-84).

23.5.2. Thrips

Kinzer *et al.* (1973) demonstrated the mechanism of resistance to thrips was due to antibiosis and non-preference.

23.5.3. Helicoverpa

Champbell *et al.* (1981). Reported non-preference and antibiosis as mechanisms of resistance in the case of *Heliothis zea*. Larval weight was reduced 3-5 fold on resistant cultivars and 100% mortality resulted when larvae were fed on some wild species.

23.5.4. Leaf miner

Anderson *et al.* (1990) analysed F_1, F_2 and F_3 progenies of 10 parental diallel crosses for different resistance traits. The F_2 was subjected to screening for incidence of Peanut Stripe Viruses (PStV), stem necrosis and damage by leaf miner. GCA effects were significant for all the traits. The Virginia parent NCAc 2821 had the best GCA effect for leaf miner resistance. They reported that resistance to ELS, LLS, rust, PStV, stem necrosis and damage by leaf miner is additive in nature.

Visalakshi (1997) reported that both morphological and anatomical characters are the contributory factor for resistance to leaf miner in the genotype ICGV 86031 and ICGV 87160. Senguttuvan and Sujatha (2000) tried to find out the biochemical basis of leaf miner resistance and reported that the less susceptible genotypes and susceptible check did not differ markedly in their chlorophyll, total protein, and total free amino acids. Thus these constitute might not have influenced the susceptibility of groundnut to leaf miner attack. However, they found marked difference between less susceptible genotypes and susceptible check in the amount of total sugar and slight difference in phenols. However, they opined that further studies are necessary for a better understanding.

S. litura: The genotypes with narrow long leaflets and profuse hairiness of the leaves and stem are less prepared by *Spodoptera* larvae for feeding compared to genotypes with broad long leaflets and slight hairiness on leaves and stem (Vasanthi and Padmavathamma, 2000).

23.6. Breeding methodology

In India groundnut is grown in two main seasons, rainy season (about 80%) and post rainy season (about 20%) under irrigated conditions. Varietal requirement of these two seasons differ because of differing disease and insect pest complexes occurring in them. Breeding methods by which new cultivars of groundnut originate are the same as those applicable to many other self pollinated crops would be equally suit for insect resistance breeding also. The general breeding methods followed in groundnut are introduction/selection, hybridization & selection, mutation breeding, multiple crosses, back cross method, etc., Earlier, introduction of released cultivars from other countries were practiced. Later single plant or mass selections were made from the introductions. Of the 103 improved cultivars released up to 1999, one cultivar (UF 70-103) has been directly introduced from USA, 46 cultivars have been developed through mass or single plant selection from the introductions, 46 cultivars through hybridization followed by selections, 6 cultivars through mutation breeding and 4 cultivars through combination of mutation and hybridization breeding (Bandyopadhyay and Manivel, 2001).

23.6.1. Introduction and selection

Undoubtedly some progress has been made through introduction and selections, but its scope is very much limited (Gibbons, 1976) and further improvement can be achieved only through artificial hybridization (Norden, 1977). It is interesting to note that till the seventies most of the groundnut cultivars were the result of selection from introductions. For example M 13, a widely adapted spreading type with very high oil content, is a reselection from the US cultivar NC 13 and TMV 10, the bold seeded Virginia type is a selection from an Argentinian cultivar.

23.6.2. Hybridization and selection

Artificial hybridization between parental lines to bring together desirable combination of genes is an integral component of any crop improvement programme. Selection of parents is the first and the most important step in hybridization for various breeding objectives. Parents are chosen for various reasons – as locally adapted cultivars, gene donors for resistance, or for specific morphological or physiological traits or quality traits. A breeder's knowledge of the parental lines and their behaviour in segregating generations helps in making a suitable choice of the parents and the direction of cross.

Briggs and Knowles (1967) have pointed that if the variety to be developed should replace the existing variety then the logical choice as one of the parents is the existing variety. The second parent chosen should complement for the characters under improvement. In NRCG and ICRISAT majority of the crosses are being made between the earlier cross derivatives designated as advanced breeding lines besides the germplasm and varieties. They have proved to be better parents than the earlier varieties. Another approach would be to select the second parent from widely different material or from a totally different region of adaptation. Several cross combinations will have to be made since wide differences exist among parents in their ability to transmit desired characters.

The reported work on groundnut regarding the choice of parents stresses on intermediate degree of diversity between the parents for achieving the best chances of useful transgressive segregants (Arunachalam *et al.*, 1984; Prasad, 1993). The usefulness of within botanical type crosses in addition to between types crosses has been pointed out by different breeders for combining productivity with earliness (Arunachalam *et al.*, 1984; Bandyopadhyay *et al.*, 1984; Durgaprasad *et al.*, 1985). The choice of parents on the basis of heterosis that is based on high general combining ability of parents has been advocated by Arunachalam *et al.* (1984).

In general the breeding for insect resistance is through pedigree breeding and the resistance sources are being crossed with the economically superior varieties. Selections are mainly made in the later generations with major emphasis on yield followed by other resistance traits including insect resistance. The resistance to insect and diseases are linked to some extent with some undesirable traits like, high reticulation, low yield, more constriction, etc. Hence the breeding for insect resistance should be aimed for tolerance/moderate resistance with higher yield rather than for fully resistance. Considerable progress in groundnut breeding through hybridization, in India has been made. To speed up cultivar development modified pedigree method of selection using single seed descent has been successfully used in North Carolina (Wynne, 1976). This method is useful when large populations are available in each generation and when dominant characters are involved. But when the resistance is governed by recessive and one, two or more genes are involved; pedigree selection will be more rewarding.

Recurrent selection through cyclic selection and crossing can be employed to increase the frequency of desirable genes and gene combinations in the populations. This method also extensively used in North Carolina. Experiments on such methods will be fruitful where extensive hybridization programme available.

23.6.3. Handling of segregating generation

One of the greatest problems of the groundnut breeder is that he has to be content with small sizes of the F_2 populations, because of difficulty in successful hybridization and very low multiplication ratio. Thus, the quantum of recombination events available to him for selection is very low. To over come this difficulty, different routes are being explored.

Fast vegetative multiplication of F_1 plants through *in-vitro* nodal culture: The protocols for in-vitro mass multiplication have been established at NRCG. Lab grown ex-plants are the best for this purpose. Interestingly 66% and 71% success in Spanish and Virginia types respectively are available. A large number of F_1 plants can be generated from one hybrid seed by nodal culture and as a result a large F_2 generation can be obtained. The possibility of increasing natural out crossing through induction of transient male sterility in the plants of the segregating generations exists. Interestingly, Dutta *et al.*, (1987) reported that low doses of radiations can be utilized to induce pollen sterility and increase the levels of out crossing in groundnut. They observed that pollen sterility induced through gamma irradiation (5 kR) could produce as high as 33% out crossing in a groundnut cultivar, M 13. Chemical sterilants can also be explored for this purpose.

Selection in groundnut is complicated owing to the fact that the produce is below the ground and the assessment based on yield has to be done only at harvest time. The evaluation of characters, which are correlated with yield, is of particular importance in this crop. Breeding approaches suggested by different workers are mentioned below.

The cultivars released so far in India are through some variant of pedigree breeding including the bulk-pedigree method in which the segregating generations are forwarded as a bulk with some culling of undesirable types till F_4 and F_5 generation and then plant-to-progeny pedigree breeding is resorted to. In single seed descent method (SSD), up to a certain generation, a pod from each plant in a segregating generations is taken and bulked, so that the bulk may represents the range of variation in that particular generation. However, whether the presumption is true can be argued against and for. In view of practical problems it is advisable to follow a kind of bulk-pedigree method rather than pure pedigree method. However, there are pros and cons to early generation selections as discussed below. It appears that early generation selection based on index involving a number of physiological characters may lead to success of a conventional pedigree breeding and hence economy of efforts in later generation. Crop simulation and analytical models will be useful tools in developing index selection, because groundnut crosses are known to throw up segregants till as late generations as F_9- F_{10} especially when inter-varietal crosses are dealt with. Reddy *et al.* (1993), however found pedigree method to be superior over bulk method.

23.6.4. Mutation Breeding

One of the earliest attempts on mutation breeding for crop improvement was by Gregory (1965) with groundnut. Groundnut is known to throw a wide range of variation through mutagens, both physical (gamma rays and x-rays) and chemical (DES, EMS, Sodium azide, etc.,). A number of cultivars have been released by Bhaba Atomic Research Centre, Trombay, Mumbai through direct mutation breeding or through use of mutants in hybridization programmes (Kale *et al.*, 2000). Some of them are TG1, TG3, TG 14, TG 17, TG 22, TG 26, TAG 24, and TKG 19A. At NRCG through mutation breeding some mutants of Girnar 1 with resistance/tolerance to different insects-pests and other streses have been identified (Mathur *et al.*, 2000). Gowda et al., 2000 also found that several late leaf resistant mutants showed resistance to insect pests also (*Spodoptera* and thrips).

23.6.5. Multiple crosses

The number of parents used for developing a cultivar in groundnut has been limited. But good variability is available in the germplasm studied. Unless variation is brought in for reshuffling the gene pool, the present stagnant level of yield cannot be over come. One of the best ways to bring in a large number of selected germplasm lines together is through use of multiple crosses. In groundnut, the most practicable and tested are the three way and double crosses (Rao, 1976). In addition to developing enhanced gene pool, such crosses can be directly utilized for developing cultivars with pyramided genes for resistance to stresses, especially biotic stresses. It has been reported by Bandyopadhyay (1985) that the frequencies of desirable selections were higher in three way crosses than in single crosses.

Double and single crosses were superior for generating desirable segregants for late leaf spot resistance as measured by green leaf area percentage. Three-way crosses followed by backcrosses were superior for releasing desirable segregants for pod yield and shelling percentage. Simultaneous selection of these characters based on an index showed that most of the desirable lines appeared in three-way and back crosses which constituted better foundation populations for further improvement (Gowda *et al.*, 1997). Multiple crosses (three-way crosses, double crosses & intermating of F2 segregants) may be more effective than conventional single crosses for the production of cultivars with acceptable levels of shelling percentage and quality traits and for incorporating rust resistance (Vindhyavarman *et al.*, 1992). It has given evidence that it is possible to have insect resistance and yield promising cultures through multiple crosses than single crosses.

Multiple crosses involving three, four or more parents each possessing a superior characteristic are expected to through greater variability and widen the genetic base. Segregants from such crosses will have wider adaptability. The most successful variety of USA, Florunner has some eight parents in its pedigree. However, In India most of the variety were developed from single cross except variety BSR 1 that was developed through three way cross (ICGS 44 x (Robut 33-1x NCAc 2821). Performance evaluation and selection of desirable plants from the multiple crosses (three way and double corsses) with resistance to biotic and abiotic resistance are in progress at NRCG.

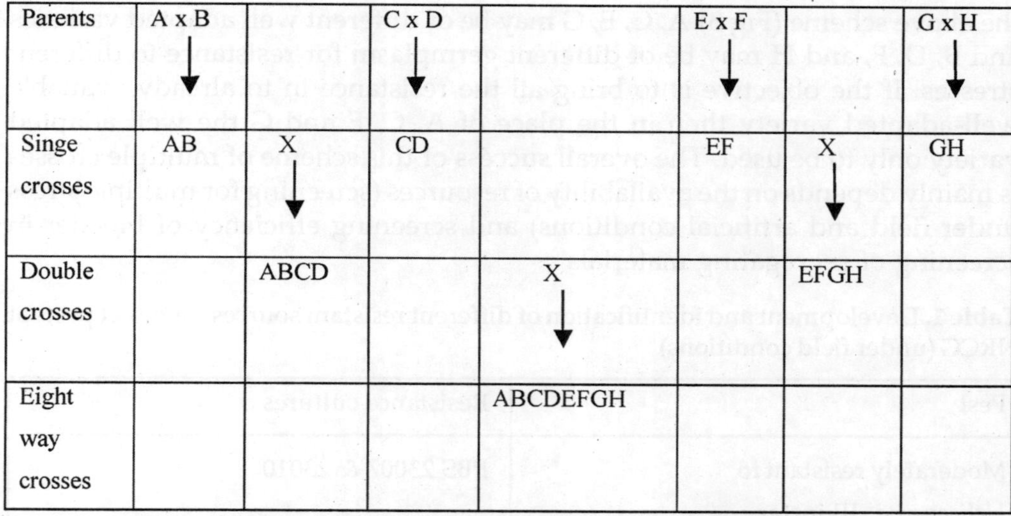

Fig 3. A model-crossing programme for multiple resistances in groundnut

Four separate single crosses for incorporation of resistance to different stresses like insect pest, diseases, drought etc. have to be made first. If the objective is only for insect resistance, the four crosses should be done for different insects pests.

Make large-scale pollination to get more number of F_1 plants.

F_2

i. Select large number of single plant with high yielding
ii. Screen the under artificial epiphytotic conditions
iii. Scoring for pest and diseases

F_3

i. Select plant with good promise for high yield, uniform maturity and acceptable pod and kernel traits.
ii. Selection should be broad based

F_6 - F_8

i. Screening under artificial epiphytotic conditions
ii. Selection with high yield and moderate level of resistance
iii. Multilocation testing

Different single crosses made for incorporation of different resistance (biotic and abiotic) could/should be bring together through various multiple crosses like four way, six way or eight way crosses. In this processes the decision of breeder will play a major role in deciding the type of crosses. In the above scheme (Fig 3) A, C, E, G may be of different well adapted varieties and B, D, F, and H may be of different germplasm for resistance to different stresses. If the objective is to bring all the resistance in to already available well-adapted variety then in the place of A, C, E and G the well adapted variety only to be used. The overall success of this scheme of multiple crosses is mainly depends on the availability of resources (screening for multiple stress under field and artificial conditions) and screening efficiency of breeder in screening of segregating materials

Table 1. Development and identification of different resistant sources for insect pests at NRCG (under field conditions)

Pest	Resistance cultures
Moderately resistant to *Helicoverpa* III instar	PBS 23007 & 23010
Highly resistant to *Spodoptera* I instar	PBS 13010, 23010, 23011, 23017, 23018, 21062 & 11050
Moderately resistant to *Spodoptera* III instar	PBS 23007, 23010, 23017, 23018, 23019, 21062 & 11050
Moderately resistant to *Spodoptera* V instar	PBS 12025, 12144, 12034, 12039, 12047, 12048, 12066, 12067, 12085, 22011, 22013, 22020, 22025 & 22035
Resistant to thrips	PBS 13010, 23003, 24005, 24006, 24030, & 24040
Resistant to leaf miner	PBS 11050 & 21062
Resistance to Bruchid beetle infestation	PBS 12006, PBS 30001, TSP 60

The above mentioned scheme is only a guidance to give the ways to attain the multiple resistance. The breeders can modify the scheme according to his convenience, objective and practability. Even though the scheme seems to be well known but following the same in groundnut is not that much easy.

23.7. Achievements in insect resistance breeding in India

Breeding projects have been initiated at the AICORPO groundnut centers during late eighties and considerable progress are have been made. Many resistance sources have been identified and being utilized in the breeding programme. The extensive screening of available germplasm for different biotic stresses have been initiated from 2000 onwards through AICRP (G) centers and is in progress. At NRCG, breeding for major insect pests and diseases was started from 1986 onwards. Earlier screening was practices and some resistance sources were identified. Hybridization was made with known sources with promising cultivars. Four advanced breeding lines bred for insect pest resistance viz., IR 12, IR 13 A (both derived from GG 11 x NCAc 2230) and IR 15 A and IR 17 (both derived from BG 2 x CGC 101) were found to have comparatively low intensity of damage due to leaf miner (under field condition). The IR 13 A was particularly notable because it showed less infestation than the known resistant varieties like BG 2 (NRCG 199-1-92). Advanced breeding lines with insect pests resistance developed at NRCG are given in the table 1. PBS 105, a multiple disease and insect resistant Virginia culture and IR 14 were highly resistant to oviposition to thrips with 4.7 egg/ leaf, 3.1 egg/leaf, respectively. A Spanish culture PBS 118 and the Virginia culture PBS 48 were identified as resistance to leaf minor. In rainy season 1995, there was severe incidence of leaf miner at Junagadh. The screening based on visual scoring (1-9 scale) and infested leaves percent, 14 entries viz., PBS 45, 170,173, 178, 183, PBDR 13, 31,32, 33A, 34, Code 7,11,Sel 28A, and IR 34 were apparently resistant (score 1-3 and infested leaves < 30%) as compared to maximum score of 8 and >95% infested leaves in susceptible ones (NRCG 1995-96). The advanced breeding lines PBS No. 24039, 29030, and 29017 were found to have resistance to leaf miner over two seasons evaluation under field condition (NRCG 1996). Two advanced breeding cultures 24001 (derived from Chandra x Chicio), 24003 (GG 11 xR 33-1) and interspecific derivatives, Code 7 (J 11 x *A. cardenasii*) and Code 11 (J 11 x 4X hybrid) showed moderate resistance to thrips infestation (< 5 eggs/plant) as compared to susceptible genotype CS 19 in which 36.5 egg/plant was recorded (NRCG 1997). During kharif 1998, heavy infestation of Helicoverpa/ Heliothis under field condition, 16 advanced breeding lines viz., PBS 11042, 12097, 12117, 12115, 21030, 21043, 21046, 21063, 22006, 22015, 23003, 24030, 24057, Code 9, Code 26 and JCA 19-B were found as resistance (< 3 score on 1-9 scale) than the most susceptible ones (9 scale). Under nylon net house conditions lab grown thrips infested plants were kept screening for resistance to thrips and aphids done. Samdur et al. 2000 reported that the advanced breeding lines PBS 23003, 24005, 24006, 24030, and 24040 showed least number of eggs (< 1.99) than the check variety JL 24 (8.9) and most susceptible one, PBS 22023 (15.01 eggs/leaf).

At the ICRISAT, germplasm accessions have been screened for resistance to major sucking pests (thrips, jassids, aphids) and several sources of resistance have been located. The sources of resistance are being used in the breeding programmes at the ICRISAT to develop high-yielding, pest resistant cultivar (Reddy and Mc Donald, 1984; Nigam et al., 1991). Results of breeding work carried out at several AICORPO centers for pest resistance have been reviewed by Reddy (1983). At Ludhiana under artificial condition c 188 and c 382 were found resistance to aphids. For jassids RMP 12 exhibited least susceptibility at Aliyarnagar. At Dharwar center 2 cultures ICG 2307 and Sel 4 were reported to be less susceptible (less than 8.8% leaf injury).

A total of 300 groundnut entries were screened for resistance to the leaf miner (*Aproaerema modicella*) in the field in Tamil Nadu, India (Mahadevan *et al.*, 1988). ICGS 50 (a cross between *Arachis cardenasii* and *A. hypogaea*) had the lowest damage and larval populations and the highest pod yield (3000 kg/ha). This genotype also had resistance against the *Spodoptera litura* and *Helicoverpa armigera* and the pathogen *Phaeoisariopsis personata*. In an another evaluation Mahadevan *et al.* (1989) screened one hundred and fifty-five genotypes for resistance to leaf miner under field condition during rainy season and eight genotypes viz., ICG 10361, 9212, 1458, 1989, 10383, 8555, 10558 and 8717 were identified with some resistance compared to the 2 control varieties JL 24 and VRI 1.

Kalaimani *et al.* (1989) screened eighteen derivatives of a cross between *Arachis cardenasii* (a wild diploid species) and cultivated groundnut (*A. hypogaea*) in unprotected field plots (4.5 sq m) and the lowest incidence of damage (4.0%) was recorded for the entry VG 101, Few entries tested combined resistance with high yield. It is concluded that other *A. cardenasii* derivatives could act as a source for improved resistance to *A. modicella*.

The influence of drought stress gradient on insect distribution in post-rainy season studied by Wheatley *et al.* (1989). The leaf hopper *E. kerri* had the reverse distribution. The thrips *Frankliniella schultzei* and *Scirtothrips dorsalis* were at first densest where drought stress was least. Their distribution subsequently became reversed and, as the condition of their hosts worsened, they again became most abundant at the wetter end of the gradient. Bud necrosis disease caused most mortality where drought stress was highest. A subsidiary experiment indicated that the intensity of overhead irrigation did not influence the density of *A. modicella*. There were marked genotype effects: JL 24 carried the highest density of insects and was most susceptible to bud necrosis disease; NC Ac 343 had the lowest leaf miner and cicadellid populations. M 13 and ICGS 11 had intermediate characteristics.

The effect of the time of sowing on the incidence of leaf miner and the effect of the pest on the yield of groundnut Raju *et al.* (1988) reported that the variety M 13 recorded the lowest (50%) incidence of leaf miner when sown on 1 June as compared to 75% when sown on 1 may. Venkateshwarulu and Swamy (1990) observed significant varietal difference in extend of leaf damage by leaf miner under green house condition. As expected, the leaf damage increased as the insect passed from one instar to another. Interaction between varieties and instar indicated no difference in damage to leaf because of varieties during first and second instar. Later instars caused significant damage to leaf. Hence these points have to be taken in to account while going for screening.

Rajgopal *et al.* (1988) preliminarily assessed 6 Virginia bunch and 18 Virginia runner accessions for resistance to *Spodaptera* in field during the rainy season. The highest resistance was shown by the Virginia bunch varieties V40 and Ah6429, and the Virginia runner varieties NCA17840, NFG79 and EC21989. Biochemical analysis of samples from accessions differing in resistance revealed that there was no relationship between nitrogen, phenolic, potash and sodium contents and leaf damage. Larval orientation was significantly correlated with leaf composition.

Dharne and Patel (2000) screened 32 genotypes under field condition for *Heliothis* resistance and eight entries viz., ICGV 86156, 86400, 86528, 87128, 87141, 87290, 87411, and 91214 recorded lowest leaf damage (<5%) than the maximum damage of 30% by the susceptible check ICG 221.

The population dynamics of *Aphis craccivora* were studied on several cultivars of groundnut in the field in Nigeria. Peaks in the aphid population occurred during the 4th and 5th weeks after planting and cultivars did not significantly influence population numbers. Aphids were not observed after the 11th week and natural enemy populations reached a peak one week after the aphid population. It is suggested that the relationship between the natural enemies and the aphids is density-dependent. Two of 6 cultivars tested were resistant to rosette disease, although aphid colonization was similar to that of susceptible cultivars (Misari *et al.*, 1987).

23.8. Resistance to storage pests

Breeding groundnut cultivars resistant to post harvest insect infestation is one possible way of reducing losses during storage. It is a strategy particularly well suited to improving storage in the Semi-Arid Tropics, where, the financial resources and technical expertise required for the effective use of chemical control methods are often lacking. Unfortunately, breeding for resistance to storage pests is often antagonistic to selection for increased yield. New high yielding varieties have frequently proved to be more susceptible to insect attack during storage than the indigenous genotypes grown by small-scale farmer. With the large of number of new varieties now being developed it is important that there is some collaboration between breeders and storage entomologists to ensure that their release will not exacerbate storage problems. Recently, some genotypes with resistance to bruchid beetle were identified at NRCG (NRCG 2000).

Genetic differences have been shown when stored groundnuts seeds were subjected to infection by red flour beetle. Of 20 cultivars examind for reduction in kernel weight from infection, responses varied from 3.9 to 22.8%. The amount of frass left by the beetle was proportional to the number of beetles that emerged. Although none of the genotypes were immune, some resistance was identified, even using small sample sizes (Kalra and Singal, 1991). Lesser corn stalk borer resistance has been found in cultivated groundnut germplasm (Smith *et al.*, 1980, 1990; Stalker *et al.*, 1984).

Chander *et al.* (1993) evaluated nine varieties of groundnut for their resistance under storage conditions to *Tribolium castaneum* by studying parameters such as initial settling and oviposition in an open-choice test, adult mortality, progeny development, weight loss, extent of kernel damage and changes in moisture, oil and free fatty acid. Considering the overall performance of the varieties, the least preferred group comprised ARK II 9001, ARK II 9005 and M 37 while JL 24 and M 13 were the most preferred. However, the induced biochemical changes did not vary much among the groundnut varieties. Mishra *et al.* (1997) observed some minimum damage on cvs ICGS 52 and OG 85-1 and maximum damage on cvs TMV-2 and AK-12-24. During a survey in the Satpura Plateau in Madhya Pradesh, India, in kharif 1987 and 1988, *Cneorane* sp. was found to infest groundnut (Sharma et al, 1989). The adult population and plant infestation ranged from 0.33 to 11.00 beetles/ sq m and 4.44-77.18%, respectively. This was the first known record of *Cneorane* sp. infesting groundnut.

23.9. Nematodes resistance

Nematode diseases are regionally important in groundnut production in India. Kalahasti malady, caused by *Tylenchorhynchus brevilineatus,* is an important disease in south India, where the disease can cause yield reduction of up to 50%. Several resistant types were found after screening 1599 genotypes. The resistant lines were, for the most part, poor yielding with undesirable pod and seed characteristics. One line showed promise for high yield (Mehan *et al.,* 1993). A seed born nematode, *Ditylenchus destructor,* has been identified in South Africa. While no resistance has been, there is genetic variability for pod yield in response to nematode pressure. Fortunately the high yielding cultivar Sellie was the most tolerant and highest yielding under infestation (Ventner *et al.,* 1993).

Part of the USDA groundnut germplasm collection has been screened in the greenhouse for resistance to root knot nematode (*Meloidogyne arenaria*) in USA, and the material with lowest gall index, egg mass index, number of eggs per plant, and number of eggs per gram of root was identified (Holbrook and Noe, 1992). When these plants were planted in the field, there was a high correlation with populations of nematodes in the rhizosphere of these genotypes. A green house screening should provide good correlation to field resistance (Noe *et al.,* 1992). In a study involving 14 genotypes, Dickson *et al.* (1990) observed that breeding lines UF 81206 showed the least nematode damage in plots without nematicide treatment and showed the least yield response to nematicide treatment. PI 203396 has been used frequently as a parent in groundnut breeding since it is a good source of resistance to several economically significant pathogens (Holbrook *et al.,* 1998).

Cho *et al.* (1986) screened 185 groundnut varieties in wooden pots and none was immune or highly resistant. However, NC 17, Schulmit, Suweon

45, Degona, Weolseong and Tainan 10 were moderately resistant. Holbrook and Noe, 1992 evaluated 1321 plant introductions and mentioned that resistance to *M. arenaria* exists in the cultivated peanut species and can be selected by rating egg-mass production on greenhouse-grown plants.

The ecology of soil nematode populations in the peanut cropping area of Senegal was studied by surveys, population dynamics and vertical distribution (Baujard and Martiny, 1995). The structure of the population corresponds to those identified in the other countries of the Sahelian zone of West Africa and might be under the dependence of two main factors: the rainfall which determines the soil moisture level and the floristic diversity, and the soil temperature which increases from the West to the East of this area. The dry season is considered to be the more appropriate sampling time for the comparison of populations since the numerical variations of population densities remained lower than during the rainy one. Sharma, and McDonald, (1990) opined that *Meloidogyne* spp. are internationally important nematode pests of groundnut and some sources of resistance have been identified against *M. arenaria, M. incognita, M. javanica* and *R. reniformis.* Facilities for resistance screening work is now exist in many countries. Information on damage thresholds of important pest species are available only from Brazil, Fiji, India and the USA.

DNA markers linked to a root-knot nematode resistance gene derived from wild peanut species have been identified. The wild diploid peanut accessions K9484 (*Arachis batizocoi*), GKP10017 (*A. cardenasii*) and GKP10602 (*A. diogoi)* possess genes for resistance to *Meloidogyne arenaria* (Burow *et al.*, 1996).

Holbrook *et al.* (1998) studied the mechanism of resistance and reported that the breeding lines UF81206-Q4, UF81206, and UF93111 yielded similarly to Florunner when grown in the field with no nematodes and significantly higher than Florunner when grown in the field with high nematode population densities. PI203396 is in the pedigree of these breeding lines and also exhibited a moderate level of resistance in a greenhouse trial. Holbrook and Noe (1990) reported that resistance to *Meloidogyne arenaria* is prevalent in both the third and fourth order gene pool of groundnut. That means, the probability of success in developing groundnut cultivars with resistance to *M. arenaria* species in the third order gene pool are cross compatible with *A. hypogaea*. Further the increased probability of resistance to *M. arenaria* in the first order gene pool is also possible and further screening in *A. hypogaea* has to be done.

Of nine groundnut varieties tested in the laboratory for resistance to *Spilosoma obliqua*, M-13 was the most resistant and M-145 and Dwarf Mutant the most susceptible (Tiwari et al 1988). Cultivars C-333, JSP-1, ICGS-26 and M-145 were the most susceptible and (JSP-2, ICGS-2, NC-4X and CGC-4) were resistant and these were thought suitable for use in future breeding resistance programmes Grewal *et al.* (1987).

23.10. Breeding for resistance to virus vectors

Virus infections transmitted by *Aphis craccivora* can be a major constraint to production. Resistance to the vector has been identified in the cultivar EC36892 (Padgham *et al.*, 1990). Field and laboratory behaviour studies in Malawi showed that this resistance takes effect only after the insect has fed for about 2 h. Electrical studies of probe penetration showed that phloem location is almost as successful in the resistant cultivar EC36892 as in the susceptible cultivar TMV2, but the mean feeding duration on the resistant cultivar was only half that on the susceptible cultivar. Chemical analysis suggests that in the resistant EC36892 an isoflavonoid-like substance may confer limited resistance at the probing stage while a high conc. of the condensed tannin procyanidin in the phloem sap may be a major constraint to prolonged ingestion. Wild *Arachis* spp. also shows evidence of resistance.

The rate of pest build-up depends on a number of factors, the more important being the availability of food, the number of generations per season, and the temperature during development. In hot dry condition pests such as aphids, leafhoppers and some caterpillars can increase very rapidly. These pests may die out naturally at the end of growing season, only to be re-established when adults migrate into the area again. In contrast, there are certain pests, e.g. white grubs, which are relatively immobile, have only one generation per year, and build up to pest proportion when suitable host plants are grown on the same land for several years.

23.11. Sources of insect resistance

Among the released cultivars, Girnar 1, TAG 24, ICGV 86590, CSMG 84-1 and BSR 1 are known to have resistance to one or more insect pests (Table 2). At NRCG, several jassid resistant cultures have been developed through intra and interspecific breeding (Table 1). Resistance to leaf eating insects is mostly confined to wild relatives although genotypic differences have been found for this trait (Table 3 and 4) Therefore cultivar development on this aspect takes some more time to yield results. Control of polyphagous species such as white grub, a menace in several groundnut-growing areas in India, may not be possible through genetic improvement alone.

23.12. Future strategies

Integrated approaches to tackle the biotic and abiotic stresses through development of multiple resistant cultivars.

Specific breeding programme to be attempted to tackle the endemic pests at particular location

Table 2. List of released groundnut cultivars with some resistance to insect pests in India.

S. No	Name of the Cultivar	Pedigree	Breeding Methods involed *	Habit group $	Year of Release	Area for which released	Originating Centre	Pod yield Kg/ha	Shelling %	Oil (%)	100 kernel wt.	Salient features
1	TG 3	Mutant of spanish improved	M	SB	1985	Kerala	BARC Trombay	1200 (R)	--	48.5	--	Spanish, medium-large pods, suitable for both rainy and summer cultivations, tolerant to pod borer
2.	Girnar 1	X 14-4-B-19-B x NCAc 17090	HS	SB	1988	Western M.H., A.P., T.N.	NRCG, Junagadh	2000 (R)	70.9	50.0	33.0	Early maturing multiple resistant cultivar suitable for rainfed conditions
3.	TAG 24	Selection from TGS 2	S	SB	1991	Maharashtra	BARC Trombay	1683	70.0	51.0	38.0	Early with 2-3 seeded; tolerant to BND leaf spot jassid and cater piller
4.	ICGV 86590	X 14-4-B-19-B x PI 259747	HS	SB	1991	Peninsular India	ICRISAT	17850	62.0	41.6	33.0	Multiple disease and insect pest resistant
5.	CSMG 84-1	Selection from MA 10	S	VR	1992	Rajasthan, U.P., and Haryana	CSAU A& T Mainpuri	2704	64.0	50.4	48.0	To lerant to thrips, termites, pod borer & foliar diseases
6.	BSR 1	ICGS 44 x (Roubut 33-1 x NCAc 2821)	HS	SB	1997	Tamil Nadu	Bhavanisagar	2845 (K) 3625 (R)	70.0	48.0		Resistant ot collar rot, stem rot, rust and leaf spot; moderatly resistant to aphids, jassids and leaf miner

* S = Selections HS = Hybridization and selection,

$ SB = Spanish bunch VL = Valencia SS = Semi-spreading VR = Virginia runner VB = Virginia bunch

Table 3. Sources of insect pest resistance in *A. hypogaea*

Insect	Resistant sources
Jassid	ICGs 156, 266, 273, 398, 411, 589, 1602, 1660, 2036, 2271, 2307, 2711, 3569, 5030, 5037, 5040, 5041, 5042, 5043, 5044, 5045, 5240, 6429, 7664, 6826, 7446, 7554, 7490, 7803, 8603,NRCGs 1844, 2197, 2828, 3818, 3960, PI 393646 BL: ICGVs 87445, 86455, 86393, 86462, 86522, 86518, 87430, 86252, 87495, 87268, CGC 101, PBS 105, 4-5-3 (IDC) CVs: Girnar 1, MH 1, Pol 2, S 206, Kisan
Leaf miner	ICGs 1697, 2248, 2271, 2465, 3053, 3806, 4508, 6544, 6862, 7237, 7404, 7881, 9116, 9893, ICGV 86031, NCAc 927, PI 390595, PI 393526, EC 24419 CV: 87160
Aphid	ICGs 5240, 5725
Thrips	ICGs 799, 2271, 2306, 2320, 2741, 5037, 5040, 5041, 5042, 5043, 5045, 6764, 7803, 8099
Spodoptera	NRCG 2845, Ah 5429, Ah 8429, NCAcs 17840, 17090, V 40, NFG 79, EC 21989 BL: IR 30, IR 17, PBSs 48, 105, 118, ICGV 86031 CVs: Kadiri 3, Girnar 1, BG 2

Table 4. Insect- pest resistance in wild *Arachis* species

Species	SP	HE	TH	JA	AP	SM	RN	PM	Pst	GR	PS
Section *Ambinervosae*											
A. sp GK 12943	-	HR	HR	I	-	S	-	-	-	-	-
A. sp GK 12946	-	HR	HR	HR	-	-	-	-	-	-	-
Section *Arachis*											
A. stenosperma HLK 410	-	HR	HR	HR	R	S	-	-	-	-	-
A. duranensis K 7988	-	HR	S	I	R	-	R	-	-	-	I
A. spegazzinii GKP 10038	-	HR	R	HR	-	S	-	-	-	-	R
A. batizocoi K 9484	-	HR	I	HR	R	-	-	-	-	-	-
A. khulmanii GK 30035	-	-	-	-	R	-	-	-	-	-	-
A. correntina GKP 9530	-	HR	HR	HR	R	S	-	R	-	-	R
A. chacoense GKP 10602	R	HR	HR	HR	R	S	-	R	R	-	R
A. cardenasii GKP 10017	-	HR	HR	HR	R	-	HR	R	I	-	S
A. villosa BU 22585	R	HR	R	HR	-	S	R	-	-	-	I
*A.*sp GK 30017	-	-	-	HR	-	-	-	-	-	I	-
Section *Erectoides*											
A. paraguariensis GKP 9646	R	S	-	R	R	-	-	-	R	-	R
A. appressipila GKP 9990	R	HR	HR	R	-	S	-	-	-	-	R
A. sp GKP 10573	-	HR	I	HR	-	MR	-	-	-	-	I
Section *Caulorhizae*											
A. repens GKP 10538	-	HR	I	HR	-	-	-	-	-	R	R

Contd...

Section *Extranervosae*											
A. macedoi GKP 10127	R	HR	I	I	R	MR	-	-	-	-	-
A. villosulicarpa	-	-	-	-	-	S	-	-	-	-	-
Section *Rhizomatosae*											
A. glabrata GKP 10127	R	HR	I	HR	-	-	HR	-	R	I	R
A. hagenbeckii HL 436	R	HR	R	I	-	-	HR	-	R	R	-
Section *Triseminalae*											
A. pusilla GKP 12922	R	HR	I	I	-	MR	-	R	S	-	-

SP-*Spodoptera*; HE-*Helicoverpa*; TH-Thrips; JA-Jassid; AP -Aphid; SM -Spider mite; RN -root knot nematode; PM -peanut mottle; Pst -Peanut stripe; GRV -Groundnut rosette; PS -Peanut stunt; I -Immune; HR - Highly resistant; R -Resistant; S -Susceptible (modified after Stalker and Moss 1987).

Knowledge on genetics of resistance to different insect pests are very much limited as compare to genetics of disease resistance. Hence, basic studies have to be initiated for all the insect pests. Simple and efficient screening methods for the entire insect pest have to be standardized. The mechanisms of resistance have to be thoroughly studied for all the insect pests. After identifying the resistance gene approaches have to standardized to incorporate through both conventional and biotechnological means. Regular survey and monitoring of pest in agro-climatic conditions. The regional importance of the major pests should be critically assessed both by survey and by appropriate on farm trials to estimate the available losses due to major pest in each region. The role of insects as vector of the virus diseases should be closely monitored to develop appropriate integrated virus-vector management strategy. The work on varietal mixture or varietal blending concepts should initiated to have multiple insect resistance multi-line varieties/blends. All the available germplasm in the country have to be screened for insect pests resistance and further more resistance sources have to be identified. Breeding for insect resistance has to be integrated with resistance to other biotic stresses and abiotic stresses. Further strengthening of research on insect pest management including breeding for insect resistance through AICRPO system has be done.

Sources of resistant:

Genotype	Catogory	Reference
Nematode		
Melloidogyne javanica		
Robut 33-1	MR	Joshi and Patel, 1990
M. arenaria		
Robut 33-1	R	Darkar and Jagdale, 1988

Leaf defoliator (*Spilosoma obliqua* and *Spodoptera litura*)

Genotype	Catogory	Reference
NCAc 343	R	Campbell *et al.* 1981;
ICGV 86031	R	Tiwari *et al.* 1988
M 13	R	
EC 36892	MR	
ICGV 86029	MR	
ICGV 86030	MR	
Chandra	MR	
NCAc 17090	MR	

Leaf miner

Genotype	Catogory	Reference
NCAc 343	R	Muthiah et al., 1992

Multiple resistances

Genotype	Category	Reference
NC 6, GP-NC 343, NCAC Nos. 01705, 02142, 02214, 02230, 02232, 02240, 02242, 02243, 02460	Resistance to more than one pests	Isleib *et al.* 1994.
ICG (FDRS) 4, ICG (FDRS) 10, ICGV 86590, GP-NC 343 and NC 6)	Tolerant to one or more insect pests besides resistance to foliar diseases.	Isleib *et al.* 1994.
ICGV87157	MR to leaf miner and leaf hopper	Nigam *et al.* 1992

References

AICORPO,. (1983-84). Annual report of all India coordinated research project on oilseeds. Directorate of Oilseeds Research, Hyderabad, India.

Amin, P.W. (1984). Methodology of screening for resistance to insect pests of groundnut. Patancheru, A.P. 502 324, India: International Crop Research Institute for Semi-Arid Tropics. 23 pp. (Limited distribution).

Amin and Mohammad, .(1980). Groundnut pest research at ICRISAT. 158-166. In Proc. Of international Workshop on groundnut, Patencheru, India.

Anderson, W.F., Beute, M.K., Wynne, J.C. and Wongkaew, S. (1990b). Statistical procedures for assessment of resistance in a multiple foliar disease complex of peanut. Phytopathology 80: 1451-1459.

Anderson, W.F., Patanothai, A., Wynne, J.C. and Gibbons, R.W. (1990a). Assessment of a diallel cross for multiple foliar pest resistance in peanut. Oleagineux 45: 373-378.

Arunachalam, V., Bandyopadhyay, A., Nigam, S.N., and Gibbons, R.W. (1984). Heterosis in relation to genetic divergence and specific combining ability in groundnut (*Arachis hypogaea* L.). Euphytica 33: 33-39.

Bandyopadhyay, A., V. Arunachanalam and K. Venkaih. (1985). Efficient selection intensity in early generation index selection in groundnut (*Arachis hypogaea* L.). Theoretical & Applied Genetics 71:300-304.

Bakhetia, D.R.C. (1980). Breeding for insects-pest resistance in Oilseeds crops. In Breeding Oilseeds Crops ed. Khem Singh Gill, 120-137.

Baujard, P., Mariny. B (1995). Ecology and pathogenicity of four trichodorid species from the semi-arid region of West Africa. Nematologica. 41: 1, 98-105.

Beck, S.D. (1965). Resistance of plants to Insects. Annu. Rev. Entomol. 10: 205-232.

Burow, M.D.; Simpson, C.E.; Paterson, A. H. and Starr, J. L. (1996). Identification of peanut (Arachis hypogaea L.) RAPD markers diagnostic of root-knot nematode (Meloidogyne arenaria (Neal) Chitwood) resistance. Molecular-Breeding. 2: 4, 369-379.

Campbell, W.V., Emery, D.A. and Wynne, J.E. (1976). Resistance of peanut tp potato leaf hopper. Peanut Science 3:40-43.

Campbell, W.V., Wynne, J.C. and Stalker, H.T.(1981). Screening groundnut for heliothis resistance. Proceeding of the International Workshop on *Heliothis* Management held at the ICRISAT center, Patancheru, Andhra Pradesh, India, during 15-20 Nov. 1981.

Chander, H.; Bakhetia, D.R.C.; Sehgal, R.C.; Berry, S.K. (1993). Evaluation of some varieties of groundnut for resistance against rust red flour beetle, *Tribolium castaneum* (Herbst) in storage. Journal of Insect Science 6(2): 219-222.

Cho, H.J., Han, S.C.; Choi, D.R. (1986). Screening peanut, pepper, cucumber, and tomato varieties for resistance to root-knot nematode, *Meloidogyne hapla.* (Ko). Research Reports, Office of the Rural Development, S. Korea, Plant Environment, Mycology and Farm Products Utilization 28(1): 94-97.

Corley Holbrook, C. (1990). Utility of early generation analysis for predicting parental potential for yield components in Peanuts. Peanut Sci. 17: 9-11.

Dharne, P.K. and Patel, S.K. (2000). Screening of groundnut genotypes for their reaction to *Spodoptera litura.* International Arachis Newsletter. No. 20:67-69.

Dickson, D.W., Oostengorp, M., Mitchell, D.J., Gommers, F.J., (ed.) Maas, P.W. (1990). Development of *Pasteuria penetrans* on *Meloidogyne arenaria* race 1 in the field. Nematology from molecule to ecosystem: Proceedings Second International Nematology Congress, 11-17 August 1990, Veldhoven, the Netherlands. 1992, 213-218.

Durgaprasad, M.M.K., V. Arunachalam and, Bandyopadhyay, A. (1985). Diversity pattern elucidating choice of parents for hybridization in varieties of groundnut (*Arachis hypogaea* L.). Trop. Agric. 62: 237-42.

Dutta, M., V. Arunachalam and, Bandyopadhyay, A. (1987). Enhanced cross-pollination to widen the scope of breeding in groundnut (*Arachis hypogaea* l.) Theory Appl. Genet. 74: 466-470.

Dwivedi, S. L. Amin, P.W. Rasheedunisa, Nigam, S.N., Nagabhushanam, G.V.S., Rao, V.R. and Gibbons, R.W. (1986). Genetic analysis of trichome characters associated with resistance to jassids (Empoasca kerri Pruthi in Peanut. Peanut Science 13:15-18.

Fowden, G.M.,Johnie,N., and William, L.P. (1972). Resistance of plants to Insects. Adv. Agron. 24:187-253.

Gallun, R.L. and Khush, G.S. (1980). Genetic inter-relationship between hosts, plants and insects. J. Environ. Quality. 1(3): 259-263.

Gowda, M.V.C., Motagi, B.N., Sheshagiri, R., and Naidu, G.K. (2000). Hyper mutation for disease resistance in groundnut. DAE-BRNS symposium on the use of nuclear and molecular technique in crop improvement Dec. 6-8, 2000. Bhabha atomic Research Centre, Mumbai. pp 175-183.

Gowda, M.V.C., R.S. Bhat, and K.M.A. Khader. (1997). Comparison of single and multiple crosses for combining resistance of late leaf spot and productivity in groundnut (*Arachis hypogaea* L.). Crop-Improvement. 24: 55-60.

Gregory, W.C. (1965). Mutation frequency, magnitude of change and the probability of improvement in adaptation. The Use of Induced Mutations in Plant Breeding. Radiat. Bot., 5 (Suppl.): 429-441.

Grewal, P.S., Chhabra, H.K., Kaul, V.K. (1987). Screening of groundnut germplasm against root-knot nematode, *Meloidogyne arenaria*. (En) Indian Journal of Nematology 17(1): 151-152.

Harber, E. (1980). Types and classification of resistance. In: Breeding Plants Resistance to Insects, (Maxwell, F.G. and Jennings, P.R. eds.) New York: John Wiley, pp. 15-21.

Harber, E. (1972). Resistance of plants to Insects. Agris. Sci. Rev. 10(2): 1-18.

Holbrook, C.C., Noe, J.P., Gorbet, D.W., Stephenson, M.G. (1998). Evaluation of peanut breeding lines with resistance to the peanut root-knot nematode. Crop Science 38 (1) : 260-262.

Holbrook, C.C. and Noe, J.P.(1990). Resistance to *Meloidogyne* in *Arachis* spp. and the implications on development of resistance peanut cultivars. Peanut Science 17: 35-38.

Holbrook-CC; Noe-JP. 1992. Resistance to the peanut root-knot nematode (*Mèloidogyne arenaria*) in *Arachis hypogaea*. Peanut-Science. 1992, 19: 1, 35-37.

Kalaimani, S., Mahadevan N.R., Manoharan V. and Sethupathi Ramalingam, R. (1989). Screening of inter-specific derivatives for leaf miner, (*Aproacerema modicella* Deventer) resistance in groundnut. International *Arachis* Newsletter No.5:15-16.

Kale, D.M., Badigannavar, A.M., and Murty, G.S.S.(2000). Forty years of mutation research in groundnut at trombay: accomplishment and potentials. DAE-BRNS symposium on the use of nuclear and molecular technique in crop improvement Dec. 6-8, 2000. Bhabha atomic Research Centre, Mumbai. pp 185-193.

Kalra, V.K., Singal, S.K. (1991). Screening of groundnut, *Arachis hypogaea Linn*. cultivars/ varieties against red flour beetle, *Tribolium castaneum* (Herbst).. (En). Journal of Insect Science 4(1): 80-81.

Kalra, V.K., Singal, S.K. (1991). Screening of groundnut, *Arachis hypogaea* Linn. cultivars/varieties against red flour beetle, *Tribolium castaneum* (Herbst). Journal of Insect Science 4(1): 80-81.

Kinzer, D.R., Pitts, J.T., Walton, R.R. and Kirby, J.S. (1973). Thrips resistance in plant introductions and in selections made for peanut improvement in Oklahoma. J. of Economic Entomology. 66:91-95.

Kogan, M. and Ortman, E.E. (1978). Antixenosis - A new term proposed to define Painter's 'nonpreference' modality of resistance. Bull. Entomol. Soc. Am. 24: 175-176.

Lorimar, N. Genetic means for controlling agricultural insects In CRC handbook of pest management Vol II.

Mahadevan, N.R., Manoharan, V., and Ramalingam, R.S. (1989). Field screening of groundnut genotypes for resistance to leaf miner, *Aproaerema modicella*. International Arachis Newsletter 5: 14-15.

Mahadevan, N.R., R. Sethupathi Ramalingam, and V. Manoharan, V. (1988). Screening for resistance to groundnut leaf-miner, *Aproacerema modicella* Deventer. International *Arachis* Newsletter No.4: 20-21.

Mathur, R.K., Manivel, P., Samdur, M.Y., Gor, H.K. and Chikani, B.M. (2000). Creation of genetic variability through mutation breeding in groundnut. DAE-BRNS symposium on the use of nuclear and molecular technique in crop improvement Dec. 6-8, 2000. Bhabha atomic Research Centre, Mumbai. pp 203-213.

McColloch, J.W. (1923). The resistance of plants to insect injury. Kans. State Hort. Soc. Bien. Rpt. 37: 196-208.

Mehan V.K.; Reddy D.D.R. and McDonald D. (1993). Resistance in groundnut genotypes to Kalahasti malady caused by the stunt nematode, *Tylenchorhynchus brevilineatus*. International-Journal-of-Pest-Management. 39: 2, 201-203.

Misari, S.M.; Ansa, O.A.; Demski, J.W.; Kuhn, C.W.; Casper, R.; Breyel, E. 1987. Reaction of peanut genotypes to the rosette virus and its vector *Aphis craccivora* Koch. (En) Proceedings, American Peanut Research and Education Society 19: 29. Abstract only.

Mishra, P.R.; Mukherjee, S.K.; Sontakke, B.K.; Dash, A.N. 1997. Screening of groundnut varieties/cultures against red flour beetle, *Tribolium castaneum*. Annals of Plant Protection Sciences 5(2): 212-213.

Mumford, E.P. (1931). Strategies in certain factors affecting the resistance of plants to insects pests. Science 73: 49-50.

Nandagopal, V., and Soni, V.C. (1994). A simple method to determine feeding and oviposition bahaviour of groundnut jassids (*Balclutha hortensis* Lindb). Journal of Oilseeds Research. 11 (1) :31-33 pp.

Nigam, S.N., Dwivedi, S.L., and Gibbons, R.W. (1991). Groundnut breeding: Constraints, achievements, and future possibilities. Plant Breeding Abstracts. 61:1127-1136.

Noe J.P., Holbrook C.C., and Minton, N.A. (1992). Field evaluation of susceptibility to *Meloidogyne arenaria* in *Arachis hypogaea* plant introductions. Journal-of-Nematology. 24: 4, Supplement, 712-716.

Norden, A.J. (1977). Genetic basis for utilizing early generation multiline cultivars in peanut breeding. *Amer.Soc.Agron. Abstr*. pp. 62-70.

Norden, A.J. (1980). Peanut. In hybridization of crop plants. Amer. Soc.Agron.-Crop Sci. Soc.. Amer., Medison, Wisconsin, pp 443-456.

NRCG. (1995). Annual Report, National Research Centre for groundnut, (ICAR), Junagadh, Gujarat 36201, India.

NRCG.(1996). Annual Report, National Research Centre for groundnut, (ICAR), Junagadh, Gujarat 36201, India.

NRCG. (1997). Annual Report, National Research Centre for groundnut, (ICAR), Junagadh, Gujarat 36201, India.

NRCG. (2000). Annual Report, National Research Centre for groundnut, (ICAR), Junagadh, Gujarat 362001, India.

Owens, J.C. (1975). An explanation of terms used in insects resistance of plants. Iowa. State J. Res. 49: 513-517.

Padgham, D.E.; Kimmins, F.M.; Barnett, E.A.; Wightman, J.A.; Rao, G.V.R.; Grayer, R.J.; Murant, A.F. (1990). Resistance in groundnut and its wild relatives to *Aphis craccivora*, and its relevance to groundnut rosette disease management. (En) Pages: 191-196 in Brighton Crop Protection Conference, Pests and Diseases - 1990. Vol. 1.. Thornton Heath, UK: British Crop Protection Council.

Painter, R.H. (1951). Insects Resistance in Crop plants. The Mcmillan Company, New York.

Painter, R.H. (1958). Resistance of plants to Insects. Ann. Rev. Entomol. 3: 257-290.

Painter, R.H. (1968). Crops that resist insects provide a way to increase world food supply. Kans. State Agric. Exp. Stn. Bull. 520: 22pp.

Prasad, M.V.R. (1993).Genetic enhancement of groundnut for higher productivity based on canopy development. (En). Pages: 2-3 in Extended summaries. National Seminar on Oilseeds Research and Development in India: Status and Strategies, Hyderabad, Andhra Pradesh, India, 2-5 Aug 1993. Hyderabad, Andhra Pradesh, India;: Indian Society of Oil Seeds Research;. (Summary: En). Summary only [633.85(540) NAT93 E 60929].

Rai, K. (1976). Pests of oilseed crops in India and their control. New Delhi, India: Indian Council of Agricultural Research. 121 pp.

Rajagopal, K.; Nandagopal, N.; Bhagat, N. R. (1988). Preliminary screening of Virginia groundnut genotypes against tobacco caterpillar, *Spodoptera litura* (F.). (En) Journal of Oilseeds Research 5(2): 162-165.

Rajgopal, K. and Bandyopadhyay.A. (1999). Elite groundnut germplasm - A ready reference. Technical Bulletin. NRCG, Junagadh. pp 35.

Raju, A.K.; Rao, P.R.M.; Rao, R.V.A; Azam, K.M.; Reddy, A.S. (1988). Influence of sowing time on the incidence of leaf miner of groundnut vs. yield. Journal of Research - APAU 16(2): 160-161.

Ranga Rao, G.V. and Wightman, J.A. (1997). Techniques for screening groundnut genotypes for resistance to insect pests , pp. 68-74 . In *Spodoptera litura* in India : Proceeding of the National Scientist Forum on *Spodoptera litura* (F.) organized from 2nd to 4th April, 1996 at International Crops Research Institute fot Semi- Arid tropics, Patancheru, Andhra Pradesh, India.

Reddy, P.S.(1983). Groundnut research under AICORPO- an over view. Paper presented at the group discussion on management of Pests and Diseases of Groundnut with special reference to Resistance Breeding during 26-29 Sept. 1983 at the ICRISAT, Patancheru, Andhra Pradesh, pp 10.

Reddy, L.J. and Mc Donald. D. (1984). 'Research strategies for increased groundnut production in India'. Paper presented at the National Seminar on 'Strategies for Increased Production of Iolseeds and Edible Oils', organized by Nastional Productive Council, New Delhi, during 6-8 February 1984, pp23.

Reddy, K.H.P., Reddy, G.L.K. Reddy, K.R. (1993). Comparison of pedigree and bulk methods in F6 generation of groundnut (*Arachis hypogaea* L.). (En). Annals of Agricultural Research 14: 119-121.

Samdur, M.Y., Nandagopal, V., Mathur, R.K., and Manivel, P. (1999). Evaluation of advanced breeding cultures of groundnut for resistance to thrips and aphids. *International Arachis Newsletter*. No. 19: 19-21.

Senguttuvan, T. and Sujata, K. (2000). Biochemical basis of resistance in groundnut against leaf miner. International Arachis Newsletter. No. 20: 69-71.

Sharma, M.L.; Singh, O.P.; Dharpure, S.R. (1989). New record of blue beetle, Cneorane sp., as a pest of groundnut in Satpura Plateau of Madhya Pradesh. Journal of Insect Science 2 (2) : 170. 2.

Sharma, S.B.; McDonald, D. (1990). Global status of nematode problems of groundnut, pigeonpea, chickpea, sorghum and pearl millet, and suggestions for future work.(ICRISAT), Patancheru 502 324, Andhra Pradesh, India.) (En) Crop Protection 9(6): 453-458. 12 .

Smith, O.D., and Boswell, T.E. (1980). Breeding for resistance to pod rot and lesion nematodes. Proceeding of the American Peanut Research and Education society, pp49-51.

Smith, C.M. (1989). (Not in text) Plant resistance to Insects: A fundamental approach. John Wiley and Sons, New York.

Smith, J.W. Jr., Posada, L. and Smith, O.D.(1980). Greenhouse evaluation of 490 peanut lines for resistance to the lesser cornstalk borer. Report of Texas Agricultural Experiment Station.; Texas; USA, 42 pp.

Snelling, R.O. (1941). Resistance of plants to insects attack. Bot. Rev. 7:543-586.

Stalker, H. T., Campbell, W. V. and Wynne, J. C. (1984). Evaluation of cultivated and wild peanut species for resistance to the lesser cornstalk borer (Lepidoptera: Pyralidae). *Journal of Economic Entomology*, 77: 1, 53-57.

Tiwari, S.N.; Rathore, Y.S.; Bhattacharya, A.K.; Sachan, G.C. (1988). (Department of Entomology, G.B. Pant University of Agriculture and Technology, Pantnagar, India).

Susceptibility of several varieties of groundnut to *Spilosoma obliqua* walker (Lepidoptera: Arctiidae). (En) Indian Journal of Entomology 50(2): 179-184.

Van der Plank, J.E. (1963). Plant diseases: Epidemic and Control. New York. Academic Press, 349pp.

Van der Plank, J.E. (1968). Disease Resistance in Plants. New York. Academic Press, 206pp.

Vasanthi R.P. and K. Padmavathamma (2000). Morphological and yield attributes of advanced breeding lines susceptible and resistance to *Spodoptera litura*. International *Arachis* Newsletter. No.20: 71-72.

Venkateswarulu, V. and Swamy, M.T. (1990). Leaf area damage by *Aproaerema modicella* Deventer - a pest of groundnut. International Arachis Newsletter 8: 20-21.

Venter,C., Waele,D.,de, Meyer,A.J., De,Waele,D. (1993). Reproductive and damage potential of Ditylenchus destructor on six peanut cultivars. Journal-of-Nematology. 25: 1, 59-62.

Varman, P.V., Kennedy, F.J.S. and Raveendran, T.S. (1989). Inheritance of leaf hopper (*Empoasca kerri* Pruthi) resistance in groundnut (*Arachis hypogaea* L.). *Journal of Oilseeds Research*, 6(2): 364-368

Visalakshi, M. (1997) Non-preference mechanism of resistance in groundnut cultures to the leaf miner *Aproaerema modicella* Deventer. J.-Oilseeds-Res. 14 (2): 265-268.

Wheatley, A. R. D., Wightman, J. A.,Williams, J. H., Wheatley, S. J. (1989). The influence of drought stress on the distribution of insects on four groundnut genotypes grown near Hyderabad, India. Bulletin of Entomological Research 79 (4) : 567-577.

Wiseman, B.R. (1985). Types of mechanisms of host plant resistance to insects attack. Insect Sci. Applic. 6: 239-242.

Wynne J.C. (1976). Evaluation of early generation testing in peanuts. Peanut Sci. 3: 62-66.

Wynne, J.C., Campbell, W.V., Emery, D.A. and Mozingo, R.W. (1977). NC 6 a southern corn root worm-resistant peanut variety. Bulletin of Agricultural Experiment Station, North Carolina State University, No.458, 14pp.

□□□

Pesticide Application Techniques

V. NANDAGOPAL, M.V.GEDIA AND P.C. RATH

24.1. Introduction

Chemical pesticides used for control of insect pests, diseases or weeds require to be applied in such a way that they should come in direct contact with the pests and distributed evenly over the plant surface to form a protective covering. The active ingredients are seldom applied as such. They are mixed with a diluents which may be solid or dust or liquid. The pest control is dependent on many factors such as pest, weather conditions prevailing, nature of pesticides, type of appliances used etc. A pesticide needs to be applied to a particular "Target" area occupied by pests. Therefore, the spray application method essentially requires:

- Thorough knowledge of the target pest
- Where, when and what amount of pesticides to be applied
- Correct droplet size, density and distribution.

There are three methods of pesticide applications:

(1) *Foliar application:* Pesticide is applied on the leaves of plant covering it thoroughly at a time.

(2) *Direct application:* Pesticide is directed on the weeds in the vicinity of plant.

(3) *Band application:* Application of pesticide to a band or strips of soil where the crop will be grown.

(4) *Spot treatment:* Application of pesticide on a particular portion of the plant. Pesticides are available in various forms. Application equipments are designed according to the type of formulations to be sprayed. The appliances are available in variety of sizes ranging from small to big keeping in view the application capacity and sources of energy.

The forms of pesticides are: Soluble powder, wettable powder, emulsifiable concentrate, soluble liquid, flowable concentrate, dust, granules, pallets etc. For the purpose of even application of these formulation there are devices designed which include 1) Sprayer 2) duster and 3) others types. The application equipments may be either manually or power operated. They are further classified as moveable or portable. The sprayer or dusters requires energy and depending upon the source of energy-either power operated and manually operated. Similarly, the energy may be classified to a) Hydraulic b) Gaseous and c) Centrifugal.

In the case of gaseous energy power sprayers which are low volume, low pressure, work on the blowing of a strong current of air towards the liquid coming out of the sprayer, which atomizes the liquid. Droplets produced are vary in size from 100-400 ml VMD (Volume median diameter), ex., Motorized knapsack sprayer. The centrifugal energy sprayer are designed to produce fine drops (50-100 ml) in the form of mists and therefore, known as mist blowers.

24.2. Nozzles

Nozzles are specially designed apertures to break up the liquid coming out of the spray tank into fine droplets. The nozzles are of different types suited to different types of work include a) Hydraulic energy b) Gaseous energy c) Centrifugal energy d) Kinetic energy and e) Thermal energy.

A typical spray nozzle consists of a base, a screw cap, a disc or orifice plate, a washer and a whirl plate and with or without a strainer. Depending upon the design of the sprayer and of the nozzle, the size of droplets varies and its forms the variable spray pattern. The nozzle are of different design for different rates of discharge and for pressure or high pressure.

The hollow nozzles are most commonly used in crop pest disease control. In this type the fluid is rotated by a whirl plate (swirl plate) by means of slanting holes in it or a screw thread on it. The liquid passes out as a conical sheet which soon beak up into droplets forming a hollow cone pattern of spray.

In Low volume nozzle, the spiral grooves in the whirl plate are very fine, orifice is smaller with a greatly reduce space between the body of nozzle and the whirl plate. Such a nozzle built-in strainer to prevent the entry of course and dust particles. Depending upon the quantity of spray fluid required per unit area the sprays are described as high volume spray, low volume spray and ultra-low volume spray. Similarly, spray liquid required to cover a hectare of field crops is 450 to 1000 liters if it were tobe a high volume sprayer, 12 to 125 liters if low volume sprayer and 0.5 to 6.0 lit./ha if it were to be a Ultra-low volume sprayer. In low volume spray the toxicant is increased by 8.25 times with corresponding decrease in total quantity of spray fluid. The droplet size varies from 70 to 150 microns, whereas in U.L.V sprays the droplet size varies from 20 to 150 microns.

Different types of spray patterns can be obtained with the use of nozzles and sprayers. The classification of pesticidal particles by their size is given in table 24.1.

Table 24.1. The classification of pesticidal particles by their size

Spray type	Equipment used	Particle size on discharge (micron)
Sprayers		
Coarser sprays	Hydraulic sprayer	More than 400
Fine sprays	Hydraulic sprayer	10 to 400
Blower & mist blowers		
Mists	Mist blowers	50 to 100
Aerosols generator		
Fog	Aerosol generator	0.1 to 50
Smokes	Smoke generators	0.001 to 0.1
Vapour	Vaporizers	Less than 0.001

The atomization devices used commercially in agriculture produce particle size ranging from 1 to 2000 micron (The droplet size is 3000 micron in case of a cloud burst and 2000 micron in heavy rain and in spray with hydraulic sprayer with large jets). The particle size may be indicated by "Mass median diameter (MMD), Numerical Average Diameter (NMD) or Surface Mean Diameter (SMD).

MMD: Mass median diameter also called mass average diameter, is the droplet diameter that satisfies the condition that half the spray volume is of large droplets and half is of smaller droplets.

NMD: Numerical average diameter is most easily obtained and is found by adding diameter of all droplets and dividing by the number of droplets. It is always less than the MMD.

SMD: The SMD is a theoretical diameter in microns of a single droplet with the same ratio of surface to volume as the entire spray. The SMD is smaller than MMD being between one half and three quarters of it depending on droplet size and distribution. The most important factor is the optimum size for spray droplets, which can be explained as that gives the maximum pest control with minimum pesticide and minimum contamination of ecosystem.

Calibration: In order to ensure uniform application of pesticides on crops, it is essential to carry out sprayer calibration exercise before under taking actual spraying work. Required spray volume can also be worked out by sprayer calibration.

The spray volume can be found out theoretically by using formula:

Application rate in litre per acre or hectare = (**Constant figure 495 british or 600 metric** x **Nozzle discharge rate in liter per minute**) / (**Effective swath width in feed or meters** x **Spraying speed in mile or K.M./ha**)

However, practical method known as area/volume (volume used or on marked area) is easier for a normal farmer to follow.

Application rate in litre/acre or hectare = (**Plot area** / **Marked area**) x **Spray volume used for marked area**

24.2.1. Some of the abbreviations used in the pesticide use

Pesticides are used in different forms i.e., liquid, powder, granules etc,. There are number of forms with which they are used. The different types of forms are given below:

AB = Grain Bait
AE = Aerosol Dispenser
AI = Active ingredient
AL = Other liquids to be applied undiluted
AP = Other powder
BB = Block bait
BR = Briquette
CB = Bait concentrate
CG = Encapsulated Granule
CR = Slow/Controlled Released granules
CS = Capsule Suspension
DC = Dispersible Concentrate
DP = Dispersible Powder
DS = Powder for Dry seed treatment
EC = Emulsifiable Concentrate
ED = Electro-chargeable Liquid
EG = Emulsifiable granule
EO = Emulsion, water in Oil
EW = Emulson, Oil in water
ES = Emulsion for Seed Treatment

FD = Smoke tin
FG = Fine granules
FK = Smoke candle
FP = Smoke cartridge
FR = Smoke rodlet
FS = Flowable concentrate for seed treatment
FT = Smoke tablet
FU = Smoke Generator
FW = Smoke pellet
G = Granuals
GA = Gas
GB = Granular bait
GE = Gas Generating Product
GC = Macrogranule
GL = Emulsifiable gel
GP = Flow dust
GR = Granule
GS = Grease
GW = Water soluble gel
HN = Hot Fogging Concentrate
KK = Combi-Pack Solid/Liquid
KL = Combi-Pack Liquid/Liquid
KN = Cold Fogging Concentrate
KP = Combi-Pack Solid/Solid
LA = Lacquer
LS = Solution for seed treatment
LV = Low volume
LVC = Low Volume Concentration
MG = Microgranule
OF = Oil miscible flowable concentrate
OL = Oil Miscible Liquid
OP = Oil Dispersible Powder
QF = Oil Miscible Flowable Concentrate
PA = Paste
PB = Plate bait
PC = Gel or paste concentrate
PO = Pour- on
PR = Plant rodlet

PS = Seed coated with a pesticide
RB = Bait
SA = Spot on
SB = Scrap bait
SC = Suspension Concentrate
SE = Suspo-emulsion
SG = Water Soluble Granule
SL = Soluble concentrate
SO = Spreading oil
SP = Water Soluble Powder
SS = Wettable Soluble Powder for Seed Treatment
SU = Ultra- low volume (ULV) suspension
TB = Tablet
TC = Technical material
TK = Technical concentrate
TP = Tracking powder
UL = Ultra Low Volume Liquid
ULV = Ultra Low Volume
VP = Vapour releasing product
WDP= Water dispersible powder
WG = Water Dispersible Granule
WP = Wettable Powder
WS = Water Dispersible Powder for Slurry Treatment
WSC = Water soluble concentrate
XX = others

24.3. Storage, handling and disposal of insecticide (Reddy *et al.*, 1990)

All insecticide are poisons. They are formulated to cause mortality/ deformity of the harmful insects. However, if misused, they may also harm animals and human being. Therefore, one should be aware of precautions to be observed in handling the insecticides. Following are the precautionary measures to be followed while handling pesticides.

24.3.1. Disposal of diluted insecticidal solutions

Any insecticidal solution that remains after spraying should be sprayed in the waste land where it will not be harmful. Never drain it into a stream or in irrigation or in drainage canals.

24.3.2. Disposal of insecticide container

Wash out the insecticide and add the washing to the spray tank before completely filling the tank with water. Store the wash container in the safe place. Punch hole in the metal container to prevent their reuse, and give or sell them to a metal merchant. Container such as card board boxes, paper sacks, polyethylene bags and other similar bags should be burnt and ashes buried.

24.3.3. Storing of insecticide

(a) Store all insecticide in original containers under lock and key, out of reach of children, pets and livestock

(b) Never transfer the insecticide to other containers

(c) Keep all insecticide away from food and feed stuff

(d) Discard any insecticide container without label or with damaged labels. Do not guess at contents

(e) Avoid exposing insecticide to extreme temperatures and to high humidity

(f) Make certain that partly used container are tightly closed and clearly labeled

(g) If spraying is to be stopped because of change in the weather, any unused dilute material should be removed from the sprayer and stored temporarily in a suitable non-corrosible container, clearly labeled with the name of the chemical, the extent to which it has been diluted and the date of preparation

24.4. Safety precautions for handling insecticide

(a) Always read the label on the chemical before using. Note warning and cautions each time before opening the container

(b) Keep sprays and dusts out of reach of children, pets and irresponsible people. They should be stored outside the home, away from food and feed. Herbicide should not be stored near seeds

(c) Always store spray and dusts in the original containers and keep them tightly closed

(d) Never smoke or chew while spraying or dusting or handling pesticides

(e) Avoid inhaling sprays or dust. If directed on label, wear protective clothing or mask

(f) Do not spill sprays or dust on the skin or clothing. If they are spilled, remove contaminated clothing immediately and wash clothing

(g) Wash hands and face, and change to clean clothing after spraying or dusting. Also wash clothing each day before reuse

(h) Use separate equipment for applying hormone type herbicides

(i) If symptoms of illness occur during or shortly after spraying or dusting, call the physician or take the patient to the hospital immediately

24.5. Emergency treatment for Poisoning By Pesticides

To using agriculture chemicals, reading the label is of utmost importance. In spite of all the warning and caution indicated on the label, poisoning of individual do occur because of mechanical accidents or carelessness. Poisoning may also be caused when the pesticide is mistaken for some harmless substance as may happen when pesticide is stored improperly. Regardless of how the poisoning occurs, remedial action should be taken immediately. When poisoning occur following measures are suggested.

- Call a physician if one can be reached easily. Describe the condition of the patient and if known, the nature of the poison and whether it was swallowed, inhaled or spilled on the skin or eyes. The doctors will advice what to do until he arrives.
- If you cannot reach the physician, the first thing to do is to remove the causes of the poisoning. If the poison has been swallowed, induce vomiting by putting the finger or a tongue depressor down the patient's throat until he vomits. Never try to make an unconscious person vomit.
- Soothe the throat and stomach and irritated membranes. Give raw eggs, milk or a thin flour paste after the stomach has been emptied. These also serve to absorb poisons.
- Give an appropriate antidote. Antidote is a substance that relieves, prevents or counter acts the insecticide concerned. Read the label for specific antidote. In general, for chlorinated hydrocarbon insecticide (eg. carbofuran, carbaryl), perform a gastric lavage and gives saline laxative. The physician may give **Phenobarbital** or **pentobarbital** to control convulsions. For organophospherous insecticides (eg. monocrotophos, dimethoate), give **atropine sulphate** or **protopam chloride** (Pralidoximechloride and 2-PAM)
- If poison has been inhaled, remove patient to open air. Give artificial respiration if patient is not breathing. Mouth to mouth sucking is quickest. Give appropriate antidote as suggested above
- If the poison has been spilled on the skin or eyes or clothing, remove the contaminated clothing and wash skin and eyes with plenty of water with or without soap.

1	✓	Aldrin																
2	✓	✓	Carbaryl															
3	✓	x	✓	Dichlorovos														
4	✓	✓	✓	✓	Dimethoate													
5	✓	?	✓	✓	✓	Endosulfan												
6	✓	✓	✓	✓	✓	✓	Fenitrothion											
7	?	✓	?	✓	✓	?	✓	Lindane										
8	✓	✓	✓	✓	✓	✓	?	✓	Malathion									
9	?	✓	?	✓	✓	?	✓	✓	✓	Metsystox								
10	✓	x	✓	✓	✓	✓	?	✓	✓	✓	Monocrotophos							
11	✓	✓	✓	*	✓	✓	?	✓	?	✓	✓	Phosphamidon						
12	?	✓	?	?	✓	?	?	!	✓	✓	?	✓	Benomyl					
13	✓	✓	✓	✓	✓	0	✓	!	0	✓	✓	✓	✓	Captan				
14	?	✓	?	?	✓	✓	?	!	✓	✓	✓	✓	✓	✓	Cabendazim			
15	✓	✓	✓	✓	✓	0	✓	!	0	✓	✓	✓	✓	✓	✓	Difolitan		
16	✓	✓	✓	✓	✓	✓	✓	✓	✓	✓	✓	✓	✓	✓	✓	✓	Moncozib	
17	✓	✓	✓	✓	✓	✓	✓	✓	✓	✓	✓	✓	✓	✓	✓	✓	✓	Thiram

✓ = Compatible; x= Non-compatible;?=Information not available;*= Not normally mixed ;0= Compatible but should be used immediately; != Use wettable powder

Fig. 1. Compatibility chart for commonly used pesticides in groundnut

Table 24.2. Compatibility of insecticides v/s fungicides

Insecicides	Benomyl	Carbendazim	Captafol	DithanM 45	DithaneZ 78	Thiophanate Metyl	Thirum	Captan	Tridemorph
Endosulfan		C			C				
Chlorpyriphos		C	C		C			C	
Dichlorvos		C			C				
Dimethoate	I		C	C	C			C	
Malathion	C	C		C	C				
Methyl demeton		C	C		C			C	
Monocrotophos		C	C	I	C	C	C	C	
Phosphamidon		C	C	C	C				
Phosalone				C	C				
Profenophos					C				
Quinalphos				C	C				
Fenvalerate	I	C	C		C				
Carbaryl			C		C	C			
Carbofuran							C	C	

The compatibility chart of commonly used pesticide in groundnut is given vide fig.1 and Tables 24.2 & 3.

24.6. Basic precautions in pesticide usage (Anon, 1997)

24.6.1. Purchase

- Purchase only just required quantity for single application in specified area.

Table 24.3. Compatibility of insecticides v/s bio-pesticides

Insecicides	*Bacillus thuringiensis*	*Beauveria bassiana*	*Helicoverpa armigera* NPV	*Spodoptera litura* NPV
Acephat	C			
Carbaryl	C	C		
Carbofuran	C			
Chlorpyriphos				
Dimethoate	C			
Endosulfan	C		C	C
Fenvalerate				C
Methyl demeton	I			
Monocrotophos	C			
Phorat				
Phosalone	C			
Phosphamidon	C			
Quinalphos	C			

- Do not purchase leaking containers, loose, unsealed or torn bags.
- Do not purchase pesticides without proper/ approved labels.

24.6.2. Storage

- Avoid storage of pesticides in house premises.
- Keep only in original container with intact seal.
- Do not transfer pesticides to other containers.
- Never keep them together with food or feed/fodder.
- Keep away from reach of children and livestock.
- Do not expose to sun light or rain water.
- Do not store weedicide along with other pesticides.

24.6.3. Handling

- Never carry/ transport pesticides along with food materials.
- Avoid carrying bulk – pesticides (dust/ granules) on head, shoulders or on the back.

24.6.4. Precautions for preparing spray solution

- Use clean water
- Always protect your nose, eyes , mouth, ears and hands
- Use hand gloves, face masks and cover your hands with cap

- Use polythene bags as hand gloves, handkerchiefs or piece of clean cloth as mask and cap or towel to cover the heads. (Do not use polythene bag contaminated with pesticides)
- Read the label on the container before preparing spray solution
- Prepare spray solution as per requirement only
- Do not mix granules with water
- Concentrated pesticides must not fall on hands etc., while opening sealed containers. Do not smell the pesticides
- Avoid spilling of pesticide solution while filling the sprayer tank
- Do not eat, drink, smoke or chew while preparing solution

  The operator should protect his bare feet and hands with polythene bags.

24.6.5 Equipment

- Select right kind of equipment
- Do not use leaky, defective equipment
- Select right kind of nozzle
- Do not blow/clean clogged nozzle with mouth. Use old tooth brush tied with the sprayer and clean with water
- Do not use same sprayer for weedicide and insecticide

24.6.6. Precautions for applying pesticides

- Apply only at recommended dose and dilution
- Do not apply on hot sunny day or strong windy condition
- Do not apply just before the rain and also after the rain
- Do not apply against the wind direction
- Emulsifiable concentrate formulations should not be used for spraying with battery operated ULV sprayer
- Wash the spryer and buckets etc. with soap water after spraying
- Containers, buckets etc. used for mixing pesticides should not be used for domestic purposes
- Avoid entry of animals and workers in the fields immediately after spraying

24.6.7. *Disposal*

- Left over spray solution should not be drained in ponds or water lines etc. Throw it in barren isolated area
- The used/empty containers should be crushed with a stone/stick and buried deep in to soil away from water sources
- Never re-use empty pesticide container for any other purpose

24.6.8. *Other Precautions*

24.6.8.1. Seed treatment

- For seed dressing, use either metal seed dresser/earthen pots or polyethylene bags
- After seed treatment, do not open lid/cover of the polybag/earthen pot immediately to avoid inhalation of pesticide/fungicide
- Do not use left over treated seeds either for human consumption or as animal feed

24.6.8.2. Harvest

- Do not harvest fruits, vegetables immediately after pesticide application
- Observe safety intervals
- Do not spray/dip harvested produce in pesticide solution

24.6.8.3. Caution

- If operator feels giddiness, uneasy, he must discontinue spraying/dusting at once taken to a Doctor
- Operator should not spray/dust more than 4 hours at a stretch in a day
- Operator should not take up spray /dusting work with empty stomach
- Appendix deals with the safety parameters inter alias classification of toxicity as per Insecticide rules, 1971 WHO classification of hazards, colour of toxicity triangle, first aid measures, symptoms of poisoning and treatment of poisoning, the extension functionaries of the state department of agriculture have to make use of this information as under:
- Basic precautions which are required to be taken as per classification of toxicity as well as hazard criteria by WHO may be seen
- The extension functionaries are to educate the farmers on safety use of pesticides with the help of colour toxicity triangles as the farming community can follow the colour and corresponding safety precautions

- The symptoms of poisoning must be known to the extension functionaries to enable them to extend first aid measures to affected persons to the extent possible
- Basically, the information on first aid measures and treatment of poisoning is required to be passed on by the extension functionaries to the doctors at Primary Health Centers as well as to the Private Doctors in the nearest vicinity of spraying of pesticides
- Extension functionaries must ensure that names of common pesticides during plant protection measures along with a copy of the leaflet which is an integral part of a pesticide container must be made available to the doctors in the vicinity of plant protection operations
- Extension functionaries are to request the doctors to intervene in procurement of antidotes for different pesticides as cited "Treatment of Poisoning" (Annexure-I).

References

Anonymous (1997). Pesticides application for plant protection, Information bulletin ASPEE Research Institute, Redg. Office Plot No.1, B.J.Patel cross road No.2 Malad (West), Bombay, pp. 40.

Reddy, P.S.; M.P. Ghewande and T. Radhakrishan (1990). Training manual on groundnut production, Directorate of Extension, Dept of Agriculture and cooperation, Ministry of Agriculture, Govt. of India, New Delhi, pp.202.

Srivastva, K.P. (1996). A text book of applied Entomology. Vol:I. Kalyani Publishers, Ludhiyana. Pp 321.

Annexure - I contd...

Treatment of Poisoning (Annexure-I).:

Name of pesticide	Classification as per insecticides Rules, 1971	Colour of toxicity triangle	WHO classification by hazard	First-aid Measures	Symptoms of poisoning	Treatment of poisoning
Insecticides						
A. Chlorinated Hydrocarbons						
1. Endosulfan	Highly toxic	Yellow	Class I moderately hazardous	Remove the person from the contaminated environment	Nausea, vomiting, restlessness tremor, apprehension convulsions, coma, respiratory failure and death	Gastric lavage with 2-4 tap water-Catharsis with 30gm (10 oz) sodium sulphate in one cup of water
Endosulfan: Agrosulfan, Agirosulfan, Bangsulfan , Endocel, Endocid, Endocrop,Endohit, Endomil, Cacosulfan, Citisulfan, Endodhan, Endosol, Endostar, Endosulfan, Entacid, Fisulfan, Hexasulfan, Hildan, Hysulphan, Kemsulfan, Kilex Endosulfan, Lusulfan, Marvel, Parasulfan, Parrysulfan, Ramfan, Sicosulfan, Slasulfan, Sujadane, Tagsulfan, Thiokill, Thiotox, Vegfro, Vikasulfan (354 EC, 4 D)						
2. Dicofol	Mode-rately toxic	Blue	Class 1:1 slightly hazardous	In case of (a) skin contact-Remove all contaminated clothings and immediately wash with lot of water and soap, (b) Eye contamination-wash the eyes with plenty of cool clean water, (c) Inhalation-Carry the person to the open fresh air. Loosen the clothings around neck and ches, and (d) Ingestion-If the victim isfully conscious, induce vomiting by ticklingback of the throat. D not administer milk, alcohol and faty substances.		Bardbiturates in appropriate dosages repeated as necessary for restlessness or convulsions. Diazepam also can be given. Watch breathing closely, administer oxygen and/or artificial respiration, if needed.
BAnmite, Delcofol, Dicofol, Hilfol, Kelthane, Micothane, Taagfol (18 EC)						
				In case the person is unconscious, make sure the breathing passage is kept clear without any obstruction. Victim's head should be little lowered and face should be turned to one side in the lying down position.		Avoid oils orally, oil laxatives. Do not give any stimulants.
						Give calcium gluconate (10% in 10 ml ampules) intravenously every four hours
B. Organophosphorous compounds						
3. Monocrotophos	Extremely toxic	Red	Class Ib Highly hazardous	In case of breathing difficulty, give mouth to mouth or mouth to nose breathing.	Mild-Anorexia headache, dizziness, weakness anexiety, tremors of tongue and eyelids, miosis impairment of visual acuity	For extreme symptoms of OP poisoning, injection of atropine (2-4mg. For adults, 0.5-1.0 for children) is recommende, repeated at every 5-10 minute intervals until signs of atropinization occur
Monocrotophos Agrocron, Agromohare, Azodrin, Balwan, Crotocel, Cor ophose, Croton, Entofos, Glorephos, Hilcron, Hycrophos, Kagrofos, Kedet, Kilex -Monocrotophos, Luphos, Macrophos, Micophos, Monohit, Monocid, Milphos, Monocil, Monocron, Monocroto, Monocrotophos, Monodhan, Monogil, Monosect, Monosil, Monosul, Monostar, Nuv acron, Paracron, Parryfos, Phoskill, Ramphos, Rasayanaphos, Sicosil, Sufos, Swadeshi Monocrotofos, Trifos, Vegfru , Vegfrokadett, Vimonfos, Vimonofos (36 WSC, 36 SL)						

4. Phospha-midon	Extremely toxic	Red	Class Ia Extremely hazardous	External cardiac(heart) massage may be given if heart action ceases	Moderate – nausea salivation, lacrimation, abdominal cramp, vomiting sweating, slow pulse muscular tremors, miosis	Atropine injection – 2 to 4 mg. Repeat 2mg. When toxic symptoms begin to recur (15-16 minute intervals). Excessive salivation is a good sign indicating need for more atropine.
Phosphamidon: Phosphamidon, Agromidon, Agriphos, Agrophos, Bangdon, Cildon, Daron, Dimecron. Entocron, Kilrodon, Phamidon, Phoskil, Phos -sul, Sicomidon, Sudon, Sum idon, Vimidon (40 SL)						
5. Phosalone	Highly toxic	Yellow	Class II Moderately hazardous	Medical aid Take the patient to a Doctor/Primary Health Centre immediately along with the original container leaflet and table.	Severe – diarrhoea, pinpoint and non-reactive pupils, respiratory difficulty, pulmonary edema, cyanosis, loss of sphinchter, control, convulsions coma and heart block	Keep airways open. Do tracheotomy and give artificial respiration respiration if needed
Agrosalone, Microzons, Phosal, Phosalone, Sugalone, Zolone (4 D, 35 EC)						
6. Methyl-o-demeton (Oxydemetonme thyl)	Highly toxic	Yellow	Class Ib Highly hazardous			
Agrotox, Dhanusystox, Hexasystox, Himax, Knockout, Metasystox, Parasystox, Zentox (25 EC)						
7. Acephate	Mode-rately toxic	Blue	Class III Slightly hazardous	If there is no specific antidote, the doctor may treat the patient symptomatically		
Acephate: Acephate, Acefecx, Acemil, Agrophate, Asataf, Dhanraj, Lucid, Orthene, Starthene, Lancer, Torpedo, Tremor (25 EC, 75 WP)						
8. Triazophos	Highly toxic	Yellow	Class Ib Highly hazardous			If ingested lavage stomach with 5% sodium bicarbonate, if not vomiting. For skin contact, wash with soap and water (eyes-wash with isotonic saline). Wear rubber gloves while washing contact area.
Hostothion, Trifos, Trizocel (40 EC)						
9. Quinalphos	Highly toxic	Yellow	Class II Moderately hazardous			In addition to atropine give 2-PAM(2-pyridine aldoxime methiodide). Lg. And 0.25 g. for infants intravenously at a slow rate over a period of 5 minutes and administer again periodically as indicated. More than one injection may be required
Quinalphos: Agroquinal, Agriphos, Agroquin, Award, Celquin, Desalux, Dhanulux, Ekalux, Entolux, Gilquin, Hyquin, Kinalux, Keterphos, Kilex, Krush, Kinalux, Milux, Phalux, Quinal, Quinalphos, Quinaltaf, Quinotox, Ramlux, Sicophos, solux, Suquin, Smash, Starbrand, Tagquin, Vazra, Vikalux (1.5 D, 25 EC, 5G)						
10. Chlorphyri-phos	Highly toxic	Yellow	Class II Moderately toxic			Avoid morphine, theophyllin, aminophyllin barbiturates or phenothiazines
Agrophos, BAngspan, Chlorofos, Classic, Coroban, Danusban, Dhanvan, Durmet, Dursban, Gilphos, Hyban, Lethal, Rdar, Ruban, Stike, Suban, Sulban, Tricel (20EC)						
11. Phorate	Extremely toxic	Red	Class Ia Extremely hazardous			Do not give atropine to a cyanotic patient. Give artificial respiration first then administer atropine
Phorate: Aelmet, Agrophorate, Cilphorate, Delsan, Dragnet, Forate, Foratox, Fortan, Glorate, Granutex, Grenade, Helmet, Kamet, Luphate, Milate, Phorate , Phorate, Ramphorate, Starphor, Starphos, Thimet, Vegfro, Volphor (10 G)						

12. Dimethoate	Highly toxic	Yellow	Class II Moderately hazardous			
Dimethoate : Agrodimet, Agromat, Banggor, Cifro, Corothiate, Cropgor, Cygon, Devigon,Diadhan, Dimethoate, Dimor, Dimer, Dinex, Entogor, Hexagor, Hygro, Kilex Dimethoate, Kemithoate, Kiltex, Micro, Milgor, Nugor, Paragor, Parrydimate, Ramgor, Rogor, Sicothate, Sulgor, Swedeshi Dimethoate, Tagor, Tara 909, Vikagor **(30EC)**						
13. Profenofos	Highly toxic	Yellow	Table -III Moderately hazardous			
Aurifos, CArina, Celcron, Curacron, Profex **(50EC)**						
14. Methtyl parathion	Extremely toxic	Red	Class Ia Extremely hazardous			
Methyl Parathion: Agropara, Agrotox, Corocid, Devithion, Dhan umar, Entocid, Folidol, Hindol, Kagrodal, Kempar, Kildet, Kilex, Kolfos, luhtion, Milon, Metacid, Metapar, Methyl parathion, Metapar, M.Devithion, Paracrop, Parahit, Paramar, Paramet, Parasul, Parataf, Parathol, Paratox, Ramcid, Ramthio, Tagpar, Vegfro, Vikacid, Vikadol **(2D, 50EC, 46.7 EC)**						
C.Carbamates						
15. Carbofuran	Extremely toxic	Red	Class-Ib Highly hazardous		Constructionof pupils, salivation, profuse sweating, lassitude muscle incoordination, nausea, vomiting, diarrhoea epigastric pain, tightness in chest	Atropine injection 1-4 mg. Repeat 2mg. When toxic symptoms begin to recur (15–50 min. interval) Excessive salivation is a good sign indicating need for more atropine
Carbofuran, Furadan, Furad, Fury, Hexafuran, Legend, Vegfrodiafuran(3G)						
16. Thiodicarb	Highly toxic	Yellow	Table-III Slighly hazardous			Keep airway open. Do tracheotomy and give artificial respiration if needed.
Larvin, Skipper (75 WP)						
17. Carbaryl	Highly toxic	Yellow	Class-II Moderately hazardous			If ingested lavage stomach with 5% sodium bicarbonate, if not vomiting. For skin contact wash with soap and water (eyes-wash with isotonic saline) Wear rubber gloves while washing contact area
Agrovin, Agroryl, Bangvin, CArbamate, Carbaryl, CArvint, Corovin, Devicarb, Hexavin, Kervin, Kildiryl, Kilexcarbaryl, Sevidol, Sevimo, Sevin (5D, 10 D, 50 WDP, 80 WDP, 85 WDP, 85 S, 40 LV, 4 G						
						Give oxygen, if needed
						Give morphine, if needed
						Avoid theoplyllin and aminophyllin or barbiturates
						2-PAM and other oximes are not harmful and in fact contra indicated for routine usage
						Do not give atropipe to a cyanotic patient. Give artificial respiration first then administeratropine
D. Synthetic pyrethroids	Highly toxic – Mode-rately toxic	Yellow-Blue	Class Ib-III Highly hazardous slightly hazardous		Irritation lacrimation, salivation, ataxia, convulsions/seizure	No specific antidotes. Treatment is essentially symptomatic

18. Fenvalerate	Highly toxic -	Yellow-	Class II Moderately hazardous			
Fenvelrate: Agrocidin, Agrofen, Bangfenn, Capvalerate, Fencel, Fencid, Fenhit, Fenkill,Fenicidin, Fensul, Fenval,Fenis, Fenoron, Fenrio, Fenvelrate, Gilfen, Hilfen, Hyfen, Karofen, Lufen, Milfen, Parafen, Parryfen, Paushafen, Pensil, Ramfen, Rasayanfen, Sicofen, Starfen, Sujafen, Sumicidin, Tagfen, Trifen, Trimpheard, Valour, Vegfru , Vikafen, (20 EC)						
E. Urea derivatives						
19. Difluben-zuron	Mode-rately toxic	Blue	Table –5 (Unlikely to present acute hazards in normal use)		Irritation of eyes, nose, throat and skin	No specific antidotes. Treatment is essentially symptomatic
Dimilin (25 WP)						
F. Inorganic						
20. Sulphur	Slightly toxic	Green	Table -5 (Unlikely to present acute hazards in normal use)		Irritation of eyes, nose, throat and skin	No specific antidotes. Treatment is essentially symptomatic
Sulphotox, Wetsulf, Devisulfan, Hexasul, Sulfex (40, 80, 85 WP)						
G.Neem products#	Slightly toxic	Green	-		Nausea, Irritation	No specific antidotes. Treatment is essentially symptomatic
ACHOOK- 0.5% Az adiradion, Nimbocinol and Epinimbocinol; APHIDIN - Soapy humus extract; AZA - 0.15% Azadirachtin; AZADIT - 0.3% Azadirachtin; BIOSOL - Oil ; BIONEEM - 0.03% Azadirachtin; BIOPEST- 0.15% Azadirachtin; ECO -NEEM- 0.3% Azadirachtin; FIELD MARSHAL - 0.3% EC Azadira chtin; FORTUNA AZA 0.15% Azadirachtin; JAI NEEM (Developed by IARI) - 0.03% Azadirachtin; JAWAN - 0.03% Azadirachtin; JEEVAN CROP PROTECTOR - 0.05% Azadirachtin; JUERKEN - 0.03% Azadirachtin; LIMONOOL - 75% EC 0.3% Azadirachtin; MARGOCIDECK- 20% EC (Azadirach tin 0.15% w/w); MARGOSAN-0- 80% EC zadirachtin; NEEMARK - 0.03% Azadirachtin; NEEMAZAL-F- 5% Azadirachtin; NEEMGOLD- 0.15% Azadirachtin; NEEM GUARD- 0.03% Azadirachtin; NEEM RICH - 0.03% Azadirachtin; NEEMOLIN - 0.15% Azadirachtin; NEEMPOURN - 0.15% Azadirac htin; NEEMOSAN- 0.03% Azadirachtin; NEEMOX - 80% EC Azadirachtin; NEEMATA - 0.15% Azadirachtin; NEETHRIN - Oil 0.03% Azadirachtin; NIMBA - Wettable powder 0.15% Azadirachtin; NIMBCIDINE - 0.03% Azadirachtin; NIMBOSOL - 0.15% Azadirachtin; NIMBASAL- 0.15% Aza dirachtin; NIMLIN- 0.03% Azadirachtin; NEEM BASE - 20% EC 20 WDPAzadirachtin; NEEM HIT - 0.15% Azadirachtin; NEEM OIL - 0.03% EC Azadirachtin; NEEM PLUS- 0.15% Azadirachtin; NEEM TOP - 0.03% Azadirachtin; NEEMOL - 0.03% Azadirachtin; NIM -76- 0.05% Azadirachti n; NEEM OIL - 2% Azadirachtin; NEEMAX - Seed Powder 300 ppm Azadirachtin; NEEMOCIDE- 0.03% Azadirachtin; NEEMACTINE- 0.15% Azadirachtin; NIMBO BAS- 0.03% Azadirachtin; NIMBO BAS- 0.15% Azadirachtin; Nimbitor (ZA – 199)- 0.03% Azadirachtin; PHYTOWIN- 0.03% Azadirachtin;						
H. Others						
21. Imidacloprid	Highly toxic	Yellow	-		Apathy, Myatonia Tremor, difficult breathing, myospasms	There is no specific antiobte. Treatment symptomatically
Gaucho, Confidor, Imidacloprid, Tatamida						
I. Fungicides (Thiocarbamates)						
22. Carbendazim	Slightly toxic	Green	Table –5 (Unlikely to present acute hazards in normal use)		Haadache, palpitation Nausea, omiting, flushed face, irritation of nose, throat, eyes & skin	No specific antidotes. Treatment is essentially symptomatic
23. Captan	Slightly toxic	Green	Table –5 (Unlikely to present acute hazards in normal use)			

24. Thiram	Moderately toxic	Blue				
25. Metalaxyl Compound	Moderately toxic	Blue				
26. Mencozeb	Slightly toxic	Green	Table 5 (Unlikely to present acute hazard in normal use)			
27. Chlorothelonil	Slightly toxic	Green	Table 5 (Unlikely to present acute hazard in normal use)			
J:Herbicide						
28. Alachlor	Mode-rately toxic	Blue	Class-III slightly hazardous		Headache, nausea, irritation, diarrhoea etc.	No specific antidotes. Treatment is essentially symptomatic
29. Pendimethalin	Mode-rately toxic	Blue	Class-III slightly hazardous			
30. Fluchloralin	Mode-rately toxic	Blue	Class-III slightly hazardous			
31. Trifluralin	Mode-rately toxic	Blue	Table –5 (Unlikely to present acute hazards in normal use)			
32. Diuron	Mode-rately toxic	Blue	Table –5 (Unlikely to present acute hazards in normal use)			
K. Rodenticides						
33. Bromodialone	Extremely toxic	Red	Class-lb Highly hazardous		Nausea, omiting diarrhoea, bleeding from nose, gums, G.I.tract	Vitamin K1 to restore bood clotting after repeated prothrombin level determination
Bromodialone (0.25 SP, SL, 0.005 CAKE, 0.005 BAIT)						
34. Zinc phosphide	Extremely toxic	Red	Class-lb Highly hazardous			
Zinc phosphide, Zincox, Ratol, Agrophos (50 WP, G)						
35. Aluminium phosphide	Fumigent					
Al-phos, Celphos, Phostoxin, Quickphos (3 g Tablet)						

Forecasting, Forewarning and Modelling of Pests of Groundnut for Integrated Pest Management

T. P. TRIVEDI, D. K. DAS,
A .DHANDAPANI AND V. NANDAGOPAL

25.1. Introduction

Known as the **poor man's almond**, groundnut contributes about 38% to the oilseed pool of India. Groundnut occupies about 6.8 million ha with a production of 6.4 million tonnes. India is the largest grower of groundnut and the second largest producer after China. About 81% of the Indian groundnut crop is rain-fed where productivity fluctuates between 500 and 1500 kg/ha with a national average at about 1000 kg/ha. Andhra Pradesh, Gujarat, Karnataka, Maharashtra and Tamil Nadu account for about 89% of the total groundnut area and contribute nearly 88% of the total groundnut production of our country. The rest of the area are mainly scattered in the states of Orissa, Rajasthan, Punjab, Uttar Pradesh, and Madhya Pradesh. The potential non-traditional areas are, in the states of West Bengal and Assam and in the northeastern hills where groundnut can be grown in both the rainy (in the uplands and terraces) and the post-rainy seasons with a productivity range of as high as 2000-3000 kg/ha.

With an increasing limitation on resources, there is hardly any scope for improvement in horizontal production of groundnut. Hence, vertical growth by increasing crop productivity and intensity is the only alternative. To attain this goal, the biotic and abiotic constraints have to be tackled in a cost effective manner. Among the biotic stresses, insect-pests and diseases cause considerable yield losses leading to non-realization of the actual yields in groundnut as well as quality of the produce. However, either pest management options are either not properly timed or the resource-poor small and marginal farmers do not follow the pest management options, as the inputs are very expensive. To get around this problem, timely adoption of integrated pest management modules that are economically viable and eco-friendly should be available with a need based management option. Under these conditions, forewarning of the pest forms an essential tool for cost-effective and eco-friendly management of pests of any crop production system.

Forecasting is a subject which deals with viewing future scenario from the present /past situations to the future. The terms 'pest' means all biotic agents which adversely affect plant/crop production. This includes insects, mites, nematodes, weeds, bacteria, fungi, viruses, parasitic seed plants and vertebrates. Therefore, pest forecasting refers to viewing future scenario of any of the above mentioned biotic agents from present/past conditions of weather/crop as well as the population of that agent. Growth and development of any living being greatly influenced by the past and the prevailing weather conditions. **Therefore, any biological activity is influenced by physical weather parameters like temperature, rainfall, humidity, radiation and atmospheric pressure.** This is the basis of pest forecasting. Several workers have developed the forecasting systems for important pests of different crops (Trivedi *et al.*, 1999; Das *et al.*, 2001).

The Inter society Consortium for Plant Protection (ICPP) has defined integrated pest management as "the use of multiple tactics in a compatible manner to maintain pest population at levels below those causing economic injury while providing protection against hazards to humans, domestic animals, plants and the environment". The term 'integrated' means it is a broad interdisciplinary approach taken using scientific principles of plant protection of variety of management strategies and tactics to fuse into a single system.. The term 'management' indicates to keep the pest population below some level based on economic criteria. It does not mean complete eradication of entire pest population.

The present day crop protection is heavily relying on synthetic chemical control. The dependence on synthetic chemicals cause serious problems to the human and the ecosystems as well.

The problems are:

a) Insects developing resistance to insecticides

b) Insect resurgence

c) Bio-magnification of certain insecticides in food chain

d) Persistence of residues of some insecticides

e) Pollution of surface and ground water causing harm to aquatic ecosystems

f) Rejection of our agricultural products at foreign market

The adoption of IPM strategies has received the attention to minimize the above problems.

The suggestions for IPM strategy

a) Use of safe synthetic pesticides and botanical pesticides

b) Minimize use of pesticides in appropriate time based on pest phenology and weather information

c) Use of biological control agents like pathogens, parasites and predators

d) Use of resistant cultivars.

e) Adoption of appropriate cultural methods to have minimum pest problems.

f) Development of transgenic plants resistant to insect pest.

In general, IPM is to use a method or combination of methods in appropriate time to manage the pests so as to reduce the pest as well as pesticide residue problems.

To implement IPM, we require some information like:

a) Why a particular pest appears in some years ?

b) If it appears, at what time the population build up start?

c) When the population will reach ETL?

d) What would be the expected population during its peak activity?

e) Mode of carry over and off season survival.

Before developing a pest forecasting system or model, it is necessary to understand the interacting factors in a crop-pest ecosystem first. Fig.1 depicted the various factors, biotic and abiotic including space and food to be considered in a general pest management programme. Among these factors, climate is the most important one which influences all others and at the receiving end, pest is the factor which is influenced by all others. In the beginning, development of pest-weather relationship is essential as weather factors are the main driving components. Later proper understanding of each factor, its inter-relationship, integration with suitable techniques will lead to a successful pest forecasting as well as management programme.

There is a close relationship between insect pests of crops and weather parameters like temperature, rainfall, humidity, cloudiness, dew duration etc. Therefore, in terms of prevailing weather conditions, the incidence of several disease and insects can be forecasted. Viral and bacterial disease are more weather-dependant than insects due to the fact that viral and bacterial pathogens remain at fixed location, consistently exposed to a particular type

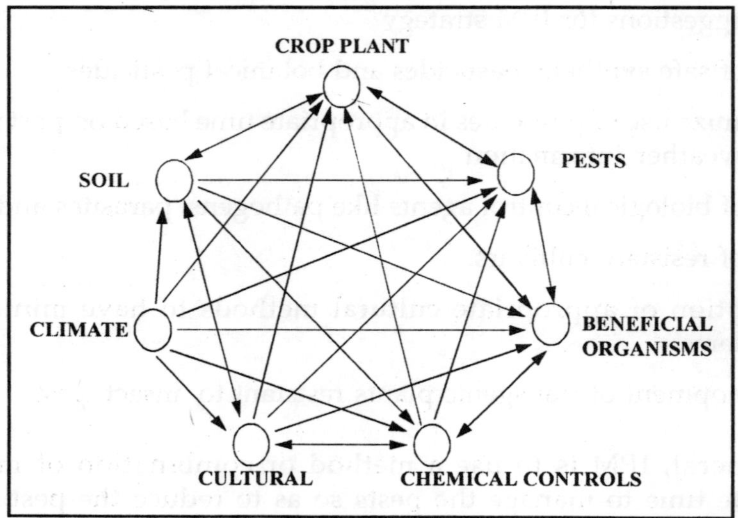

of weather conditions for a sufficiently long period. Insects, however, are mobile and move from place to place at different times and hence are less affected by uniformity of weather. It is certain that most diseases are more dependent on weather than insects are; however, spread of certain insect pests is also directly weather dependent.

25.2. Development of forecasting system for insect Pests

The prerequisite of an economically viable, environmentally sound and easily adoptable pest management programme may encompass a precise forewarning of the pest occurrence, population dynamics. Forewarning techniques should be simple, based on the knowledge of quantitative seasonal studies on biology and ecology over several years, including, geographical distribution, key mortality factors and weather. The following basic information and methods should be followed for development of dynamic pest prediction system.

25.2.1. Quantitative seasonal studies

Sampling of population, spatial and temporal distribution of pest abundance, seasonal occurrence in relation to weather and phenophase of the crop.

25.2.2. Life History and Biology of the pest

The length of life cycle, number of generations in a year, growth index, life span, survival rate, food, habitat, over wintering, collateral hosts and intrinsic growth rate in field conditions are to be collected in a systematic way. The effects of biotic and abiotic factors on development of pests, spread and mode of carry over needs to be worked out accurately.

25.2.3. Ecological Studies of the pest

Life-table studies of pests are important for better understanding of pest population build-up. Life-table of a pest can be used to find the key mortality factors operating in different stages of the insects. Besides these, the spatial pattern can also provide clues to find the suitable sampling schemes to be used for collecting pest population samples. The flux of pests along with mode of carry over and spread can be used for forecasting of pests.

25.3. Surveillance of pests

Emergence of soil pests or migrating pest through regular monitoring of population as direct count on crops plants, light traps, pheromone traps, wind traps and also by drawing soil samples for visual observation of immature stages of pests indicates the pattern of population build up. Even emergence traps fixed in the soil and crops plants also work as a tool. The population of the pests should be also counted on alternate or collateral hosts on periodical intervals for their occurrence on major host plants. The use of satellite for ground images of the crops when mono-cropped over large area is also one of the tools to study the pest infestation.

25.4. Modelling of pest population

Based on the data available on the other factors such as weather, natural enemies present, crop biology, various modelling options are available. Some of the successful modelling techniques used in pest forecasting include statistical, mathematical, simulation models and geographical information system. Reliable, genuine data on pest population for long period on definite time intervals, weather records of the study sites, crop phenology, population data of natural enemies are the requirements for developing comprehensive pest forecasting models. Different aspects of modelling are discussed in detail below.

25.4.1. What is Model?

A model is a physical or mathematical representation of an object, system or process within the limits of reality. Physical models are the models of proportionate shape and size which also called 'iconic' models. Model of a building, mannequins in the departmental stores are the examples of physical models. The mathematical models range from deterministic models, simple regression type statistical model to complex multicomponents computer simulation model. In ecological and agricultural entomology, models are widely used. (i) to seek general conclusions about how organisms interact with each other and their environment; and (ii) to predict how populations or communities will behave in relation to some expected changes in their existing relationships or in the prevailing environmental conditions. The later will often include man's interventions with biocides or introduction of exotic organism (Holt and Cheke, 1997). There are a number of issues which need to be considered when embarking on model building or commonly known as modeling.

The different aspects of modeling needs to be considered before a model building are:

- Deterministic or stochastic (Type of model)
- Modeling over time or space or both
- Time frame - Discrete or continuous
- Units to be considered
- Deductive or inductive

25.4.2. Type of model – stochastic or deterministic

Deterministic models produce identical results each time (for the same starting state and set of parameter values). Variability is fundamental to biological processes, however, the parameter values may be better represented as having a mean and variance, not simply a fixed value. For example, daily fecundity for the individuals in a population may be somewhere in the range of 10-30 eggs with a mean of 20. In a deterministic model, the fecundity may be set at 20. In a stochastic model, explicit account is taken of the variability. The actual fecundity used is determined by selecting value from the frequency distribution of possible values. Thus, each time different unpredictable results are obtained from a stochastic model. Stochastic models are important when dealing with a small population size. However, the deterministic models forms the basic of modeling. For example, the disease progression may be directly proportional to the time. The mathematical model to denote this would be exponential model.

This model can be made stochastic to count the reality by adding an error term which follows some statistical distribution.

25.4.3. Space

Most models are concerned with the dynamics of processes overtime but spatial dynamics are also important. By choosing a non-spatial model it is assumed that the use of mean representation of the dynamics of the system. However, if it is known that different parts of the system are not spatially homogenous, then some partitioning of the model to represent this heterogeneity may be useful.

25.4.4. Time frame – discrete or continuous

The model concerns generation changes or annual population estimates, different equations with discrete time steps may be appropriate. If the model variables can be represented as a continuous process, a model in continuous time can be appropriate. Typically, for each variable in the model, an equation is specified for its rate of change. It is sometimes possible to find solutions to such models algebraically and they would then properly be called analytical models.

25.4.5. Units

The question of variable resolution also relates to the units of population size to be used. All the individuals in the field could be treated as a single population. Alternatively, the individuals on each plant or even each individual could be modeled separately.

25.4.6. Deductive or inductive

Analytical models are usually deductive in the sense that the model is specified on arguments from the general to the particular. The inductive approach is to argue from the particular to the general by starting with the available data and attempting to induce pattern or relationships. Empirical models based on statistical relationships are usually inductive, where as simulation models are usually built using a mixture of deductive and inductive methods.

25.4.7. Modeling Techniques

25.4.8. Statistical models

Most of the biological activities are interrelated. For example, the insect population, among other things, may be related to the prevailing weather factors. The fluctuations in the pest population may be explained by the fluctuation in the weather factors. Statistical models try to assign the variability in one factor (called independent variables) to other factors (called independent/explanatory variables) to other factors whatever left unexplained by the independent/explanatory variables are attributed to the error or random term. Thus, the statistical models are stochastic in nature and the stochastic models account the variability in any biological process. The statistical models are of different kinds based on how the explanatory and explained variables are related in the models. If the relationship between explained and explanatory models is linear, the models are known as linear models and if it is non-linear, non-linear models.

Most of the time, statistical models explain the empirical relationship found in the data and no biological reasons may be assigned to these models. But, one can develop, statistical models which can be biologically explained. The models such as exponential, Gumpertz etc. used in disease forecast are some examples of such models. Besides these, one can use time-series models in which the dependent variable is related to the previous values of itself. Some examples of such time-series models are AROMA, ARIMA etc. In the subsequent section, we discuss various kinds of statistical models.

25.4.9. Regression models

Here, the dependent variables are related to a set of independent variables linearly. If there is only one independent variable, then the model is called simple linear regression models. If there are more than one independent variables then it is called multiple regression model. An example will be useful in understanding the regression models.

Ex. 1. The population of jassids (.......) of groundnut is related to the prevailing weather factors. Thus, if we denote jassids as γ and minimum temperature (min. temp) and maximum temperature as maximum temp then the multiple regression model would be

$$\gamma = a + b \text{ (min. temp)} + c \text{ (max. temp)} + e$$

Here, *a* is known as the intercept term and *b* and *c* are called regression coefficients. *e* is the error or random term. The random term accounts for the unexplained variation in jassid population.

Thus, in general, a multiple regression models is written in the form

$$Y = b_0 + b_1 x_1 + b_2 x_2 + \ldots\ldots + b_n x_n + e$$

(dependent) (intercept)

wehre $b_i'x$ (I = 1, ..., n) are known as regression coefficients.

The unknown parameters are estimated from the data using the procedure called Leant square method. Most of the standard statistical programs and popular spread sheets such as Ms-Excel, Lotus-123 etc. have options to extract regression coefficient.

The part of the variation as explained by the independent variables is usually known as R^2. R^2 may be used as a yard stick in choosing a model. Before we end, some cautions are listed.

1. Never include two independent variables which are themselves related. For example, if min. temp is included in the model, never include (min. temp)^2 as another variables. If two independent variables are related to each other, the regression equation is unstable and this problem is usually known as multi-co linearity.
2. Regression models may be highly influenced my few observations. Such observations are called outliers. Diagnostic tests are available to check outliers.
3. One can use step-wise methods for choosing a subset of independent variables.

25.4.10. Non-linear models

If the dependent variable and the independent variables are not related in a form which in not a linear then the models are known as non-linear models. For example,

$$Y = e^{(elogb.a)t}$$

Non-linear models, in general are difficult to fit. There are several statistical packages are available by which one can fit a non-linear models. Most of the computer programs require the initial values of the parameters have to be given and the program find the best value of the parameters by iteration. Some cautions in using the non-linear model are

(i) the iteration may not converge if proper initial values are not chosen

(ii) the final parameteric values obtained may not be the best.

Another popular modelling technique is by simulation.

25.4.11. Simulation

With the advent of computer and its application in the field of agriculture the modeling of the total process of development can be done step by step with the time. A set of rate variables and state variables are used. It mimics the total system behavior with proper understanding of biological, ecological process involved into it. In a discrete model, population of pest of 't + 4' th week can be estimated from 't' th week population. But in simulation the output will be continuous in nautre i.e., estimated population of every week will come as output. Two types of input data are required for simulation model – one time input i.e., initial adult or larval or egg population, crop stage, presence or absence of natural enemies etc. other is continuous input like daily weather parameters like, temperature, humidity, rainfall etc. Thus using simulation models, one can develop a system by which the pest population may be predicted.

25.4.12. Degree day model

The development rate of insects is dependent on temperature. Thus it is possible to relate insect development not to calendar time, but to temperature. It is common assumption in such models that above a certain temperature threshold, development rate is proportional to temperature. Below this threshold, development is assumed to be zero. For example, if the development threshold is 3°C, and the mean temperature on a particular day was 15°C, then the degree day elapsed during that day is 12°C. This degree day is also called physiological time. Incorporating the degree days in the model helps in studying the pest dynamics more clearly at micro-level.

25.4.13. Rule based model

It is possible to represent a system as a set of linked logical relationships. In such models, the relationships have taken the form of rules of the structure.

'IF some condition(s) THEN some conclusion(s)

ELSE other conclusion(s)'

Such Boolean conditions were incorporated into conventional models. A rule based model is composed entirely of Boolean logic. Rule based models are qualitative equivalent of simulation models and provide a means by which quite crude or subjective knowledge about a system can be used. In rule based models, system components are represented by a small set of discrete states and changes are described by rules. These models however, may be used as the starting point in model building.

25.4.14. Expert System

An expert system is a model that mimics the decision-making of a human expert. Most applications in entomology which use the term expert system refers to rule based models. A rule based model was used above to simulate the dynamics of processes over time through computer software. In an expert system, rules are used to represent the decision-making associated with a particular problem. The main aims of this expert system are to diagnosis of problems and prescription for treatment.

The expert system may also include statistical models or simulation models. Thus, expert system would be the logical outcome of a modeling activity for field level application of models.

25.5. Pest-Weather Relationship

The population dynamics of the organisms responsible for insect pest and disease outbreak in crops is influenced by a number of factors like quantum of food supply, availability of intermediary hosts are carriers, presence or absence of natural enemies, competition for shelter and suitability of weather condition for development or successful resting. It is a possible that under certain conditions a pest may be free from the effects of food supply, shelter and natural enemies. But it is very rare for a pest to remain immune from weather influences. In fact, weather or climatic factors have direct bearing on pest development as well as indirect bearing on the other factors affecting the population dynamics of the pests (Venkataraman, 1992).

25.5.1. Pest-weather Relationships in groundnut

As it is known the pest-weather relationship study is very important in developing pest forecasting models and formulation of effective management practices. A good number of studies have been conducted in this aspect. Among the many insect-pests, tobacco caterpillar (*Spodoptera litura*) and the groundnut leaf miner (*Aproaerema modicella*), are economically important as they cause severe yield losses across groundnut growing regions of the country, (Amin, 1988).

At Maharashtra, high maximum and minimum temperatures and low humidity in the pre-monsoon period, followed by continuous rain for about 3 weeks with maximum and minimum temperatures in the range of 27.0 - 32.7°C and 21.3 - 23.7°C respectively in the period mid June to mid July, followed by monsoon break conditions for a week or two favoured an outbreak of leaf miner (*Aproaerema modicella*) (Dubey and Ballal, 1995). Madhumati and Thippaswamy (1996) calculated net reproductive rate and different rates of increase and found that weekly rate of multiplication was highest at 31.9°C and 64.16 mm rainfall. Shanower *et al.* (1993) calculated different degree days requirement for different phonological stages of this pest.

At Anantapur, it was observed that a sharp rise in maximum temperature by more than 2°C and their persistence for few days (heat wave conditions) associated with dry atmospheric conditions results in outbreak of leaf Webber in groundnut in 3-4 days. Thus, it is possible to predict the incidence of leaf Webber in groundnut in 3-4 days and the prediction models could be developed.

On the other hand, emergence of Red Hairy Caterpillar was found to be closely associated with the amount and occurrence of rainfall. The population reached a peak on the third to fourth day after the heavy rain event. However, the dry spells following rain event had a controlling effect on pest population. Outbreak of pest population can thus be predicted 3 to 4 days well in advance (Anon, 1998).

Jena and Kuila (1997) reported a good correlation between the weather variables and information of leaf miner in groundnut. It was observed that leaf miner infested plants were negatively correlated with maximum temperature (-0.606), minimum temperature (-0.345) mean temperature (-0.760), sunshine hours (-0.593) and positively correlated with rainfall (0.90) and relative humidity (0.622). Shekh *et al.* (1999) observed in Anand that the population of aphids, leaf hoppers, thrips and whitefly were negatively correlated with most of the weather parameters. The vapour pressure and relative humidity had favourable effect on the population build of leaf miner, tobacco leaf eating caterpillar and gram pod borer. While sunshine hours, maximum temperatures and wind speed had detrimental effect. The population of lady bird beetle was adversely affected by most of the parameters except sunshine hours and temperature. Besides weather parameters, the age of the crop has also significant correlation on build up of leaf miner, leaf eating caterpillar, pod borer and lady bird beetle.

In case of leaf spot disease of groundnut, Patel *et al.* (1997) reported that three days mean diurnal temperature range (6.3 to 15.0°C) and three days mean night temperature of 25.2 to 21.4°C were found to be congenial to accelerate the spore production.

The pest weather relationships study is must for any pest modeling, particularly pest forecast modeling.

Temperature

Sherasiya and Butani (1998) showed that maximum temperature had a significant and positive influence on infestation levels of leaf miner during the summer. During the rainy season, relative humidity and rainy days were correlated with high infestation levels. Lewin *et al.* (1979) reported that temperature was also positively correlated with GLM incidence and accounted for more of the variation than rainfall. Correlation analysis of the variables showed that mean temperature had a positive correlation with the populations of leaf miner (Bagmare *et al.*, 1995).

Nandihalli *et al.* (1989) studied the population dynamics of *Spodoptera litura* in the field in Karnataka, India, observed a significant negative correlation between number of catches and average minimum and maximum temperatures for both traps, and a positive correlation with average morning relative humidity for light trap catches only. All the weather parameters in combination recorded, influenced the catches in light and pheromone traps by 40.6 and 42.0%, respectively.

Rainfall

The abundance of insect pest in an area is affected not only by the total rainfall, but also its distribution (Khan and Reodeo 1987), also claimed that high rain fall was key factor regulating leaf miner population. Amin (1987) suggested that heavy rain fall reduces leaf miner population. There was no linkage between the rain fall and the appearance of leaf miner male moth in pheromone traps (Anon 1993, 1994). Lewin *et al.* (1979) found a significant negative correlation between GLM incidence and rainfall: higher GLM incidence was correlated with lower rainfall. Amin (1987) and Khan and Rao Deo (1978) have suggested that high rainfall reduces leaf miner populations even though data from the latter do not support this conclusion. They observed the high populations in August-September during a high rainfall period, and populations declined in March when no rain was recorded (Khan and Raodeo, 1978).

Relative humidity

Mishra and Sontakke (1992) recorded highly significant correlations between pheromone trap catches of *S. litura* and both maximum and minimum humidity. Similarly Rajaram *et al.* (1999) reported relative humidity had a significant positive relation with the catches of *S. litura* in water pan pheromone traps. Jayanthi *et al.*, (1993) reported that populations of *Caliothrips indicus* had a positive correlation with temperature, sunshine, rainfall and morning relative humidity and a negative correlation with evening relative humidity and wind speed.

Sunshine hours

Bright hours of sunshine showed significant and positive correlation with leaf miner population Bagmare *et al.* (1995) whereas Sherasiya and Butani (1998) showed that maximum temperature and sunshine hours were negatively correlated with the leaf miner population.

Congenial conditions for the pest build-up

An epidemic of the pest occurs only when the conditions are congenial for the rapid multiplication and spread of the pest over an agro-climatic location. Such information helps in development of good forewarning systems/thumb rules for prediction of the outbreak of epidemics. The literature was surveyed and the congenial conditions for the build-up of the pests have been tabulated and presented in table 26.1.

Table 25.1. Information on period of incidence and climate for development of major insect pests of groundut

Pest	Period of occurrence	Congenial climate
Leaf miner	March-October	High temperature
Tobacco caterpillar	March-October	Hot and humid climate
White grub	August-October	Temperature coupled with humidity
Jassid	March-October	Dry weather, bright sun shine and occasional rains
Aphids	July-September	Succulence in plants
Hairy caterpillars	June-October	Long drought period followed by rains
Helicoverpa armigera	March - October	High rain followed by drought for few years
Termites	September-October	Warm humid weather soil temperature above 30^0 C passively

25.5.2. Importance and field application of prediction model

As mentioned earlier, the biotic stress on groundnut put great threat to stable yield. The indiscriminate uses of pesticide also pose serious health hazards in humans, animals and put the entire eco system in great peril. The dependency on chemical pesticides can be reduced if avoidable sprays are eliminated. To avoid unnecessary sprays, monitoring of pests, pest forecasting models are helpful. If the pest is controlled in the initial stages, the losses can be reduced to considerable extent. Thus the pest forecasting models are very important.

Moreover, the IPM strategy in plant protection will improve the environment but at the same time, the yield levels are sustained in longer run An IPM module coupled with inputs from a pest-forecasting models is badly required.

The field application of pest-forecasting model may pose some hardship as it may be difficult to reach farmers within a short-period of time. However with the recent advent in information technology, this may be reduced considerably.

25.5.3. Scopes and limitations of pest modeling

Pest modeling is an emerging field with coordination of entomology, environmental sciences, statistics and computer sciences. It enable the planners, researches, students and farmers to understand the crop-pest environment in an effective way. Pest forecasting through modeling, effective action plan through knowledge based expert systems and yield loss modeling due to pest attack are the main fields in which proper attention is necessary.

Although the pest modeling is needed for effective management of pests, this pest modeling is heavily dependent upon huge, intensive and reliable data on pest population dynamics, corresponding weather parameters, crop phenology and natural enemy population. Availability of the data of the above aspects in a uniform data sheet is very rare. Although some scattered works on forecasting or modeling have been done at different places, the results are not of consistent type. Availability of uniform data is the main constraint. The tendency observed in pest forecasting in India is that even though lost of information collected on pest-weather relationships were collected and various types of models were developed, almost no effort had been so far taken by any agency in applying these in field condition. Due to this, pest forecasting models are not so successful. Trained manpower is also very limited in this field.

Apart from these, the modeling simply an attempt to describe what goes on in Nature and there is always a big gap in human understanding of the Nature. This gap, though shrinking in the recent time, put greater limitations of modeling, particularly pest modeling in which so many complex system interacting with one another.

25.5.4. Modeling work on groundnut pests

Several types of modeling work have been carried out on groundnut pests. These are

a) Modeling for forewarning insect pests and diseases;
b) Modeling on crop weed competition and yield losses;
c) Modeling of pesticide effectiveness or dynamics; and
d) Expert system or decision support system on groundnut pest management

25.5.5. Forewarning of insect pests and diseases of groundnut.

The forewarning models range from simple regression model to computer based simulation for early predictions. A simple regression model used by Mack *et al.* (1991) to determine whether the adults of *Elasmopalpus lignosellus* in week 'i' could explain the various in larval abundance in week 'I+1' over the other years of observation. The mean number of larval from week 'I+1' increased linearly with an increase in adult flush count from week 'i', indicating that the larval density could be predicted by adult abundance.

Like most soil insects, termites the pests of groundnut are influenced by soil moisture. Johnson *et al.* (1981) derived a relationship between annual rainfall (x) and the percentage of groundnut stand with tap root invaded by *Microtermes* sp. (y).

$$Y = (24493 / x) - 20.6$$

In Nigeria where the annual rainfall is more than 800 mm, less than 10% of stands are attacked.

In India, a stepwise regression model was used by Sanadhikar *et al.* (1989) to predict rust epidemic in groundnut and one biological and seven meteorological variables were identified as factors for rust epidemic.

The stepwise regression technique was followed to develop the prediction equation for population buildup of groundnut pests in Anand, Gujarat (Shekh *et al.*, 1999). The results were presented in table 26.2.

The coefficient of determination (R^2) thus obtained, ranged between 0.132 to 0.508. It might further be noted that weather parameters altogether accounted for only 13.2. to 32.3. per cent. Variation in the population buildup in aphid, jassid, thrips and whitefly (Eq. 1 to 4). However, if crop was incorporated in the regression analysis, the coefficient of determination was higher which ranged between 0.323 to 0.508 (Eq. 5 to 8). These equations may be used for prediction of pest population.

The degree day model used by Lu (1998) to forecast the egg hatch of Holotrichia formosana a major pest of ground nut in china. The developmental zero (base temperature) for egg, larval and nymphal stages were 13.6 0.7, 13.9 0.7 and 9.3 0.6oCand the effective accumulated temperature were 126.3 7.8, 842.4 51.5 and 227.4 13.6 day degree C, respectively.

Table 25.2. Prediction equations for different pest population in groundnut crop.

Pest	Equation	R^2	Eq. No.
Aphid	$Y = 0.1102 - 0.0050 * X_3 - 0.0010 * X_9 + 0.0319 * X_{10} + 0.0067 * X_{12}$	0.1896	1
Jassid	$Y = 1.3450 - 0.0030 * X8 - 0.0506 * X12$	0.1318	2
Thrips	$Y = 3.4713 - 0.0761 * X2 + 0.0826 * X3 - 0.0383 * X5 - 0.0152 * X8 - 0.0575 * X12$	0.3237	3
White fly	$Y = 0.8590 + 0.0259 * X3 - 0.0368 * X6 + 0.0445 * X10 - 0.0442 * X12$	0.2073	4
Leaf miner	$Y = 0.0869 + 0.0159 * X1 + 0.0453 * X5 - 0.1605 * X12$	0.3333	5
Tobacco leaf eating caterpillar	$Y = 6.0696 + 0.0123 * X1 - 0.1295 * X2 - 0.0897 * X6 + 0.2376 * X7 - 0.0460 * X8 - 0.0760 * X10 - 0.0648 * X11 - 0.1167 * X12$	0.5075	6
Gram pod borer	$Y = 3.6114 + 0.0056 * X1 - 0.0762 * X2 - 0.0388 * X6 + 0.1280 * X7 - 0.0262 * X8 - 0.0635 * X10 - 0.0367 * X11 - 0.0869 * X12$	0.4884	7
Lady bird beetle	$Y = - 0.9106 + 0.0463 * X1 + 0.1827 * X3 - 0.2161 * X7 + 0.0192 * X8 + 2.1649 * X13$	0.3226	8

Where : Y is population of respective pests, X1: crop age (week after sowing), X2: maximum temperature (°C), X3: minimum temperature (°C), X4: mean temperature (°C), X5: vapour pressure (morning) (mm of hg), X6: vapour pressure (afternoon) (mm of hg), X7: mean vapour pressure (mm of hg), X8: relative humidity (%), X9: rainfall (mm), X10: number of rainy days, X11: sunshine hours (hrs/day), X12: wind speed (km/hr), and X13: aphid population

A type of indexing based on minimum temperature and hours of RH > 90% was used in USA by Johnson (1989) to forecast the Cercospora leaf spot on groundnut. The index numbers for 2most recent 24h periods were summed and this is 3.5, the groundnut growers in that area were given forewarning for disease development. Pedro Junior *et al.* (1994) described a technique of forecasting Cercospora leaf spot of groundnut based on correlations between minimum daily temperature and leaf wetness duration and development of Cercosporidium personata on groundnut. This was tested in USA and Argentina and using these methods the number of fungicide sprays needed to control the disease was reduced to 1-3 per season.

Regarding simulation a dynamic simulation of groundnut rust was reported by Savary *et al.* (1990) from Ivory Coast. The model involves 2 units: crop growth and development and rust epidemic. The simulated output were found fairly similar to observed rust severity and crop growth data from Adiopdoume, southern Ivory Coast.

A simulation model of the population dynamics of Aproaerema modicella was developed in the early stages of IPM research project in India. It was linked to a dynamic programming procedure which indicated the number of insecticide applications needed to optimize income over ranges of natural mortality and host plant resistance and extended to include the initial level of pest population, mortality caused by insecticide to natural enemies, the effectiveness of insecticide application and groundnut prices. Model results indicated that future research should evaluate the effectiveness of farmers' insecticide applications. The initial population density had a significant influence on subsequent levels of damage.

Forewarning of groundnut insect-pests

Among the insects, *S. litura* and the GLM are important insects causing severe yield losses to groundnut crop through out the groundnut growing zones of India. To know the very presence of the insects it is essential to locate the adults first. This is possible through monitoring of the insects using trap with many attractants. Earlier Scientists have used light trap data to find out the probable occurrence of the insects in a season. Subsequently, using weather parameters prediction models were developed and the formulae are presented in table 26.3.

Table 25.3. The prediction equations for insect pests reported by various workers

Insect pest	Formula	Reference
S. litura	Y = -68.7 +0.92X1 + 7.2X2 4.4X3 9.3X4 + 1.1X5 + 1.0X6 + 1.0 X7 where X1=Tmax, X2=Tmin, X3=VP(M), X4 = VP(E), X5= RH(M), X6=RH(E), X7=SSH	Nandagopal *et al.*, 2001
S. litura	Y = 6.0696 + 0.0123X1 0.1295 X2 0.0897 X6 + 0.2376 X 70.0460 X8 0.0760 X10 0.0648 X11 0.1167 X12 where X1=crop age, X X2=Tmax, X6=VP(AN), X7= Mean VP, X8=RH%, X10= No. of rainy days, X11= SSH, X12= Wind speed	Sheikh *et al.*, 1989
GLM	1. Y = 200.4 2.7470X1 3.4938 X2 where X1= Tmax, X2=Tmin 2	Logeswaran *et al.*, 1982
	2. Y = 886.02-4.6139X1 14.1182X2 where X1= Rh, X2=Tmax 3	Bagmare *et al.*, 1995
	3. Y = 827.05 2.933X1 22.6155X2 where X1=Rh, X2=Tmin	
	4. Y = -2.59 + 0.11X1 - 0.0007x3 where X1=T Mean, X3=Rainfall,	
	5. Y = -2.49 + 0.10X1 - 0.005X3 where X1=T Mean, X3 Rainfall	
	6. Y = -0.10 + 0.04X1 - 0.006X3 where X1= T Mean, X3= Rainfall,	
	7. Y = 1.44 + 0.03X1 + 0.04X3 - 0.024 where X1=T Mean, X3=Rainfall, X4=RH	

Approaches followed

The common practice of Indian farmers is to apply pesticides when the infestation is fairly visible, which is usually 3-4 weeks after initiation of the infestation and hence they may be least beneficial. Subsequent development of light traps for monitoring has helped to some extent to avoid such situations.

With the advent of female sex pheromone of *S. litura* (Tamaki *et al.* 1973) and for GLM (Hall *et al.* 1993), it is now possible to precisely monitor the activity of the pest even before the damage caused.

Experiments were carried out to establish the prediction modules for the target pest using sex pheromones. The correlations between male moth catches in traps with weather parameters in three different dates of sowing at four centres were established. Monitoring was done at six centres for the target pests by staggered planting during kharif seasons and setting up standard sex pheromone trapping (Ranga Rao *et al.*, 1991) for insects and spore traps for diseases. The traps were fixed 1 M above the ground level. Two pheromone traps were used in each field. The pheromone vial was replaced at and interval of three weeks.

Weather data was collected from the meteorological observatory. These weather parameters were correlated with moth catches and the damage by insects on the foliage using stepwise regression and based on the relationships prediction formulae were worked out. An effort was made to develop relationship between moth catch and percent damage due to insect-pests so that the pheromone trap data could be effectively used for estimation of losses due to pest infestation. The derived weather parameters such as growing-degree-days (GDD), relative temperature difference (RTD), relative humidity difference (RHD), diurnal variation (DV) calculated using the following formulae were also used for the development of prediction models. In addition, the already published models from the literature were applied to the data generated to validate these models.

Historical data analysis

By utilizing the historical data, detailed analysis was done to develop prediction models for Spodoptera with the help of IASRI, New Delhi. The results are as follows

Models were developed for forecasting pest population for weeks 28 to33.

Weather variables lagged by 3,4 and 5 weeks were used to develop the model.

The model was of the form :

$$Y_t = a_0 + \sum_{i=1}^{p} a_i z_i + \sum_{i \neq j}^{p} b_{ij} z_{ij} + cI_y + dL + e$$

where

$$z_i = \sum_{w=t-5}^{t-3} r_{iw} x_{iw} \qquad z_{ij} = \sum_{w=t-5}^{t-3} r_{ijw} x_{iw} x_{jw} \qquad I_y = \sum_{w=t-5}^{t-3} r_{yw} Y_w$$

Y_t is pest population in t[th] week, L is pest population of the previous year (42nd week), x_{iw} is value of ith weather parameter in wth week, r_{iw} is value of correlation coefficient between y_t and i^{th} weather parameter in wth week, r_{ijw} is correlation coefficient between y_t and product of x_i and x_j in wth week, r_{yw} is correlation coefficient between y_t and $y_{w;}$ p is number of weather variables; e is error term

Models were developed for forecasting *Spodoptera* population trapped using the pheromone trap. Weather variables as well as pest population lagged by 3,4 and 5 weeks were used to obtain indices. These indices along with pest population of previous year (42nd week) were used as regressors in the model. The results are given in 26.4

Table 25.4. Models developed for forecasting *Spodoptera* population

Data used up to	Week	Model	R2
1999	28	$Y_{28} = 83.42 + 7.36\ Z_1 - 0.03\ L$	0.79
	29	$Y_{29} = 19.82 + 0.78\ Iy - 2.06\ Z_1 + 0.001\ Z_{14}$	0.94
	30	$Y_{30} = 74.71 + 0.80\ Iy + 1.50\ Z_1 - 0.60\ Z_3 + 0.03\ Z_{13} + 0.001\ Z_{34}$	0.98
	31	$Y_{31} = 57.93 + 0.52\ Iy + 1.97\ Z_1 + 0.03\ L$	0.87
	32	$Y_{32} = 803.99 + 0.20\ Z_{23}$	0.48
	33	$Y_{33} = 698.39 - 0.59\ Z_4 + 0.20\ Z_{13}$	0.78
1998	28	$Y_{28} = 83.35 + 7.31\ Z_1 - 0.03\ L$	0.80
	29	$Y_{29} = 1.20 + 0.72\ Iy$	0.89
	30	$Y_{30} = 47.69 + 0.78\ Iy + 1.53\ Z_1$	0.87
	31	$Y_{31} = 86.77 + 0.47\ Iy + 2.10\ Z_1 + 0.03\ Z_{23}$	0.93
	32	$Y_{32} = 813.34 + 0.21\ Z_{23}$	0.49
	33	$Y_{33} = 698.85 - 0.51\ Z_4 + 0.20\ Z_{13}$	0.80
1997	28	$Y_{28} = 83.38 + 7.33\ Z_1 - 0.03\ L$	0.80
	29	$Y_{29} = 1.48 + 0.71\ Iy$	0.90
	30	$Y_{30} = 71.98 + 0.79\ Iy + 1.29\ Z_1 - 0.43\ Z_3 + 0.03\ Z_{13}$	0.96
	31	$Y_{31} = 88.56 + 0.48\ Iy + 2.14\ Z_1 + 0.03\ Z_{23} + 0.02\ L$	0.94
	32	$Y_{32} = 364.66 + 0.26\ Z_{13} + 0.36\ L$	0.81
	33	$Y_{33} = 728.4 - 0.58\ Z_4 + 0.21\ Z_{13}$	0.85

X_1 : Av temp.; X_2 : Av soil temp.; X_3 : Av R.H.; X_4 : Av rainfall

As seen from the above table the predications for the low populations were close to actual incidence. However as the incidence increased suddenly there was a wide gap between actual and predicted values. Similarly, based on correlation studies, contribution of different weather parameters to the build up of leaf miner was worked out.

Development and validation of prediction modules
Insect-pests

Groundnut leaf miner and tobacco caterpillar are major biotic stresses that cause significant yield losses in groundnut production systems across the Country. Based on the information already available as well as data generated through experimentation at different groundnut production systems, weather based forewarning systems were developed for leafminer and *Spodoptera litura*. Using regressions, prediction equations were developed for *S. litura* for different centres viz. Junagadh (Y = 87.68 + 4.3 Tmax – 6.88 Tmin – 0.35 Rhe; R^2=0.28), Jalgaon (Y = -706 + 0.23Tmax – 0.243Rhe; =0.28), Dharwad (Y = -208.1 + 19.49Tmax -10.06Tavg; R^2=0.28), and Kadiri (Y = 240.75 – 6.42Tmax; R^2 =0.47) with significant goodness of fit. Using the experimental data generated across six centres over two years, prediction equations were developed and validated. Real-time surveys were conducted in farmers' fields to validate the prediction models developed. Correlations were established between moth catches in the traps and crop damage due to pest. Analysis of eight years of data only showed significant and positive correlation between maximum temperature and moth catches of leafminer. Using multiple correlation and regression analysis, historical data as well as the data generated under the project over three years, prediction models were fitted and validated. Also, from the published literature, the formulae were tried for different regions to assess their suitability. For developing models against diseases, regular as well as derived weather parameters were used. Validation of the published prediction models for the target pests against experimental data showed that either the models were directly not suitable or could predict the trends only during early stages of pest build up.

Validation of published modules

The published models shown in table 2 were validated using the results of field trials laid out during kharif seasons of 2001, and 2002. The equations The predicted and actual values for Vriddhachalam are presented in the table26.5.

Table 25.5. The predicted and actual values of incidence of leaf miner for Vriddhachalam centre

DAS	Actual incidence	Predicted values as per equation (moth catches)						
		1	2	3	4	5	6	7
II-68	36.0	39.3	81.7	80.6	0.3	0.1	0.9	0.9
II-87	24.0	27.2	117.8	105.6	0.6	0.4	1.0	1.0
II-88	27.5	31.0	106.2	123.2	0.5	0.3	1.0	1.0
II-90	20.5	24.8	92.1	100.5	0.6	0.4	1.1	1.0
III-68	36.0	39.3	81.7	80.6	0.3	0.1	0.9	0.9
III-88	25.5	31.0	106.2	123.2	0.5	0.3	1.0	1.0
IV-69	40.5	42.7	141.7	141.4	0.3	0.1	0.9	0.9
IV-87	22.5	27.2	117.8	105.6	0.6	0.4	1.0	1.0

II, III, IV indicate the dates of sowing; 1-7 equations shown in Table 25.2 for leafminer

Validation of the data for Tirupati location is shown in Table 6. The results showed similar trend as at Vriddhachalam.

Table 25.6. The predicted and actual values of incidence of leafminer for Tirupati centre

DAS	Actual	1	2
I-81	37	33.6	208.4
I-94	38	33.7	193.9
I-100	32	38.7	100.3
II-78	60	33.7	55.4
II-86	115	189.9	109.5
II-79	108	35.7	102.1
I-34	80	12.4	164.5
I-81	37	33.6	91.6
II-56	138	32.6	128.2

I and II indicate the dates of sowing; 1 and 2 equations shown in table 2 for leafminer

At Anantpur location at 35 days after sowing when crop was planted at recommended date of sowing, the prediction could be validated successfully. For both leafminer and Spodoptera, the prediction equations could hold good for the early stages of pest incidence especially when pest populations were low.

25.5.6. Modeling on crop weed competition

The effect starbur (*Acanthospermum hispidium*) competition in peanuts, were studied by Walker et al 91989) in USA using regression technique. Full season interference from 8, 16, 32 and 64 briestly starbun plants/7.5 m row reduced the seed yields by 14, 26, 43 and 50% respectively. Weed interference for 2 weeks after crop emergence reduced by an average of 4%. A model of competition for light between peanut and broad leaved weeds was developed by Barbour and Bridges (1995) to run with the PNUTGROW model. The model simulates shading of the groundnut canopy by reducing the daily photo synthetically active radiation (PAR) received by the groundnuts in a manner that realistically represents timing and quantity of light capture by the weeds. The weed Florida beggar weed overtopped the groundnut canopy 52 days after sowing (DAS) and reduced PAR reaching the groundnuts45% by 73 DAS. Sickle pod overtopped the crop canopy 42 DAS. Wild poinsettia overtopped the groundnut canopy 44 DAS and had an attenuation value of 39% 85 DAS. It was suggested that models will play an important role in reducing the expenditure of time and resources required to document yield losses due to weeds in groundnut.

Modelling of pesticide effectiveness

Use of simulation model to explore fungicide strategies for control of *Cercospora* leaf spot of peanut was done by Knudsen et al (1988). The persistence and efficacy of chlorothalonil was incorporated as a submodel into a computer simulation model of *Cercospora* leaf spot of groundnut caused by *C. arachidicola*. The model was found to be a good estimator of the combined effects of weather and chlorothalonil treatments on disease progress.

25.5.7. Expert system or decision support system

In IPM programme this are the modeling for developing strategies for specific purposes. A decision support system (PEET, the pesticide economic and environmental tradeoff) was developed to evaluate both economic and environmental impacts of the different pest management practices.

25.6. Future Strategies

1. A network programme for collection of data for each of the above pests to be established. This can be done through AICRP on groundnut.
2. A detailed format/schedule for collection of daily/weekly pest data, weather data, crop phenology, crop husbandry, predator, parasites, alternate hosts and pest migration to be developed, circulated to the coordinating centres in the network programme. Data to be collected, strictly according to the schedule to maintain uniformity, throughout the year/crop season.
3. The main centre will be collecting data from the coordinating centres daily/weekly thorough phone/fax/e-mail and will be maintaining a database in their computer. All centres have to be equipped with the above facilities These facilities will be used later for giving pest forewarning as well as pest management tips.
4. A multidisciplinary team comprises of entomologist, plant pathologist, meteorologist, statistician, computer scientists will be engaged for synthesizing, testing and validation of model. The coordinating centers with entomologists, pathologists, agronomics and economists will be required for on farm trial and economic analysis.
5. Required manpower to be developed and knowledge of present workers to be enhanced to international standard through training.
6. The recent advances in IT industry may pave ways to remove the important bottle-neck of spreading the information to the farmers without any loss. Software developed in Indian Languages/web/Internet based information centres are required.

25.7. Conclusions

Modeling and forecasting of pests and diseases are complex approaches and need integration of various disciplines and multi institutional support. Besides knowledge of biology; ecosystem, behavior and epidemiology are the bases for development of prediction models. The approach has become important and useful to forewarn the farmers for preparedness to apply various pest management options to reduce the load of pesticides in the environment. This will be helpful in cutting down the expenditure on pest control, increase profitability and quality of food and feed.

References

AICRPAM (1995). All India Coordinated Research Project on Agrometeorology. *Annual Report*. 1994-95. pp. 34.

Amin, P.W. (1983). Major field and storage insect pests of groundnut in India and their control. Presented at the Summer Institute on *'Recent Advances in Groundnut Productivity'*, Junagadh, 9-30, April 1983.

Anonymous (1998). The regional impact of climate change – An assessment and vulnerability (Eds. Robert T. Watson, M.C. Zinyowera and Richard Moss). IPCC, Cambridge University Press.

Atwal, A.S. (1976). Agricultural pests of India and south east Asia. New Delhi:Kalyani Publishers, 502 p.

Bagmare, A., Deepesh Sharma and Ajay Gupta (1995). Effect of weather parameters on population build up of various leaf miner species infesting different host plants. *Crop Research* 10 (3): 344 –352.

Bailey, J.B., Johnson , G.L. and Toth, S.J. (1994). Evolution of a weather-based peanut leafspot spray advisory in Nortern Carolina. *Plant Disease*. 78: 530-535.

Bainbridge Fletcher, T. (1914). Some south Indian insects and other animals of importance – considered especially from an economic point of view. Government Press, Madras – 565 Pages.

Barbour, J.C. and Bridges, D.C. (1995). A model of competition for light between peanut (*Arachis hypogaea*) and broad leaf weeds. *Weed Science* 43(2):247-257.

Bhama, K.S. (1972). A rust on groundnut leaves near Madras. *Current Science* 41 (6) : 188-189.

Boyle, L.W. (1964). Low-gallonage application of fungicides for control of peanut leafspot. University of Georgia Agriculture. *Experimental Station Mimeograph Serve*. 203, 20p.

Castellani, E. (1959). La ruggine dell'arachide. *Olearia* 13: 261-270.

Chahal, D.S., and Chohan, J.S. (1971). *Puccinia* rust on groundnut. *FAO Plant Protection Bulletin* 19 (4): 90.

Cornell, J.A and Berger, R.D. (1987). Factors that influence the value of coefficient of determination in simple linear and non linear regression models. *Phytopathology* 7:63-70.

Cummins, D.G. and Smith, D.H. (1973).Effect of *Cercospora* leaf spot of peanuts on forage yield and quality on seed yield, *Agronomy Journal.*, 65: 919.

Das, D. K. Trivedi, T. P. and Srivastava, C.P. (2001). Simple rule to predict *Helicoverpa armigera* on the crops growing at Andhra Pradesh. *Indian Journal of Agricultural Sciences.* 71(6): 421-423.

Das, Srikanta., Raj, S.K. and Das, S. (2000). *Indian Phytopathology.* 53: 71-75.

Dubey, R.C., Ballal, A.S. (1995). Influence of weather factors on the groundnut leaf miner attack. *Agro-ecosystem Management.* 94-96.

Frezzi, M. J. (1960). Enfermedades del mani in la provincia de Cordoba (Argentina), Rev. Invest. Agr. 14: 113-155.

Frohlich, G. and Rodewald, W. (1963). Planzenschutz in den Tropen. Special Volume 3. Karl Marx University, Leipzig. Pages 136-141. (In German).

Garren, K.H. and Jackson C.R. (1973). Peanut disease in Peanut-culture and Uses, American Peanut Research and Education Association, Inc., Stillwater, Okhlahoma, 429.

Gerlagh, M. and Bokdman, M. (1974). Yield of groundnut in relation to the reduction of photosynthesis by *Cercospora* leaf spot or by its simulation, Review of *Plant Pathology* 53: 2008.

Gujrati, J.P., Kapoor, K.N. and Gangrade, G.A. (1973). Biology of soybean leafminer, *Stomopteryx subsecivella* (Lepidoptera: Gelechiidae). *The Entomologist,* 106: 187-191.

Hall, D.R., Beevor, P.S., Campion, D.G., Chamberlain, D.J., Cork, A., White, R., Almetre, A., Henneberry, T.J., Nandagopal,V., Wightman, J.A. and Ranga Rao, G.V., (1993). Identification and synthesis of new pheromones. *IOBC Wprs Bulletin,* 16: 1-10.

Higgins, B.B. (1956). Les Maladies de l'arachide aux Etats-Unis. *Oleagineux* 11: 213-230.

Holt, J. and Cheke, R.A. (1997). Modelling. Methods in Ecological and Agricultural Entomology. (Eds. Dent, D.R. and Walton, M.P.). CAB International pp. 387.

Jayaraj, S. and Shantharam, G. 1985. Ecology based Integrated control of *Spodoptera litura* (F). on Cotton,pp 256 –264 in Microbial Control and Pest Management (ed,S.Jayaraj)., Tamil Nadu Agricultural University,Coimbatore,Tamil Nadu.

Jayanthi, M., Singh, K. M. and Singh, R. N. 1993. Population build up of insect pests on MH 4 variety of groundnut influenced by abiotic factors. *Indian Journal of Entomology.* 55: 109-123.

Jena, B.C. and Kuila, B. (1997). Seasonal incidence of leaf miner *Aproaerema modicella* in groundnut and its chemical control. *Ind. J. of Ent.* 59(1): 27-33.

Jenkins, W.A. (1939). The development of Mycosphaerella berkeleyii, Journal of Agricultural Research, 58: 617.

Johnson, G. (1989). A operational system for prediction of peanut leaf spot disease. *Nineteenth Conf. Agril. And Forest Meteorology and Ninth Conf. Biometeorology and Aerobiology.* Charleston South Carolina, USA, 7-10 March 1989, 67-69.

Johnson, R.A.., Lamb, R.W. and Wood, T.G. (1981). Termites and crop loss studies in Nigeria – a survey of damage to groundnuts. *Tropical Pest Management,* 27: 325-342.

Knudsen, G.R., Johnson, C.S., Spurr, H.W. (Jr.) (1988). Use of a simulation model to explore fungicide strategies for control of *Cercospora* leaf spot of peanut. *Peanut Science*. 15 (1): 39-43.

Kulkarni, S.A. 1989. Bio-ecology and management of *Spodoptera litura* (F). (Lepidoptera: Noctuidae). on groundnut (*Arachis hypogaea* L.). Ph.D.Thesis, UAS, Dharwad, India, pp 270.

Leefmans, S. (1927). Zeikten en plagen der culturregewassen in Nederlandsch - Indie in 1926. Mededelingen van het Institut voor Plantenzeikten. Buitenzorg No. 73.pp 60.

Logeswaran, G., Madhava Rao, S., Vasudevan, G., and Kannan, V. (1982). Influence of time of sowing and weather factors on the infestation of leaf miner , *Aproaerema modicella* Deventer and yield in rain-fed groundnut. *Madras Agriculture Journal* 69 (6): 359-363.

Lu, D.Y. (1988). The developmental zero and effective accumulated temperature for *Holotrichia formosana* and their application in forecasting. *Insect knowledge*. 25(4): 207-208.

Mack, T. P., Davis, D. P., Backman, C.B. (1991). Predicting lesser cornstalk borer (Lepidoptera: Pyralidae) larval density from estimates of adult abundance in peanut fields. *Journal of Entomological Sciences*. 26(2): 223-230.

Mack, T. P.; Davis, D. P. and Backman, C. B.1991. Predicting lesser cornstalk borer (Lepidoptera: Pyralidae) larval density from estimates of adult abundance in peanut fields. *Journal of Entomological Science*, 26 (2) : 223-230.

Madden, L.V. (19860). Statical analysis and comparison of disease progress curve. In *Plant Diseases Epidemiology :population Dynamics and Management*. (Ed.) K. J. Leonardand W.E. Fry) Vol. 1. pp. 55-83. Macmillan publishing Company. New York.

Madhumathi, T. and Thippeswamy, M. (1996). Age specific fecundity tables and growth rate of leaf miner, *Aproarema modicella* Deventer (Lepidoptera:Gelechiidae) on groundnut during different seasons in Andhra Pradesh (India). *J. of Entomological Research*. 19(4): 367-371.

McVery, D. V. (1965). Inoculation and development of rust on peanut grown in the greenhouse . *Plant disease Reporter*. 49: 191-192.

Meyrick, E. (1906). Description of Indian microlepidoptera. *Journal of Bombay Natural History Society*, 17: 139-140.

Mishra, P.R. and Sontakke, B.K. 1992. Monitoring of *Spodoptera litura* F. by pheromone trap catches. *Indian Journal of Plant Protection*. 20: 188-190

Mohammad, A. (1981). The groundnut leafminer (*Stomopteryx subsecivella* Zeller). (Lepidoptera : Gelechidae). A review of world literature. Occasional paper 3, Groundnut Improvement Program, International Crops Research Institute for Semi Arid Tropics (ICRISAT), Patancheru, A.P., India. Pp 21.

Nath, V. R. and Indira, S. (1975). Screening of germplasm of groundnut gaainst rust. *Indian Phytopathology* 28: 121-3.

Nandihalli, B. S., Patil, B.V., Somasekhar and Hugar, P. (1989). Influence of weather parameters on the population dynamics of *Spodoptera litura* (Fb.) in pheromone and light traps. *Karnataka Journal of Agricultural Sciences*. 2: 62-67.

Patel, H. R., Valand, G. B., Shekh, A. M. and Patel, J. G. (1997). Weather in relation to late spot disease development in the two groundnut cultivars. *An. Agric. Res.* 18 (1): 16-20.

Pedro Junior, M. J., Moracs, S. A. de., Godoy, I. J. de., De Moraes, S.A., De Godoy, I. J. (1994). Agrometeorological forecasting method for Cercospora leaf spot in peanuts. *Fitopatologia Brasileira* 19(1): 69-73.

Rajaram, V., Janarthanan, R and Ramamurthy, R. 1999. Influence of weather parameters and moon light on the attraction of light trap and pheromone trap catches of cotton pests under dry farming system. *Annals of Agricultural Research*. 20: 282-285.

Ranga Rao, D.V., Singh, K.N., Wightman, J.A and Ranga Rao,G.V. (1991). Economic status of neem cake mulch for termite control in groundnut. *International Arachis Newsletter.*, 9: 12-13.

Sandhikar, R. N., Bulbule, S.V. and Mayee, C. D. (1989). A linear model to predict incubation period of groundnut rust. *Indian Phytopathology* . 42: 347.

Sandhikar, R.N., Bulbule, S.V., Mayee, C.D. (1989). Prediction model for rust epidemic in groundnut. *Indian J. Mycol. Pl. Pathology*. 19(1): 60-67.

Savary, S., Jong, P.D. De, Rabbinge, R., Zadoks, J.C. (1990). Dynamic simulation of groundnut rust: a preliminary model. *Agricultural systems* 32(2): 113-141.

Shanower, T.G., Gutierrez, A.P. and Wightman., J.A. (1993). Effect of temperature on development rates, Fecundity, longevity of groundnut leaf miner, *Aproeaema madicella* (Lepidoptera: Gelechiidae) in India. *Bulletin of Entomology Research*. 83(3): 413-419.

Shekh, A. M., Patel, H. R. and Pandey Vyas (1999). Predicting the outbreak and spread of pests and diseases using weather parameters. *Proceedings of the National Workshop on Dynamic Crop Simulation Modelling for Agrometeorological Advisory Services*. National Centre for Medium Range Weather Forecasting, Department of Science and Technology: Jan 1999: 249-256.

Sherasiya, R.A. and Bhutani, P.G. (1998). Population dynamics of groundnut leaf miner in relation to weather parameters. *Gujarat Agriculture University Research Journal*. 223 (2): 39-44.

Singh, S.P. and Jalali, S.K. (1997). Management of *Spodoptera litura* (F) Lepidoptera:Noctuidae). In Whiteman, J.A. and Ranga Rao, G.V. (Eds.) *Spodoptera litura* in India: Proceedings of National Scientists Forum on *Spodoptera litura* (F), 2-4, April, 1996. India: ICRISAT Asia Centre, 27-66.

Spegazzini, C.L. (1884). Fungi guarantici. Anales, Sociedad Cientifica Argentina, 17, 69-96 and 119-134, *Puccinia arachidis* Speg. (n.sp.), 90.

Subrahmanyam, P., Reddy, D.V.R., Gibbson, R.W., Rao, V.R., and Garren, K.H. (1979). Current distribution of groundnut rust in India PANS (UK). (1979). v. 25(1) p. 25-29.

Tamaki, Y., Noguchi, H and Yashima,T., (1973). Sex pheromones of the *Spodoptera litura* (Lepidoptera:Noctuidae): Isolation, identification and synthesis. *Applied Entomolgy and Zoology*, 8 : 200-203.

Trivedi, T.P.; Puri, S.N; Paul Khurana,S.M; Bhar,L. M; Mehta, S.C; Jain, R.C and Singh, G. (1999). Modelling for forecasting of Aphid (*Myzus persicae*) in potato. *Global Conference on potato*. December 1999, IARI, New Delhi.

Venkataraman, S. (1992). Weather in the incidence and control of pests and diseases. In Crops and Weather (Eds. S. Venkataraman and A. Krishnan). ICAR pp. 586.

Walker, R.H., Wells, L.W., McGuire, J.A. (1989). Bristly starbur (*Acanthospermum hispidum*) interference in peanuts. (*Arachis hypogea*). *Weed Science*. 37(2): 196-200.

Wang, Z.Z. and Lin K.H. (1986). A long –term forecasting equation for groundnut rust. (in chinses with english summary), *J. South ChinaAgri. Univ*. 7 (4): 41-44.

Wightman, J.A. and Ranga Rao, G.V. (1993). A Groundnut Insect Identification Handbook for India. Information Bulletin No. 39 ICRISAT. Patancheru, Andhra Pradesh, India, pp. 28–37.

Wightman, J.A., Dick, K.M., Ranga Rao, G.V., Shanower, T.G., and Gold, C.S. (1990). Pests of groundnut in the semi-arid tropics. In: *Insect Pests of Food Legumes* (Ed. by S.R.singh, John Wiley & Sons, New York. pp.243-322

Yang, C. and Liu, H. (1966). Biological observations on *Stomopteryx subsecivella* Zell. in Den Bei District, Kwantung. (In Chinese). *Acta Entomol. Sin.*, 15, 39-46.

Zadoks, J. C. (1971). Systems analysis and dynamics of epidemics. *Phytopathology* 61: 600-610.

□□□

26

Transfer of Pest Management Technologies

G. D. SATISH KUMAR

26. 1. Introduction

Technology is the application of knowledge for practical purposes (Galbraith, 1967; Swanson, 1997). Generally technology is used to improve the human condition, the natural environment, or to carry out other socio-economic activities. Technology can be classified into two major categories i. material technology, where knowledge is embodied into a technological product such as sprayers, pesticides, pheromone traps and ii. Knowledge based technology such as technical knowledge, management of insect pests to successfully grow the crop. The transfer of material technology to farmers generally involves the production, distribution, and sale of seeds, sprayers, pesticides and other production inputs. Therefore the transfer of material technology is simpler than training and disseminating technical knowledge and management skills to large number of poorly educated farmers who operate in diverse agro ecological situations. Also, the delivery systems needed for these are generally different. In most cases, the private sector is best suited to produce and distribute material technology. On the other hand, most knowledge based technologies such as the integrated pest management are generally taught through vocational training programs for rural young people or disseminated through a publicly funded technology transfer system.

At the same time most material technology requires technical knowledge so that these products or tools can be used effectively. For example to properly use an agrochemical in pest management, farmers need to know the proper application rates, the time and conditions for application, safety procedures, and so forth. In addition, if farmers use a sprayer to apply the chemical, then they need to know how to operate, adjust, calibrate and clean the equipment to achieve the best results. Therefore, material and knowledge based technologies are generally closely intertwined. Private sector firms have very limited technical capacity to train farmers in these product related skills and knowledge, also it may not be economical to them; therefore, the transfer of most knowledge based technologies is by design or by default, left to the national technology transfer system (Swanson, 1997).

Transfer of technology in its crude sense is a process by which the recommended practices produced by research and development are transmitted through extension agents to producers. It implies active role for research and sees the farmer as little more than a target organism who has to be persuaded somehow to adopt the new practices. In real situations, the farmers have an initiating role; it is their purpose, socio-economic conditions and available resources to which the technology has to be appropriate. One of the many purposes of the extension system is to identify those purposes, socio-economic conditions and resources, to prescribe known technology suited to the farmers, where possible, and to refer to researchers those difficulties for which new knowledge is required. Thus, transfer of technology is a two way process in which the extension system has a central role to play, not only as the intermediary for new knowledge but also advisor of the farmers and their window into the knowledge system.

26.2. Origin of transfer of technology approach in Indian oilseeds/ groundnut scenario

India's oilseed scenario showed distinct transformation from its long inertia, thanks for setting up of the Technology Mission on Oilseeds (TMO) in May, 1986 by the GOI and series of synergistic and innovative farmers-oriented policies it has launched, backed by better availability of inputs, services, attractive and guaranteed price support operations. The country's aggregate output of oilseeds, which in the past never surpassed 13 million tonnes scaled an all time record of 18 mt during 1988-89. Simultaneously the quantum of imports of vegetable oils and out flow of scarce foreign exchange reserves on account of it came down from 18 lakh tones (920 crores, Foreign Exchange Equivalent) in 1987-88 to as low as 4.0 lakh tones (250 crores) by 1989-90. These amazing developments in the country's most sensitive and vital sector namely oilseeds have not only capitulated it from a position of one of the major sources of drain on the exchequer to a net foreign exchange earner in recent years.

Encouraged by the tremendous potential of Oilseeds Indian Council of Agricultural Research (ICAR) for the first time launched a special A.P Cess Fund supported project on "On-farm Research on Annual oilseed crops from July 1988. The scheme was unique of its kind. It provided hither to unavailable opportunities to All India Coordinated Research Project on Oilseeds (AICORPO) researchers actively engaged in generation of location specific technologies. It has given space to demonstrate to farmers what they have been preaching for many years. It provided scientists an opportunity to test the feasibility and viability of their technologies and obtain feed back from the farmers. Based on the feed back scientists may modify the technology to improve their acceptability and suitability to real farm situations. This adhoc scheme covered all the mandate crops of the project and its network of 65 centres spread over 17 states spanning over all important and potential oilseed growing areas/situations in the country (Ranga Rao, 1991).

The AICRPO network has developed several location specific, viable and feasible agro production and protection technologies, improved varieties and hybrids for raising the yield and income of farmers. Though TMO made efforts for transferring the technologies to farmers, the transfer of technology (TOT) programmes in oilseeds have not received the attention they deserve. One of the reasons attributed for poor implementation of TOT projects in oilseeds according to Prasad (1993) were lack of finance to take up the On-farm researches and demonstration in agro-production and protection technologies generated from time to time under real farm situations. Hence, in order to remove the financial and other constraints which were hindering the spread of technologies already developed by AICRPO a special project on "Front Line Demonstrations in Oilseed" (FDO) under the support of Ministry of Agriculture & Cooperation, GOI came into operation from kharif 1990-91. This project is a continuation of the earlier project on On-farm research on annual oilseeds crops, which was discontinued from 1990. According to Reddy and Kiresur (1995) the new project made a distinct improvement over the A.P. Cess Fund supported programme and provided adequate financial support for various extension activities such as production of publicity material, farmers meetings, farmers days, travel by the scientists for monitoring and implementation of the programme besides enhanced contingencies.

Even though an exclusive mechanism does not exist for transferring pest management technologies in oilseeds/ groundnut the above programmes discussed include them as a major component in their ambit of operations.

The Frontline Demonstrations (FLD's) on groundnut conducted during 1987 to 1997 had helped in increasing the productivity from 842kg/ha to 1047 kg/ha during the same period. However not withstanding the above positive changes in the output and the increasing resilience and strength the groundnut acquired in recent years, the general per hectare yields still far short of those realized else where in the world and even the research farms. In order to increase the yield of groundnut there are limited options available and one of which is reducing the losses in production and realizing the full yield potential of the crop. It was estimated that the avoidable yield loss due to insect pests alone was upto 39.5% in groundnut (Sherasiya, 1998). Studies by Central Research Institute for Dryland Agriculture (CRIDA) in three states Gujarat, Andhra Pradesh and Tamilnadu showed that adoption of the plant protection component of the technology input raised productivity form 72.5 to 85% (Anon, 2000). This clearly shows the achievable yield potential of transferring the pest management practices in groundnut.

26.3. Extent of adoption of plant protection measures by groundnut farmers

Bhalara (1979) observed that only 24.6% of groundnut farmers had high knowledge, whereas 38.4% and 37% of farmers had low and medium level of knowledge about plant protection measures in groundnut. He further revealed that majority of them (43%) were low adopter of plant protection measures followed by medium (29.2%) and high adopters (27.6%). Meti *et al.* (1997) revealed that 33.3% of groundnut farmers interviewed were aware of pest and disease management and only 27.5% actually adopted the recommended pests and diseases management practices in groundnut cultivation in Karnataka. Verma *et al.* (2003) reported a technological gap of 65% in adoption of plant protection measures by groundnut farmers.

These studies indicate that there exists a gap in the adoption of pest management technologies in groundnut. Hence, the transfer of technology system in groundnut has to be redefined to speed up the rate of adoption among farmers. For this it is important to understand the existing transfer of technology system for groundnut in India and the inherent constraints associated with it.

26.4. Systems involved in the transfer of technology

Four set of basic components are involved in groundnut technology transfer system. Each set of activities is performed basically by a system, which is inter-linked with others for implementing the process (Jaiswal and Arya, 1981; Peterson, 1997). In India, though, there is no separate organization especially involved in groundnut pest management activities; the system segments shown in Figure 2 have included it as a component in their gamut of activities. The organizations involved in these systems and their linkages are explained in Figure 3.

The four systems are

1. Research/Technology generation system to evolve technology
2. Extension/Dissemination/Communication System to transfer the technology from research station to the farmer's field
3. Technology Utilisation system/Farmers system which adopts the technology and takes benefit from it and
4. Support system which supports the technology transfer process by providing necessary inputs required for the use of technology and also by providing necessary facilities for processing and marketing of the output

26.4.1. Research/Technology generation system

The research system takes care of the technology production; its output becomes the input of the technology utilization system. A process of feed back operates at all levels. The research system consists of Agricultural Universities, Research Institutes and other organizations which create new pest management technologies through applied research.

26.4.1.1. National Research Centre for Groundnut (NRCG)

The National Research Centre for Groundnut (NRCG) was established in 1979 by the Indian Council of Agricultural Research (ICAR). The mandate of the centre is to conduct basic and strategic research to enhance production, productivity, and quality of groundnut, to act as the national repository of working collection of groundnut germplasm and information on groundnut research, to establish relevant institutional linkages, offer consultancy and training and to provide logistic support and coordination mechanism for generation of location specific technology through the All India Coordinated Research Project on Groundnut (AICRPG) productivity and production including the pest management technology.

26.4.1.2. National Centre for Integrated Pest Management (NCIPM)

National Centre for Integrated Pest Management (NCIPM), New Delhi became functional in 1988 to develop and promote IPM technologies for major crops to sustain higher crop yields with minimum ecological implications.

26.4.1.3. Indian Institute of Chemical Technology (IICT)

The Indian Institute of Chemical Technology (IICT) is collaborating with the NRCG in the development of pheromone traps for the monitoring and control of the leaf miner in groundnut.

26.4.1.4. International Crops Research Institute for Semi Arid Tropics (ICRISAT)

The International Crops Research Institute for Semi Arid Tropics (ICRISAT) was established in 1972 at Patancheru, Hyderabad in Andhra Pradesh with the mandate crops Sorghum, Pearl millet, Chickpea, Pigeon pea and Groundnut. The mission is to conduct research that can lead to enhanced sustainable production of these crops and to improved management of the limited natural resources of the Semi Arid Tropics. ICRISAT communicates information on technologies as they are developed through workshops, networks, training, library services and publishing.

26.4.1.5. State Agricultural Universities (SAUs)

The main function of the SAUs is research, teaching and extension. The different research stations of SAUs collaborate with AICRP-G in developing and popularizing improved production and protection technologies.

However their primary method of extension continues to be through institutional training programmes to extension personnel of the State Department of Agriculture. They also participate in radio talks of All India Radio, farmer's discussions on Door Darshan and organize kisan melas

26.4.2. Extension system

Evaluates and adapts research outputs. Widely disseminates the knowledge and inputs to different target adopters, farmers, and private companies and so on from the research system and obtains feedback. It includes government agencies and input manufacturing, credit, export and marketing agencies as well.

26.4.2.1. State Department of Agriculture

The major functions of the state departments of agriculture are i. Overall responsibility for extension work ii. Study field problems and formulate new pest management technologies through a process of deliberations with the agricultural universities iii. Share with agricultural universities the responsibility of conducting field trials and demonstrations iv. Establishing state information cell in cooperation with the university v. Organise training programmes on Groundnut Pests Management and Integrated Pest Management (IPM) technologies. These organizations transfer pest management technologies effectively by utilising individual, group and mass contact methods. In India, Training and Visit (T&V) system is followed in almost all states, where the emphasis is on the contact farmers (progressive farmers) so that through them the technology is to tickle down to the other small and marginal farmers. Regular training to staff involved in the transfer of technology was also given more emphasis in the system.

26.4.2.2. State Agricultural Universities (SAUs)

One of the major functions of agricultural universities in addition to teaching and research is to collect, synthesise, and communicate relevant research findings to farmers and extension workers. Directorate/Department of Extension Education is responsible for the farm advisory, training and transfer of technology activities in the entire state. Some universities maintain communication centre for production of information material such as printed literature, audio-visuals, radio and television programmes and exhibits. However their primary method of extension continues to be through institutional training programmes to extension personnel of the State Department of Agriculture.

26.4.2.3. National Research Centre for Groundnut (NRCG)

The AICRPG with its headquarters at NRCG, functions under the supervision of NRCG. AICRP-G is concerned with planning, monitoring and conducting location specific groundnut research at various centers in the country. The details of demonstrations conducted at different centres from 1998-2000 are provided in table 1.

Table 26.1. FLDs conducted under the AICRPG from 1998-2000.

Year	No. of Front Line Demonstrations	
	Plant protection	Total
1998	12	150
1999	6	204
2000	24	205
Total	42	559

It is observed that only 7.51 % of the total FLDs were related to pest management (AICRPG, 1999-2001). Hence there is a need to increase the number of demonstrations on plant protection especially that of Integrated Pest Management.

NRCG performs first line transfer of technology through innovative projects like Technology Assessment and Refinement through Institution Village Linkage Programme (TAR-IVLP). The details of Integrated Pest Management (IPM) On-farm trials are given in table 2.

Table 26.2. Effect of IPM Components on the control of insect pests, diseases and yield of Groundnut

S. no.	Biotic Factors	Insect, Pests/ Diseases Score		% Control Over
		IPM Components	Farmers Practice	Farmers Practice
A. Insects				
1.	Jassids	1.29	2.04	29.09
2.	Thrips	1.90	3.0	35.55
3.	De-foliators	1.50	2.19	35.7
B. DISEASES				
4.	ELS	1.92	3.29	40.84
5.	LLS	1.91	3.21	48.15
6.	Rust	2.41	4.15	34.94
7.	Collar rot (%)	6.81	13.8	43.22
8.	Stem rot (%)	7.09	15.35	56.08
9.	Aflaroot (%)*	3.24	4.03	26.14
10.	PBND (%)	2.38	4.38	39.19
11.	Pod yield of Groundnut (kg/ha)	2416	2050	17.84
12.	Fodder yield of Groundnut (kg/ha)	3462.85	3114.34	9.6

Source: Annual reports TAR-IVLP.

It was observed that there was reduction in defoliators by 35.7%, early leaf spot by 41 %, late leaf spot by 48%, rust by 35 %, collar rot by 43%, stem rot by 56%, aflarot by 26% and PBND by 39 % where IPM components were used over farmers' practice (Table 4). The male moths of *Spodoptera* and *Heliothis* trapped were 17/trap/week and 14/trap/week respectively in IPM treatment. This resulted in increased pod yield of groundnut by 18% (2416 kg/ha) and fodder yield by 10% (3462.85kg/ha) over the farmers' practice (2049 kg/ha and 3114 kg/ha).

NRCG plays a catalytic role in the transfer of technology through organising krishi melas and farmer's day. It participates in Krishi melas of national importance and educates the farmers on the improved practices in groundnut.

26.4.2.4. Input and other agencies

These agencies make individual contacts, organize kisan melas, farmers meetings to educate the farmers on the concepts of the pest management in groundnut. The primary goal of input agencies is to promote their products through extension activities as mentioned above. They help to accentuate the successful suppression of some of the major pests of groundnut.

Commodity organizations like Indian Oil Products Export Authority (IOPEA), Agricultural Produce Export Development Authority (APEDA), banks and NGOs are also engaged in transfer of technology process mainly in collaboration with the state or central institutes. Their method of transfer is creating awareness among farmers, export groups and oil processing units.

26.4.3. Technology utilization system

Encompasses the users of the agricultural technology, mainly farmers. User awareness, adaptation and adoption of improved pest management technologies from various sources affect farm-level productivity and profitability. Interaction and feedback between users, researchers and transfer systems improve cooperation and the relevance of technology.

26.4.4. Supporting system

The task of technology transfer is not complete by merely passing information on pest management technologies from research stations to farmers. The actual adoption of improved practices by the farmers needs supporting services to make inputs available as well as to make suitable arrangements for output storage and marketing. Supporting system includes credit institutions like commercial banks, cooperative societies and farmers service societies, input supply agencies, marketing and processing agencies.

26.5. Constraints in transfer of pest management technologies in groundnut

Development of pest management technologies are the result of applied research. The prime mission of institutions producing this research is to develop technologies superior to the existing ones. The following are the attributes that have to be incorporated into a technology before it can be transferred to the farmers.

Relative advantage, is the degree to which a technology is percervied as better than the idea it supersedes in terms of pest control, profitability, ease of operation, etc. *Compatibility* is the degree to which the technology is consistent with the agro-physical, existing culture and past experiences of the farming community where the technology is to be transferred. *Simplicity* is the degree to which the technology is readily and easily understandable and usable by the farmers. *Divisibility* is the degree to which the innovation could be tried on a limited scale and *Communicability* is the degree to which the efficient use of technology is easily communicable. However, several problems make it difficult to achieve the total incorporation of all the above elements in the technology. One of the problems is the lack of communication between farmers and scientists. Scientists are supposed to be aware of the field problems through the feedback given by the farming community directly or through extension agencies. However, this feed back is extremely meager and, as a result the research system functions in some isolation. This prevents scientists from developing adequate sensitivity to the conditions and requirements of the ultimate users. Quite often, the technologies are not oriented to field conditions and the information developed by the research are mostly "publication-oriented rather than application oriented".

It is a well known fact that the extension system has not been completely successful in transferring pest management technologies to groundnut farmers. Some of the problems responsible for this are the lack of effective linkage between research and extension, lack of farmers participatory programmes, lack of educational approach, ineffective use of different extension methods, lack of knowledge among extension workers on groundnut pest management technologies.

The T and V system has an inherent bias towards the progressive farmers, but the experience suggests that the small and marginal farmers have their own farming systems, which are complex, diverse, and risk prone. Hence the common strategy for all the farmers doesn't suffice the need. There should be client oriented extension network based on the farmers need. Also the effective pest management system requires adequate manpower and inputs such as biocontrol agents, pesticides, application equipments, and transportation and communication facilities. Though presently the infrastructure for the transfer of technology is available, there is a need to strengthen the extension system as a whole. There is also wide spread lack of understanding on the actual type of training needed.

It is evident that the educational level and socio-economic conditions of the farmers greatly affect the implementation of pest management strategies mainly because it essentially involves the knowledge of the pest, suitable methods of control and correct technique(s) of their application. Some of the important constraints identified at farmers level are lack of knowledge on pest management, timely non-availability of critical inputs (pesticides) to farmers and high cost of the pesticides, which increases the cost of cultivation (Mathukia, 1981; Mayani, 1987; Patel, 1988).

Farmers do not think about Integrated Pest Management (IPM), agro-ecosystem, pest status, side effects of pesticide application on the natural enemy complex and host-parasite relationship. Concept of economy threshold level and the harmonious blending of several compatible measures appears complicated to the farmers. They also do not take cognizance of the available pest management tactics especially if the pest damage does not occur periodically or when it is less severe. Sometimes even though farmers are aware of pest management techniques but the timely non-availability of inputs in their pure form hinders the process of adoption.

Extension scientists have done a great deal of work on barriers to the adoption of new technologies, trying to explore why farmers accept a certain technology not other. The major acceptance motives identified were economic profitability, relative advantage, attainment of goal, security, problem solving. Reasons for rejection were found to be lack of information, timely non-availability of critical inputs (pesticides) to farmers and high cost of the pesticides, financial limitations, and labour. In additions to these factors, personal and environment characteristics may facilitate or block adoption. The factors can be grouped into client-centered factors (caste, age, education, knowledge, attitudes, etc.), client-environmental factors (education, social factors, infrastructure, etc.), technology variables (profitability, compatibility, etc.) and communication variables (contact with extension agency, media exposure, etc.). Other factors include price policies relating to price of pesticides and prices of groundnut and government policies with respect to the crop itself. Thus, diverse set of factors act as constraints to transfer of pest management technologies. All the above facts donot act independently or in isolation. They interact with each other to form a barrier-complex. This complex of resistance, with manifold intensity and strength acts, as an impediment to transfer of technology from research system to farmers.

26.6. Future thrust

The basic challenge in augmenting the transfer of pest management technologies is to improve the relevance and develop technologies with the desirable attributes like relative advantage, profitability and compatibility with the farmers socio-economic conditions.

Multidisciplinary, farmer participatory research approach for evolving pest management technologies will reduce the time lag between developing a technology and its adoption by farmers. For this to happen training may be imparted to the scientists actively engaged in research on participatory approaches and their usefulness.

The innovative approaches of TAR-IVLP clearly indicated the potential of IPM under farmer's fields, such efforts should be increased and spread to more villages.

Setting up of a national centre for training the trainers on groundnut extension will be useful. Such a centre can plan, coordinate and provide high level training to all the trainers. Establishing of groundnut pest forecasting centre and advance dissemination of the information regarding pest incidence through modern communication media such as TV, Internet, e-mail and interactive computer aided systems.

Strengthening the linkage between research system and the extension system is the need of the hour. The state department of extension personnel are to be involved in the research projects and they should provide an effective feed back to the research system.

Extension personnel involved in the transfer of pest management technologies are concerned with the diagnosis, prognosis and control of different pests. These extension personnel have varying educational backgrounds, which may include graduation in Agriculture or in other sciences, diploma in Agriculture or just matriculation with some experience in extension work. Many of the extension personnel might have received their formal education/training at a time when great emphasis was placed on the use of pesticides for pest control. Moreover, shift from subsistence agriculture to intensively managed production systems as well as introduction of new high yielding varieties have resulted in differential situations in pest-host relationships, which has led to changes in the pest status (Harris, 1972 and Anon, 1972). This makes it imperative to impart extensive education on the principles of pest management by way of intensive refresher and reorientation courses to the extension personnel.

The extension personnel should take advantage of all possible methods and techniques of extension teaching: Farmers are generally contacted through mass, group, and individual methods. However studies on the impact of various extension methods and media have indicated that use of number of methods in combination are highly effective as compared to the effect obtained when single method is used. This may be taken into consideration while planning the extension programmes.

Timely availability of information is important for adoption of pest management technologies, which can be achieved through efficient use of mass media communication. New millennium communication media Internet, e-mail can be used for effective and efficient dissemination of information regarding pests and diseases of groundnut. The new ad-age mantra "Information, Communication combined with Entertainment (ICE)" can be effectively utilized in the transfer of pest management technologies.

It is widely recognized that some form of organization (NGOs) is essential to reach wider spectrum of people within less time. The indigenous ones in particular have cohesiveness and trust of the local farmers, which can be utilized in the transfer of the pest management technologies.

Farmers should be encouraged to form into groups and cooperative organizations based on the criteria like land, irrigation, export etc. Groups can have effective linkages with other sub-systems of Transfer of Technology (TOT). Groups can exert pressure on the other systems for effective functioning, thus can develop a demand driven transfer of technology system.

Making the farmers participate in research process right form the beginning (Farmers Participatory Research Approach) for evolving pest management technologies will not only reduce the time lag between developing a technology and its adoption but also empowers the groundnut farmers.

Establishment of Local Resource Centres (LRCs) in each taluk and can be connected to district head quarters. Location specific expert systems on pest management for groundnut along with other crops may be developed and installed on a touch screen system at LRCs. So that the farmers can get the required information with clear animation effects with in few seconds.

The major groundnut growing states Gujarat, Andhra Pradesh, Tamilnadu, Karnataka and Maharashtra which accounts for 85% of kharif groundnut production (Reddy *et al.*, 1997) can be linked through net and can share the information between themselves and the other partners involved in the research, development and extension of the groundnut pest management technologies.

In these maneuvers the development of Pest Management Information Systems (PMIS) as was developed by the NCIPM will be very handy. PMIS is user-friendly software to identify pests of cotton and their management. Like wise the PMIS for groundnut can be developed. It can be used as a training tool by any agencies to train master trainers, training organizers and progressive farmers. The software gives detail information on all the pests of groundnut, their symptoms of damage and control measures.

All the efforts will go waste if critical inputs are not available on time. The linkage between research, extension system and inputs agencies should be further strengthened and timely availability of inputs should be ensured for adoption of pest management technologies.

References

Annual report. (2000). National Centre for Integrated Pest Management. New Delhi.

Anonymous, (2000). A comprehensive report of the work done under the pilot project for Technology Assessment and Refinement through Institution Village Linkage Programme from September 1995 to march 1999. National Research Centre for Groundnut, Junagadh. pp 4.

Arulraj, S. (1986). Transfer of pest management technologies In: Sugarcane entomology in India. (Ed) David H, Easwaramoorthy S and Jayanthi R. Sugarcane Breeding Institute, ICAR. pp 475-482.

Bhalara, V.G. (1979). An analysis of groundnut growers knowledge about plant protection measures in groundnut. M.Sc (Agril.) Thesis (Unpublished). Gujarat Agricultural Univeristy, Sardar Krushinagar.

Galbraith, J.K. (1967). The new Industrial State. Boston.

Ghewande, M.P., Nandagopal, V., Devi Dayal, Satish Kumar, G.D., Sojitra, V.K. and Chavda, V.N. (2001). On-farm assessment of integrated pest management technology in rainfed groundnut (*Arachis hypogaea* L.). Paper presented at National Symposium on Oilseeds and Pulses, Tamilnaddu Agricultural University, Coimbatore.

Glass, E.H. (1975). Integrated Pest Management: Rationale, potential needs and implementation. E.S.A. Special Publ. 75-2. Ent. Soc. Am., Washington D.C.

Harris, D.R. The origins of agriculture in the tropics Am. Sci. 60: pp.180-193.

Jaiswal, N.K., and Arya, H.P.S. (1981). Transfer of Farm technology in India. *In*: Management of transfer of farm technology. National Institute of Rural Development, Hyderabad. 13-48.

Meti, S.K, Hanchinal,S.N, and Benaji, V.I. (1997). Maharashtra J of Extension Education. XVI. 258-262.

Peterson, W. (1997). Context of Extension in Agricultural and Rural Development. In: Improving Agricultural Extension-A Reference manual. (eds) Swanson , B.E., Bentz, R.P., and Sofranko , A.J., 1997. Food and Agricultural Organization of the United Nations. Rome.pp 21-26.

Prasad, M.V.R. (1993). Review and Reorganisation of Frontline Demonstrations in Oilseed Crops. Proceeding of Group meeting. DOR, Hyderabad. 21-27.

Progress report,. (1998-2000). All India Coordinated Research Project on Groundnut (AICRP-G). National Research Centre for Groundnut, Junagadh. pp A 24-26.

Ranga Rao,V. (1991). Improved technologies in oilseeds performance and promise under real farm situations, Directorate of Oilseeds, Hyderabad. pp137.

Reddy, P.S., Ghewande M.P., and Devi Dayal. (1997). Groundnut In: Efficient Management of Dryland Crops in India- Oilseeds. (eds) Singh, R.P., Reddy, P.S, Kiresur,V.1997. Indian Society of Oilseeds Research, Hyderabad. pp 21-58.

Reddy, P.S., and Kiresur,V. (1995). Transfer of improved oilseeds production technologies through Frontline Demonstrations: Significant features and Achievements (1993-94). Directorate of Oilseeds Research, Hyderabad. pp 1-6.

Report. (2000). A brief report on research and technological achievements during 1986 to 1999 in groundnut, National Research Centre for Groundnut, Junagadh. pp 4.

Sherasiya, R.A., Butani, P.G. (1998.) Assessment of avoidable yield loss due to insect pests in groundnut. Journal of Oilseeds Research., 15: 2, 390-392.

Swanson, B.E. (1997). Strengthening Research-Extension-Farmer Linkages In: Improving Agricultural Extension-A Reference manual. (eds) Swanson , B.E., Bentz , R.P., and Sofranko , A.J., 1997. Food and Agricultural Organization of the United Nations. Rome.pp 171-178.

Verma, P.D., Munshi, M.A., and Popat, M.N. (2003). International Arachis Newsletter. 23: 30-32.

□□□